Lecture Notes in Artificial Intelligence 1640

Subseries of Lecture Notes in Computer Science
Edited by J. G. Carbonell and J. Siekmann

Lecture Notes in Computer Science
Edited by G. Goos, J. Hartmanis and J. van Leeuwen

Springer

Berlin
Heidelberg
New York
Barcelona
Hong Kong
London
Milan
Paris
Singapore
Tokyo

William Tepfenhart   Walling Cyre (Eds.)

Conceptual Structures: Standards and Practices

7th International Conference
on Conceptual Structures, ICCS'99
Blacksburg, VA, USA, July 12-15, 1999
Proceedings

Series Editors

Jaime G. Carbonell, Carnegie Mellon University, Pittsburgh, PA, USA
Jörg Siekmann, University of Saarland, Saarbrücken, Germany

Volume Editors

William M. Tepfenhart
Software Engineering Department, Monmouth University
West Long Branch, NJ 07764-1898, USA
E-mail: btepfenh@moncol.monmouth.edu

Walling Cyre
Virginia Tech, Automatic Design Research Group
The Bradley Department of Electrical and Computer Engineering
Blacksburg, VA 24061, USA
E-mail: cyre@vt.edu

Cataloging-in-Publication data applied for

Die Deutsche Bibliothek - CIP-Einheitsaufnahme

Conceptual structures : standards and practices ; proceedings / 7th International Conference on Conceptual Structures, ICCS '99, Blacksburg, VA, USA, July 12 - 15, 1999. William M. Tepfenhart ; Walling Cyre (ed.). - Berlin ; Heidelberg ; New York ; Barcelona ; Hong Kong ; London ; Milan ; Paris ; Singapore ; Tokyo : Springer, 1999
(Lecture notes in computer science ; Vol. 1640 : Lecture notes in artificial intelligence)
ISBN 3-540-66223-5

CR Subject Classification (1998): I.2, G.2.2, F.4.1, F.2

ISBN 3-540-66223-5 Springer-Verlag Berlin Heidelberg New York

Typesetting: Camera-ready by author
SPIN 10703464 06/3142 – 5 4 3 2 1 0 Printed on acid-free paper

Preface

With all of the news about the Internet and the Y2K problem, it is easy to forget that other areas of computer science still exist. Reading the newspaper or watching the television conveys a very warped view of what is happening in computer science. This conference illustrates how a maturing subdiscipline of computer science can continue to grow and integrate within it both old and new approaches despite (or perhaps due to) a lack of public awareness.

The conceptual graph community has basically existed since the 1984 publication of John Sowa's book, "Conceptual Structures: Information Processing In Mind and Machine." In this book, John Sowa laid the foundations for a knowledge representation model called conceptual graphs based on semantic networks and the existential graphs of C.S. Peirce. Conceptual graphs constitutes a very powerful and expressive knowledge representation scheme, inheriting the benefits of logic and the mathematics of graphs.

The expressiveness and formal underpinnings of conceptual graph theory have attracted a large international community of researchers and scholars. The International Conferences on Conceptual Structures, and this is the seventh in the series, is the primary forum for these researchers to report their progress and activities. As in the past, the doors were open to admit alternate representation models and approaches.

These proceedings include papers that illustrate the adaptivity of the conceptual graph community and the degree to which this area has matured. First, John Sowa has consented to have a proposed draft standard for conceptual graphs made part of these proceedings. The adoption of a standard within the community is a major landmark in maturity by which a community has reviewed and agreed upon a core set of concepts and practices. Second, there are a number of papers in which conceptual graph based systems compete against other technical approaches to solve the same kind of problem. The ability of this community to provide more than one application is a significant accomplishment. Third, one will notice a number of papers describing applications for very different kinds of problem domains that are based upon this technology. The presence of these papers provides a certain credibility that what has been a research area is now producing systems that can actually be used. Finally, the theory papers, which have long been a standard at this conference, are still present. In our opinion this indicates that researchers still find this to be a fertile field and a valuable approach for investigating tough computational problems.

These proceedings contain 34 papers and are organized into seven major sections: Conceptual Graph Modeling; Natural Language; Applications; SISYPHUS-I; Contexts, Logic, and CGs; Logic; and Position Papers. A new feature was introduced in this conference, position papers, which allow investigators entering into the field to present the basic direction in which their laboratory is heading without having to present the results of what may be rather immature work. In addition, a special track was incorporated into the program this year. The SISYPHUS-I track reports the work of conceptual graph researchers as measured against a standardized testbed application – a resource allocation problem.

We would like to thank the authors, editorial board members, and program committee members for helping to making this conference possible. Without the help of Bheemeswara Reddy Dwarampudi and Dwyna M. Macdonald, these proceedings wouldn't have been produced. In addition, we would like to thank Bob Bekefi for his help in setting up the computer accounts for this effort.

Finally, on behalf of the organizing committee, we would like to thank the organizations that have sponsored this conference: The Bradley Department of Electrical and Computer Engineering at Virginia Tech, the Division of Continuing Education at Virginia Tech, and Monmouth University. This conference has been held in cooperation with the American Association for Artificial Intelligence.

William Tepfenhart and Walling Cyre

Organizing Committee

Honorary Chair

General Chair

Program Chair

Editorial Board

Program Committee

Michel Leclère	IRIN, Université Nantes, France
Graham A. Mann	University of New South Wales, Australia
Philippe Martin	Griffith University, Australia
Rokia Missaoui	Université du Québec à Montréal, Canada
Jens-Uwe Moeller	University of Hamburg, Germany
Bernard Moulin	Université Laval, Canada
Maurice Pagnucco	Macquarie University, Australia
Mike P. Papazoglou	University of Tilburg, The Netherlands
Heather Pfeiffer	New Mexico State University, Las Cruces, USA
Anne-Marie Rassinoux	Geneva Hospital, Geneva, Switzerland
Daniel Rochowiak	University of Alabama in Huntsville, USA
Eric Salvat	I.N.A.P.G., Paris, France
Leroy Searle	University of Washington, USA
Gerd Stumme	Technische Universität Darmstadt, Germany
Eric Tsui	CSC Financial Services, Australia
Mark Willems	Cycorp, USA
Vilas Wuwongse	Asian Institute of Technology, Thailand

Table of Contents

Invited Talk

CG Modeling

Understanding Natural Language

Applications

SISYPHUS-I

Context, Logic, and CGs

Logic

Position Papers

Conceptual Graphs: Draft Proposed American National Standard

John Sowa

SUNY at Binghamton, USA
sowa@west.poly.edu

Abstract. This is a copy of the draft proposed American National Standard (dpANS) as established May 3, 1999. It is a draft, with many parts remaining to be completed. The most curent version (updated as the standard approaches completion) is always available at: http://concept.cs.uah.edu/CG/Standard.html.

Foreword

This draft proposed American National Standard (dpANS) was developed by the NCITS.T2 Committee on Information Interchange and Interpretation. During that development, the following persons, who were members of NCITS.T2, commented on and/or contributed to this dpANS:

- J. Lee Auspitz, TextWise
- Roger Burkhart, Deere & Company
- Murray F. Freeman, FOSI
- Michael R. Genesereth, Stanford University
- Nancy Lawler, DoD
- Sandra Perez, Concept Technology
- Anthony Sarris, Ontek
- Mark Sastry, Compuware
- John F. Sowa, Concept Technology
- Robert Spillers, IBM

In addition to the NCITS.T2 committee members, the following users and implementers of conceptual graph systems commented on early drafts of this dpANS, tested its implementability in whole or in part, and/or suggested revisions or extensions to the specifications:

- John Black, Deltek Systems, Inc.
- Stephen Callaghan, Peirce Holdings International Pty. Ltd.
- Harry S. Delugach, Univerity of Alabama in Huntsville
- Peter W. Eklund, Griffith University
- Gerard Ellis, Peirce Holdings International Pty. Ltd.
- John W. Esch, Lockheed Martin
- Gil Fuchs, UC Santa Cruz
- Olivier Gerbé, DMR Consulting Group
- Steve Hayes, Peirce Holdings International Pty. Ltd.
- Pavel Kocura, Loughborough University
- Fritz Lehmann, Cycorp
- Robert Levinson, UC Santa Cruz
- Philippe Martin, Griffith University
- Russell Matchan, Peirce Holdings International Pty. Ltd.
- Guy Mineau, Université Laval
- Heather Pfeiffer, New Mexico State University
- Doug Skuce, University of Ottawa
- Finnegan Southey, University of Guelph
- William M. Tepfenhart, AT&T Laboratories
- David J. Whitten, US Department of Veteran's Affairs

1 Scope

This dpANS specifies the syntax and semantics of conceptual graphs (CGs) and their representation as machine-readable character strings in the conceptual graph interchange form (CGIF). CGs have been developed as a conceptual schema language, as specified by ISO/IEC 14481 on Conceptual Schema Modeling Facilities (CSMF). CGIF has been designed for interchange between CSMF systems or between other IT systems that require a structured representation for logic. This Standard also provides guidance for implementers of IT systems that use conceptual graphs as an internal representation or as an external representation for interchange with other IT systems. The external representations are readable by humans and may also be used in communications between humans or between humans and machines.

2 Conformance

An IT system is in conformance with this dpANS if it can accept information expressed in the conceptual graph interchange form (CGIF) defined in Chapter 7, translate that information to some internal representation, and then translate its internal representation to CGIF in a form that is equivalent to the original input according to the criteria defined in Chapter 9. The level of conformance of the IT system shall be specified by a *conformance pair*, which consists of two identifiers called the *style* and the *expressiveness*:

1. The style specifies how much nonsemantic information is preserved by the IT system that accepts and generates information represented in CGIF.
2. The expressiveness specifies the largest subset of semantic features defined by the abstract syntax of Chapter 6 that can be represented in CGIF that is accepted and generated by the IT system.

Chapter 9 specifies the identifiers that shall be used to represent the style and expressiveness of a conformance pair.

3 Normative References

The following normative documents contain provisions, which, through reference in this text, constitute provisions of this dpANS. For dated references, subsequent amendments to, or revisions of, any of these publications do not apply. However, parties to agreements based on this dpANS are encouraged to investigate the possibility of applying the most recent editions of the normative documents indicated below. Members of ISO and IEC maintain registers of currently valid International Standards. ANSI maintains a register of currently valid American National Standards.

ISO/IEC 10646-1:1993, Information Technology (IT) - Universal Multiple-Octet Coded Character Set (UCS).

ISO/IEC 14481:1998, Information Technology (IT) - Conceptual Schema Modeling Facilities (CSMF).

ANSI X3.42:1990, Information Technology (IT) - Representation of numerical values in character strings for information interchange.

4 Terms and Definitions

For the purpose of this dpANS, the terms and definitions given in ISO/IEC 10646-1, ISO/IEC 14481, ANSI X3.42, and the following list apply. Some terms may be shortened to acronyms or abbreviations, which are listed in parentheses after the term.

The term “abstract syntax”, for example, may be abbreviated “AS”, and “conceptual relation” may be abbreviated “relation”.

abstract syntax (AS).

A formal specification of all semantically significant features of conceptual graphs, independent of any concrete notation.

actor.

A conceptual relation whose semantics may be computed or otherwise determined by an IT system external to the current knowledge base. It may affect or be affected by external entities that are not represented in the abstract syntax.

arc.

An ordered pair <r,c>, which is said to link a conceptual relation r to a concept c.

blank graph.

An empty CG containing no concepts or conceptual relations.

bound label.

An identifier prefixed with the character “?” that marks a bound concept of a coreference set.

catalog of individuals.

A context of type CatalogOfIndividuals whose designator is a CG that contains a unique concept for every individual marker that appears in a knowledge base. It may also contain additional concepts and relations that describe the individuals and the relationships among them.

compound graph.

A conceptual graph that is not simple. It contains one or more contexts or defined quantifiers.

concept.

A node in a CG that refers to an entity, a set of entities, or a range of entities.

conceptual graph (CG or graph).

An abstract representation for logic with nodes called concepts and conceptual relations, linked together by arcs.

designator.

A symbol in the referent field of a concept that determines the referent of the concept by showing its literal form, specifying its location, or describing it by a nested CG.

display form (DF).

A nonnormative concrete representation for conceptual graphs that uses graphical displays for better human readability than CGIF.

encoded literal.

An encoded representation of a literal other than a number or a character string. It may be used to specify strings in a language other than conceptual graphs or MIME encodings of text, images, sound, and video.

entity.

Anything that exists, has existed, or may exist.

extended syntax (ES).

Extensions to the CGIF syntax defined in terms of the basic features of CGIF.

first-order logic (FOL).

A version of logic that forms the foundation of many IT languages and systems, including SQL. See Chapter 9 for further information.

formal parameter (parameter).

A concept in a lambda expression whose referent is not defined until the concept is linked to some coreference set whose referent is defined.

higher-order logic (HOL).

An extension to first-order logic that is represented by conceptual graphs and KIF. See Chapter 9 for further information.

identifier.

A character string beginning with a letter or the underscore character "_" and continuing with zero or more letters, digits, or underscores. Case is not significant: two identifiers that differ only in the case of one or more letters are considered identical.

indexical.

A designator represented by the character "#" followed by an optional identifier.

conceptual graph interchange form (CGIF).

A normative concrete representation for conceptual graphs.

conceptual relation (relation).

A node in a CG that has zero or more arcs, each of which links the conceptual relation to some concept.

conformance pair.

A pair of identifiers, called the style and expressiveness, which specify the subset of CGIF that may be accepted or generated by an IT system.

conformity operator.

A symbol "::" that relates types to individual markers. If t::i is true, the entity identified by the individual marker i is said to *conform* to the type t.

context.

A concept that contains a nonblank CG that is used to describe the referent of the concept.

coreference label.

A defining label or a bound label used to identify the concepts that belong to a particular coreference set.

coreference set.

A set of concepts in a CG that refer to the same entity or entities. In CGIF and LF, the members of a coreference set are marked with coreference labels that have matching identifiers; in DF, they may be linked by dotted lines called coreference links.

defined quantifier.

Any quantifier other than the existential. Defined quantifiers are defined in terms of conceptual graphs that contain only existential quantifiers.

defining label.

An identifier prefixed with the character "*" that marks the defining concept of a coreference set.

It is used by implementation-dependent methods for locating the referent of a concept.

individual marker.

A designator, represented by the character "#" followed by an integer, used as an index of some entry in the catalog of individuals. Individual markers can be used as unique identifiers, since two distinct individual makers cannot refer to the same entry in the catalog.

knowledge base (KB).

A context with four nested contexts: a type hierarchy, a relation hierarchy, a catalog of individuals, and an outermost context of type Assertion.

Knowledge Interchange Format (KIF).

A language for representing logic that has a model-theoretic semantics equivalent to the semantics of conceptual graphs.

lambda expression.

A CG with zero or more concepts marked as formal parameters. There is no semantic distinction between a CG with no formal parameters and a zero-adic lambda expression.

layout information.

Specification of the shapes and positions of the nodes and arcs of a CG when it is displayed on a two-dimensional surface. The layout information is not represented in the abstract syntax, and it has no effect on the CG semantics.

linear form (LF).

A nonnormative concrete representation for conceptual graphs intended for better human readability than CGIF, but without requiring graphical displays.

negation.

A context of type Proposition to which is attached a monadic conceptual relation of type Neg or its abbreviation by the character "~" or "¬".

nested conceptual graph.

A CG that is used as a designator in the referent field of some concept.

outermost context.

A context of type Assertion containing a nested CG that asserts the facts and axioms of some knowledge base.

quantifier.

A symbol in the referent field of a concept that determines the range of entities in the referent of the concept. The default quantifier is the existential, which is represented by a blank; all others are defined quantifiers.

referent.

The entity or entities that a concept refers to.

referent field.

The area in a concept where the referent is specified.

relation hierarchy.

A context of type RelationHierarchy that specifies a partial ordering of relation labels, some or all of which are defined by lambda expressions.

relation type label (relation label).

An identifier that specifies a type of conceptual relation.

scoping context.

A special context with the type label SC, which is used to delimit the scope of quantifiers. It has no attached conceptual relations.

signature.

For any conceptual relation r with valence n, a list of n types that specify the type of concept that may be linked to each of the n arcs of r.

simple graph.

A conceptual graph that contains no contexts, negations, or defined quantifiers.

singleton graph.

A CG that contains a single concept and no conceptual relations.

special context.

A kind of context defined by the extended syntax in Section 8.1. Special contexts include negations and contexts whose type label is If, Then, Either, Or, and SC.

star graph.

A CG that contains a single conceptual relation and the concepts that are attached to each of its arcs.

type field.

The area in a concept node where the concept type is specified.

type hierarchy.

A context of type TypeHierarchy that specifies a partial ordering of type labels, some or all of which are defined by monadic lambda expressions.

type label.

An identifier that specifies a type of concept.

valence.

A nonnegative integer that specifies the number of arcs that link a conceptual relation to concepts. All conceptual relations with the same relation label have the same valence. A conceptual relation or relation label of valence n is said to be *n-adic*; *monadic* is synonymous with 1-adic, *dyadic* with 2-adic, and *triadic* with 3-adic.

5 CG Representations

5.1 Interrelationships

A conceptual graph (CG) is a representation for logic specified by the abstract syntax (AS) defined in Chapter 6 of this dpANS. Informally, a CG is a structure of concepts and conceptual relations where every arc links a concept node and a conceptual relation node. Formally, the abstract syntax specifies the CG structures and constraints without commitments to any concrete notation or implementation. Chapter 6 presents the abstract specifications of conceptual graphs in ten normative sections; each section concludes with an informative note that illustrates and explains the abstract details. CGs may be implemented in any machine-readable representation or any humanly readable style that preserves the information specified by the abstract syntax.

For communication between machines, Chapter 7 specifies a concrete syntax called the *conceptual graph interchange form* (CGIF), which has a simplified syntax and a restricted character set designed for compact storage and efficient parsing.

Chapter 8 specifies syntactic extensions, which can be translated to the basic CGIF form that maps to the abstract syntax. For communication with human users, two concrete notations have become traditional: a graphic notation called the *display form* (DF) and a more compact notation called the *linear form* (LF). Although the CGIF notation and the extensions are specified in the normative Chapters 7 and 8, the DF and LF notations are described only by recommended guidelines in the nonnormative Appendix A. Chapter 9 defines the semantics of conceptual graphs by a normative translation to predicate calculus.

5.2 Example

To illustrate the abstract syntax and concrete notations presented in this dpANS, Figure 1 shows the display form of a conceptual graph that represents the propositional content of the English sentence *John is going to Boston by bus*.

In DF, concepts are represented by rectangles: [Go], [Person: John], [City: Boston], and [Bus]. Conceptual relations are represented by circles or ovals: (Agnt) relates [Go] to the agent John, (Dest) relates [Go] to the destination Boston, and (Inst) relates [Go] to the instrument bus. For dyadic relations, the arcs that link relations to concepts are represented by arrows. For n-adic relations, the arcs are numbered from 1 to *n*.

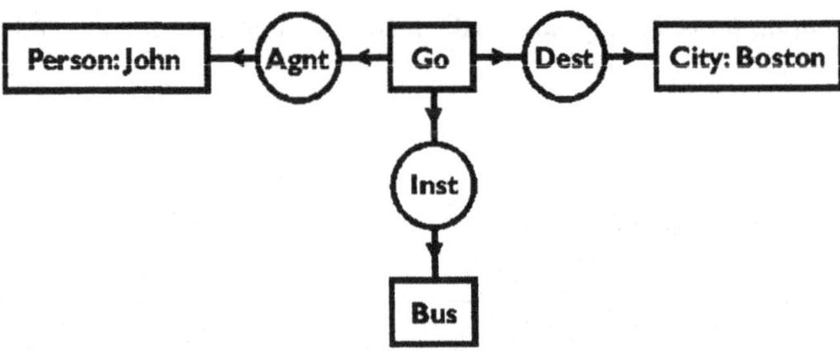

Fig. 1. CG Display Form for "John is going to Boston by bus."

As a mnemonic aid, an arrow pointing toward the circle may be read “has a(n)”; an arrow pointing away may be read “which is a(n)”; and any abbreviations may be expanded to the full forms. With this convention, Figure 1 could be read as three English sentences:

- Go has an agent which is a person John.
- Go has a destination which is a city Boston.
- Go has an instrument which is a bus.

This English reading is a convenience that has no normative status. The numbering or direction of the arcs takes precedence over any such mnemonics.

The linear form for CGs is intended as a more compact notation than DF, but with good human readability. It is exactly equivalent in expressive power to the abstract syntax and the display form. Following is the LF for Figure 1:

```
[Go]-
(Agnt)-[Person: John]
(Dest)-[City: Boston]
(Inst)-[Bus].
```

In this form, the concepts are represented by square brackets instead of boxes, and the conceptual relations are represented by parentheses instead of circles. The hyphen on the first line indicates that the relations attached to [Go] are continued on subsequent lines.

Both DF and LF are designed for communication with humans or between humans and machines. For communication between machines, the conceptual graph interchange form (CGIF) has a simpler syntax and a more restricted character set. Following is the CGIF for Figure 1:

```
[Go *x] (Agnt ?x [Person 'John']) (Dest ?x [City
'Boston']) (Inst ?x [Bus])
```

CGIF is intended for transfer between IT systems that use CGs as their internal representation. For communication with systems that use other internal representations, CGIF can be translated to the Knowledge Interchange Format (KIF):

```
(exists ((?x Go) (?y Person) (?z City) (?w Bus))
(and (Name ?y John) (Name ?z Boston)
(Agnt ?x ?y) (Dest ?x ?z) (Inst ?x ?w)))
```

Although DF, LF, CGIF, and KIF look very different, their semantics is defined by the same logical foundations. Any semantic information expressed in any one of them can be translated to the others without loss or distortion. Formatting and stylistic information, however, may be lost in translations between DF, LF, CGIF, and KIF.

6 Abstract Syntax

6.1 Conceptual Graph

A *conceptual graph* g is a bipartite graph that has two kinds of nodes called *concepts* and *conceptual relations*.

- Every *arc* a of g must *link* a conceptual relation r in g to a concept c in g. The arc a is said to *belong* to the relation r; it is said to be *attached* to the concept c, but it does not belong to c.
- The conceptual graph g may have concepts that are not linked to any conceptual relation; but every arc that belongs to any conceptual relation in g must be attached to exactly one concept in g.
- Three kinds of conceptual graphs are given distinguished names:
  1. The *blank* is an empty conceptual graph with no concepts, conceptual relations, or arcs.
  2. A *singleton* is a conceptual graph that consists of a single concept, but no conceptual relations or arcs.
  3. A *star* is a conceptual graph that consists of a single conceptual relation and the concepts that are attached to its arcs.

Comment.

To illustrate this definition, consider the conceptual graph in Figure 1 for the sentence *John is going to Boston by bus*. The term *bipartite* means that every arc of a CG links a conceptual relation to a concept; there are no arcs that link concepts to concepts or relations to relations. Two of the six arcs in the CG belong to (Agnt), two belong to (Dest), and the remaining two belong to (Inst).

A conceptual graph g with n conceptual relations can be constructed from n star graphs, one for each conceptual relation in g. Since Figure 1 has three conceptual relations, it could be constructed from the following three star graphs, which are represented in the linear form (LF):

```
[Person: John]<-(Agnt)<-[Go]

[Go]-(Dest)-[City: Boston]

[Go]-(Inst)-[bus].
```

These three star graphs constitute a disconnected CG, which does not indicate whether the three copies of [Go] refer to the same instance or different instances of going. If they refer to the same instance, the three identical concepts of type [Go] could be overlaid to form the connected CG in Figure 1.

6.2 Concept

Every concept has a *concept type* t and a *referent* r.

Comment.

This abstract definition does not say how the type and referent are represented. In computer storage, they may be represented by a pair of pointers, one pointing to a specification of the concept type and the other pointing to a specification of the referent. In the concrete notations, the type field is on the left, and the referent field is on the right. In the concept [Bus], "Bus" is the type, and the referent field contains a blank, which represents an existential quantifier; the actual referent is a physical entity of type Bus that exists somewhere in the world. In the concept [Person: John], "Person" specifies the type, and the name "John" designates some person who is the referent.

6.3 Conceptual Relation

Every conceptual relation r has a *relation type* t and a nonnegative integer n called its *valence*.

- The number of arcs that belong to r is equal to its valence n. A conceptual relation of valence n is said to be n-*adic*, and its arcs are numbered from 1 to n.
- For every n-adic conceptual relation r, there is a sequence of n concept types $t_1,...,t_n$, called the *signature* of r. A 0-adic conceptual relation has no arcs, and its signature is empty.
- All conceptual relations of the same relation type t have the same valence n and the same signature s.
- The term *monadic* is synonymous with 1-adic, *dyadic* with 2-adic, and *triadic* with 3-adic.

Certain conceptual relations, called *actors*, may have side effects that are not represented in the abstract syntax; formally, however, actors are treated like other conceptual relations.

Comment.

In Figure 1, Agnt, Dest, and Inst are dyadic relation types. Examples of monadic relation types include Psbl for possibility and Past for the past tense. A 0-adic conceptual relation is logically equivalent to the proposition stated by its defining conceptual graph. Figure 2 shows the between relation (Betw) as an example of a triadic relation (valence 3), whose first two arcs are linked to two things that are on either side of a third. That graph may be read *A person is between a rock and a hard place.*

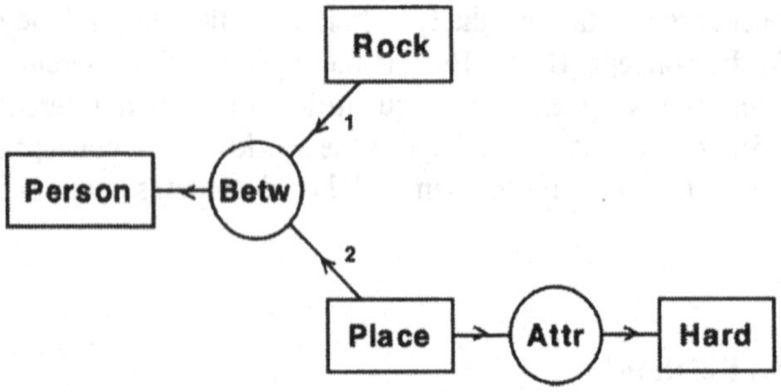

Fig. 2. A conceptual relation of valence 3

The signature of a relation represents a constraint on the types of concepts that may be linked to its arcs. For Agnt, the signature is (Act,Animate), which indicates that the type of the concept linked to its first arc must be Act or some subtype, such as Go, and the type of the concept linked to its second arc must be Animate or some subtype, such as Person. For Betw, the signature is (Entity,Entity,Entity), which shows that all three concepts must be of type Entity, which is the most general type that imposes no constraints whatever.

For a conceptual relation with n arcs, the first n-1 arcs have arrows that point towards the circle, and the n-th or last arc points away. In LF, Figure 2 may be represented in the following form:

```
[Person]<-(Betw)-
<-1-[Rock]
<-2-[Place]-(Attr)-[Hard].
```

The hyphen after the relation indicates that its other arcs are continued on subsequent lines. The two arcs that point towards the relation are numbered 1 and 2. The arc that points away is the last or third arc; the number 3 may be omitted, since it is implied by the outward pointing arrow. For monadic relations, both the number 1 and the arrow pointing towards the circle are optional. For dyadic relations, the arcs are either numbered 1 and 2, or the first arc points towards the circle and the second arc points away.

6.4 Lambda Expression

For any nonnegative integer n, an n-adic *lambda expression* e is a conceptual graph, called the *body* of e, in which n concepts have been designated as *formal parameters* of e.

- The formal parameters of e are numbered from 1 to n.

- There is a sequence $(t_1,...,t_n)$ called the *signature* of e, where each t_i is the concept type of the i-th formal parameter of e. Since a 0-adic lambda expression has no formal parameters, its signature is the empty sequence.

Comment.

This abstract definition does not specify how the formal parameters are designated. The traditional notation, which is used in LF and DF, is to mark a parameter with the Greek letter lambda λ. If n is greater than 1, the parameters may be marked λ$_1$, λ$_2$, ..., λ$_n$. As an example, the conceptual graph for the sentence *John is going to Boston* could be converted to the following dyadic lambda expression by replacing the name John with the symbol λ$_1$ and the name Boston with λ$_2$:

```
[Person: &lambda;1]<-(Agnt)<-[Go]-(Dest)-[City:
&lambda;2].
```

Since this is a dyadic lambda expression, its signature is a pair of types <Person,City>, which come from the type fields of the formal parameters. This lambda expression may be used to define a conceptual relation that relates a person to a city.

To simplify the parsing, the CGIF notation avoids the character λ and represents lambda expressions in a form that shows the signature explicitly:

```
(lambda (Person*x, City*y) [Go *z] (Agnt ?z ?x)
(Dest ?z ?y))
```

For every n-adic lambda expression, the keyword **lambda** is following by a list of n pairs that specify the formal parameters. Each pair consists of a type, such as Person, and a defining label, such as *x. The list of n types is the signature of the lambda expression.

6.5 Concept Type

A *type hierarchy* is a partially ordered set T whose elements are called *type labels*. Each type label in T is specified as *primitive* or *defined.*

- For any concept c, the type of c is either a type label in T or a monadic lambda expression.
- The type hierarchy T contains two primitive type labels Entity, called the *universal type*, and Absurdity, called the *absurd type*. The character "⊤" is synonymous with Entity, and the character "⊥" is synonymous with Absurdity.
- For every defined type label, there is a monadic lambda expression, called its *definition.*

- A defined type label and its definition are interchangeable: in any position where one may occur, it may be replaced by the other.
- The partial ordering over T is determined by the *subtype* relation, represented by the characters “≤” for *subtype*, “<” for *proper subtype*, “≥” for *supertype*, and “>” for *proper supertype*. If t is any type label, Entity≥t and t≥Absurdity; in particular, Entity>Absurdity.
- The partial ordering of type labels must be consistent with the rules of inference defined in Chapter 9.

Comment.

The type hierarchy starts with some set of primitive type labels, which includes at least Entity and Absurdity. The definitional mechanisms introduce new type labels, whose place in the hierarchy is determined by their definitions. As an example, the following equation defines the type label MaineFarmer by a lambda expression for a farmer located (Loc) in Maine:

```
MaineFarmer = [Farmer: &lambda;]-(Loc)-[State:
Maine].
```

The character “λ” indicates that the concept [Farmer] is the formal parameter, and the sequence (Farmer) is the signature of the lambda expression. The type label of the formal parameter is always a supertype of the newly defined type: Farmer ≥ MaineFarmer. As an alternate notation, type labels can be defined with the keyword type and a variable:

```
type MaineFarmer(*x) is [Farmer: ?x]-(Loc)-[State:
Maine].
```

Either the type label MaineFarmer or its defining lambda expression could be placed in the type field of a concept. The following two conceptual graphs are equivalent ways of saying *Every Maine farmer is laconic*:

```
[MaineFarmer: &forall;]-(Attr)-[Laconic].

[ [Farmer: &lambda;]-(Loc)-[State: Maine]:
&forall;]-(Attr)-[Laconic].
```

The second graph may be read *Every farmer who is located in the state of Maine is laconic*. Either graph could be converted to the other by interchanging the type label and its defining lambda expression.

6.6 Relation Type

A *relation hierarchy* is a partially ordered set R whose elements are called *relation labels*. Each relation label is specified as *primitive* or *defined*, but not both.

- For every relation label in R, there is a nonnegative integer n called its *valence.*
- For every n-adic conceptual relation r, the type of r is either a relation label in R of valence n or an n-adic lambda expression.
- For every defined relation label of valence n, there is exactly one n-adic lambda expression, called its *definition.*
- A defined relation label and its definition are interchangeable: in any position of a CG where one may occur, it may be replaced by the other.
- The partial ordering over R is determined by the *subtype* relation, with the symbols ≤ for *subtype*, < for *proper subtype*, ≥ for *supertype*, and > for *proper supertype.*
- The partial ordering of relation labels must be consistent with the rules of inference defined in Chapter 9.
- If r is an n-adic relation label, s is an m-adic relation label, and n is not equal to m, then none of the following is true: r<s, r>s, r=s.

Comment.

As an example, the relation type GoingTo could be defined by a CG that makes GoingTo a synonym for a dyadic lambda expression:

[Relation: GoingTo]-(Def)-

-[Person: λ$_1$]<-(Agnt)<-[Go]-(Dest)-[City: λ$_2$].

This definition says that the relation GoingTo is defined by a lambda expression that relates a person (marked by λ$_1$), who is the agent (Agnt) of the concept [Go], to a city (marked by λ$_2$), which is the destination (Dest) of [Go]. With this relation, the graph for the sentence *John is going to Boston* could be represented by the following CG:

[Person: John]-(GoingTo)-[City: Boston].

This graph can be expanded to a more detailed graph by replacing the relation type label GoingTo with its definition:

[Person: John]-([Person: λ$_1$]<-(Agnt)<-[Go]-

-(Dest)-[City: λ$_2$])-[City: Boston].

The next step is to remove the lambda expression from inside the circle or parentheses, to join the first parameter [Person: λ$_1$] with the concept attached to the first arc, and to join the second parameter [City: λ$_2$] with the concept attached to the second arc:

[Person: John]<-(Agnt)<-[Go]-(Dest)-[City: Boston].

This graph says that the person John is the agent of going and that the city Boston is the destination of going. Each step of this derivation could be reversed to derive the original graph from the expanded graph.

6.7 Referent

The referent of a concept is specified by a *quantifier* and a *designator*.

- A quantifier is one of the following two kinds:
    1. **Existential**. An *existential* quantifier is represented either by the character “∃” or by the absence of any other quantifier symbol or expression.
    2. **Defined**. A *defined* quantifier is a symbol or expression in the extended syntax that may be translated to conceptual graphs that contain only existential quantifiers.

  A referent whose quantifier is existential is called an *existential referent*.

- A designator is one of three kinds:
    1. **Literal**. A *literal* is a representation of the form of the referent. The three kinds of literals are numbers, character strings, and *encoded literals*, which are specified by a pair consisting of an identifier and a string.
    2. **Locator**. A *locator* is a symbol that determines how the referent may be found. The three kinds of locators differ in the way the referent is determined: an *individual marker* specifies a unique concept in the catalog of individuals of a knowledge base; an *indexical* is a symbol that determines the referent by an implementation-defined search that is not defined by this dpANS; and a *name* is a symbol that determines the referent by some conventions that are independent of the current knowledge base.
    3. **Descriptor**. A *descriptor* is a conceptual graph that is said to *describe* the referent.

Comment.

The quantifier and the designator determine the connections between the abstract formalism and the physical referents. Those connections are implementation dependent because they go beyond the notation to the actual people, things, actions, and events. An existential quantifier declares that at least one instance of the type exists. A defined quantifier may specify other quantities or amounts. A designator specifies the referent by showing its form (literal), by pointing to it (locator), or by describing it (descriptor). Following are some examples:

- A literal shows the form of a referent, as in the concept [String: "abcdefg"]. For a multimedia system, an encoded literal may represent sound, graphics, or full-motion video.

- A locator is either a name like Boston or symbol that begins with the character "#". In the concept [Tree: #23846], the locator #23846 is an individual marker that identifies some tree in a catalog of individuals. Examples of indexicals include [Person: #you] and [Book: #this]. Names include symbols like 'John Q. Public' or 'ISBN-0-534-94965-7', whose resolution to a particular referent is independent of the IT system.

- A descriptor is represented by a conceptual graph nested in the referent field of a concept, as in the following example:

```
[Proposition: [Cat]<-(Agnt)<-[Chase]-(Thme)-
[Mouse]].
```

This graph may be read *There exists a proposition, which states that a cat is chasing a mouse.* A concept with a completely blank referent, such as [Cat], has an implicit existential quantifier and a blank conceptual graph as descriptor. Since a blank graph does not say anything about the referent, the concept [Cat] by itself simply means *There exists a cat.*

- An encoded literal begins with the character "%" followed by an identifier and a literal string. The identifier might be the name of a language, such as English, or it might be the name of some encoding such as GIF for images or WAV for sounds. The following concept represents a situation that is described by an English sentence:

```
[Situation: %English"A plumber is carrying a pipe."]
```

In each of the above concepts, there is an implicit existential; the concept [String: "abcdefg"], for example, may be read *There exists a string, whose form is represented by the literal* "abcdefg". Two or more designators in the same concept represent different ways of indicating the same referent, as in the concept [Cat: Yojo #5549], which says that the cat named Yojo is cataloged as individual #5549.

The defined quantifiers include the *universal quantifier*, which is represented by the character "∀" or the string "@every", and *collections* such as {1, 2, 3} or {Tom, Dick, Harry}. The translation rules specified in Section 8.2 translate the universal quantifier to an existential quantifier and a pair of negations. They translate collections to multiple concepts, each containing an existential referent.

To illustrate descriptors and literals in the referent field, Figure 3 shows a concept [Situation] with a nested conceptual graph that may be read *A plumber is carrying a pipe.* In the nested graph, the agent relation (Agnt) indicates that the plumber is the one who is doing the action, and the theme relation (Thme) indicates that the pipe is the theme of the action. That nested CG is a descriptor that describes the situation.

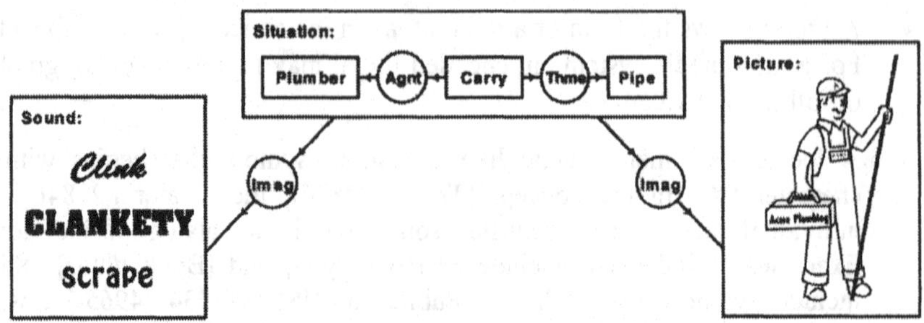

Fig. 3. A conceptual graph with a descriptor and literal referents

The situation node in Figure 3 is linked via the image relation (Imag) to a concept of type Picture, whose referent field contains an encoded literal of the image. In CGIF, the character "%" marks an encoded literal, which is followed by an identifier that specifies the format and a string that represents the encoding of an image. The situation node is also linked via an Imag relation to a concept of type Sound, whose referent field might contain an encoded literal that represents the sound. For better readability, the tools that display DF may translate the CGIF representation to a graphic form, as in Figure 3.

6.8 Context

A *context* C is a concept whose designator is a nonblank conceptual graph g.

- The graph g is said to be *immediately nested* in C, and any concept c of g is also said to be immediately nested in C.
- A concept c is said to be *nested in* C if either c is immediately nested in C or c is immediately nested in some context D that is nested in C.
- Two concepts c and d are said to be *co-nested* if either c=d or there is some context C in which c and d are immediately nested.
- If a concept *x* is co-nested with a context C, then any concept nested in C is said to be *more deeply nested* than *x*.
- A concept d is said to be *within the scope* of a concept c if either d is co-nested with c or d is more deeply nested than c.

Comment.

A context is a concept with a nested conceptual graph that describes the referent. In Figure 3, the concept of type Situation is an example of a context; the nested graph describes the situation as one in which a plumber is carrying a pipe. Figure 4 shows a CG with two contexts; it expresses the sentence *Tom believes that Mary wants to marry a sailor*.

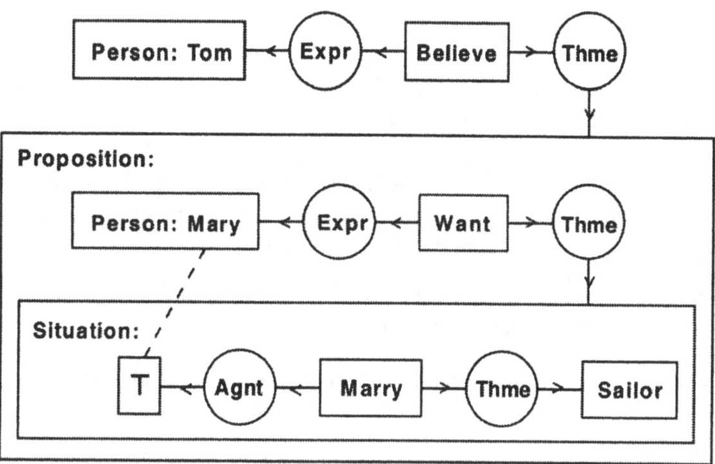

Fig. 4. A conceptual graph containing a nest of two contexts.

In Figure 4, Tom is the experiencer (Expr) of the concept [Believe], which is linked by the theme relation (Thme) to a proposition that Tom believes. The proposition box contains another conceptual graph, which says that Mary is the experiencer of [Want], which has as theme a situation that Mary hopes will come to pass. That situation is described by another nested graph, which says that Mary (represented by the concept [⊤]) marries a sailor. The dotted line, called a *coreference link*, shows that the concept [⊤] in the situation box refers to the same individual as the concept [Person: Mary] in the proposition box. Following is the linear form of Figure 4:

```
[Person: Tom]<-(Expr)<-[Believe]-(Thme)-
[Proposition:  [Person: Mary *x]<-(Expr)<-[Want]-
(Thme)-
[Situation:  [?x]<-(Agnt)<-[Marry]-(Thme)-[Sailor]
]].
```

Both the display form and the linear form follow the same rules for the scope of quantifiers. The outermost context contains three concepts: [Person: Tom], [Believe], and the proposition that Tom believes. Any graph in the outermost context has the effect of asserting that the corresponding proposition is true in the real world. Inside the proposition box are three more concepts: [Person: Mary], [Want], and the situation that Mary wants. Since those three are only asserted within the context of Tom's belief, the graph does not imply that they must exist in the real world. Since Mary is a named individual, one might give her the benefit of the doubt and assume that she also exists; but her desire and the situation she supposedly desires exist in the context of Tom's belief. If his belief is false, the referents of those concepts might not exist in the real world. Inside the context of the desired situation are the concepts [Marry] and [Sailor], whose referents exist within the scope of Mary's desire, which itself exists only within the scope of Tom's belief.

6.9 Coreference Set

A *coreference set* C in a conceptual graph g is a set of one or more concepts selected from g or from graphs nested in contexts of g.

- For any coreference set C, there must be one or more concepts in C, called the *dominant nodes* of C, which include all concepts of C within their scope. All dominant nodes of C must be co-nested.
- If a concept c is a dominant node of a coreference set C, it may not be a member of any other coreference set.
- A concept c may be member of more than one coreference set $\{C_1,C_2,...\}$ provided that c is not a dominant node of any C_i.
- A coreference set C may consist of a single concept c, which is then the dominant node of C.

Comment.

In the display form, the members of a coreference set C may be shown by connecting them with dotted lines, called *coreference links*. In Figure 4, the dotted line connecting [Person: Mary] to [⊤] is an example of a coreference link. The concept [⊤] is within the scope of [Person: Mary], which is the dominant node. Since the two concepts represent the same individual, any information in the dominant node may be copied to the other node. The type label Person or the referent Mary, for example, could be copied from the dominant node to the concept [⊤].

In CGIF, which is defined in Chapter 7, the members of a coreference set are marked by *coreference labels*. One of the dominant nodes is marked with a *defining label*, prefixed with an asterisk; all the other nodes, are marked with *bound labels*, prefixed with a question mark.

6.10 Knowledge Base

A *knowledge base* is a context of type KnowledgeBase whose designator is a conceptual graph consisting of four concepts:

1. *Type hierarchy*. A context of type TypeHierarchy whose designator is a CG *T* that specifies a partial ordering of type labels and the monadic lambda expressions for each defined type label.
2. *Relation hierarchy*. A context of type RelationHierarchy whose designator is a CG *R* that specifies a partial ordering of relation labels, the valences of the relation labels, and the lambda expressions for each defined relation label.
3. *Catalog of individuals*. A context of type CatalogOfIndividuals whose designator is a CG *C* that contains exactly one concept for each individual

marker that appears in any concept in the knowledge base. The designator may also contain other concepts and relations that describe the individuals.

4. *Outermost context.* A context of type Assertion whose designator is a conceptual graph *O*.

   The contents of the knowledge base must satisfy the following constraints:

- The type labels in any conceptual graph or lambda expression in *T, R, C,* and *O* must be specified in *T.*
- The relation labels in any conceptual graph or lambda expression in *T, R, C,* and *O* must be specified in *R.*
- The individual markers in any conceptual graph or lambda expression in *T, R, C,* and *O* must be specified in *C.*

The type labels KnowledgeBase, TypeHierarchy, RelationHierarchy, CatalogOfIndividuals, and Assertion are metalevel labels that need not be specified in *T.* However, if a knowledge base contains information about a knowledge base, including knowledge bases that may be nested inside itself, then its type hierarchy must specify the metalevel labels.

Comment.

The outermost context of a knowledge base B corresponds to what Peirce called the *sheet of assertion.* Its designator is a CG that asserts the information that is assumed to be true in B. The type hierarchy and the relational hierarchy define the labels that appear in the concepts and conceptual relations of any conceptual graphs in any of the contexts nested in B. The catalog of individuals contains global information about the referents identified by individual markers in any concepts in B. As an example, the following catalog specifies individual markers for four entities: a cat, a dog, a village, and a state.

```
[CatalogOfIndividuals:
[Cat: #5539]-
(Name)-[Word: "Yojo"]
(Attr)-[Black]
(Loc)-[Village: #3711]-(Name)-[Word: "Croton-on-
Hudson"] ]
[Dog: #7603]-
(Name)-[Word: "Macula"]
(Attr)-[Spotted]
(Loc)-[State: #2072]-(Name)-[Word: "New Hampshire"]
]]
```

Four concepts in the nested CG have individual markers: #5539 marks a cat named Yojo; #3714 marks a village named Croton-on-Hudson; #7603 marks a dog named Macula; and #2092 marks a state named New Hampshire. The other concepts and relations specify additional information about the four cataloged individuals. The outermost context of the same knowledge base may make further assertions about them, but the problem of finding the individuals themselves is outside the scope of the formalism.

Individuals that are not cataloged might be implied by existential quantifiers in some concepts of a knowledge base. As an example, O might assert that for every human being x, there exists exactly one father y of x and exactly one mother z of x. Any finite knowledge base, however, must eventually stop with some persons whose parents cannot be identified, even though their existence is implied.

The structures that can be represented in CGs are more general and more highly organized than the information in most commercial databases and knowledge bases. Most such systems can be represented as special cases of a CG knowledge base:

- *Relational database.* The names of the SQL tables map to relation labels in the relational hierarchy R, and the names of the columns in the relations map to type labels in the type hierarchy T. The rows of the tables map to star graphs stored in the outermost context O. An SQL table with n rows and m columns is translated to n star graphs, each of which contains an m-adic conceptual relation whose label is the same as the label of the table; each of the n arcs of the relation is attached to a concept whose type label is the same as the name of the corresponding column and whose referent contains the data in the table element of the corresonding row and column. Integrity constraints on the database are translated to axioms represented by CGs that are also stored in the outermost context O.
- *Object-oriented database.* Since O-O databases may have more structure than relational databases, their representation in a CG knowledge base may have more complex graphs. The O-O class hierarchy determines the partial ordering of the type hierarchy T, and the O-O class definitions map to lambda expressions associated with the type labels in T. The data in an O-O DB map to CGs that may be larger than the star graphs that result from a relational database.
- *Expert system knowledge base.* Since there are many different kinds of expert system tools, the mapping to a CG knowledge base may be defined differently for each kind of tool. Frame definitions, for example, would usually be represented in the type hierarchy while instance frames would be represented in the outermost context. For Prolog, the signatures of the predicates would be represented in the type and relation hierarchies while the rules and facts would be asserted in the outmost context.

7 Conceptual Graph Interchange Form (CGIF)

The *conceptual graph interchange form* (CGIF) is a representation for conceptual graphs intended for transmitting CGs across networks and between IT systems that use different internal representations. All features defined by the abstract syntax of Chapter 6 and the extensions of Chapter 8 are representable in CGIF. The CGIF syntax ensures that all necessary syntactic and semantic information about a symbol is available before the symbol is used; therefore, all translations can be performed during a single pass through the input stream. The CGIF syntax is specifed in terms of the Extended BNF (EBNF) rules and metalevel conventions, which are defined in Annex B.

7.1 Lexical Categories

The lexical categories of CGIF are defined as sequences of characters as specified in ISO/IEC 10646-1. Characters outside the 7-bit subset of ISO/IEC 10646-1 occur only in strings that represent identifiers, literals, and comments. By using escape sequences in such strings, it is possible to encode and transmit CGIF in the 7-bit subset of ISO/IEC 10646-1. Numerical values are represented by the conventions of ANSI X3.42. The following lexical categories may be used in EBNF rules in this chapter or in the translation rules in chapters 8 and 9:

EscapeSequence.

An *escape sequence* is a sequence of characters beginning with a backslash that is used to represent certain characters. Nine escape sequences are said to be *valid*:

1. Backspace: "\b".
2. Horizontal tab: "\t".
3. Linefeed: "\n".
4. Form feed: "\f".
5. Carriage return: "\r".
6. Double quote: "\"".
7. Single quote: "\'".
8. Backslash: "\\".
9. Unicode: "\uXXXX", where each X represents one of the following hexadecimal digits: "0", "1", "2", "3", "4", "5", "6", "7", "8", "9", "a", "b", "c", "d", "e", "f", "A", "B", "C", "D", "E", "F".

Unicode escapes may be used to represent any CGIF character, including those in the 8-bit subset of ISO/IEC 10646-1. To ensure a consistent interpretation, the CGIF syntax is designed to allow parsing programs to process Unicode escapes before applying other syntactic rules.

Identifier.

An *identifier* is a string beginning with a letter or underscore "_" and continuing with zero or more letters, digits, or underscores.

```
Identifier  ::=  (Letter | "_") (Letter | Digit |
"_")*
```

Case is not significant: two identifiers that differ only in the case of one or more letters are said to *match.* As an example, the identifier "CarStop" matches "CarsTop", "carstop", "car\u0053top", and "car\u0073top".

Identifiers beginning with "_" followed by one or more digits are generated by the Gensym rule defined in Section 8.2. Such identifiers should be avoided unless they have been generated by a call to the Gensym rule.

Number.

A *number* is a sequence of characters that represents an integer or a floating-point number as defined in ANSI X3.42.

String.

A *string* is a sequence of zero or more characters as specified in ISO/IEC 10646-1. The first eight escape sequences must be used to represent the characters they define; a Unicode escape sequence must be used for all characters outside the 8-bit subset of ISO/IEC 10646-1. Characters in the 8-bit subset may be represented by a Unicode escape sequence. The backslash "\\", double quote "\"", or single quote "\'" or their Unicode representations "\u005c", "\u0022", and "\u0027" may only occur in valid escape sequences.

WhiteSpace.

White space is a sequence of one or more characters that represent a space, a horizontal tab, a carriage return, a line feed, or a form feed.

Comment.

These definitions allow CG tools to process Unicode escape sequences that may occur in CGIF before parsing or otherwise processing a CGIF stream. A tool that uses only 8 bits to represent each character may convert any escape sequence of the form "\u00XX" to an 8-bit character while leaving the other escape sequences unchanged.

7.2 Syntactic Categories

Every syntactic category of CGIF is first described in English and then defined formally by an EBNF rule, as defined in Annex B. Each EBNF rule is followed by an English statement of possible constraints and implications. Context-sensitive constraints, which are determined by the abstract syntax defined in Chapter 6 or by the extended syntax defined in Chapter 8, are stated with a reference to the governing section in Chapter 6 or 8. In all questions of interpretation, the EBNF rules take precedence over the English descriptions and statements. The following syntactic categories may be used in EBNF rules and translation rules:

Actor.

An *actor* begins with a less-than symbol "<" followed by a relation type label or a lambda expression. It continues with zero or more arcs, a vertical bar "|", zero or more arcs, and an optional semicolon ";" and actor comment. It ends with a greater-than symbol ">".

```
Actor  ::=  "<" (RelTypeLabel | LambdaExpression)

Arc* "|" Arc* (";" ActorComment)? ")" ">"
```

The arcs that precede the vertical bar are called *input arcs*, and the arcs that follow the vertical bar are called *output arcs*. If the relation type label or lambda expression has valence n, the total number of input and output arcs must be equal to n.

ActorComment.

An *actor comment* is a sequence of zero or more characters that appears in a actor after a semicolon ";" and before the final ">". Any occurrence of ">" or the Unicode escape "\u003e" in an actor comment is assumed to be the final ">" of the actor.

Arc.

An *arc* of a conceptual relation is specified by either a concept or a bound label.

```
Arc  ::=  Concept | BoundLabel
```

Formally, an arc is a pair consisting of one concept and one conceptual relation. In CGIF, the concept attached to the arc is determined by the concept or bound label in its specification; the relation to which the arc belongs is the one in whose specification the arc is contained.

BoundLabel.

A *bound label* (BoundLabel) consists of a question mark "?" followed by an identifier.

```
BoundLabel   ::=   "?" Identifier
```

A bound label is said to match a defining label if their identifiers are identical, ignoring possible differences between uppercase and lowercase letters.

CG.

A *conceptual graph* (CG) is a list of zero or more concepts, conceptual relations, actors, special contexts, or CG comments.

```
CG  ::=   (Concept | Relation | Actor |
SpecialContext | CGComment)*
```

The alternatives may occur in any order provided that any bound coreference label must occur later in the CGIF stream and must be within the scope of the defining label that has an identical identifier. The definition permits an empty CG, which contains nothing. An empty CG, which says nothing, is always true.

CGComment.

A *CG comment* is a sequence of zero or more characters enclosed in the strings "/*" and "*/".

```
CGComment   ::=   "/*" Character* "*/"
```

Any occurrence of "*/" including the Unicode escape "\u002a\u002f" terminates a CG comment.

Concept.

A *concept* begins with a left bracket "[" followed by an optional type, an optional colon followed by a referent, optional coreference links, and an optional semicolon ";" followed by a concept comment. Finally, the concept is terminated by a right bracket "]".

```
Concept  ::=   "[" Type? (":" Coreflinks? Referent)?
Coreflinks?

("; " ConComment)? "]"
```

If the type is omitted, the default type is Entity. This rule permits the coreference labels to come before or after the referent. If the referent is a CG that contains bound labels that match a defining label on the current concept, the defining label must precede the referent.

In Figure 4, for example, the concept [Person: Mary] could be written in CGIF as [Person:'Mary'*x]; the coreferent concept [⊤] could be written [?x], and its implict type would be Entity.

ConComment.

A *concept comment* (ConComment) is a sequence of zero or more characters that appears in a concept after a semicolon ";" and before the final "]". Any occurrence of "]" or the Unicode escape "\u005d" in a concept comment is assumed to be the final "]" of the concept.

ConTypeLabel.

A *concept type label* (ConTypeLabel) is any identifier other than the five special context labels "If", "Then", "Either", "Or", and "SC" or the reserved label "Else".

CorefLinks.

Coreference links (CorefLinks) are either a single defining coreference label or a sequence of zero or more bound labels.

```
CorefLinks  ::=  DefLabel | BoundLabel*
```

If a dominant concept node, as specified in Section 6.9, has any coreference label, it must be either a defining label or a single bound label that has the same identifier as the defining label of some co-nested concept.

DefLabel.

A *defining label* (DefLabel) consists of an asterisk "*" followed by an identifier.

```
DefLabel  ::=  "*" Identifier
```

The concept in which a defining label appears is called the *defining concept* for that label; a defining concept may contain at most one defining label and no bound coreference labels. Any defining concept must be a dominant concept as defined in Section 6.9.

Every bound label must be resolvable to a unique defining coreference label within the same context or some containing context. When conceptual graphs are imported from one context into another, however, three kinds of conflicts may arise:

1. A defining concept is being imported into a context that is within the scope of another defining concept with the same identifier.

2. A defining concept is being imported into a context that contains some nested context that has a defining concept with the same identifier.

3. Somewhere in the same knowledge base there exists a defining concept whose identifier is the same as the identifier of the defining concept that is being imported, but neither concept is within the scope of the other.

In cases (1) and (2), any possible conflict can be detected by scanning no further than the right bracket "]" that encloses the context into which the graph is being imported. Therefore, in those two cases, the newly imported defining coreference label and all its bound labels must be replaced with an identifier that is guaranteed to be distinct. In case (3), there is no conflict that could affect the semantics of the conceptual graphs or any correctly designed CG tool; but since a human reader might be confused by the similar labels, a CG tool may replace the identifier of one of the defining coreference labels and all its bound labels.

Designator.

A *designator* is a literal, a locator, or a conceptual graph.

```
Designator  ::=  Literal | Locator | CG
```

EncodedLiteral

An *encoded literal* is the character "%" followed by an identifier and a string enclosed in double quotes.

```
EncodedLiteral  ::=  "%" Identifier "\"" String "\""
```

The identifier specifies some implementation-dependent method for interpreting the following string.

FormalParameter.

A *formal parameter* consists of a type and a defining label.

```
FormalParameter  ::=  Type DefLabel
```

Indexical.

An *indexical* is the character "#" followed by an optional identifier.

```
Indexical  ::=  "#" Identifier?
```

IndividualMarker.

An *individual marker* is the character "#" followed by an integer.

```
IndividualMarker  ::=  "#" Integer
```

LambdaExpression.

A *lambda expression* begins with "(" and the keyword "lambda", it continues a signature and a conceptual graph, and it ends with ")".

```
LambdaExpression  ::=  "(" "lambda" Signature CG ")"
```

A lambda expression with n formal parameters is called an n-adic labda expression. The simplest example, represented "(lambda ())", is a 0-adic lambda expression with a blank CG.

Literal.

A *literal* is a number, a string enclosed in double quotes, or an encoded literal.

```
Literal  ::=  Number | "\"" String "\"" |
EncodedLiteral
```

Any double quote that occurs in a literal must be preceded with a backslash.

Locator.

A *locator* is a name, an individual marker, or an indexical.

```
Locator  ::=  Name | IndividualMarker | Indexical
```

Name.

A *name* is a string of one or more letters or a string of one or more arbitrary characters enclosed in single quotes “\'”.

```
Name  ::=  Letter+ | ("\'" Character+ "\'")
```

The quotes are optional for one-word names like John, but they are required for names that contain nonletters, such as ‘John Q. Public’. A string in double quotes, such as “John Q. Public”, is a literal.

Negation.

A *negation* begins with a tilde “~” and a left bracket “[” followed by a conceptual graph and a right bracket “]”.

```
Negation  ::=  "~[" CG "]"
```

A negation is an abbreviation for a concept of type Proposition with an attached relation of type Neg. It has a simpler syntax, which does not permit coreference labels or attached conceptual relations. If such features are required, the negation can be expressed by the unabbreviated form with an explicit Neg relation.

In CGIF, the tilde “~” or “\u007e” is the only symbol used to abbreviate Neg. In LF or DF, the not sign “¬” or “\u00ac” may be used as an alternative to “~”, but “¬” is avoided in CGIF because it is outside the 7-bit subset of ISO/IEC 10646-1.

Quantifier.

A *quantifier* is an existential or a defined quantifier.

```
Quantifier  ::=  Existential | DefinedQuantifier
```

An existential quantifier is represented by an empty string "". A defined quantifier is defined in Chapter 8 by an expansion of the containing CG to an equivalent CG that does not contain that quantifiers.

Referent.

A *referent* consists of a quantifier followed by a designator or a designator followed by a quantifier.

```
Referent  ::=  (Quantifier Designator | Designator
Quantifier)
```

The referent of a concept contains information about the entity or entities that the concept refers to. A blank referent, which implies a blank quantifier and a blank designator, indicates that the concept refers to something that conforms to the concept type, but without giving any further information that might help to identify it.

Relation.

A *conceptual relation* (Relation) begins with a left parenthesis "(" followed by a relation type label or a lambda expression. It continues with zero or more arcs and an optional semicolon ";" and relation comment. It ends with a right parenthesis ")".

```
Relation  ::=  "(" (RelTypeLabel | LambdaExpression)

Arc* (";" Comment)? ")"
```

If the relation type label or the lambda expression has a valence n, the relation must have a sequence of exactly n arcs.

RelComment.

A *relation comment* is a sequence of zero or more characters that appears in a conceptual relation after a semicolon ";" and before the final ")". Any occurrence of ")" or the Unicode escape "\u0029" in a relation comment is assumed to be the final ")" of the concept.

RelTypeLabel.

A *relation type label* (RelTypeLabel) is any identifier other than "lambda".

Signature.

A *signature* is a parenthesized list of zero or more formal parameters separated by commas.

```
Signature  ::=  "(" (FormalParameter (","
FormalParameter)*)? ")"
```

SpecialConLabel.

A *special context label* (SpecialConLabel) is one of the five identifiers "If", "Then", "Either", "Or", and "SC".

```
SpecialConLabel  ::=  "If" | "Then" | "Either" |
"Or" | "SC"
```

The identifier "Else" is reserved for possible future use as a special context label.

SpecialContext.

A *special context* is a negation or a left bracket, a special context label, a CG, and a right bracket.

```
SpecialContext  ::=  Negation | "[" SpecialConLabel
CG "]"
```

In Chapter 8, every special context is a defined by a translation to a context as defined in Section 6.8. Therefore, it is a kind of concept. But unlike other concepts, a special context may not have a defined quantifier, coreference labels, or attached conceptual relations.

Type.

A *type* is a concept type label or a monadic lambda expression.

```
Type  ::=  ConTypeLabel
|  "(lambda" "(" FormalParameter ")" CG ")"
```

Comment.

As an example, the DF representation in Figure 1 was translated to LF, CGIF, and KIF in Section 5.1. Following is a translation of Figure 2 from DF to CGIF:

```
(Betw [Rock] [Place *x1] [Person]) (Attr ?x1 [Hard])
```

For more compact storage and transmission, all white space not contained in comments or enclosed in quotes may be eliminated:

```
(Betw[Rock][Place*x1][Person])(Attr?x1[Hard])
```

This translation takes the option of nesting all concept nodes inside the conceptual relation nodes. A logically equivalent translation, which uses more coreference labels, moves the concepts outside the relations:

```
[Rock *x1] [Place *x2] [Person *x3] (Betw ?x1 ?x2
?x3)

[Hard ?x4] (Attr ?x2 ?x4)
```

The concept and relation nodes may be listed in any order as long as every bound label follows the defining node for that label.

Following is a translation of Figure 3 from DF to CGIF:

```
[Situation *x1 (Agnt [Plumber] [Carry *x2]) (Thme
?x2 [Pipe])]

(Imag ?x1 [Sound: %WAV"..."; The literal string
encodes the audio. ])

(Imag ?x1 [Picture: %JPEG"..."; The literal string
encodes the image. ])
```

This example shows how literals of any kind may be represented in the referent field of a concept. To specify the method of encoding, the type label, such as Sound, could be specialized to a subtype, such as AU. In DF, the sound might be represented by a transcription such as "Clink Clankety Scrape" or it could be converted to audio when someone clicks a mouse on the concept node. To avoid storing multiple copies of large literals, such as sound or video, a single copy might be stored outside the CG system with only a locator in the referent field.

Following is a translation of Figure 4 from DF to CGIF:

```
[Person: Tom *x1] [Believe *x2] (Expr ?x2 ?x1)

(Thme ?x2 [Proposition

[Person: Mary *x3] [Want *x4]

(Thme ?x4 [Situation

(Agnt [Marry *x5] ?x3) (Thme ?x5 [Sailor]) ]) ])
```

Note that the concept [⊤] in Figure 4, which may be represented [?x3] in LF or CGIF, may also be omitted completely in CGIF since the coreference label ?x3 inside the conceptual relation of type Agnt is sufficient to show the connection. As these examples illustrate, CGIF is not as readable as DF, but it contains the equivalent semantic information. To reconstruct an exact image of the original DF, the comment fields in the concept and relation nodes may contain additional layout information to specify the style and placement of the nodes.

7.3 Mapping AS to CGIF

Whenever a feature of the abstract syntax (AS) specified in Chapter 6 has the same name as some CGIF category of Section 7.2, the AS feature is represented by a string defined by that CGIF category. The mapping of AS features to CGIF categories is not one to one because of the following exceptions:

1. **Comments.** No comments are represented in the abstract syntax. Therefore, the CGIF categories ActorComment, CGComment, ConComment, and RelComment do not correspond to anything in AS.

2. **Lexical Categories.** Since the AS features are independent of any notation or implementation, the lexical categories of Section 7.1, which are defined as character strings, do not have a direct mapping to AS. Some of them, such as WhiteSpace, do not correspond to anything in AS.
3. **Noncontiguous Constituents.** Every CGIF category defines a class of contiguous character strings. Some AS features, such as coreference sets, cannot be represented by a single contiguous string. Therefore, they must be mapped to a noncontiguous collection of strings, such as the defining and bound labels.

Each of the following subsections from 7.3.1 to 7.3.10 specifies the CGIF strings that represent the AS features defined in the corresponding sections from 6.1 to 6.10.

7.3.1 Conceptual Graph.

Every conceptual graph (CG) is represented by a string of category CG.

- Every arc *x* of a CG is a pair <r,c consisting of a relation r and a concept c that is linked to r. The Relation string that represents r contains the Arc string that represents *x*. That Arc string may either be the Concept string that represents c or a BoundLabel string whose identifier matches a DefLabel or BoundLabel in the Concept string that represents c.

- A CG string may contain Concept strings that represent concepts that are not linked to any conceptual relation.

- Three kinds of conceptual graphs are given distinguished names:

    1. The *blank* is represented by an empty string or a string that contains nothing but strings of category WhiteSpace or CGComment.

    2. A *singleton* is represented by a string of category CG that contains exactly one string of category Concept, but no strings of category Relation.

    3. A *star* is represented by a string of category CG that contains exactly one string of category Relation. Every Concept string in the CG string must represent one of the concepts attached to the conceptual relation of the star graph.

7.3.2 Concept.

Every concept c is represented by a Concept string that contains a Type string that represents the concept type of c and a Referent string that represents the referent of c.

7.3.3 Conceptual Relation.

Every conceptual relation r is represented by a Relation string or an Actor string. The relation type of r is the RelTypeLabel or the LambdaExpression contained in the Relation string or the Actor string. The valence of r is the valence of the RelTypeLabel or the LambdaExpression.

- The number of Arc strings in the Relation string or the Actor string is equal to the valence of r.
- The signature of r is the signature of the RelTypeLabel or the LambdaExpression contained in the Relation string or the Actor string.

A conceptual relation that is represented by an Actor string may have side effects that are not represented in the abstract syntax or the translation to predicate calculus defined in Chapter 8.

7.3.4 Lambda Expression.

Every lambda expression e is represented by a LambdaExpression string that contains n FormalParameter strings that represent the formal parameters of e and a CG string that represents the body of e.

- For each i from 1 to n, the i-th formal parameter of e is represented by the i-th FormalParameter string.
- The signature of e is the sequence of Type strings contained in the FormalParameter strings.

7.3.5 Concept Type.

Every type hierarchy T may be represented by a concept of type TypeHierarchy that contains a nested CG that defines the concept type labels of T and the partial ordering over T:

```
TypeHierarchy  ::=  "[" "TypeHierarchy"
(TypeDefinition | TypeLabelOrdering)* "]"
```

A type definition is a star graph containing a conceptual relation of type Def that relates a type label to a lambda expression:

```
TypeDefinition  ::= "(" "Def"
"[" "TypeLabel" "\"" ConTypeLabel "\"" "]"
"[" "LambdaExpression" "\"" LambdaExpression "\""
"]" ")"
```

A type label ordering is a star graph containing a conceptual relation of type EQ, GT, or LT that relates two type labels:

```
TypeLabelOrdering  ::= "(" ("EQ" | "GT" | "LT")
```

```
"[" "TypeLabel" "\"" ConTypeLabel "\"" "]"

"[" "TypeLabel" "\"" ConTypeLabel "\"" "]" ")"
```

For two type labels s and t, the type label ordering has a conceptual relation of type EQ iff s=t, of type GT iff st, and of type LT iff s<t.

The type definitions and type label orderings that define a type hierarchy must obey the following constraints:

- A type label that does not appear in a type definition is said to be *primitive.*
- Two primitive type labels are Entity and Absurdity. Every type hierarchy contains the star graph

```
(GT [TypeLabel "Entity"] [TypeLabel "Absurdity"]).
```

- A type label that appears in a type definition is said to be *defined*, and no type label may appear in more than one type definition.

For more concise storage and transmission, multiple star graphs for the type label ordering may be abbreviated by using a collection as defined in Section 8.2. As an example, the following star graph asserts that the type label "Entity" is related to each type label in the collection by the relation GT:

```
(GT [TypeLabel "Entity"]

[TypeLabel @Col{"Physical", "Abstract",
"Independent",

"Relative", "Mediating", "Continuant",
"Occurrent"}]))]
```

The translation rules in Section 8.2 would expand this star graph to seven separate star graphs—one for each type label in the collection.

7.3.6 Relation Type.

Every relation hierarchy R may be represented by a concept of type RelationHierarchy that contains a nested CG that defines the relation type labels of R and the partial ordering over R:

```
RelationHierarchy  ::=  "[" "RelationHierarchy"

(RelationDefinition | ValenceSpec |
RelationLabelOrdering)* "]"
```

A relation definition is a star graph containing a conceptual relation of type Def that relates a relation label to a lambda expression:

```
RelationDefinition  ::= "(" "Def"
```

```
"[" "RelationLabel" "\"" RelTypeLabel "\"" "]"

"[" "LambdaExpression" "\"" LambdaExpression "\""
"]" ")"
```

A valence specification is a star graph containing a conceptual relation of type Has that relates a relation label to its valence:

```
ValenceSpec  ::= "(" "Has"

"[" "RelationLabel" "\"" RelTypeLabel "\"" "]"

"[" "Valence" Integer "]" ")"
```

A relation label ordering is a star graph containing a conceptual relation of type EQ, GT, or LT that relates two type labels:

```
RelationLabelOrdering  ::= "(" ("EQ" | "GT" | "LT")

"[" "RelationLabel" "\"" RelTypeLabel "\"" "]"

"[" "RelationLabel" "\"" RelTypeLabel "\"" "]" ")"
```

For two relation labels r and s, the relation label ordering has a conceptual relation of type EQ iff r=s, of type GT iff rs, and of type LT iff r<s.

The relation definitions, valence specifications, and relation label orderings that define a relation hierarchy must obey the following constraints:

- Every primitive relation label must have exactly one valence specification and no relation definition.

- Every defined relation label must have exactly one relation definition and no valence specification.

For more concise storage and transmission, multiple star graphs for relation label orderings and valence specifications may be abbreviated by using collections. As an example, the following star graph asserts that each of the four type labels in the collection has the valence 2:

```
(Has [TypeLabel @Col{"EQ", "GT", "LT", "Has"}]
[Valence 2])
```

The translation rules in Section 8.2 would expand this star graph to four separate star graphs—one for each type label in the collection.

7.3.7 Referent.

The referent of a concept is represented by a quantifier string and a designator string.

- A quantifier string is one of the following two kinds:

1. **Existential**. An existential quantifier is represented by the empty string "" or by the string "@exist".

2. **Defined**. A defined quantifier is represented by the character "@" followed by an identifier or expression that is defined as a quantifier in Section 8.2.

- A designator is one of three kinds:

1. **Literal**. A literal is represented by a Literal string.

2. **Locator**. A locator is represented by an Indexical string, an IndividualMarker string, or a Name string.

3. **Descriptor**. A descriptor is represented by a CG string containing at least one Concept string or Relation string.

7.3.8 Context.

A context is represented by a SpecialContext string or by a Concept string that contains a CG string containing at least one Concept string or Relation string.

7.3.9 Coreference Set.

Every coreference set C is represented by an identifier i that matches the coreference labels of every Concept string that represent concepts in C.

- One of the dominant nodes d in C shall be selected as the *defining node* of C, and the Concept string that represents d shall contain the defining label *i and no other coreference labels.

- Every Concept string that represents a concept in C other than d shall contain a bound label ?i and no defining labels.

- No Concept string that represents any dominant concept in C may contain a coreference label that does not match i.

- Any Concept string that represents a nondominant concept in C may contain one or more bound labels other than ?i.

- If d is the only concept in C, the defining label *i is optional and may be omitted.

- No concept within the scope of d that is not in C may be represented by a Concept string that contains a coreference label that matches i.

7.3.10 Knowledge Base.

A *knowledge base* KB is represented by a context, whose structure is defined by the following translation rule (see Annex B.2):

```
KB  ::=  "[KnowledgeBase:"
"[TypeHierarchy:" CG.T "]"
"[RelationHierarchy:" CG.R "]"
"[CatalogOfIndividuals:" CG.C "]"
"[OutermostContext:" CG.O "]" "]"
```

1. *Type hierarchy*. The conceptual graph T contains one concept of the form "[Type:" Identifier.*t* DefLabel "]" for each type label *t*. The partial ordering is defined by a collection of star graphs of the form "(Subt" BoundLabel BoundLabel ")", which specifies that the first bound label refers to a type label that has as subtype (Subt) the type label referred to by the second bound label. For each defined type label, there is a star graph of the form "(Def" BoundLabel "[LambdaExpression:" "\"" LambdaExpression.e "\"" "])", which specifies that the bound label refers to a type label that is is defined (Def) by the quoted lambda expression in the following concept.

2. *Relation hierarchy*. A relation hierarchy R together with the n-adic lambda expressions that define some or all of the relation labels in R.

3. *Catalog of individuals*. A concept whose designator is a conceptual graph *g* that contains global information about the individuals identified by individual markers in B. For each individual marker *i* that occurs in any concept of B, there is exactly one concept *c* immediately nested in *g* whose designator is *i*. The graph *g* may also contain other concepts and conceptual relations that describe the individuals in the catalog and the relationships among them.

4. *Outermost context*. A context O called the *outermost context of* B whose type label is Assertion and whose nested CG contains only type labels in T, relation labels in R, and individual markers in C.

8 Extended Syntax

The abstract definitions in Chapter 6 define the core structures of conceptual graphs, which are sufficient to represent all of first-order and higher-order logic. To represent logic more concisely, however, the extended syntax supports special contexts and defined quantifiers, which can all be translated to the core structures. The translation rules defined in Section B.2 provide a mechanism for mapping the extended syntax to CGIF, whose mapping to the abstract syntax is defined in Section 7.3.

8.1 Special Contexts

All Boolean operators can be represented in terms of conjunction and negation. The conjunction of two CGs is represented by including them in the same context. The negation of any CG is represented by including it in a negation. For convenience and readability, some combinations of conjunction and negation are defined as special contexts, whose syntax is defined by EBNF rules and whose semantics is defined by translation rules that convert them to the basic CGIF notation.

The first two special context labels are "If" and "Then", which are used to represent material implication. The syntactic category IfThen is defined as a context with the label "If" whose designator is any CG and a context with the label "Then":

```
IfThen  ::=  "[" "If" CG "[" "Then" CG "]" "]"
```

Any CGIF string of this form can be translated to a basic CGIF string by the TrIfThen translation rule:

```
TrIfThen  ::=  "[" "If" CG.x "[" "Then" CG.y "]" "]"
=  "~[" x "~[" y "]" "]"
```

The first line of this rule matches any IfThen string; during the match, it assigns the string that represents the first CG to the variable x and the string that represents the second CG to y. Then the second line generates a string that encloses the two Concept strings x and y in nested negations.

As an example, the following CGIF string represents the sentence *If Sam has a car, then Sam drives the car*:

```
[If (Has [Person 'Sam'] [Car]
[Then [Drive *x] (Agnt ?x [Person 'Sam']) (Thme ?x
[Car #]) ]]
```

The TrIfThen rule translates this string to the following:

```
~[ (Has [Person 'Sam'] [Car]
~[ [Drive *x] (Agnt ?x [Person 'Sam]) (Thme ?x [Car
#]) ]]
```

This string is corresponds to the sentence It is false that Sam has a car, and Sam does not drive the car.

The Either and Or contexts represent disjunction. The syntactic category EitherOr is defined as a context that has the special label "Either" that contains a conceptual graph and zero or more contexts with the special label "Or":

```
EitherOr  ::=  "[" "Either" CG
("[" "Or" CG "]")* "]"
```

Any CGIF string of this form can be translated to the basic CGIF form by the TrEither translation rule and the TrOrs rule:

```
TrEither  ::=  "[" "Either" CG.x
("[" "Or" CG "]")*.y "]"
=  "~[" x TrOrs(y) "]"
```

The first line of this rule matches a context with the label "Either" and assigns the first CG, which may be blank, to the variable *x*. The second line matches zero or more contexts with the label "Or" and assigns all of them to the variable y. Then the third line generates a negative context containing x and the result of translating y by the TrOrs rule.

```
TrOrs  ::=  ("[" "Or" CG.x "]")?.y String.z RightEnd
=  (y="" ? "" | "~[" x "]" TrOrs(z))
```

The first line of this rule matches an optional context with the special label "Or"; during the match, it assigns the nested CG to *x*, the possibly nonexistent context to y, and the remainder of the input string to z. Then the second line has a conditional that tests to see whether y is empty; if so, it returns the empty string; otherwise, it returns a negation of x followed by the result of applying the TrOrs rule to the string z.

As an example, the following CGIF string has two nested Or contexts to represent the sentence *Either Sam has a car, or Sam rides a bus*:

```
[Either
[Or (Has [Person Sam] [Car]) ]
[Or [Ride *x] (Agnt ?x [Person Sam]) (Inst ?x [Bus])
]]
```

The TrEither and TrOrs rule translate this string to the following:

```
~[
~[ (Has [Person Sam] [Car]) ]
~[ [Ride *x] (Agnt ?x [Person Sam]) (Inst ?x [Bus])
]]
```

Since the Either context contains the Or contexts within its scope, it is the place to put a concept that has coreference links to all the options. As an example, consider the following sentence *Either a cat is on a mat, or it is eating, or it is chasing a mouse.* The concept [Cat] must be placed in a context outside all three disjunctions:

```
[Either [Cat *x]
[Or (On ?x [Mat]) ]
[Or (Agnt [Eat] ?x) ]
[Or [Chase *y] (Agnt ?y ?x) (Thme ?y [Mouse]) ]]
```

This graph is logically equivalent to the graph that represents the sentence Every cat is on a mat, is eating, or is chasing a mouse or to the graph for the sentence If there exists a cat, then it is on a mat, it is eating, or it is chasing a mouse. In all these sentences, the cat is a hypothetical or generic cat. If the speaker has a particular cat in mind, then the concept [Cat: *x] would have to be moved outside the context of Either:

```
[Cat *x]
[Either
[Or (On ?x [Mat]) ]
[Or (Agnt [Eat] ?x) ]
[Or [Chase *y] (Agnt ?y ?x) (Thme ?y [Mouse]) ]]
```

This graph corresponds to the sentence There exists a certain cat, which is on a mat, is eating, or is chasing a mouse.

8.2 Defined Quantifiers

[This section is still under construction.]

As an example, consider the following CGIF representation the sentence *Every cat is on a mat*:

```
(On [Cat @every] [Mat])
```

The occurrence of the symbol "@every" in the referent field of the concept [Cat @every] triggers a translation rule that generates the following basic CGIF representation:

```
~[ [Cat: *_5219]
~[ (On ?_5219 [Mat]) ]]
```

The translation rule generates a new coreference label whenever it is invoked. In this case, it generated the defining label *_5219 with the corresponding bound label ?_5219. The nest of two negations in the translated CGIF corresponds to an implication. Therefore, it can be read *If there exists a cat, then it is on a mat.*

Following are some other constructions that can be introduced by translation rules:

- *Collective designator*. In the concept [Candle: {*}@50], the *generic set symbol* {*} is a collective designator that represents a set of unspecified elements of type Candle, and the quantifier @50 is the *count* or cardinality of the set of candles.
- *Set* and *sequence*. In the concept [Person: {Bill, Mary, Sam}], the collective designator represents a set of three people, and the order is not significant. In [Person: <Bill, Mary, Sam>], the collective designator represents a sequence

of the same three people, but the order is significant: Bill is the first, Mary is second, and Sam is third.

- *Quantifier*. The existential quantifier, which is represented by a blank, is the default. The universal quantifier is represented by the character "∀", as in [Person: ∀], read *every person*. A count is represented by the symbol @ followed by an integer, as in [Person: {*}@5], read *five persons*. A measure is represented by the symbol @ followed by a number and a measuring unit, as in [Interval: 15 sec], read *an interval of 15 seconds*.
- *Collective quantifier*. The generic set symbol {*} and the generic sequence symbol "<*>" are collective designators, which may be combined with a quantifier that indicates the number of elements in the set or sequence: [Farmer: {*}] may be read *a set of farmers*; [Letter: <*>@3] may be read *a sequence of three letters*. The referent {*}∀ implies a set of all elements of the type specified in the type field of the concept, as in [Farmer: {*}∀], read *the set of all farmers*.
- *Lambda expressions as types*. Since type labels are synonyms for lambda expressions, a lambda expression may occur anywhere that a type label may occur. Consider the following concept:

```
[ [Farmer: &lambda;]-(Loc)-[State: Maine]:
{*}&forall;].
```

The type field contains a lambda expression for farmer located (Loc) in the state of Maine, and the referent field contains the generic set symbol {*} and a universal quantifier ∀. Altogether, this concept may be read *the set of all farmers located in the state of Maine*.

The macros allow the full power of logic to be used for defining generalized quantifiers. The character "¬" in the referent field represents the universal negative quantifier, which is read as the English word *no*. The qualifier @>18 may be read as *more than 18*. In combination, they can represent the sentence *No trailer truck has more than 18 wheels*:

```
[TrailerTruck: &not;]-(Part)-[Wheel: {*}@>18].
```

The first step of the macro expansion produces a graph for the sentence *It is false that a trailer truck has more than 18 wheels*:

```
&not;[ [TrailerTruck]-(Part)-[Wheel: {*}@>18] ].
```

This graph, in combination with Figure 4.8, can be used to say that every trailer truck has at least 18 and no more than 18 wheels. The qualifier @=18 can be defined by the conjunction of both these forms to say that every trailer truck has exactly 18 wheels.

Precedence levels for quantifiers.

In predicate calculus, the scope of quantifiers is determined either by parentheses or by the order in which the quantifiers are written, as in the following two formulas:

```
(&forall;x:Woman)(&exist;y:Man)married(x,y).

(&exist;y:Man)(&forall;x:Woman)married(x,y).
```

The first formula may be read *For every woman x, there exists a man y, where x married y* or more simply *Every woman married some man.* That formula does not guarantee that each woman found a unique man, but it leaves the possibility open. The second formula, however, could be read *Some man married every woman.* If there is more than one woman, the man would be a bigamist.

In EGs and CGs, the scope is normally shown by context enclosures, which correspond to the parentheses in predicate calculus. To reduce the number of contexts, another way of showing scope is to assume *precedence levels* for the various kinds of quantifiers. By convention, the universal quantifier ∀ has higher precedence than the existential quantifiers in the same context. With that convention, the two sentences may be represented by the following conceptual graphs:

```
(Past)-[Situation: [Woman: &forall;]<-(Agnt)<-
[Marry]-(Thme)-[Man] ].

[Man: *x]  (Past)-[Situation: [Woman: &forall;]<-
(Agnt)<-[Marry]-(Thme)-[?x] ].
```

The first graph may be read *Every woman married some man*, but the second is more contorted: *There exists a man that every woman married.* The PAST relation attached to the contexts shows the past tense explicitly.

In English, scope is often shown by special words and stylistic conventions. The word *certain*, in the sentence *Every woman married a certain man*, suggests that the existential quantifier for *a certain man* has higher precedence than the universal quantifier for *every woman.* Therefore, the CG symbol @certain may be defined as an existential quantifier with a higher precedence than ∀:

```
(Past)-[ [Woman: &forall;]<-(Agnt)<-[Marry]-(Thme)-
[Man: @certain] ].
```

In this graph, the quantifier @certain has highest precedence, ∀ has middle precedence, and the existential in [Marry] has lowest precedence. That means that for each woman, there is a separate instance of marrying, but the same man is the bridegroom in each instance. Following are the three precedence levels for CG quantifiers:

- *High precedence.* The symbol @certain in the referent field represents an existential quantifier whose scope includes any universal quantifier in the same context.

- *Middle precedence.* The universal quantifier ∀ has lower precedence than @certain, but it has higher precedence than the existential quantifier shown by a blank or by the character "∃".
- *Low precedence.* The basic existential quantifier, represented by ∃ or by a blank, has the lowest precedence. It is within the scope of all other quantifiers in the same context.

When new quantifiers are defined, they are assigned to one or another of these levels. The symbol @1, for example, represents the quantifier ∃!, which means there exists exactly one. It has the same low precedence as the implicit existential. For complex mixtures of quantifiers, the scope can be delimited by transparent contexts (marked by context brackets [ ] with no type label).

9 Logical Foundations

9.1 Translation to Predicate Calculus

When a conceptual graph is represented in CGIF, the grammar rules permit several different options, all of which are logically equivalent. As an example, consider the following CGIF representation for Figure 1, which was discussed in Chapter 5:

```
[Go *x] (Agnt ?x [Person: John]) (Dest ?x [City:
Boston]) (Inst ?x [Bus])
```

As another option, the grammar permits all concept nodes to be moved to the front with all arcs represented by bound variables:

```
[Go *x] [Person: John *y] [City: Boston *z] [Bus *w]

(Agnt ?x ?y) (Dest ?x ?z) (Inst ?x ?z)
```

This representation takes somewhat more space and requires more coreference labels. However, it has a more direct mapping to predicate calculus:

```
(&exist;x:Go)(&exist;y:Person)(&exist;z:City)(&exist
;w:Bus)

(Name(y,'John') &and; Name(z,'Boston') &and;

Agnt(x,y) &and; Dest(x,z) &and; Inst(x,w))
```

In CGIF, the names John and Boston are represented in the referent fields of concepts. In predicate calculus, the Name predicate is added to indicate that they are names. In KIF, however, the absence of a question mark indicates that they are constants, and the Name predicate may be omitted. Following is the corresponding statement in KIF:

```
(exists ((?x Go) (?w Bus))
```

```
(and (Person John) (City Boston)
(Agnt ?x John) (Dest ?x Boston) (Inst ?x ?w)))
```

For a given abstract CG, all the variations of CGIF permitted by the grammar are logically equivalent, and they map to statements in predicate calculus or KIF that are logically equivalent. For output, an IT system may generate any variation of CGIF permitted by the grammar; for input, it must be able to recognize all of them.

The translation from CGIF to predicate calculus is defined by the function φ. If u is any CG, then φ(u) is the corresponding predicate calculus formula. As examples, following are five English sentences and their represetations in CGIF, predicate calculus, and KIF:

1. Every cat is on a mat.

```
(On [Cat: @every] [Mat])
(&forall;x:Cat)(&exist;y:Mat)On(x,y)
(forall (?x cat) (exists (?y mat) (on ?x ?y)))
```

2. It is false that every dog is on a mat.

```
~[(On [Dog: @every] [Mat])]
~(&forall;x:Dog)(&exist;y:Mat)on(x,y)
(not (forall (?x dog) (exists (?y mat) (on ?x ?y))))
```

3. Some dog is not on a mat.

```
[Dog *x]  ~[(On ?x [Mat])]
(&exist;x:Dog)~(&exist;y:Mat)On(x,y)
(exists (?x dog) (not (exists (?y mat) (on ?x ?y))))
```

4. Either the cat Yojo is on a mat, or the dog Macula is running.

```
[Either
[Or (On [Cat: Yojo] [Mat])]
[Or (Agnt [Run] [Dog: Macula])] ]
((&exist;x:Cat)(&exist;y:Mat)(Name(x,'Yojo') &and;
On(x,y)) &or;
((&exist;z:Dog)(&exist;w:Run)(Name(z,'Macula') &and;
Agnt(w,z)))
```

```
(or (exists (?y mat) (and (cat Yojo) (on Yojo ?y)))
(exists (?w run) (and (dog Macula) (agnt ?w
Macula))))
```

5. If a cat is on a mat, then it is happy.

```
[If (On [Cat *x] [Mat])
[Then (Attr ?x [Happy])] ]
(&forall;x:Cat)(&forall;y:Mat)(On(x,y) &implies;
(&exist;z:Happy)Attr(x,z))
(forall ((?x cat) (?y mat))
(= (on ?x ?y)
(exists (?z happy) (attr ?x ?z))))
```

9.2 Canonical Formation Rules

All operations on conceptual graphs are based on combinations of six *canonical formation rules*, each of which performs one basic graph operation. These rules are fundamentally graphical: they are easier to show than to describe.

The first two rules, which are illustrated in Figure 5, are *copy* and *simplify*. At the top is a CG for the sentence "The cat Yojo is chasing a mouse." The down arrow represents the copy rule. One application of the rule copies the Agnt relation, and a second application copies the subgraph -(Thme)-[Mouse]. Both copies are redundant, since they add no new information. The up arrow represents two applications of the simplify rule, which performs the inverse operation of erasing redundant copies. The copy and simplify rules are called *equivalence rules* because any two CGs that can be transformed from one to the other by any combination of copy and simplify rules are logically equivalent. The two formulas in predicate calculus that are derived from the CGs in Figure 5 are also logically equivalent. The top CG maps to the following formula:

- (∃x:Cat)(∃y:Chase)(∃z:Mouse)(name(x,'Yojo') ∧ agnt(y,x) ∧ thme(y,z)),
- which is true or false under exactly the same circumstances as the formula that corresponds to the bottom CG:
- (∃x:Cat)(∃y:Chase)(∃z:Mouse)(∃w:Mouse)(name(x,' Yojo') ∧ agnt(y,x) ∧ agnt(y,x) ∧ thme(y,z) ∧ thme(y,w) ∧ y=z).

By the inference rules of predicate calculus, either of these two formulas can be derived from the other.

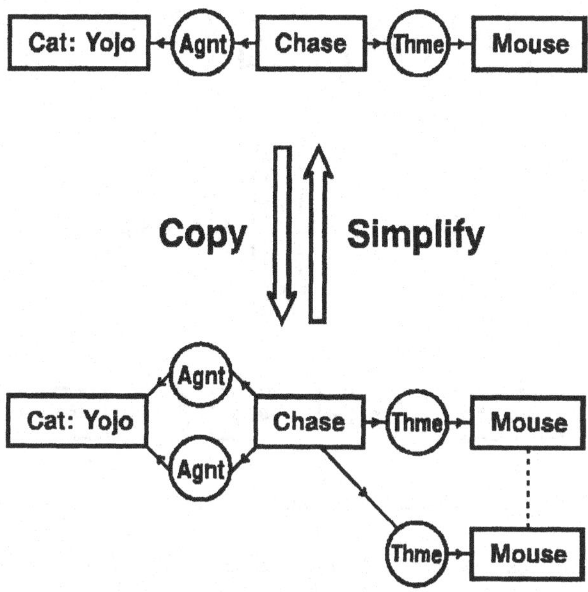

Fig. 5. Copy and Simplify rules.

Figure 6 illustrates the restrict and unrestrict rules. At the top is a CG for the sentence "A cat is chasing an animal." By two applications of the restrict rule, it is transformed to the CG for "The cat Yojo is chasing a mouse." The first step is a *restriction by referent* of the concept [Cat], which represents some indefinite cat, to the more specific concept [Cat: Yojo], which represents an individual cat named Yojo. The second step is a *restriction by type* of the concept [Animal] to a concept of the subtype [Mouse]. Two applications of the unrestrict rule perform the inverse transformation of the bottom graph to the top graph. The restrict rule is called a *specialization rule*, and the unrestrict rule is a *generalization rule*. The more specialized graph implies the more general one: if the cat Yojo is chasing a mouse, it follows that a cat is chasing an animal. The same implication holds for the corresponding formulas in predicate calculus. The more general formula

- (∃x:Cat)(∃y:Chase)(∃z:Mouse)(agnt(y,x) ∧ thme(y,z)) is implied by the more specialized formula:

- (∃x:Cat)(∃y:Chase)(∃z:Mouse)(name(x,'Yojo') ∧ agnt(y,x) ∧ agnt(y,x) ∧ thme(y,z)).

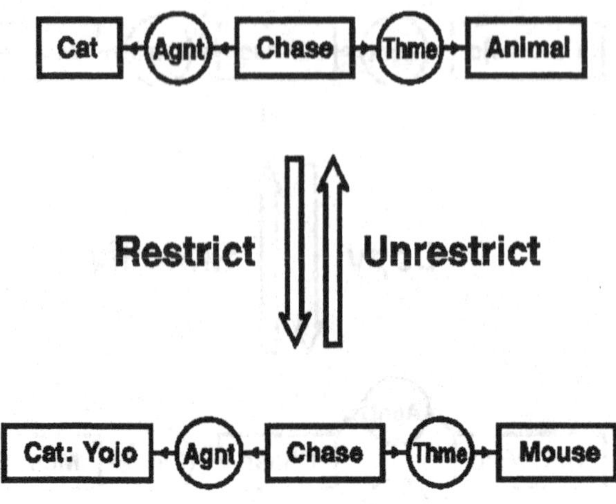

Fig. 6. Restrict and Unrestrict rules

Figure 7 illustrates the *join* and *detach* rules. At the top are two CGs for the sentences "Yojo is chasing a mouse" and "A mouse is brown." The join rule overlays the two identical copies of the concept [Mouse] to form a single CG for the sentence "Yojo is chasing a brown mouse." The detach rule performs the inverse operation. The result of join is a more specialized graph that implies the one derived by detach. The same implication holds for the corresponding formulas in predicate calculus. The conjunction of the formulas for the top two CGs

- ((∃x:Cat)(∃y:Chase)(∃z:Mouse)(name(x,'Yojo') ∧ agnt(y,x) ∧ thme(y,z)) ∧ ((∃w:Mouse)(∃v:Brown)attr(w,v))

is implied by the formula for the bottom CG:

- (∃x:Cat)(∃y:Chase)(∃z:Mouse)(∃v:Brown)(name(x,' Yojo') ∧ agnt(y,x) ∧ thme(y,z) ∧ attr(z,w)).

Although the canonical formation rules are easy to visualize, the formal specifications require more detail. They are most succinct for the simple graphs, which are CGs with no contexts, no negations, and no quantifiers other than existentials. The following specifications, stated in terms of the abstract syntax, can be applied to a simple graph u to derive another simple graph w.

1. *Equivalence rules.* The copy rule copies a graph or subgraph. The simplify rule performs the inverse operation of erasing a copy. Let v be any subgraph of a simple graph u; v may be empty or it may be all of u.

- *Copy.* The copy rule makes a copy of any subgraph v of u and adds it to u to form w. If c is any concept of v that has been copied from a concept d in u, then c must be a member of exactly the same coreference sets as d. Some conceptual relations of v may be linked to concepts of u that are not in v; the copies of those conceptual relations must be linked to exactly the same concepts of u.

- *Simplify.* The simplify rule is the inverse of copy. If two subgraphs v_1 and v_2 of u are identical, they have no common concepts or conceptual relations, and corresponding concepts of v_1 and v_2 belong to the same coreference sets, then v_2 may be erased. If any conceptual relations of v_1 are linked to concepts of u that are not in v_1, then the corresponding conceptual relations of v_2 must be linked to exactly the same concepts of u, which may not be in v_2.

2. *Specialization rules.* The restrict rule specializes the type or referent of a single concept node. The join rule merges two concept nodes to a single node. These rules transform u to a graph w that is more specialized than u.

- *Restrict.* Any concept or conceptual relation of u may be *restricted by type* by replacing its type with a subtype. Any concept of u with a blank referent may be *restricted by referent* by replacing the blank with some other existential referent.
- *Join.* Let c and d be any two concepts of u whose types and referents are identical. Then w is the graph obtained by deleting d, adding c to all coreference sets in which d occurred, and attaching to c all arcs of conceptual relations that had been attached to d.

3. *Generalization rules.* The unrestrict rule, which is the inverse of restrict, generalizes the type or referent of a concept node. The detach rule, which is the inverse of join, splits a graph in two parts at some concept node. The last two rules transform u to a graph w that is a generalization of u.

- *Unrestrict.* Let c be any concept of u. Then w may be derived from u by unrestricting c either by type or by referent: unrestriction by type replaces the type label of c with some supertype; and unrestriction by referent erases an existential referent to leave a blank.
- *Detach.* Let c be any concept of u. Then w may be derived from u by making a copy d of c, detaching one or more arcs of conceptual relations that had been attached to c, and attaching them to d.

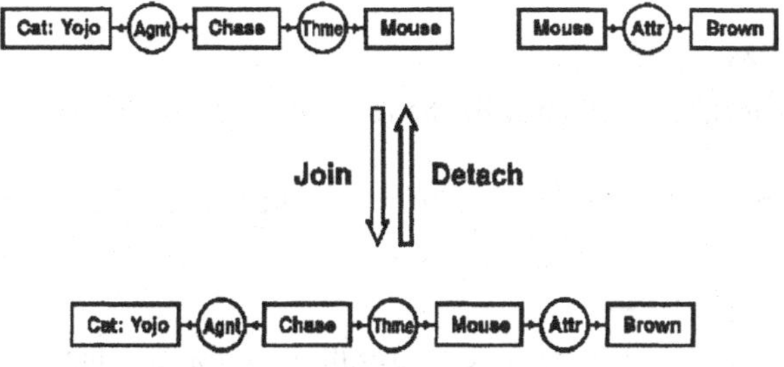

Fig. 7. Join and Detach rules

Although the six canonical formation rules have been explicitly stated in terms of conceptual graphs, equivalent operations can be performed on any notation for logic.

For nested contexts, the formation rules depend on the level of nested negations. A positive context (sign +) is nested in an even number negations (possibly zero). A negative context (sign -) is nested in an odd number of negations.

- *Zero negations.* A context that has no attached negations and is not nested in any other context is defined to be positive.
- *Negated context.* The negation relation (Neg) or its abbreviation by the ~ or ¬ symbol reverses the sign of any context it is attached to: a negated context contained in a positive context is negative; a negated context contained in a negative context is positive.
- *Scoping context.* A context C with the type label SC and no attached conceptual relations is a scoping context, whose sign is the same as the sign of the context in which it is nested.

Let u be a conceptual graph in which some concept is a context whose designator is a nested conceptual graph v. The following canonical formation rules convert u to another CG w by operating on the nested graph v, while leaving everything else in u unchanged.

1. Equivalence rules.

- If v is a CG in the context C, then let w be the graph obtained by performing a copy or simplify rule on v.
- A context of type Negation whose referent is another context of type Negation is called a *double negation.* If u is a double negation around that includes the graph v, then let w be the graph obtained by replacing u with a scoping context around v:

```
[Negation: [Negation: v]]  =  [SC: v].
```

A double negation or a scoping context around a conceptual graph may be drawn or erased at any time. If v is a conceptual graph, the following three forms are equivalent:

```
~[ ~[ v]],   [v],   v.
```

2. Specialization rules.

- If C is positive, then let w be the result of performing any specialization rule in C.
- If C is negative, then let w be the result of performing any generalization rule in C.

3. Generalization rules.

- If C is positive, then let w be the result of performing any generalization rule in C.
- If C is negative, then let w be the result of performing any specialization rule in C.

In summary, negation reverses the effect of generalization and specialization, but it has no effect on the equivalence rules.

9.3 Rules of Inference

The canonical formation rules are the foundation for all logical operations on CGs. For each rule, there is a corresponding inference rule for predicate calculus. If some rule transforms a CG u to a CG v where u implies v, then the corresponding formula φ(u) in predicate calculus implies the formula φ(v). The graph rules, however, are usually simpler than the corresponding rules in predicate calculus.

- *Equivalence rules.* The equivalence rules may change the appearance of a graph, but they do not change its logical status. If a CG u is converted to a CG v by these rules, then u implies v, and v implies u.
- *Specialization rules*:. The specialization rules transform a CG u to another CG v that is logically more specialized: v implies u.
- *Generalization rules.* The generalization rules transform a CG u to another CG v that is logically more generalized: u implies v.

By handling the syntactic details of conceptual graphs, the generalization and specialization rules enable the rules of inference to be stated in a general form that is independent of the graph notation.

- *Erasure.* In a positive context, any graph u may be replaced by a generalization of u; in particular, u may be erased (i.e. replaced by the blank, which is a generalization of every CG).
- *Insertion.* In a negative context, any graph u may be replaced by a specialization of u; in particular, any graph may be inserted (i.e. it may replace the blank).
- *Iteration.* If a graph u occurs in a context C, another copy of u may be drawn in the same context C or in any context nested in C.
- *Deiteration.* Any graph u that could have been derived by iteration may be erased.
- *Equivalence.* Any equivalence rule (copy, simplify, or double negation) may be performed on any graph or subgraph in any context.

Coreference links.

Lambda conversions.

Both Peirce and Frege recognized the need to define new relations as *abstractions* from other logical expressions. But with the *lambda calculus*, Alonzo Church (1941) was the first to formalize the rules for operating on those abstractions. Definitions 7.1.5 and 7.1.6 authorize the fundamental equivalence rule for a lambda calculus of CGs: the label for a defined concept or relation type is interchangeable with its defining lambda expression. By itself, however, this rule is not sufficient; additional *conversion rules* are necessary to transform the CG in which the lambda expression occurs. The two rules of *lambda expansion* remove the defining graph from the type field of a concept or relation node and join its formal parameters to other concepts in the current context:

- *Type expansion.* Let g be a conceptual graph containing a concept c that has a defined type t and an existential referent (i.e. a blank or ∃ as its quantifier). Then the operation of *type expansion* performs the following transformations on g:
    1. The definition of t must be a monadic lambda expression whose body is a conceptual graph b and whose formal parameter p is a concept of b. Let the signature of t be the sequence (s), where s is the type of p.
    2. Remove the type t from the concept c of g, and replace it with s.
    3. Remove the λ marker from the concept p, and replace it with the designator of c.
    4. At this point, the concepts c and p should be identical. Combine the graphs g and b by joining p to c.
- *Relational expansion.* Let g be a conceptual graph containing an n-adic relation r that has a defined type t. Then the operation of *relational expansion* performs the following transformations on g:
    1. The definition of t must be an n-adic lambda expression whose body is a conceptual graph b and whose formal parameters $<p_1,...,p_n>$ are concepts of b. Let the signature of t be the sequence $(s_1,...,s_n)$, where each s_i is the type of p_i.
    2. Remove the relation r from g, and detach each arc i of r from the concept c_i of g.
    3. Remove the λ marker from each parameter p_i, and replace it with the designator of c_i.
    4. For each i from 1 to n, restrict the types of c_i and p_i to their maximal common subtypes. Then combine the graphs g and b by joining each p_i to c_i.

The corresponding rules of *lambda contraction* are the inverses of the rules of lambda expansion: if any conceptual graph g could have been derived by lambda expansion from a CG u, then g may be *contracted* to form u. These rules are the graph equivalents of Church's rules for lambda calculus. Like Church's version, they can be used as equivalence rules in proofs.

9.4 Conformance Levels

Annex A (Informative) Presentation Formats

The abstract syntax (AS) is independent of any notation. It allows a concept to be identified with a block of computer storage or with multimedia presentations that use colors, three dimensions, specialized shapes, sound, and motion. The display form (DF) is intended as a readable two-dimensional representation that can be drawn on paper or computer screens. The linear form (LF) is convenient for keyboard entry of CGs or for printing CGs in a more compact form than DF. Neither DF nor LF is a normative notation, and implementations of CG tools that conform to this dpANS may extend or modify DF or LF to adapt them to specialized applications.

Unlike the abstract form, the concrete notations must deal with the limitations of some physical medium: material such as paper, plastic, or glass; images displayed on a two-dimensional surface or in a three-dimensional space; or structures of bits and pointers in computer storage. The physical medium may introduce features that are not part of the formal definition:

- **Location**. Every concept and relation node in a concrete representation must have some location, described either by spatial coordinates or by some kind of address.
- **Visual phenomena**. The nodes of a concrete representation may differ in size, position, brightness, and color.
- **Storage limitations**. When graphs and collections of graphs become large, they may exceed the size of a sheet of paper, a display screen, or computer storage. Various techniques may be used to move, display, scroll, or zoom in and out of graphs.

These constraints introduce extraneous features that may be important for human factors or computer efficiency, but they are irrelevant to the abstract syntax, the semantics, and the formal operations.

A.1 Display Form

The *display form* (DF) is a concrete representation for CGs that can be printed or displayed on two-dimensional surfaces. It is designed to enhance readability while representing the abstract syntax as closely as possible. Any information about the display form that is not explicitly derived from AS or CGIF is called *layout information.* Conformance to this dpANS does not require the preservation of layout information across translations to and from CGIF or to and from other conceptual schema languages such as KIF. Following are the basic principles for mapping the abstract form to the display form:

1. Concept nodes are drawn as rectangles, which are also called *boxes*.
2. Conceptual relation nodes are drawn as circles, ovals, or ellipses; they may be called *circles*, even though they may be elongated to ovals or ellipses to accommodate long identifiers.
3. The arcs that link a conceptual relation node to a concept node are drawn as solid lines, preferably straight. If necessary, the arcs may cross other arcs, coreference links, and nodes; but to minimize the crossing, the arcs may be curved or bent.
4. The size and position of the nodes, the orientation of the arcs, and the point where an arc is attached to a node has no semantic significance. Sizes and positions may be changed to enhance readability or save space.
5. To distinguish the arcs of an n-adic relation, an integer from 1 to n is drawn next to each arc. For improved readability, other conventions for distinguishing the arcs are permissible:

- For monadic relations, the integer 1 is optional. It may either be omitted entirely or be replaced by an arrowhead pointing away from the relation.
- For dyadic relations, the integer 1 may be replaced by an arrowhead pointing towards the relation, and the integer 2 may be replaced by an arrowhead pointing away from the relation.
- For n-adic relations where n>2, the integer n may be replaced by an arrowhead pointing away from the relation. For the arcs from 1 to n-1, each arc may have both an integer and an arrowhead pointing towards the relation
- If a dyadic relation is declared to be *symmetric*, the integers or arrowheads are optional.
- If an n-adic relation where n>2 is declared to be *symmetric* in the arcs from 1 to k for 1<k<n, the integers from 1 to k may be replaced by arrowheads pointing towards the relation.
- If an n-adic relation is declared to be *functional*, its n-th arc may be drawn as a double line with an arrowhead pointing away from the relation.
- An arrowhead on an arc may be placed either in the middle of an arc or at the end towards which it points.

6. The coreference links that connect concepts to concepts are drawn as dotted or dashed lines that are lighter in weight than the arcs that link conceptual relations to concepts. Any coreference link may be replaced by coreference labels with the same conventions used for CGIF.
7. The type and referent fields of a concept box are separated by a vertical or horizontal line of two or more dots; a vertical line of exactly two dots represents the colon character ":", which is the normal separator in LF.
8. When the separator is a vertical line, the type field is on the left, and the referent field is on the right. When the separator is a horizontal line, the type field is above, and the referent field is below the line. Figure 8 shows the five possible orientations of the type and referent fields. The form in the upper

left-hand corner is the preferred one; the others are used to provide more space for large type or referent expressions.

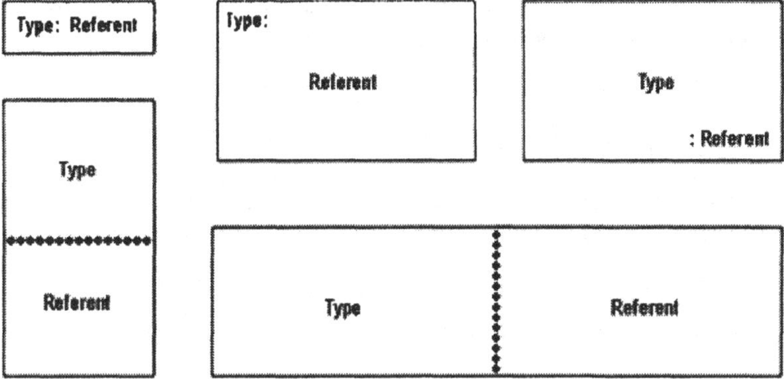

Fig. 8. Five possible orientations of the type and referent fields

9. The contents of the type and referent fields, which are defined in Sections 6.5 and 6.7, are the same for both the display form and the linear form.

10. A context drawn in DF may contain a nested CG in DF or LF. A context in LF, however, may not contain a nested CG in DF.

11. To minimize line crossing, any concept node may be split in two or more nodes in the same context. The first node contains all the type and referent information of the original concept node. It must be marked with a defining label such as *x or a bound label such as ?x. All the other nodes may be drawn as boxes containing nothing but a corresponding bound label such as [?x]. Any coreference links attached to the original node must remain attached to the first node of the split. Any arc of a conceptual relation attached to the original node must be attached to one of the split nodes, but which one is semantically irrelevant.

Since DF is not normative, a CG tool that conforms to this dpANS may modify or extend DF for particular applications or I/O devices. An application to poultry farming, for example, might display a concept of type Chicken in the shape of a chicken.

A.2 Linear Form (LF)

The following EBNF rules define a recommended context-free syntax of the linear form (LF). The starting symbol is CG, which represents a linear representation of a conceptual graph. The referent field, which is common to both the linear form and the display form, is defined in Section 7.2. To minimize dependencies on implementation-dependent encodings, these rules contain no character constants. The symbol BeginConcept, for example, which is normally represented by the left bracket

"[", could be redefined as some other symbol if necessary. The recommended print forms are listed in Section 4.5.

- A CG is either a ConceptBranch or a Relation followed by either a ConceptLink or a ConceptList:

```
CG  ::=  ConceptBranch | Relation (ConceptLink |
ConceptList)
```

- A ConceptBranch is a Concept, which is optionally followed by either a RelationLink or a RelationList:

```
ConceptBranch  ::=  Concept (RelationLink |
RelationList)?
```

- A RelationBranch is a Relation, which is optionally followed by either a ConceptLink or a ConceptList:

```
RelationBranch  ::=  Relation (ConceptLink |
ConceptList)?
```

- A ConceptLink is an Arc followed by a ConceptBranch:

```
ConceptLink  ::=  Arc ConceptBranch
```

- A RelationLink is an Arc followed by a RelationBranch:

```
RelationLink  ::=  Arc RelationBranch
```

- A ConceptList is a BeginLinks marker, a ConceptLink, zero or more pairs of LinkSep and ConceptLink, and an EndLinks marker:

```
ConceptList  ::=  BeginLinks ConceptLink (LinkSep
ConceptLink)* EndLinks
```

- A RelationList is a BeginLinks marker, a RelationLink, zero or more pairs of LinkSep and RelationLink, and an EndLinks marker:

```
RelationList  ::=  BeginLinks RelationLink (LinkSep
RelationLink)* EndLinks
```

- A Concept is a BeginConcept marker, a Type, an optional pair of a FieldSep and a Referent, and an EndConcept marker:

```
Concept  ::=  BeginConcept Type (FieldSep Referent)?
EndConcept
```

- A Relation is a BeginRelation marker, a Type, and an EndRelation marker:

```
Relation  ::=  BeginRelation Type EndRelation
```

- An Arc is an optional Integer followed by either a LeftArc or a RightArc:

```
Arc  ::=  Integer? (LeftArc | RightArc)
```

- A Type is either a TypeLabel or a LambdaExpression:

```
Type  ::=  TypeLabel | LambdaExpression
```

Annex B (Normative) Metalevel Conventions

Annex B represents the metalevel conventions for EBNF rules and translation rules. None of the notation in this annex appears in CGIF, and none of it is required to be implemented in any CG tools. Annex B is normative only because the metalevel conventions it defines are used to define CGIF in the normative chapters 7, 8, and 9 of this dpANS.

B.1 Extended BNF Rules

The syntax of CGIF and LF is specified by metalevel rules written in Extended Backus-Naur Form (EBNF). Each EBNF rule is a context-free grammar rule that has a single *category symbol* on the left, a *defining marker* "::=" in the middle, and a *defining expression* on the right. The following EBNF rule is a metametalevel rule that specifies the syntax of every EBNF rule, including itself:

```
EBNFRule  ::=  CategorySymbol DefiningMarker
DefiningExpression
```

Each EBNF rule defines the category symbol on the left as the name of a class of strings specified by the defining expression on the right. Each defining expression is a regular expression constructed from the following *metalevel categories*:

Category.

A class of strings named by an identifier called a *category symbol*. Every category is a metalevel category, a lexical category, or a syntactic category.

DefiningExpression.

A *defining expression* is a sequence of one or more terms separated by a concatenator (white space) or an alternator (vertical bar with optional white space on either side).

```
DefiningExpression  ::=  Term
  ((WhiteSpace | WhiteSpace? "|" WhiteSpace?) Term)*
```

The concatenator indicates sequence: each term must match a string of the corresponding type in the order in which they occur. The alternator, which may include optional white space on either side, indicates options: either the term on its right or the term on its left, but not both, must match a string of the corresponding type. If a defining expression contains both concatenators and alternators, the concatenators have higher precedence (tighter binding) than the alternators.

DefiningMarker.

A *defining marker* is a string consisting of two colons and an equal sign "::=" with optional white space on either side.

```
DefiningMarker  ::=  WhiteSpace? "::=" WhiteSpace?
```

In an EBNF rule, it indicates that the category on its left is defined by the expression on its right.

MetalevelCategory.

A *metalevel category* is any category whose symbol is defined in Annex B of this dpANS. Metalevel categories are used only to define the syntax of EBNF rules or translation rules.

LexicalCategory.

A *lexical category* is any category whose symbol is defined in Section 7.1 of this dpANS. Lexical categories may be used in EBNF rules or translation rules.

SyntacticCategory.

A *syntactic category*is a category whose symbol appears on the left side of some EBNF rule in Section 7.2 of this dpANS.

Term.

A *term* is a category symbol, a literal string, or a defining expression enclosed in parentheses followed by an optional iterator "*", repeater "+", or option "?" symbol.

```
Term  ::=  (CategorySymbol | Literal | "("
DefiningExpression ")" )
              ("*" | "+" | "?")?
```

The iterator indicates zero or more occurrences of the preceding string, the repeater indicates one or more occurrences, and the option indicates zero or one occurrence.

No category symbol that appears in an EBNF rule or a translation rule in this dpANS may contain embedded white space. In an English sentence, however, a category symbol whose identifier contains an embedded uppercase letter may be written with white space inserted before that letter, and uppercase letters may be translated to lowercase. For example, the category symbol written "WhiteSpace" in an EBNF rule may be written "white space" in an English sentence.

B.2 Translation Rules

The EBNF rules described in Section B.1 specify the syntax of a language, but not its semantics. To specify semantics, the syntactic categories of a *source language* may be translated to some *target language*, such as predicate calculus, whose semantics is independently defined. This section defines *translation rules*, which augment the EBNF rules with *variable assignments*, *tests*, and *translation sequences*. Each translation rule is a sequence consisting of a rule name, a defining marker "::=", a defining expression, a translation marker "=", and target.

```
TranslationRule   ::=   RuleName "::="
DefiningExpression "=" Target
```

A translation rule may be *called* by passing a source string to a program called a *parser*, which matches the defining expression to the source string. If the match is successful, the value of the target is returned as the value of the *rule call*. If the match is unsuccessful or some rule call in the target fails, the rule call is said to *fail*. The parts of a translation rule are constructed from the metalevel categories of Section B.1 and the following additional categories:

Conditional.

A *conditional* is a translation sequence and test followed by a true option and a false option.

```
Conditional  ::=   "(" TranSequence Test "?"
TranSequence ":" TranSequence ")"
```

The value of the first translation sequence is tested for equality "=" or nonequality "!=". If the test is true, the value of the conditional is the value of the second translation sequence; otherwise, it is the value of the third translation sequence.

DefiningExpression.

A *defining expression* of a translation rule is a sequence of one or more translation terms separated by a concatenator (white space) or an alternator (vertical bar with optional white space on either side).

```
DefiningExpression  ::=  TranTerm
((WhiteSpace | WhiteSpace? "|" WhiteSpace?)
TranTerm)*
```

Any defining expression of an EBNF rule as specified in Section B.1 is valid in a translation rule. The only difference is that the translation terms may have an optional variable assignment and an optional test.

GenerationTerm.

A *generation term* is a literal, a variable, a rule call, or a conditional.

```
GenerationTerm  ::=  Literal | Variable | RuleCall |
Conditional
```

The value of the generation term is the value of the literal, variable, rule call, or conditional.

LeftEnd.

A *left end* is an empty string that is assumed to precede any string. In a defining expression, the left end matches the start of any source string, and its value is the empty string "".

RightEnd.

A *right end* is an empty string that is assumed to follow any string. In a defining expression, the right end matches the ending of any source string, and its value is the empty string "".

RuleCall.

A name of a translation rule followed by a translation sequence enclosed in parentheses.

```
RuleCall  ::=  RuleName "(" TranSequence ")"
```

When a rule is called, the value of the translation sequence is parsed by the defining expression of the rule. Then the value of the target sequence of the rule is returned as the value of the rule call. A failure of a rule call during parsing causes the current option to fail. If all parsing options fail or the target sequence of the rule fails, then the rule call fails.

Target.

A translation sequence that follows the translation marker "=" of a translation rule. When the translation rule is called, the value of the target sequence is the value returned by the rule call.

Test.

A *test* consists of an equal sign "=" or an nonequal sign "!=" followed by a generation term or a translation sequence enclosed in parentheses.

```
Test ::= ("=" | "!=") (GenerationTerm | "("
TranSequence ")")
```

If the preceding string matches the value that follows, the equal test is true and the nonequal test is false. Otherwise, the nonequal test is true and the equal test is false.

TranTerm.

A *translation term* (TranTerm) has the same options as a term in an EBNF rule with the addition of an optional variable assignment and an optional test.

```
TranTerm ::= (CategorySymbol | Literal | "("
DefiningExpression ")" )
("*" | "+" | "?")? VarAssignment? Test?
```

If a test occurs at the end of any term, a copy of the string matched by the term is tested. If the test is true, the parser continues; otherwise, it backtracks to another option in the defining expression.

TranSequence.

A sequence of one or more generation terms separated by white space.

```
TranSequence ::= GenerationTerm (WhiteSpace
GenerationTerm)*
```

The value of the translation sequence is the concatenation of the values of all the generation terms in the order in which they occur.

VarAssignment.

A *variable assignment* consists of a period "." followed by an identifier called a *variable*.

```
VarAssignment ::= "." Variable
```

The string of the source language matched by the immediately preceding term is assigned as the value of the variable. Any value previously assigned to the variable must be saved in case a subsequent mismatch in the parsing of the source string causes the parser to backtrack and try another option. A variable that has never been assigned a value has the empty string "" as its default value.

Comment.

As an example, the following translation rule named Paragraph matches any text string and replaces all blank lines by the HTML paragraph tag “<p>”.

```
Paragraph  ::=  LeftEnd String.Front
("\n" WhiteSpace? "\n" String.Back)? RightEnd
=   Front (""=Back ? "" : "\n<p>" Paragraph(Back))
```

When this rule is called to parse any string, the category named LeftEnd matches the start of the source string. Then the category named String matches any string of zero or more characters, which it assigns to the variable named Front. In this case, the parser can scan ahead to look for the first occurrence of a line feed “\n” followed by optional white space, another line feed, and any string up to the right end, which it assigns to the variable Back. If the parser fails to find a sequence of two line feeds separated by white space, it assigns the entire source string to Front and leaves Back with the default value “”.

After the parser finishes matching the defining expression to the source string, the target sequence is evaluated to generate the value of the rule call. In this case, it generates the value of Front concatenated to the value of the conditional. If Back is the empty string, the conditional returns the empty string. Otherwise, it returns the value “\n<p>” concatenated to the result of calling Paragraph to continue processing Back.

Annex C (Informative) Bibliography

This dpANS is based on the CG syntax and semantics as published in book form [12,14], in three special issues of journals [1,13,16], and in the proceedings of a series of international conferences and workshops [2,3,6,7,8,9,11,15]. Conceptual graphs have been recommended as a conceptual schema language [10], and they have been implemented as an intermediate language for relating different conceptual models [5]. Conceptual graphs are semantically equivalent to the Knowledge Interchange Format (KIF), as defined in the dpANS [4]; any information expressed in either KIF or CGs can be automatically translated to the other without loss or distortion.

1. Chein, Michel, ed. (1996) *Revue d'Intelligence artificielle*, Special Issue on Conceptual Graphs, vol. 10, no. 1.
2. Eklund, Peter W., Gerard Ellis, & Graham Mann, eds. (1996) *Conceptual Structures: Knowledge Representation as Interlingua*, Lecture Notes in AI 1115, Springer-Verlag, Berlin.
3. Ellis, Gerard, Robert A. Levinson, William Rich, & John F. Sowa, eds. (1995) *Conceptual Structures: Applications, Implementation, and Theory*, Lecture Notes in AI 954, Springer-Verlag, Berlin.
4. Genesereth, Michael R., & Richard Fikes, eds. (1998) *Knowledge Interchange Format (KIF)*, draft proposed American National Standard, NCITS.T2/98-004.

5. Hansen, Hans Robert, Robert Mühlbacher, & Gustaf Neumann (1992) *Begriffsbasierte Integration von Systemanalysemethoden*, Physica-Verlag, Heidelberg. Distributed by Springer-Verlag.
6. Lukose, Dickson, Harry Delugach, Mary Keeler, Leroy Searle, & John Sowa, eds. (1997) *Conceptual Structures: Fulfilling Peirce's Dream*, Lecture Notes in AI 1257, Springer-Verlag, Berlin.
7. Nagle, T. E., J. A. Nagle, L. L. Gerholz, & P. W. Eklund, eds. (1992) *Conceptual Structures: Current Research and Practice*, Ellis Horwood, New York.
8. Mineau, Guy W., Bernard Moulin, & John F. Sowa, eds. (1993) *Conceptual Graphs for Knowledge Representation*, Lecture Notes in AI 699, Springer-Verlag, Berlin.
9. Mugnier, Marie-Laure, & Michel Chein, eds. (1998) *Conceptual Structures: Theory, Tools, and Applications*, Lecture Notes in AI 1453, Springer-Verlag, Berlin.
10. Perez, Sandra K., & Anthony K. Sarris, eds. (1995) *Technical Report for IRDS Conceptual Schema*, Part 1: Conceptual Schema for IRDS, Part 2: Modeling Language Analysis, X3/TR-14:1995, American National Standards Institute, New York, NY.
11. Pfeiffer, Heather D., & Timothy E. Nagle, eds. (1993) *Conceptual Structures: Theory and Implementation*, Lecture Notes in AI 754, Springer-Verlag, Berlin.
12. Sowa, John F. (1984) Conceptual Structures: Information Processing in Mind and Machine, Addison-Wesley, Reading, MA.
13. Sowa, John F., ed. (1992) *Knowledge-Based Systems*, Special Issue on Conceptual Graphs, vol. 5, no. 3, September 1992.
14. Sowa, John F. (1999) Knowledge Representation: Logical, Philosophical, and Computational Foundations, PWS Publishing Co., Pacific Grove, CA.
15. Tepfenhart, William M., Judith P. Dick, and John F. Sowa, eds. (1994) *Conceptual Structures: Current Practice*, Lecture Notes in AI 835, Springer-Verlag, Berlin.
16. Way, Eileen C., ed. (1992) *Journal of Experimental and Theoretical Artificial Intelligence* (JETAI), Special Issue on Conceptual Graphs, vol. 4, no. 2.

Constraints on Processes: Essential Elements for the Validation and Execution of Processes

Guy W. Mineau

Dept. of Computer Science, Université Laval
Quebec City, Quebec, Canada
mineau@ift.ulaval.ca

Abstract. A process is often described as a sequence of actions that changes the state of a system. To make sure that it is semantically valid, it must abide by semantic constraints defining its proper behavior. These constraints are called *behavioral constraints*. In the past, [1] presented how to describe and structure constraints on conceptual graphs in a declarative yet operational way; and [2] presented a framework to describe and execute processes using conceptual graphs. This paper combines these two approaches to show how processes can be constrained. It also gives two examples showing why constrained processes are needed in real applications. The first example is a database application where migration constraints must be enforced; the second example shows how agent systems must use behavioral constraints in their interaction. By adding behavioral constraints to conceptual graphs based tools, this paper proposes the CG theory as a powerful modeling language not only for data but also for process modeling.

1 Introduction

There are applications for which processes must be represented in a declarative manner. For instance, whenever one wants to make inferences from a process, either for validation, explanation or simulation purposes, a declarative representation is required. At last year's conference, [2] showed how such a CG representation could be obtained automatically from any non-ambiguous well-formed syntactical procedural representation. Consequently, it offered a declarative representation for processes which could easily be obtained from other representations, and which is useful for the modeling of some application domain where processes are objects of interest.

One additional step towards the full modeling of processes is to allow the representation of constraints on processes, i.e., of *behavioral constraints*. The inferences made from processes must take these constraints into account if the validation or execution of these processes is sought. Because such constraints define semantically valid behavior, they allow the definition of *who is allowed to do what* in the system. The most common use for these constraints is to define the scope and applicability of transactions. In agent systems for instance, message passing is defined as a process involving agents. This can be seen as a transaction occurring between agents. Since agents must communicate through a particular interface, there is a validation that must take place to that effect. The message is sent from the source agent to the destination agent if this transaction is validated. Behavioral constraints are at the heart of this type of applications.

When update operations are represented as processes, then the update of the system, and therefore its evolution, can be monitored and constrained. This is useful in all database and knowledge base applications in order to prevent any update operation that would cause a violation of the semantics of the model. Since the domain and the transactions are described in a declarative format, explanations about the failure of the update operation could provide the user with valuable knowledge about this failure with regard to the actual state of the system. Since the CG formalism can be used to describe database applications, this type of application is also of interest to us.

Consequently, this paper presents a framework that allows the declarative description of behavioral constraints under the CG formalism. Section 2 summarizes the representation of processes under the CG formalism as proposed in [2]. Section 3 reminds the reader about the framework proposed in [1] to represent semantic constraints. Section 4 combines the two previous notions to illustrate how semantic constraints on processes can be represented under the CG formalism. To that effect, the migration constraint of database literature is used. Section 5 exemplifies in more details what is presented in Section 4 by applying it to behavioral constraints in agent systems.

2 The Representation of Processes

The representation of processes, i.e., of procedural knowledge, falls under two different representation paradigms, either the procedural or declarative paradigm. The former provides an algorithmic representation that is useful for the execution of the process; while the latter allows inferences to be drawn from the description of the process since it is then available to an inference engine. In our previous work [2], we proposed a declarative representation for processes under the CG formalism, that would also allow its execution. Based on state transition machines, this representation describes a state change from one state to some other state where some new knowledge is asserted and/or old knowledge retracted whenever some conditions are met. Thus, we proposed to use pairs of pre and postconditions to describe a process. Figure 1 below shows the process of delivering a diploma to a student who completed an undergraduate program. It states that when an enrolled student has completed the total amount of credits required by his undergraduate program, then he is awarded the corresponding diploma, is then considered to be a graduate student (is no longer considered to be undergraduate, if it was the case), and is no longer enrolled in this program. In Figure 1, the precondition states what is required in order for the process to fire (to execute); the postconditions state what happens as the result of the process. The overall execution mechanism, including parameter passing, is described in more

3 Semantic Constraints

Since for a particular canon the set of all syntactically correct graphs can be structured into a partial order of generality ( ), called generalization hierarchy in [4], our previous work on constraints [1] proposes to identify valid and invalid subspaces of this generalization hierarchy. The idea is to identify parts of this space which are semantically invalid according to the semantics of the domain. To that purpose, we proposed to use non-validity intervals such as [u,v[, where u and v are two graphs such that u v. The interpretation of this interval states that any syntactically correct graph x such that x u is semantically invalid unless x v. This interval could also be represented using a **false-unless** construct: **false** u **unless** v, which is equivalent to [u,v[, and which is the default interpretation of this statement. Of course, the interval may exclude u and/or include v in the interval, and the **unless** part may be omitted[2]. Figure 2 shows a constraint that ensures that a student can not be considered graduate if he has not completed his entire undergraduate program. This prevents the first graph to be acquired without any basis to support this assertion (e.g., as could be the case through direct manipulation of the knowledge base). Figure 3 shows a constraint that states that it is impossible for a student to have both statuses: undergraduate and graduate, at the same time.[3]

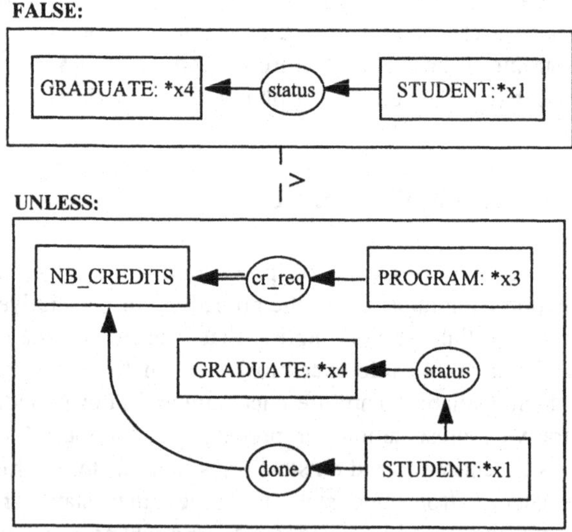

Fig.2: Constraint C1 = **false** u_1 **unless** v_1, i.e., $[u_1,v_1[$.

---

[2] The syntax of a constraint statement is: **false {exclude}** u **{unless {include}** v**}**. The default interpretation being [u,v[, the **exclude** and **include** keywords inverse the associated bracket when present. When omitted, the **unless** part is equivalent to **unless included** [⊥], where [⊥] is the absurd graph (always semantically invalid).

[3] The => symbol represents a functional dependency between the input concepts and the output concept of a relation, as defined in [5].

details in [2, 3]. The double diamond box gives the name of the process which is called when the conditions are met. Different subsumption relations among processes can be defined (see Appendix A). Specifically, the process/subprocess ordering will be useful for what follows. Informally, process P1 is said to be a subprocess of process P2, written P1 $_s$ P2, iff the triggering of P2 always entails the triggering of P1 and the resulting state from applying P1 onto some data is also reached at some point when P2 is applied onto the same data (though not necessarily the final state of P2). The most general process is represented by a double diamond box labeled Link which does not have statement boxes (assertions and negations) attached to it.[1]

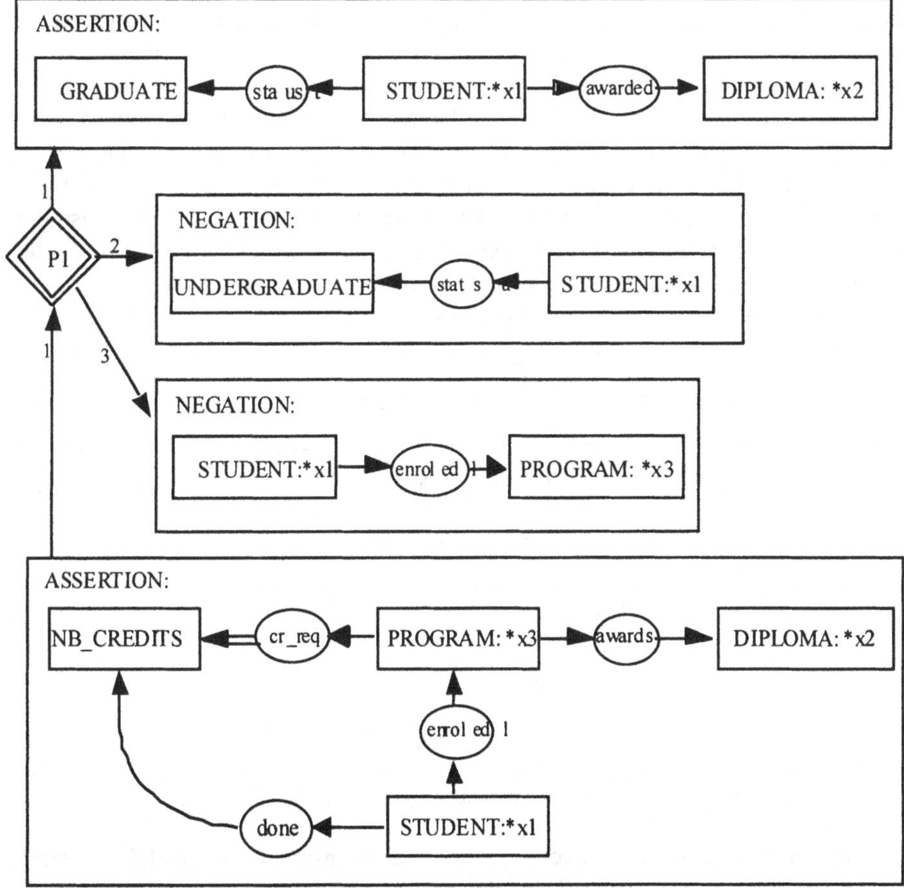

Fig.1: A graphical representation of P1: delivering a diploma to a graduating student.

[1] Throughout this paper, we assume that coreferencing is global, i.e., there is an implicit outer context where all quantified variables are known.

FALSE:

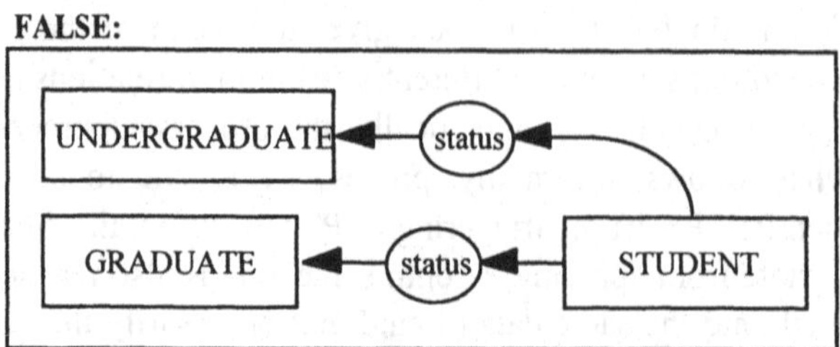

Fig.3: Constraint C2 = **false** u_2, i.e., $[u_2,[\perp]]$: a student can not have both statuses.

Both constraints can be integrated into the space of all syntactically correct graphs in order to identify invalid subspaces. A graphical representation of such a structure, called *validity space*, is shown in Figure 4. The nodes of this structure are: the universal (top) and absurd (bottom) graphs and the graphs describing the two intervals: u1, v1 and u2. The links explicit the relation between each pair of nodes, when needed. The labels - and + on the arcs and nodes indicate whether the corresponding graphs are known to be semantically invalid according to the constraints (-) or not, i.e., are semantically plausible (+) (since they do not violate any constraint). This structure is automatically produced from a set of constraints expressed under **false-unless** constructs. Any asserted graph is classified in this structure. If any – label is associated with it, then the graph can not be asserted because it violates some constraint; its acquisition fails.

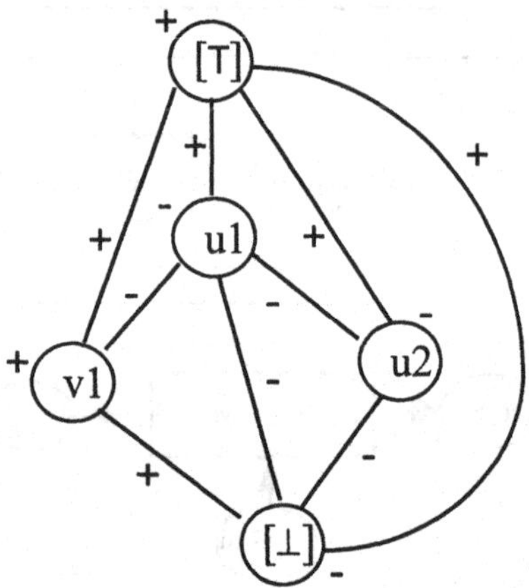

Fig.4: The validity space associated with constraints C1 and C2.

Already the reader should notice that since these constraints (C1 and C2) identify what graphs may be acquired and stored in the knowledge base, they already constrain the execution of a process. For instance, constraint C1 states that the graduate status of a student can not exist independently from the fact that this student has completed some (undergraduate) program. Any process asserting the former fact could not do so if the latter has not previously been asserted. In terms of an implementation, if a postcondition of a currently executing process would result in a violation of a constraint, the process would be blocked and the system would produce a fatal error statement warning the system's administrator of this situation.

4 Constraints on Processes

However, constraints as expressed in Section 3 above are not sufficient; one may still want to block any process that would carry some illegal operation. For instance, we may have the following constraint to enforce: a graduate student can never lose his graduate status once acquired. This is really a constraint on the possible update operations that the system may carry: it is a *behavioral* constraint. It may be expressed as: there should not be any process whose postcondition deletes the graduate status of a graduate student. Figure 5 shows process P2, which would delete the graduate status of a graduate student if executed.

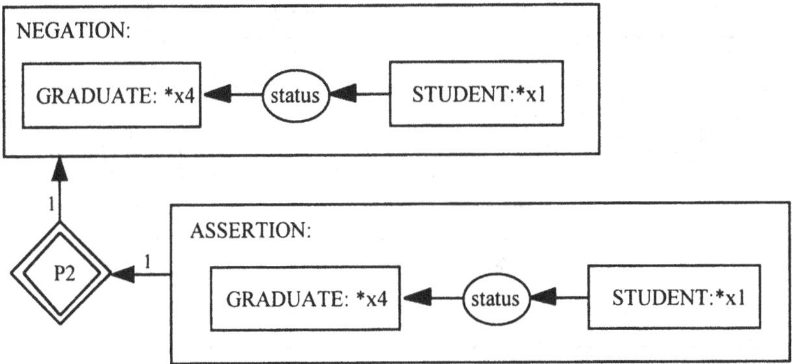

Fig.5: Process P2 should be blocked as the result of the behavioral constraint above.

Using the subprocess partial order relation defined over processes $_s$ (see Appendix A), and using the same scheme as presented in Section 3, one could give the following constraint C3 = **false** u_3, where u_3 is the graph of Figure 5, in order to block any process whose execution would resemble that of P2, that is, would include P2 as a subprocess. In fact, any process which is a specialization of P2 according to the $_s$ relation would fall underneath it in a validity space, and would be considered semantically invalid. Consequently, it could not be asserted in order to trigger its execution (see [2]) without violating constraint C3, and thus would not be executed.

As another example, let us say that an *undergraduate* status can not be deleted from a student's record (process P3) unless it is replaced by a *graduate* status (process P4). In that case, the graph of Figure 6 would be semantically invalid, but the one in Figure 7 would be valid. This new constraint C4 could be expressed as: **false** g_4 **unless** g_5. Any process falling underneath P3 in a validity space would be semantically invalid and would be blocked unless it also falls under P4 in which case it could fire since its behavior would then be specific enough to meet the constraint.

The reader should notice that together, constraints C1, C2, C3 and C4 enforce a *migration* constraint as defined in database literature [6]. In our example, once an *undergraduate* status is given to a student, it can never be deleted; it can only be replaced by a *graduate* status (C4) and this claim most be supported (C1).

Furthermore, once a *graduate* status is given to a student, it can never be deleted (C3). Finally, a student can never carry both status at the same time (C2). Any update operation must conform to these constraints, which prevent the evolution of the database (or knowledge base) to ever violate the semantics of the model. Behavioral constraints confine the evolution of a database (knowledge base) to semantically plausible states; they constantly monitor the evolution of the system, catching violations as they occur. Providing a declarative model that describes the evolution of a system is a simple yet powerful modeling tool that helps maintain the consistency of the system throughout its use and evolution without having to give explicit hard-coded procedures that must all be triggered each time consistency must be checked.

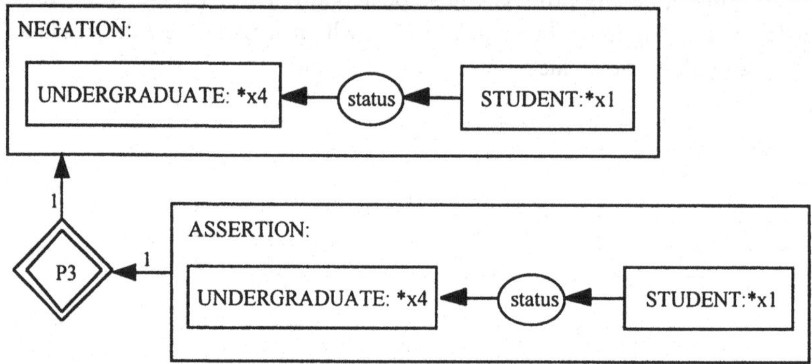

Fig.6: Process P3 = graph g_4.

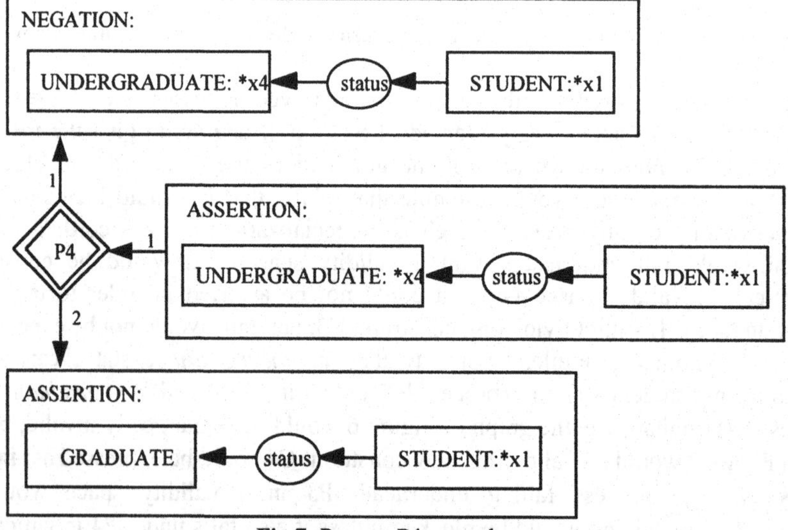

Fig.7: Process P4 = graph g_5.

5 Behavioral Constraints in Agent Systems

As demonstrated above on a simple example, behavioral constraints are powerful in the sense that they improve the expressiveness of the modeling language so that update operations can be constrained within the semantics of the model. For other applications like agent systems, since the interaction between agents follows some well-established protocol, behavioral constraints are essential. If not modeled, some operations (behaviors) may have side effects that could jeopardize the soundness of the model. In effect, without enforced behavioral constraints, the soundness of the system relies solely on the commitment and ability of the agents to act according to the agreed-upon protocol of interaction. Section 5.3 gives examples of current operations in agent systems where behavioral constraints, as presented in this paper, significantly improve the consistency of the system because they enforce the interaction protocol. Sections 5.1 and 5.2 state basic assumptions needed to represent agent systems under the CG formalism.

5.1 The Memory Structure

Agents each model the world independently from each other. Under the CG formalism, they must have sheets of assertion of their own to represent assertions that: they know are true, they think are true, they hope are true, they wish were true, and so on. Through these sheets of assertion, each agent also models what it knows about other agents [7]. These sheets of assertion being private to an agent, they compose its *private memory*. Also, agents may need to share some sheets of assertion for communication purposes or for collective work such as planning and coordinating [8,9]. In our simple model, we state that these sheets of assertion compose a *public memory* accessible to all agents.[4] Provided that time frames are sufficient to correctly model the domain and the events that change it[5], we propose to use contexts as defined in [10] to represent and structure the different sheets of assertion required by the application.

In brief, any assertion must be done in relation to a context. Figure 8 below shows such an assertion which states that agent Guy believes that there is some agent in charge of coordination. This assertion is made in Guy's private memory. Thus the assertion (which is the *extension* of the context) is made in the context of Guy's beliefs in his private memory (which is the *intention* of the context). The truth-value of the assertion holds only in this context. Based on Formal Concept Analysis [11], our previous work on contexts [10] proposes a way to structure and relate contexts like those of Figure 9, both inferred from Figure 8. The interested reader should refer to it.

---

[4] Of course we could complexify this model and identify what agents share what sheet of assertion, but that is not the purpose of this paper.

[5] In this paper, again for simplicity reasons, we assume that the time structure of the application can be modeled using time frames, i.e., can be mapped onto discrete snapshots of the world, and thus can be predicated.

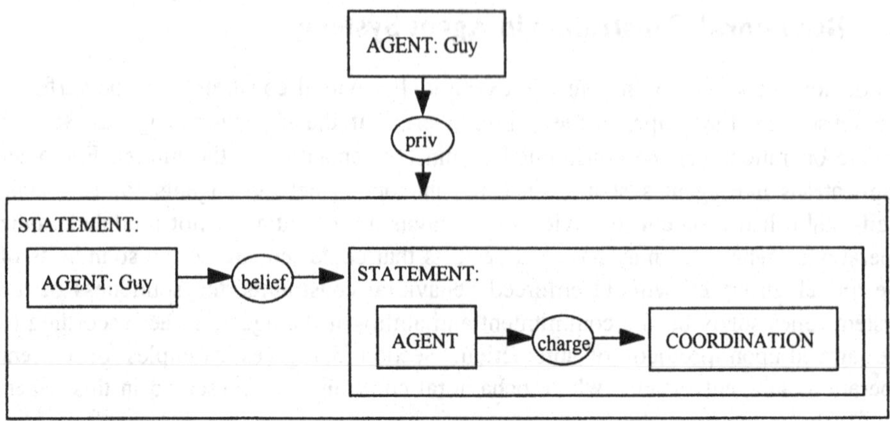

Fig.8: An assertion made in relation to some sheet of assertion.

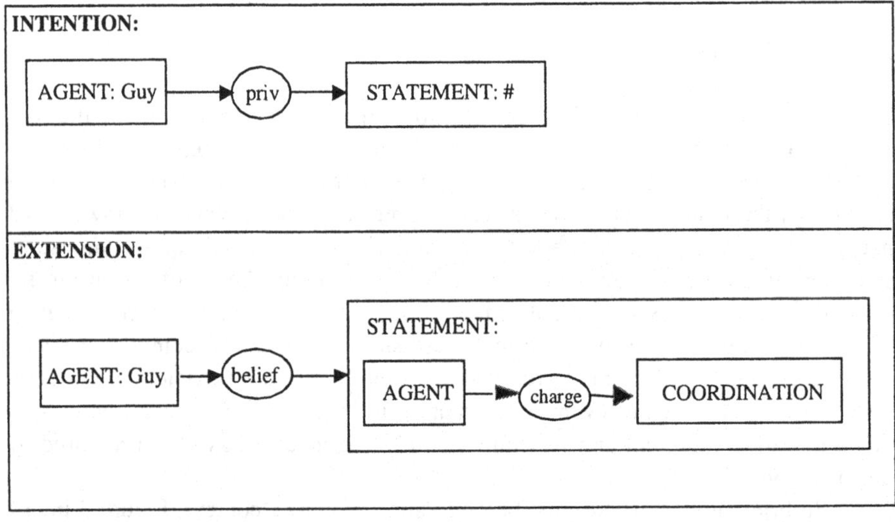

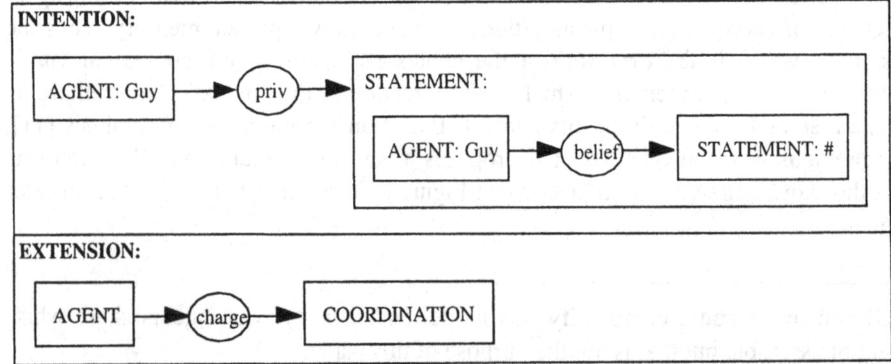

Fig.9: The two contexts inferred from the assertion of Figure 8, according to [10].

5.2 Asserting New Facts

The acquisition of new facts asserted by agents can be modeled through actors. For example, if Guy is asked to make a statement, the graph of Figure 10 should be asserted. Here, the **in** actor is a procedural attachment which, once the graph is asserted, calls on an input function which will ask Guy to input a statement (represented as a conceptual graph, referent of the STATEMENT concept). When Guy makes a statement, the output concept of the actor will be instantiated with this statement, a conceptual graph, which will then be asserted in the system. Section 5.3 below shows how such assertions can be validated against some behavioral constraints. Such constraints are mandatory to ensure that agents can access only the sheets of assertion that they own, i.e., access only their own memory.

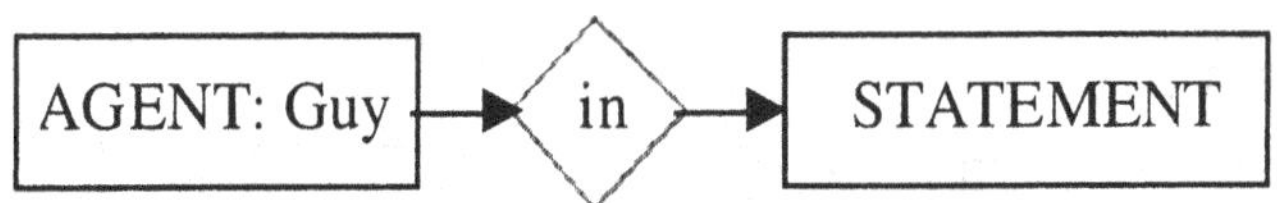

Fig.10: Asking Guy to input some statement.

5.3 Behavioral Constraints to Enforce Interaction Protocols

In this section, we show how some basic operations of agent systems can be validated against behavioral constraints which enforce the interaction protocol between them. For instance, we show how to restrict access to private memories (for asserting new facts), how to send messages, to access public memory, to transfer assertions between sheets of assertion owned by the same agent, and to access private sheets of assertion owned by other agents. These examples are sufficient to support the claims that this paper makes: a) behavioral constraints are essential to agent systems because they help to constrain the actions of the agents according to the interaction protocol, and thus help to improve the soundness and consistency of the system, and b) having behavioral constraints represented in a declarative formalism makes them available to inference processes.

Restricting Access to Private Memories. With the memory structure described in Section 5.1, any agent other than the one owning it should never alter a private memory. In other words, no agent should be allowed to make an assertion in anybody else's private memory. Figure 11 shows the corresponding behavioral constraint.

Constraint C5 verifies that any statement made by any process in some agent's private memory has originated from this agent. In our example, in order for our first precondition to be met, there must have been some prior action which resulted in having this agent input some statement. Once acquired, this precondition is denied so that this process does not fire repeatedly over the same data.

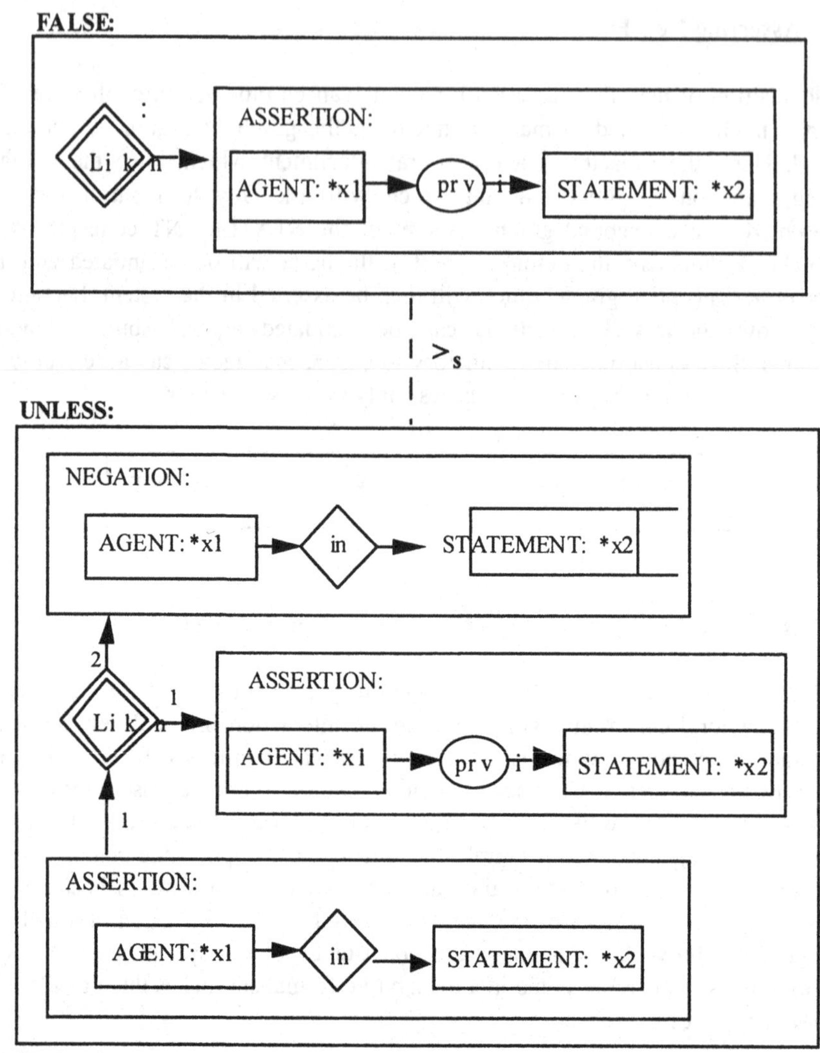

Fig.11: Constraint C5: any assertion in a private memory must come from its owner.[6]

[6] Any triggered process must disable some triggering condition once it is executed. Therefore, there should always be some postcondition which is a negation of a precondition in the description of a process.

Sending Messages. As an exception to constraint C5 (of Figure 11), constraint C6 states that an assertion made in the *in-box* of some private memory could be the result of the arrival of a message from another agent (see Figure 12).[7]

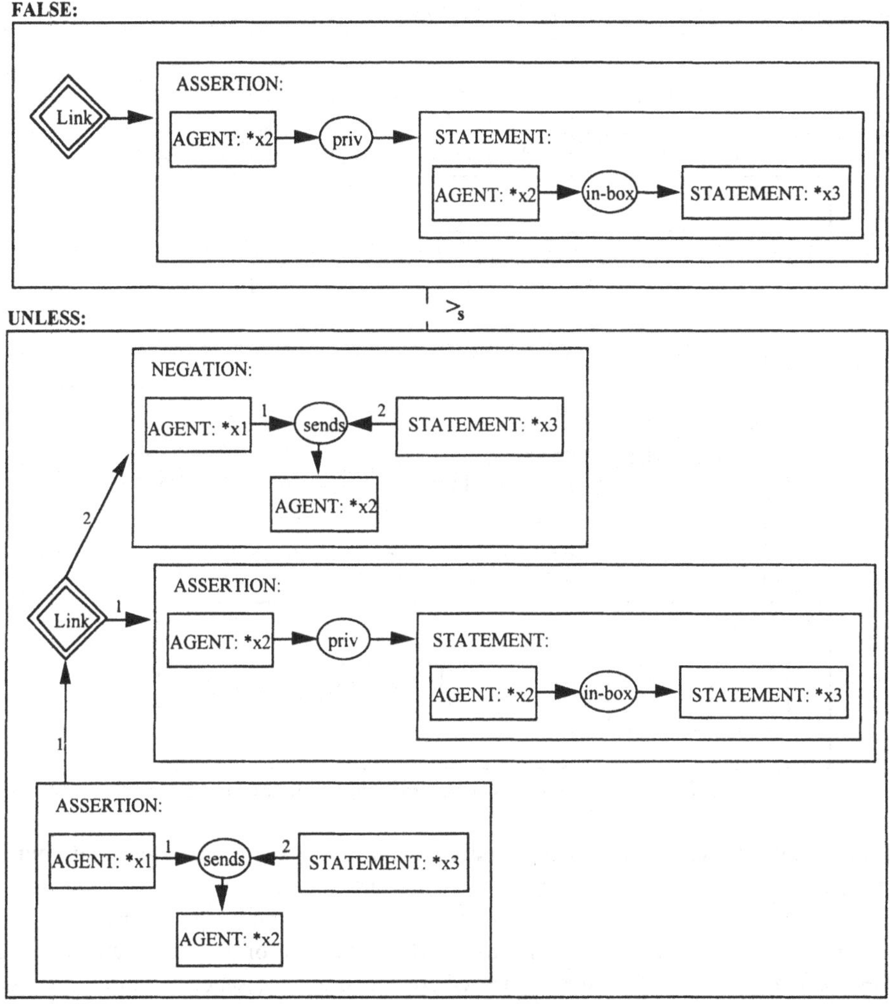

Fig.12: Constraint C6: sending a message to an agent.

Access to Public Memory. In our example application, we chose to have one sheet of assertion for each agent, in which public information would be asserted. This would result in a public sheet of assertion owned by each agent. For instance, if g_6 is some

[7] Of course, the choice of the label **in-box** is arbitrary. These decisions must be done when the memory structure of the agent is defined. Also, in this example, we did not choose to keep track of who sent the message, for simplicity reasons, although we could have done so.

conceptual graph, asserting g_6 to Guy's public sheet of assertion results in the assertion shown in Figure 13. In our application, public information is not censored; so we chose not to constrain assertions made to public sheets of assertion. Finally, we chose to implement a broadcast mechanism which copies everybody's public information to everybody else's (see Figure 14).

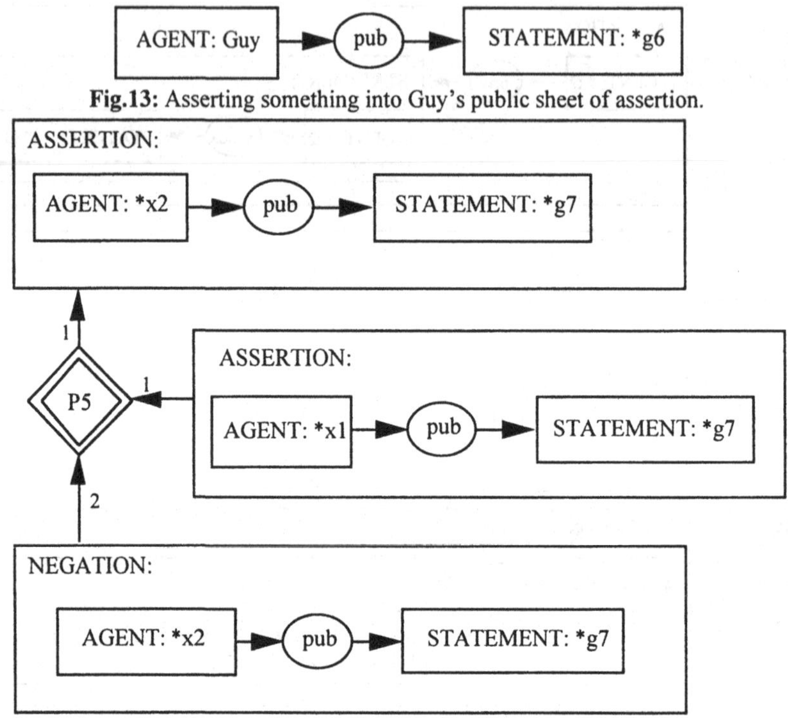

Fig.13: Asserting something into Guy's public sheet of assertion.

Fig.14: Process P5: broadcasting information between agents.

Transferring Statements Between Contexts Owned by the Same Agent. Of course an agent must be allowed to shift assertions around between his own sheets of assertion (contexts). For instance, if honesty is assumed among agents, then received statements (messages from other agents) could be thought of as true statements, and then believed. Figure 15 shows a case where incoming messages are moved to the agent's beliefs, and how in-boxes are emptied (Process P6). This process implies an assertion that does not come from either a message or the acquisition of a new assertion. Therefore, in order to remain valid, we should have some other constraint C7 stating that any assertion in a context may come from some other context owned by the same agent (not shown).

Accessing Information by Rightful Owner. As achieved by constraints C5, C6 and C7, assertions in private memories can only be carried out by their rightful owners. However, the content of private memories may not be for everybody's eyes. Consequently, we must also constrain the query mechanism upon which processes are based. So we chose to add the following constraint (C8) to the system (see Figure 16). This constraint states that if a process accesses information in some private memory, it must request permission to do so from the owner of the memory. The access actor used in this constraint makes sure that only the rightful owner of this information allows access to it. Once x1 is instantiated, the actor will ask agent x1 for access permission. If denied, the resulting Boolean value False will not match that of the asserted graph, blocking the process [3]. According to C8, all previously described processes should be updated to include this *request for permission to access data* as an additional precondition.

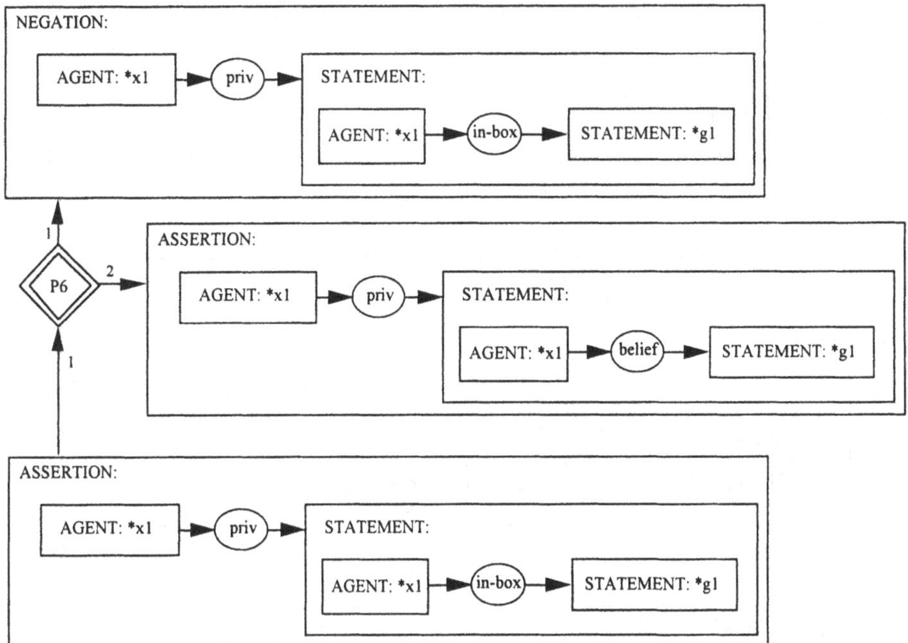

Fig.15: Process P6: moving assertions out of the in-box of an agent.

6 Conclusion and Future Developments

This paper combines our previous work on semantic constraints and on processes in order to propose a mechanism that allows the description of constraints on processes, i.e., of behavioral constraints. When update operations are expressed as processes, they can then be validated. Consequently, such constraints monitor and control the evolution of a data or knowledge base. As a matter of fact, this paper showed how migration constraints, which are still a problem in database environments, could be represented in a declarative format and enforced using CGs.

In order to prove its usefulness in some real application, this paper addressed a more complex problem: the representation of agent systems under the CG formalism, where the representation of behavioral constraints is essential to the enforcement of the interaction protocol. By doing so, we wanted to show how behavioral constraints help improve the soundness and consistency of a system, which would be otherwise left to the good will of the agents. The examples given above should be sufficient to convince the reader that the validation of processes against behavioral constraints is mandatory in agent systems where the commitment of agents towards the interaction protocol is not fully guaranteed.

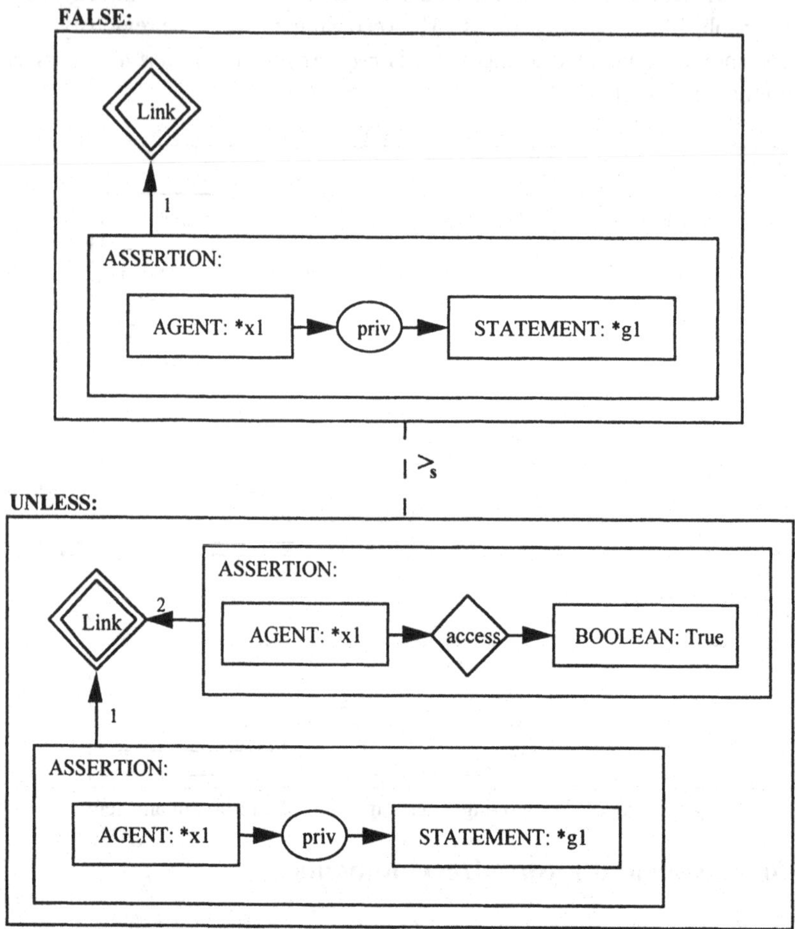

Fig.16: Constraint C8: restricting access to information.

Furthermore, though not explicitly argued for in this paper, the declarative representation of behavioral constraints allows their integration to the knowledge base, and thus their availability to the inference engine. Consequently, explanations about these constraints, faulty processes, and the relations between the two could be generated as support to the knowledge engineer or the user. Once step further, we will eventually address the problem of constraint validation. This activity clearly needs inferences to be drawn from a set of constraints, thus requiring their representation under a declarative format.

Our future work will explore alternative subsumption relations among processes. In this paper, it is clear that the computation of the $\geq_s$ relation is done for very simple conceptual graphs as operands. From an implementation point of view, because processes are usually more complex than those presented here, and because they can be broken into a set of non-trivial subprocesses, other relations, easier to compute, will be needed.

Providing the CG formalism with mechanisms to represent objects, processes and constraints on objects and processes, makes it a very powerful though simple representation framework applicable to the representation of agent systems, of OO systems, and of other complex applications such as real time systems. Using simple constructs like the double diamond box for processes and **false-unless** constructs to describe constraints, we hope to provide knowledge engineers with simple-to-use yet powerful tools which will prompt them to choose the CG formalism as a universal conceptual modeling language for a vast spectrum of applications.

References

1. Mineau, G.W. & Missaoui, R.: The Representation of Semantic Constraints in CG Systems. In: Lukose, D., Delugach, H., Keeler, M., Searle, L. & Sowa, J.F. (eds): Lecture Notes in Artificial Intelligence, vol. 1257. Springer-Verlag, (1997) 138-152
2. Mineau, G.W.: From Actors to Processes: The Representation of Dynamic Knowledge Using Conceptual Graphs. In: Mugnier, M.-L, Chein, M. (eds): Lecture Notes in Artificial Intelligence, vol. 1453. Springer-Verlag, (1998) 65-79
3. Mineau, G.W. & Missaoui, R.: Semantic Constraints in Conceptual Graph Systems. DMR Consulting Group, R&D Division. Internal Report #960611A. (1996)
4. Sowa, J. F.: Conceptual Structures: Information Processing in Mind and Machine. Addison-Wesley, (1984)
5. Mineau, G.W.: Views, Mappings and Functions: Essential Definitions to the Conceptual Graph Theory. In: Tepfenhart, W.M., Dick, J P., Sowa, J.F. (eds): Lecture Notes in Artificial Intelligence, vol. 835. Springer-Verlag, (1994). 160-174
6. Bertino, E. & Martino, L.: Object-Oriented Database Systems, Concepts and Architectures. Addison-Wesley, (1993)
7. Moulin, B.: A Logical Framework For Modeling A Discourse From the Point of View of The Agents Involved in it. Technical Report #DIUL-RR-9803. (1998)
8. de Moor, A. & Mineau, G.W.: Handling Specification Knowledge Evolution Using Context Lattices. In: Mugnier, M.-L, Chein, M. (eds): Lecture Notes in Artificial Intelligence, vol. 1453. Springer-Verlag, (1998) 416-430
9. Wing, H., Colomb, R. & Mineau, G.W.: Using CG Formal Contexts to Support Business System Interoperation. Mugnier, M.-L, Chein, M. (eds): Lecture Notes in Artificial Intelligence, vol. 1453. Springer-Verlag, (1998) 431-438
10. Mineau, G.W. & Gerbé, O.: Contexts: A Formal Definition of Worlds of Assertions. In: Lukose, D., Delugach, H., Keeler, M., Searle, L. & Sowa, J.F. (eds): Lecture Notes in Artificial Intelligence, vol. 1257. Springer-Verlag, (1997) 80-94
11. Wille, R., Restructuring Lattice Theory: an Approach Based on Hierarchies of Concepts. In: I. Rival (ed.): Ordered Sets. (1982) 445-470

Appendix A

This section provides two formal definitions: a *generality* relation (written $_g$) and a *subprocess* relation (written $_s$) among processes. The latter is used to construct a validity space from constraints on processes and is thus of immediate interest to us; the former is a stepping stone towards understanding the latter. In order to define these two partial order relations, let us first give some general definitions based on our previous work on processes [2].

a) Let a system S be described in terms of truth-valued statements. At one point in time, the set of such statements describing S is said to describe the *state* S is in at that time.
b) Let S* be the set of all such states. All states in S* are part of a partial order of generality. Letting s_1 and s_2 be states in S*, either $s_1 = s_2$ (same state), $s_1 > s_2$ (s_1 is more general than s_2), $s_1 < s_2$ (s_1 is more specific than s_2), or s_1 and s_2 are not comparable. If $s_1 > s_2$ then all models of s_2 (truth values associated with the statements describing s_2 and which make s_2 true) are models of s_1.
c) A process changes the state of a system by applying transformation operators on it, unless it is a null process (that is, its execution has no effect: its initial and final states are the same).
d) A process can be described in terms of the sequence of states resulting from these transformations. At a higher level of abstraction, a process can be described by its initial and final states. The set of initial states is called T; the set of final states is called R.
e) Let process P be a deterministic process. Then P can be described as a function from T to R, i.e., $P: T \rightarrow R$, or in other words, $\forall t \in T, \exists! r \in R$, such that $r = p(t)$, where p is the transformation (function) from t to r. Consequently, $R = \{r \mid r = p(t)\ \forall t \in T\}$. Then P can be defined as: $<T,p>$.
f) Let S*(P,t) be the set of all states generated by the application of process P from the initial state t, including t and $r = p(t)$, for $t \in T$.
g) Let P_1 and P_2 be two processes described as: $P_1 = <T_1,p_1>$ and $P_2 = <T_2,p_2>$.

Definition 1: ($_g$)

P_1 is said to be more general than P_2, written $P_1\ _g\ P_2$, iff: $T_2 \subseteq T_1$ and $r_2\ \ r_1\ \forall t \in T_2$, where $r_1 = p_1(t)$ and $r_2 = p_2(t)$.

Definition 2: ($_s$)

P_1 is said to be a subprocess of P_2, written $P_1\ _s\ P_2$, iff: $\exists t \in T_2$ such that $\exists t' \in T_1$, $t' \in S^*(P_2,t)$ and $r_1 \in S^*(P_2,t')$, where $r_1 = p_1(t')$.

User Modelling as an Application of Actors

Ani Nenkova and Galia Angelova

Bulgarian Academy of Sciences, Linguistic Modeling Lab,
25A Acad. G. Bonchev Str., 1113 Sofia, Bulgaria
{ani,galja}@lml.bas.bg

Abstract. Many AI systems define, store, and manipulate a user model (UM) by knowledge representation means different or separate from the system's knowledge base (KB). This paper describes a UM strategy in a system for generation of NL explanations. The idea is to track the user's requests and to modify the declarative patterns for information retrieval by actors. This allows for dynamic tailoring of the explanations to the user's previous knowledge.

1 Introduction and General Framework for Development of the User Model

Many AI systems support a user model (UM). Usually it is defined, stored and manipulated by knowledge representation means different or separate from the system's knowledge base. However, due to the growing importance of agent-oriented design, it becomes more and more desirable to integrate the UM within the system knowledge, thus allowing its processing by the same inference engine.

This paper presents work-in-progress in the field of modelling users[1] within the system DBR-MAT which generates explanations from a KB of conceptual graphs (CG). DBR-MAT explains in several natural languages (NL) *static* domain facts, encoded as predefined CGs (see [1], [2]). The current version of DBR-MAT supports dynamically a model of the user who interacts with the system. By monitoring the user's requests, the system collects information about individual characteristics of the user (*beginner, medium level* or *domain expert*, already *familiar* with some notion or *non-familiar*) and generates NL explanations whose detailness and focus are tailored according to the current status of the UM. The internal user representation is changed between different sessions as well as within a single session, since the user's characteristics evolve along the interaction. We see two ways for dynamic adaptation of the system to a particular user: (*i*) to adapt the strategy for extracting knowledge relevant to the user's request, (*ii*) to support more dynamic views to the KB, so that any user looks at a "different" KB whenever an answer is extracted for her. The first possibility—*dynamic changes of the way relevant knowledge is retrieved*—concerns user- and request-specific updates of the predefined patterns for knowledge extraction by

[1] We prefer to talk about *users* since at present our system does not recognize itself as an *agent* and therefore it supports only a model for its dialogue partners.

projection and join. The idea is to keep static the menu-based organization of the main user interface but to provide user-dependent modifications of the extraction patterns, in order to conduct the extraction of more or less (detailed) knowledge about focused concepts. Therefore, we aim at changes of the declarative knowledge *what* is to be extracted, always applying the same procedures of *how* it should be extracted.

The second possibility—*(simulation of a) dynamic KB update*—looks easy to implement, given the fact that in every multi-agent model the system practically supports records about what each particular user knows. So the KB turns to a collection of static facts encoding domain knowledge plus individual, dynamically stored users' KBs keeping information about each user's competence.

Section 2 briefly reviews related work in UM. Section 3 presents ideas and examples of how our user and his behaviour are modelled by actors. Section 4 discusses the benefits of this approach, future work and concluding remarks.

2 User Modelling in Systems Generating Help-Texts and Explanations

Two very important problems should be solved by systems generating explanations oriented to an *individual* user (not a group of users). The first is *how to gather* reliable information about the characteristics of the user and then decide *how to reason* about it. The second is *how to apply* the already built UM for tailoring the generated text to the user's competence. Many researchers have concentrated on only one of those aspects of user modelling; typical examples are GUMS (general UM shell, [3]) and TAILOR [6]). GUMS aims at building a UM without any consideration how it will later be integrated in a particular application, while TAILOR takes a convenient UM as given and discusses the possible variations of generated explanations, which are due to the activation of different presentation strategies depending on the UM status.

Most often, the UM applications have built-in schemes of possible reactions to changes in the UM status. In fact these systems support their UM procedurally. Up to our knowledge, none of the UM approaches changes dynamically the system's strategy for formulation of the answer depending on the user's behaviour. On the other hand, sophisticated UMs are often misleading and useless, because after some level of complication they start differ from the agent's view of herself. To summarize, the field of agent modelling contains mostly successful examples in particular dialogue settings rather than ultimate and stable design solutions of how an agent is to be modelled in an adequate manner.

DBR-MAT is a system for naive users, which provides complicated explanations of linguistic and domain information behind a very simple and standard interface. There are many ways to approach the UM task in such a system. We can build hypotheses whether the user is a *beginner* or an *expert* depending on the sequence of queries about the type hierarchy. Many specific details in the UM concern the procedural part of the system but in this paper we discuss only users' characteristics, relevant to the declarative knowledge.

3 CG and Dynamic Knowledge

In section 3.1 we briefly mention some relevant, most recent discussions on modelling events and object transformations by CG. We consider only those ideas which directly influenced our UM. In section 3.2 we discuss our current solution for supporting a dynamic UM in the DBR-MAT system.

3.1 Dynamic Modelling by Actors and Processes in CG

Papers and recent discussions about the CG standard show slightly differing views on actors. All authors agree that *(i)* it is reasonable to model dynamic behaviour, by extending the FOL-kernel of the CG; *(ii)* actors are functional conceptual relations, so formally, the terms actor and conceptual relation are synonymous; *(iii)* the semantics of an actor may be computed or otherwise determined by an IT system external to the KB.

The differences seem to concern the question what procedural semantics is attached to the actors. One possible view is that "the actor is an active relation between two states (positively nested graphs), getting triggered when some set of conditions is satisfied. The actor applies when the *from*–graph evaluates to true and changes to the consequent graph, so in another sense the actor is a graph rewrite operator because of its effect on the nested graph of the state"[2].

Paper [4] says that "actors are procedural attachments, they compute values, while processes assert and retract graphs. Actors implement computations for infinite domains such as mathematical computations, which would require an infinite number of assertions if represented by processes". This paper shows definition of processes and their parameters and how inference on processes becomes possible in order to produce explanations e.g. about processes.

Paper [7] considers dynamic KB changes and formalises actor definitions and assertion/retraction of graphs. Examples in [9] show that actors simplify the control of external conditions. All papers advocating the use of actors in CG present elegant declarative solutions for reaction to external (procedural) changes. In our considerations we apply the notion of assertion events and the ways for drawing the coreference links between types in the *from*–graphs and the consequent graph (we need coreference links to control specific individuals).

3.2 User Modelling by Actors

To provide more flexible UM, we need tools for transformations of objects as well as mechanisms to assert and retract graphs. We apply them when: (*i*) user's knowledge evolves and the system needs to remember what this particular user knows; (*ii*) a "CG-external" condition concerning the user's competence is true (e.g. *the system decides that the user is familiar with concept X because an explanation about it has been generated three times during the current session*).

[2] Compilation of William Tepfenhart's messages to the CG mailing list (1998).

Modification of Projection Patterns. The important point here is to explain why we need to extract knowledge fragments at all — can't we explain graphs as a whole, as they are encoded in the KB? The problem is that Sowa's utterance path for NL generation [8] can be computed successfully for relatively simple graphs only. Even when one applies a complex and perfect grammar to conduct the incremental building of a sentence from the utterance path, it is clear that very large graphs (say, 20 concepts and 15 relations plus contexts) cannot be verbalized as one sentence. So the system has to *view* them as several "sub-"graphs with appropriate coreference links and to generate several coherent sentences. Thus the NL generation from complex graphs turns into a very problematic issue. Some aspects of this task are concidered in [5] but we are far from ultimate solutions. To overcome the KB complexity, DBR-MAT extracts relatively simple graphs, which fit to the grammar's potential to maintain correct sentences. The users become familiar with the domain knowledge by asking follow-up clarification questions. Practically every KB fact might be verbalised, if the correct sequence of requests is posed.

The DBR-MAT user initiates the generation of NL explanations by highlighting text (let us assume a single term) and then selecting items from a menu, similar to the well-known interfaces of text processors. Each menu selection causes projection from the (static) KB graphs, using predefined projection patterns. Table 1 gives the menu items and the conceptual relations, retrieved "around" the focused concept after each request type (one of q1–q14). Fig. 1 exemplifies simple graphs, applied as projection patterns [1].

In Table 1 and Fig. 1 some conceptual relations and their corresponding patterns for extraction of answers overlap. For instance, if questions q6 or q7 have been answered, and immediately after this q1 is posed, the generated explanation should not include characteristics, so the graphs causing these explanations should be retracted either from the list of projection patterns, or from the list of already extracted temporary graphs underlying the generation itself. Similarly, if the user is an expert in the domain, she might not need a complete verbalization of the type hierarchy in q1, so there is no need to include the whole q2's answer as a subset of the q1's answer - the information from q3 will suffice.

Simple examples below illustrate the actors we apply at present. Fig. 2 shows an elementary KB consisting of four graphs. EXPR denotes the thematic role Experiencer. For simplicity we don't discuss the individual KB of different users, which are separated via the internal identifiers of the corresponding graphs. Thus the current user is the only one to be treated in a sample UM.

Let us consider the customization of projection patterns (Fig. 1) by retraction of query graphs within a single dialog session. Table 1 overlaps for certain conceptual relations to allow for more complete answers when the user asks for explanation *for the first time* or *only once*. But after some dialog progress, Table 1 produces more and more overlapping answers, so the corresponding projection patterns in Fig. 1 need to be updated dynamically. This effect is obtained by retractions of query graphs from the lists shown in Fig. 1. For example, if the user asks q7–q11, immediately subsequent request for q6 and the same concept

Table 1. A static user menu and relations maintaining the extraction of graphs

#	Menu	Submenu Item	Corresponding Conceptual Relations to be extracted	Inheritance of characteristics
q1	What is?		[type-def.]+q2 (Types of ... / All)+ CHAR, PART-OF, FUNCT, RSLT	Yes
	Types of...			
q2		All	super- + sub- + sister-concept	
q3		General	All superconcepts from the hierarchy	
q4		Concrete	All subconcepts from the hierarchy	
q5		Similar	All sister-concepts from the hierarchy	
	Characteristics			
q6		All	ATTR+CHAR+WHO+OBJ+ HOW+WHERE+THME+PTNT...	Yes
q7		Attribute	CHAR	Yes
q8		Who	AGNT, EXPR	
q9		Object	THME, OBJ, PTNT	
q10		How	INST, MANR	
q11		Where	LOC, DEST, FROM, IN, TO	
q12	More...		All remaining relations which were not extracted by the other menu items	
q13	Examples		Individuals	
q14	Want All		All from above without repetitions	

will explain "other, additional characteristics" instead of "all". The retract operators are actors, activated when predefined conditions about the query sequence within a single session are satisfied.

The query graphs in Fig. 1 also cause overlapping answers for "neighbour" concepts which are highlighted one after another. Suppose the user asks q8 for *oil particles*. The answer is obtained from the KB graphs 1, 2 and 3 (see Fig. 2) by projection with the query graphs given in Fig. 1 for q8. In English (which is not the original result, since our generators work in German and Bulgarian), the obtained explanation looks like *The oil particles, which are lighter than water, swim up. The oil particles, which are heavier than water, precipitate. The swimming-up oil particles stick*. (Note that it would be much better if the third sentence could appear at second position, but at present DBR-MAT has no means to order the discourse clauses with such precision).

If now the user asks q8 for *precipitate*, after q8 for *oil-particles* the resulting explanation again contains (among others) the sentence *The oil particles, which are heavier than water, precipitate*. In this case the query graph yielding the repetitive answer is retracted from the list in Fig. 1.

In general, the customization of Fig. 1 proceeds as follows: if the user query activates an *update*-actor, because of the specific request sequence and the truth value of the actor preconditions, the query graphs are updated accordingly. Otherwise, an *recovering*-actor is activated to copy the initial query graphs. To be precise, we'll mention that the real query graphs are more complicated than

Query	Graphs Applied as Projection Patterns (Incomplete Lists)
q7	[highlighted concept]←(CHAR)← [T: ?]
	[highlighted concept]→(CHAR)→ [T: ?]
q8	[T:?] ←(ATTR)←[highlighted concept]←(AGNT)←[T: ?]→(ATTR)→[T:?]
	[T:?] ←(ATTR)←[highlighted concept]←(EXPR)←[T: ?]→(ATTR)→[T:?]
	[T:?] ←(ATTR)←[highlighted concept]→(AGNT)→[T: ?]→(ATTR)→[T:?]
q9	[highlighted concept]←(OBJ)←[T: ?]→(ATTR)→[T: ?]
	[highlighted concept]→(OBJ)→[T: ?]→(ATTR)→[T: ?]
	[highlighted concept]←(THME)←[T: ?]→(ATTR)→[T: ?]
	[highlighted concept]←(PTNT)←[T: ?]→(ATTR)→[T: ?]

Fig. 1. Query graphs for projection, for extracting the answer of q7, q8 and q9.

those in Fig. 1. There are different query graphs for concepts attached to nouns, to verbs, etc. due to the fact that the query graphs have to conduct the extraction of the proper fillers for thematic roles, when concepts corresponding to verbs are focused (otherwise the correct construction of sentences fails with the available grammar). That is why asking q1 for *stick* produces the answer *The swimming-up oil particles stick, which leads to a dense oil layer*. So in the real system we maintain more query graphs to allow for correct verbalization of verb valences, but this fact does not concern the ideas of UM summarized here.

Dynamic KB of user's competence. A simple actor ≪ASS_FAM≫ conducts the assertions of familiar facts in the user's KB. For instance, if the user asks q1, q6 or q7 about *density*, she receives an explanation containing (among the others) the sentence *Density is a characteristic of substance*. The corresponding graph, extracted from KB's graph 4, is asserted in the user's KB: [[DENSITY]←(ABOUT)←[USER]→(ASKING)→[QUERY_TYPE:{q1|q6|q7}]] →≪ASS_FAM≫→[[DENSITY]←(CHAR)←[SUBSTANCE]].
The actor ≪ASS_FAM≫ is initiated by the system interface monitor; the asserted graph is copied from the temporary graphs. Thus DBR-MAT remembers what the users (are expected to) know. To avoid repetition, another actor ≪RETR_TEMP≫ retracts temporary graphs:

[[USER]→(FAM)→[*graph*]]→≪RETR_TEMP≫→[*graph*] .

Thus, if the user asks q1, q6 or q7 for *substance* after having asked it for *density*, the explanation is *Substance is characterized by relative weight*. Otherwise the answer would be *Substance is characterized by density and relative weight*.

For simplicity we do not consider all the details of passing parameters (DBR-MAT supports parameters for distance of the generated explanations).

3.3 Conclusion

This paper deals with actors which assert and retract graphs under certain conditions, declared in their *from*-states. These actors are a compilation and an adaptation of several ideas from in earlier works. The goal is to encode declaratively knowledge about users and corresponding system behaviour, which is

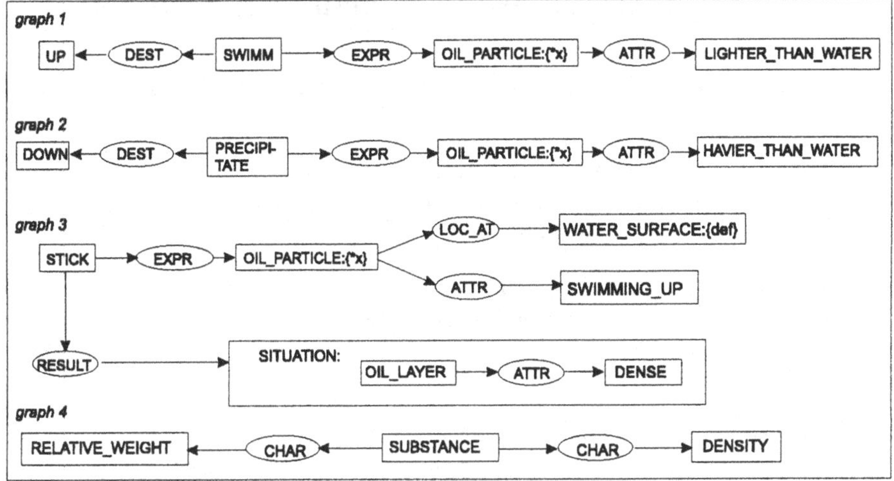

Fig. 2. Static domain knowledge, from where fragments are extracted by projections.

typically treated in a procedural way. We should admit, however, that a substantial part of the UM is supported outside the declarative approach.

We find very important the fact that CGs are employed for maintenance of dynamic collections of graphs, because this shows how the KBs of CGs move towards the classical paradigm in multi-agent environments. They allow for the support of many users and their corresponding KBs, and for reasoning about user's competence with respect to the static domain knowledge.

References

1. Angelova, G., K. Bontcheva. *DB-MAT: Knowledge Acquisition, Processing and NL Generation using Conceptual Graphs.* Proc. ICCS-96, LNAI 1115, pp. 115–129.
2. Angelova, G., K. Bontcheva. *Task-Dependent Aspects of Knowledge Acquisition: a Case Study in a Technical Domain.* Proc. ICCS-97, LNAI 1257, pp. 183-198.
3. Finin, T. *GUMS—A General User Modeling Shell.* In Kobsa and Wahlster (eds.): User Models in Dialog Systems, pp. 411–430, Springer-Verlag, Berlin.
4. Mineau, G. *From Actors to Processes: The Representation of Dynamic Knowledge using Conceptual Graphs.* Proc. ICCS-98, LNAI 1453, pp. 65–79.
5. N. Nikolov, Ch. Mellish, G. Ritchie. *Sentence Generation from Conceptual Graphs.* Proc. ICCS'95, LNAI 954, pp. 74–88.
6. Paris, C. *The Use of Explicit User Models in a Generation System for Tailoring Answers to the User's Level of Expertise.* In Kobsa and Wahlster (eds.): User Models in Dialog Systems, pp. 200–230, Springer-Verlag, Berlin.
7. Raban, R. and Delugach, H. *Animating Conceptual Graphs.* Proc. ICCS'97, LNAI 1257, pp. 431–445.
8. J. Sowa. *Conceptual Structures: Information Processing in Mind and Machine.* Addison-Wesley, Reading, MA, 1984.
9. Mann, G. *Procedural Renunciation and the Semi-automatic Trap.* Proc. ICCS'98, LNAI 1453, pp. 319-333.

Spatial Universals as the Human Spatial Notion

Jane Brennan

Department of Artificial Intelligence
School of Computer Science and Engineering
University of New South Wales
Sydney 2052
Australia
jbrennan@cse.unsw.edu.au

Abstract. Peirce (CP 2.753) suggests that humans possess an inherited notion of space, which amongst other notions allows them to adapt to the environment. This paper discusses a conceptual graph approach to defining such spatial notions (i.e. spatial universals) as a finite set of canonical graphs and suggests its use to derive a potentially infinite number of canonical graphs. Our approach is illustrated by a bilingual example.

Introduction

Peirce (CP 2.75 as quoted in Gärdenfors 1993) notes that "[all humans and animals] have from birth some notions, however crude and concrete, of force, matter, space, and time". This, in addition to "some notion of what sort of objects their fellow-beings are" adapts them to their environment. This paper suggests an implementation of the human inherited[1] notion of space, in the following called spatial universals, using conceptual graphs.
Language is our way of communicating our internal representation of the real world (i.e. our thoughts) to others. It is therefore not surprising that many AI approaches representing spatial and temporal knowledge are strongly related to natural language. A very common approach is to analyse natural language constructs describing spatial scenes or temporal situations and then derive formalisations, models or concepts from these constructs. Some examples of such approaches can be seen in Herskovits (1986), Habel (1988) and Mann (1996).

However, since the seventeenth century, it has frequently been speculated from a philosophical and linguistic perspective that the analysis of vocabularies of all human languages can be performed in terms of a finite set of semantic components (i.e. universals). These universals are independent of the semantic structure of any given language. Combining the separate universals in language specific ways would result in language specific concepts, which would, by analysis, still be identifiable as the original universal (Lyons 1968). We believe that universals can be defined as

[1] This is to be understood in terms of human physiology.

canonical graphs and "[a]lthough the number of canonical graphs may be infinite, the formation rules can derive all of them from a finite [definable] set." (Sowa 1984, p. 94)

Katz (1967) suggested that there are "language invariant but language linked components of a cognitive structure of the human mind." Together with Fodor (Katz and Fodor 1963), he introduced primitives, which could be combined by operations called projection rules. Sowa (1984) strongly criticised Katz and Fodor's approach by pointing out the weaknesses of this theory based on primitives. For instance, there is a lack of linguistic and psychological evidence for the existence of a truly universal set of primitives.

However, we believe that the spatial domain is a special case due to its physically very constraining nature. Therefore, we would like to introduce an approach combining both views in suggesting that there are basic spatial concepts which are language-independent and the foundation for language dependent concepts of space. Space is physically far more constraining than any other aspect of the real world and could be thought as the physical reality we live in. "Human categorisation is constrained by [this] reality, since it is characterised in terms of natural experiences that are constantly tested through physical and cultural interaction." (Lakoff and Johnsons 1980, p. 181). Even though disparate people or cultures perceive the real world in very different terms given that they are living in disparate physical environments, there are still some physical constraints such as gravity or basic human physiology, which are almost identical across the planet. This gives rise to the claim that universal spatial concepts are reinforced by these identical physical constraints.

The paper will first discuss previous research in the area of qualitative temporal and spatial reasoning, then it will outline our approach of universal spatial concepts using spatial reasoning and conceptual graphs. The conclusions will discuss possible applications of our approach and future research directions.

Previous Research

The qualitative representation of space has been a primary focus of previous research in the area of spatial reasoning [e.g. Hernández 1994] because such an approach is close to the human way of representing and reasoning about space.

Allen (1983) introduced an interval calculus for dealing with qualitative temporal information by describing comparative relations between the time intervals. This idea was extended to represent binary topological relations between spatial regions [e.g. Guesgen 1989, Egenhofer 1991]. Conceptual structures have been used as spatial representation to enable reasoning about space to solve spatial problems such as navigation (Mann 1994, 1996). Another interesting application of conceptual graphs is the representation of temporal intervals. Esch (1992) combines Allen's temporal interval calculus and Matuszek's (1988) end point relations on temporal intervals into a set of 12 base relations on intervals. Matuszek's end point relations consider the endpoints of the intervals rather than the duration of them. For example, Allen's relations START, EQUALS and STARTED BY can be combined to form SWS (i.e.

starts when starts) using the end point approach. This allows a much greater freedom in defining relations between time intervals as there is no necessity to consider both end points of the intervals. Although, this makes the relations more ambiguous , for our purpose this greater degree of freedom is desirable.

Only two of Esch's base relations are used in this paper, the relation *F=S* and *F)S*, where *x F=S y* means that the x interval finishes when the y interval starts and *x F)S y* means that the x interval finishes before the y interval starts.

As we discussed before, temporal representations have repeatedly been used to represent spatial aspects. It is therefore an obvious choice to use Esch's base relations as conceptual relations for spatial descriptions.

The universal spatial concepts

The spatial concepts which we will be looking at work within a deictic frame (i.e. everything is related to oneself). The deictic perspective is the natural perspective of humans in their early stages of cognitive development, when the self is the centre of the world and everything is related to this centre. Because all humans generally have the same physiology (i.e. eyes at the front of the head, ears on the sides etc.), and are subject to the same basic physical environment (i.e. gravity) and receive similar perceptual sensations, psychological evidence [e.g. Pick 1993] suggests that all humans develop similar basic spatial concepts which we will call universal spatial concepts. We do not claim that this is the true and ultimate mental model of space, but argue that such a model could be very useful for various applications such as robot navigation or language understanding.

Our model of universal spatial concepts will be a formalisation based on Landau and Jackendoff's (1993) axes model. The formal description of their psychological model will give the opportunity of concept expansion in order to derive language specific spatial concepts based on the spatial universals.

Most objects have a top and a bottom, a front and a back, and sides and/or ends. As for humans, the front of the body can be determined by the fact that one's eyes, nose, feet and navel point in the same direction; and the fact that arms are attached opposite one another and orthogonal to the front determines the sides (Landau and Jackendoff 1993). Landau and Jackendoff differentiate three layers of axes, the generating axis, orienting axes and directed axes. The generating axis is the principal axis, which is vertical in case of a human. The orienting axes are orthogonal to the generating axis and each other, they represent the front-to-back and side-to-side axes. Finally, the directed axes indicate inherent regularities that distinguish one end from the other (i.e. top from bottom, front from back). Fig. 1 illustrates the axes model.

The universal spatial concepts can be defined along these axes by using one of Esch's base relations between two intervals; the relation *F=S* describes the fact that the first interval finishes at the start of the second interval. As base relations describe time intervals, which are directed (from start to finish), we can use those unchanged to define the directed axes in Landau and Jackendoff's model.

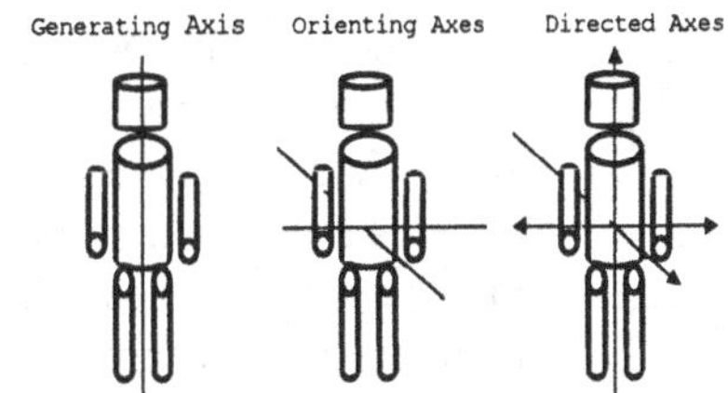

Fig. 1. Landau and Jackendoff's Axes (1993)

Consider the extent of our reference object (i.e. the human body) by visualising the body surface to be interval start or interval end. The spatial concepts are defined along the directed axes assuming an erected position of the human body.

The overall basic spatial concept can be defined by the three axes and their relations to each other. The conceptual relations between the axes are in this case bi-directional and the arrows are therefore omitted.

The generating axis can be defined as a combination of directed intervals, the bottom interval starting at an undefined point before the end point of the bottom interval (i.e. body surface under the feet), the body interval is limited by the body surface as its end points, ending at the body surface which also starts the top (i.e. over the head). The front-to-back axis is very similar to the generating axis apart from the fact that the body extent concerned is horizontal.

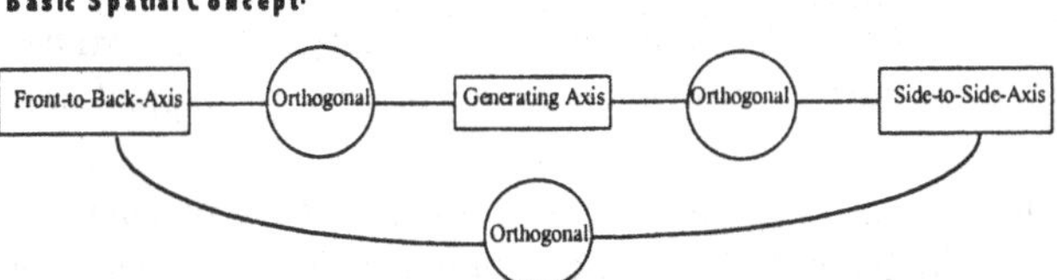

The side-to-side axis consists of two opposite directed intervals each starting from the centre of the horizontally considered body (i.e. where the generating axis symmetrically divides the horizontal body).

The body extent differs for the different axes, they can be defined as seen in Fig. 2. Assuming that our understanding of the spatial environment - including the relationships between objects but excluding ourselves- is based on the above described concepts, we are now facing the problem that all of the defined concepts are unique to one particular object and do therefore not allow us to perform any JOIN operations between conceptual graphs of different objects.

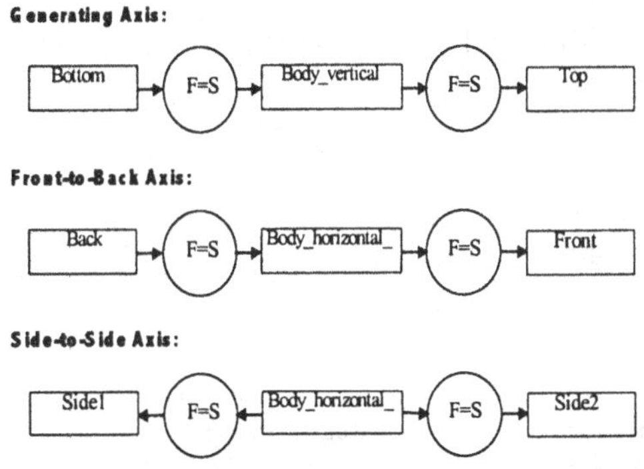

Fig. 2. Concepts of the Axes

We need to introduce a further axis, which is universal to all objects, to be able to relate different objects to each other. The ground axis does fulfil this purpose. It represents the ground we stand on, which would, in respect to the human body, generally be seen as orthogonal to the generating axis and parallel to the orienting

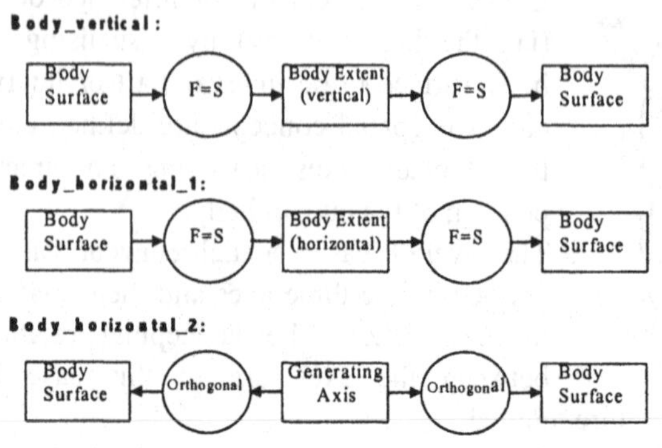

Fig. 3. Concepts of the Body Extent

axes.[2] This means, additionally to the deictic[3] character of the three axes already defined, the ground axis adds a new component to the basic spatial concepts in order to account for relations between objects out of our deictic range. The ground axis is not directed and will not be defined any further.

We will now look at a bilingual example, which embraces different concepts for the same spatial situation. This example compares concepts of *above* and *on* situations in English and in the Mexican language Mixtec (Regier 1996). Mixtec uses a body-part system which is metaphorically mapped onto the spatial system.

The example is illustrated in Fig. 4. In English, the circle is *above* the rectangle, if their surfaces (i.e. bottom of circle, top of rectangle) do not touch each other and it is *on* the rectangle, if their surfaces do touch. In Mixtec, there are two prepositions for describing the circle being vertically on or above the rectangle, but surface contact is not considered. Instead, the extent of the rectangle is important. The preposition *siki*, meaning "animal-back", is used to describe the circle above or on a horizontally extended rectangle. While the preposition *šini*, meaning "head", is used to describe the circle above or on a vertically extended rectangle.

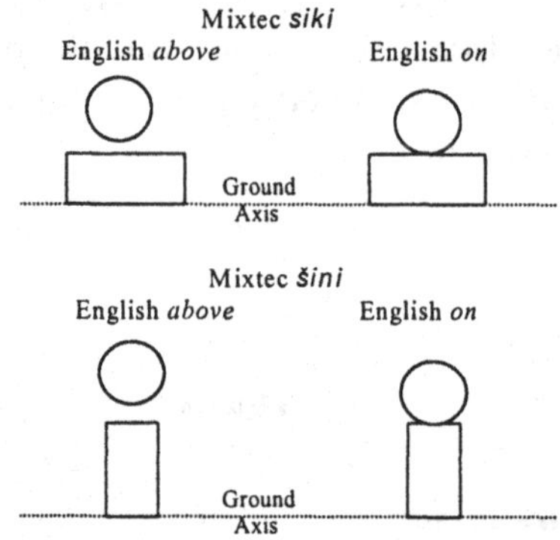

Fig .4. Language dependent concepts (Regier 1996)

Applying the defined basic concepts to the language concepts *siki* and *above*, we can see that for *siki* the generating axis of the rectangle is parallel to the ground axis and orthogonal to the generating axis of the circle. The contact information is not important in the usage of this concepts. In English however, the generating axes of both the rectangle and the circle are orthogonal to the ground axis and parallel to each other and the non-contact information differentiates this concepts from *on*.

The conceptual graphs for *siki* and *above* can be seen in Fig. 5a, 5b respectively. JOIN operations can then be performed on the graphs (see Fig. 5c, 5d). Esch's *F)S*

[2] Only even ground is considered for the purpose of this paper.

[3] egocentric reference frame

relation could replace the *above* non-contact implying conceptual relation in the graph for a more specific description. For a better conceptual understanding however,

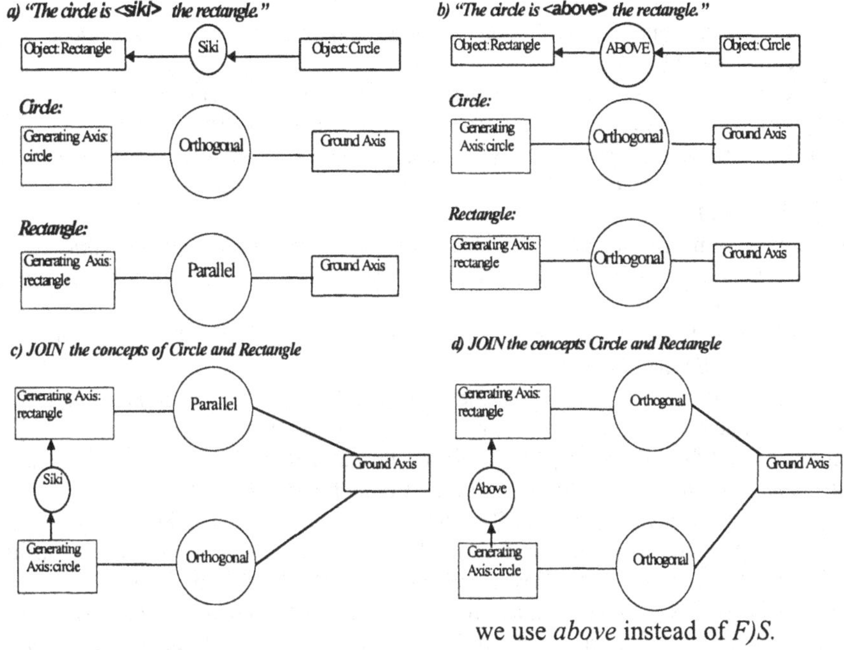

Fig. 5. Concept Definition and the Application of the JOIN Formation Rule

As can be seen in the example, the set of basic (i.e. universal) concepts can be seen as a finite set of canonical graphs (Sowa 1984) which can be used to derive a potentially infinite number of canonical graphs. In our case, all possible spatial situations could be covered by applying formation rules appropriately to the basic concepts and creating more complex concepts (i.e. conceptual graphs).

Conclusion and Outlook

The universal spatial concept model allows the representation of basic spatial concepts using structures, which can be used for various applications such as robot navigation or language understanding. For example, a robot could be initially equipped with an implementation of the model as its simple concept of space and then expand the concept by learning from experience with the environment (i.e. sensory input). Future work could include the creation of an algorithm applying formation rules sufficiently to the discussed set of spatial concepts (i.e. the universal spatial concepts) to allow real world linguistic or navigational problems to be solved. Also, the concepts defined in this paper need to be extended to include more topological information (e.g. contact information) by including more of Esch's base relations in order to describe more complex spatial situations.

References

Allen J.F.: "Maintaining knowledge about temporal intervals" in *Communications of the ACM*, 26(11), pp832-843, 1983.

Egenhofer M.J. and Franzosa R.: "Point-set topological spatial relations" in *International Journal of Geographical Information Systems*, 5(2), pp161-174, 1991.

Esch J.W.: "Temporal Intervals", In Nagle T.E., Nagle J.A., Laurie L.G. and Eklund P.W. (eds) *Conceptual Structures: Current Research and Practice*, pp363-380, Ellis Horwood, Chichester, 1992.

Gärdenfors P.: "Induction and the Evolution" in *"Charles Peirce and the Philosophy of Science IX"*, D. Prawitz, B. Skyrms, D. Westerstahl (eds.); pp. 423-449, Elsevier Science, Amsterdam, 1993.

Guesgen H.-W.: "Spatial reasoning based on Allen's temporal logic", Technical Report TR-89-049, ICSI, Berkley/CA, 1989.

Habel C.: "Repräsentation räumlichen Wissens" in "*Wissensrepräsentation in Expertensystemen*", G. Rahmsdorf (eds.), Springer-Verlag, Berlin, 1988.

Hernández D.: "Qualitative Representation of Spatial Knowledge" in *Lecture Notes in Artificial Intelligence (804)*, Springer-Verlag, Berlin, Heidelberg, 1994.

Herskovits A.: "Language and Spatial Cognition", Cambridge University Press, 1986.

Katz J.J.: "Recent Issues in Semantic Theory", In *Foundations of Language*, International journal of language and philosophy, 3, pp124-194, Dordrecht, Holland, 1967.

Katz J.J., Fodor J.A.: "The structure in semantic theory", In *Language*, Journal of the Linguistic Society of America, 39, pp170-210, Baltimore, 1963. (Reprinted in Fodor J.A. and Katz J.J. (eds): "The Structure of Language: Readings in the Philosophy of Language", Englewood Cliffs, N.J.: Prentice-Hall, 1964.)

Lakoff G., Johnson M.: "Metaphors we live by", University of Chicago Press, Chicago,1980.

Landau B., Jackendoff R.: " 'What' and 'where' in spatial language and spatial cognition", In *Behavioral and Brain Sciences*, 16, pp217-265, Cambridge University Press, 1993.

Lyons J.: "Introduction to theoretical linguistics", Cambridge University Press, 1968.

Mann G.A.: "A Relational Goal-Seeking Agent using Conceptual Graphs", In Tepfenhart W.M., Dick J.P. and Sowa J.F.(eds), *Conceptual Structures: Current Practices*, Lecture Notes in AI 835, pp113-126, Springer-Verlag, Berlin, 1994.

Mann G.A.: "Control of a Navigation Rational Agent by Natural Language", PhD dissertation, University of New South Wales, Sydney, Australia,1996.

Matuszek D., Finin T., Fritzson R., Overton C.: "Endpoint Relations on Temporal Intervals", In *Proceedings of the Third Annual Rocky Mountain Conference on Artificial Intelligence*, pp182-188, 1988.

Peirce C.S.: "Collected Papers of Charles S. Peirce", Hartshorne, Weiss and Burks (eds.), Harvard University Press, Cambridge, MA, 1932.

Pick H.J. Jr.: "Organization of spatial knowledge in Children", In Eilan N., McCarthy R., Brewer B. (eds.), *Spatial Representation: Problems in Philosophy and Psychology*, Oxford: Blackwell, pp31-42, 1993.

Regier T.: "The Human Semantic Potential - Spatial Language and Constrained Connectionism", A Bradford Book, The MIT Press Cambridge, 1996.

Sowa J.F.: "Conceptual Structures", Addison-Wesley Publishing Company, Menlo Park, CA, 1984.

Knowledge Engineering with Semantic and Transfer Links

Karima MESSAADIA, Mourad OUSSALAH

LGI2P/EMA-EERIE
Parc scientifique Georges Besse - 30000 Nimes, France
Email :messaadi@eerie.fr, oussalah@eerie.fr.

Abstract. Recent methodologies in knowledge engineering (K.E) are oriented towards the construction of component libraries. The concepts commonly used in describing Knowledge Base Systems are tasks, PSMs (problem solving methods) and domains. Developers have to select them from a library, adapt and link them so that they fit their specific needs. In order to help developers to quickly understand, find, and configure the components[1] best suited to their applications, we need to specify languages for describing tasks, PSMs and domains plus the different interactions between them. We propose to clarify the different interactions using inter- and intra-concept links and to describe the different concepts languages using ontologies improving thus their reusability and sharing.

Keywords: Knowledge Engineering; Ontologies; Knowledge representation.

Introduction

Recent approaches in knowledge engineering (K.E) are oriented towards the construction of component libraries similar to the component-based approach used in software engineering. Constructing an application from reusable components is a promising way to minimise development time and facilitate evolution and maintenance. In order to help developers to quickly understand, find, and configure the components best suited to their applications, component- oriented approaches need to specify languages for describing components, and reuse-methodologies defining how to construct a reusable component and how to structure and index the components in a library. Existing K.E approaches often use task, PSM and domain concepts to describe a knowledge base system (KBS). This separation between the task to solve, the reasoning used to solve a task and the domain knowledge, permits to see the construction of an application as a combination of these three components.

---

[1] We use components referring to task, PSM, domain, and links concepts and their specialisation.

Different libraries have been described, generally PSM- libraries [4, 5, 8, 16, 25]. PSMs in these libraries are often described using task-specific or domain-specific terms. Task specific and domain-specific PSMs are less reusable than generic ones. This is known as the *reusability/usability* trade-off [11]. Other studies focus on ontology portability in order to construct libraries of reusable PSMs, tasks and domains. Ontologies were initially used to *formally* describe generic models of domains so that they can be reused to describe different specific domains sharing the same structure. As we can reuse tasks and PSMs, the notion of task ontology and PSM ontology appeared in recent works [5] [8] [15] referring to independent descriptions of them. The problem of these libraries is the lack of : a clear description of the different relationships between the three concepts, at different levels of abstraction. Comparative studies can be found in [15] [19]. We have described in [16], the structure of a KBS model integrating the three main concepts: task, PSM, domain, plus semantic and transfer links, used to describe interactions between the three main concepts. In this article we will first introduce the Y model then, we will focus on the ontology of semantic and transfer links. This ontology enhance the reusability of the overall model since we can reuse the task, the PSM and the domain components, plus the links used to bind them. The separation between the semantic and the transfer notions enable us to have a fine grained reuse.

Description of the Y- model life-cycle development

We have called our model the Y-shaped model (Figure1). This model represents the three concepts used to construct a KBS: task, PSM and domain. In order to clarify the different kinds of relationships between these three concepts, we use inter-concept links and intra-concept links.

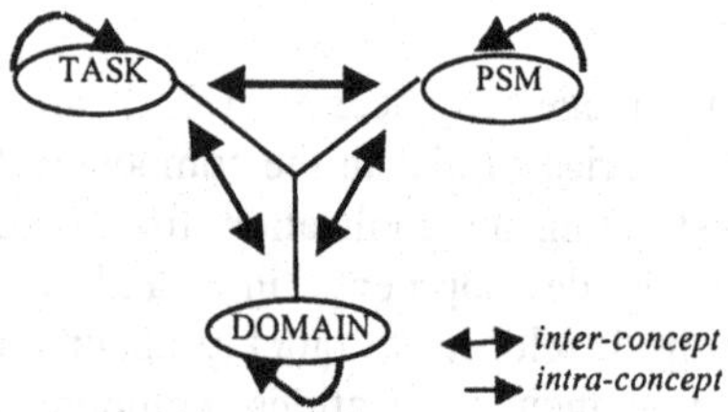

Fig. 1. The conceptual level

An inter-concept link describes relationships between concepts of different sorts: task/PSM, PSM/domain or task/domain. An intra-concept link describes the ones between two concepts of the same sort: task/task, PSM/PSM and domain/domain. For example, the literature shows that a generic PSM becomes more specific either by linking it to a task or a domain, or by specialising it through inheritance [8]. The inter-concept link can be used for relating a PSM with a task or a domain, and an intra-concept link for specialising it. Inter-concept links are bi-directional but, for simplification, we use undirected links considering for example task/PSM and PSM/task links as being the same.

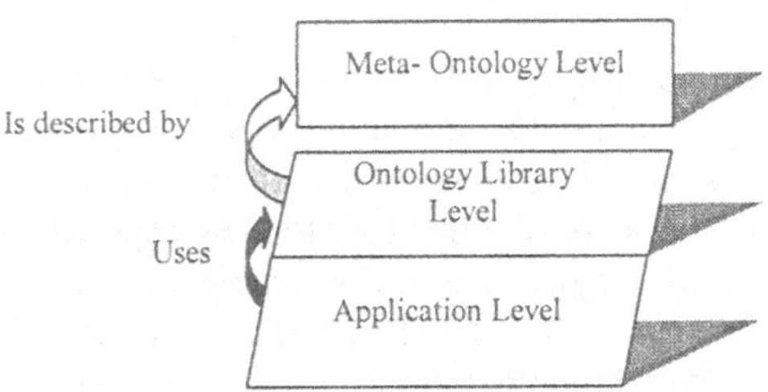

Fig. 2. Abstract levels

We can use different abstract levels to describe The Y-model. For our case we have identified three levels of abstraction (fig. 2): meta- ontology, ontology library and application.

1. The *meta- ontology level* describes the basic concepts needed to describe a KBS. Thus, at this level, we identify generic types of concept: task, PSM, domain knowledge plus inter- concept link and intra- concept link.
2. The *ontology library level* enables the description of different sorts of ontology - task, PSM, domain and link. It can be decomposed into different layers according to their degree of specialisation.
3. The *application level* is concerned with describing applications. An application is constructed using task-, PSM- and domain-ontologies specialised, if needed, using intra-concept links, and linked using inter-concept links.

In this paper, we will see first the meta-ontology of semantic and transfer links then, we will go on describing the ontology of semantic and transfer links. For the other components, readers can refer to [21].

Meta-Ontology of semantic links

Definition of a semantic link: A semantic link [12] describes a relation between two concepts. Its semantic can express an association (logical, physical, etc.), a composition, a specialisation, etc. It has its own semantics and behaviour allowing the concepts it relates to communicate and collaborate. It is often used in data modelling (data base systems, or CAD) so that semantic information is not distributed among related concepts, but defined in an independent entity, thus enabling modularity and reusability. For example, in an access telecommunication network, a semantic link *connected to* permits to describe a physical connection of a terminal to a concentrator, using for example a cable. The link cost characteristic is proper to the physical cable type. Its description in the link without referencing it in the two concepts, facilitate manipulation. If we change the cable or some of its characteristics, this, will not affect the related concepts.

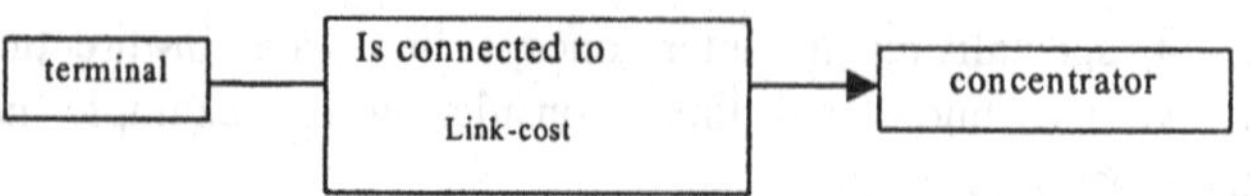

Fig. 3. Example of a semantic relation

A semantic link is defined by:
from and *to* : link attributes define the source and destination concepts.
The *roles* determine the roles, source and destination play in the link. These roles can help to identify the semantic link best suited to a specific case.
The *semantic attributes* describing the semantic properties of the links. These properties permit to enhance their representation and simplify their manipulation[16]. These properties are:

1. *Exclusivity/ sharing*: expressing that a concept referenced by an exclusive semantic link cannot be referenced by another semantic link of the same family. A sharing link is the opposite case. A concept referenced by an exclusive link, can be referenced simultaneously by other semantic shared links. The notions of exclusivity and sharing are limited to the semantic links of the same family. Thus, if a component is referenced by an exclusive composition link, it can not be referenced by another composition link but, it can be referenced by other sorts of links such as association.
2. *Dependence/ independence :* The dependence specifies that the existence of a destination component is dependent of the source component. The dependence notion permits to define that the destruction of a source component of a semantic link implies the destruction of the destination component referenced by this link. The independence is the opposite case of the dependence.
3. *Predominance / no- predominance* : the semantic of predominance and no predominance is symmetric to the dependence and independence. Predominance permits to specify the case where the source concept is dependent of the destination one. The no-predominance is the opposite case.
4. *Cardinality/ Inverse cardinality*: *Cardinality* expresses the number of target concepts that can be associated with a source concept and *inverse Cardinality* expresses the number of source concepts that can be associated with a target concept.
5. *Transfer link* : this attribute defines the transfer link associated to the semantic link.

name : <semantic-link-name>
from : <from-concepts>
to : < to-concepts>
Roles: <from-role; to-role>
Transfer < associated transfer name>
Semantics attributes

Meta-Ontology of transfer links

Definition of a transfer link : A transfer link[20] expresses the transferability of information between concepts. It may also have transfer functions used to transform the information it transports between source and target concepts. A transfer link may be associated with a semantic link. The commonly used semantic link is specialization (sort of). There is an implicit mechanism permitting to transfer (copy) the attributes from the super class to the sub class (transfer with an identity function). The transfer is composed of at least one translator, defining such a function.
Translator definition A transfer is composed of a set of translators describing information propagation between the related concepts. Each translator defines information propagation between the attributes of linked concepts and specifies the transfer function used to translate the information being sent.

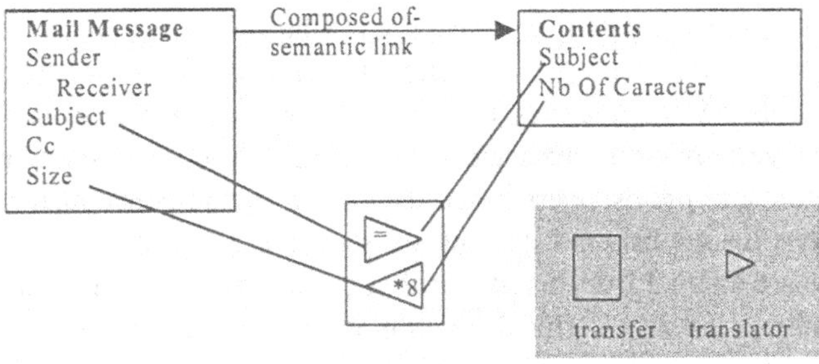

Fig. 4. Transfer link example

The transfer has to respect some principles:

1. The existence of a transfer link attached to a semantic link is dependent of the semantic link.
2. The transfer can be uni-directional, describing the transfer of information from source concept to destination concept, or bi-directional, describing an information flux in the two directions.
3. The transfer is selective. The different source and destination attributes have to be clearly specified. A concept can have intrinsic attributes which can not be shared, so, not propagated to its linked concepts.
4. The transfer is not transitive. It is limited to its source and destination concepts.
5. Like semantic links, a transfer link is defined at class level but operates effectively at the instance level.

Name :	<transfer-name>
Semantics :	<associated semantic link name>
source :	<source-concepts>
destination :	<destination-concepts>
translator :	<translator-names>

Name:	<translator-name>
Input:	< source -attributes >

Output: < destination-attributes > *Transfer-functions*

Semantic link Ontology

At the ontology level, we focus on the different sort of[2] semantic and transfer links, which can be either an inter or an intra- concept link according to the is-a[3] input and output concepts they bind.

Many disciplines studied semantic relations (links), such as linguistics, logic, psychology, information systems, and artificial intelligence (Winston & Al 88, Herrmann &Chaffin 87, Iris &Al 88, Woods 91, Dahlberg 94, Priss 96). The reader can see [22] for a comparison of these studies.

These different studies lead to the specification of different classifications of semantic links. For example, psychologists like Winston and Chaffin & Herrmann based their work on the lexical level. They use "relational elements" which characterises a relation in terms of its defining features, to define a classification of composition links [28]. Their classifications differ because they haven't used the same relational elements. Logicians, and philosophers based their work on the conceptual level. For example, Dahlberg distinguishes 4 classes of semantic relations. In information systems, the links are studied from a functional point of view distinguishing between dependent and independent links. This distinction is studied at the instance level then generalised at the class level. This leads to different hierarchies, where links can have the same name but referring to different semantics because there is no explicit formalism defining them. According to Priss [22], if a study concerns few basic semantic links, claiming that other links are based on these basic ones, we can describe the different links at the conceptual level. But if it concerns a big amount of links like describing a dictionary, the work can be done on the lexical level. The reader can find in [22] a comparison of these studies. For our case, we have focused our work on two big categories, often used in object modelling, and used in the Y-model. These two categories are the inclusion where inheritance and composition are the most known sort of, and association.

---

[2] We use sort of refering to the specialisation link (which we also call inheritance) used in object oriented domain.

[3] We use is-a refering to the intanciation mechanism used in object oriented domain.

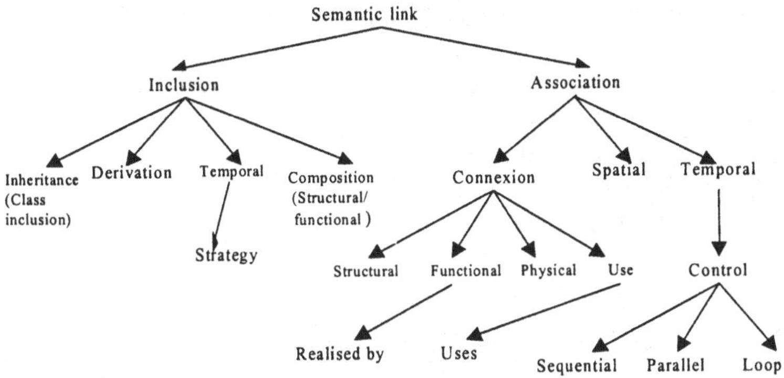

Fig. 5. Some semantic links (relations)

Inclusion

The inclusion link specifies that a concept includes other concepts. This inclusion is then refined as structural, functional, spatial, temporal and class inclusions.

name : <Inclusion-name>
sort of: <semantic relation>
from <from-concept>
to : <to-concept>
Roles: < sub set, set>
Transfer <transfer- inclusion>

Inheritance (Class inclusion)

This link defines the inheritance link. The main characteristic of this link is the similarity, permitting to differentiate it from the other sorts of inclusion link. The source concept is similar to the destination concept.

name : <sort-of>
sort of: <inclusion>
from : <from_concept>
to : <to_concept>
Roles: <subtype , super-type >
Transfer <transfer sort-of>

Composition (structural and functional inclusion)

The composition link connects a source concept called composite to another concept (s) called component. Fig 4 is an example of such links.

name :	<Composed of-name>
sort of:	<inclusion>
from :	<from_concept>
to :	<to_concept>
Roles:	< composite, component >
Transfer	<transfer composition>

Derivation

Often confused with inheritance, this link states that a destination concept is derived from a source concept. It is often used for evolution management in object design.

name :	<sort-of>
sort of:	<derivation>
from :	<from_concept>
to :	<to_concept>
Roles:	<versioned, version >
Transfer	<transfer version>

Spatial and temporal Inclusion

The spatial inclusion link defines when a concept is surrounded by another concept (*containment link)*. The temporal inclusion defines when a concept is included in another according to time. The *Strategy* we used in the Y-model[16], is a sort of temporal inclusion.

name :	<inclusion-spatial name>
sort of:	<inclusion>
from :	<from-concept>
to :	<to-concept>
Roles:	< contains; contained >
Transfer:	<transfer inclusion spatial- temp>

Association

The association link relates two or many concepts. It contrasts with the inclusion which consists of constructing a concept from other concepts. The association can be structural, defined under assembling constraints, functional, spatial, or temporal.

name	<association-name>
Sort of	<semantic link>
from	<from-concept>
to	<to-concept>
Roles	< from-role; to-role >
Transfer	<transfer- association>

Connection

Association links are undirected links. Connection [20] is a directed association. It can be functional, structural, physical, or usable. Fig 3 is an example of a physical connection.

name	< connection-name>
sort of	<association>
from	<from-concept>
to	<to-concept>
Roles	< connected; connected >
Transfer	< connection transfer>

Functional Connection

This link defines a functional connection where the linked concepts are associated to realise a specific relation. The *realised by* link used in the Y-model [16] is a sort of functional connection link.

name :	<*realised by*-name>
sort of:	<functional-connection>
from :	<from-concept>
to :	<to-concept>
Roles :	< problem; resolution method>
Transfer	<transfer-functional>

Spatial and temporal association

this defines an association of concepts in a special order (in space, or in time). The *control* we used in the y-model is a sort of temporal association.

Name	< temporal-association -name>
sort of	<association >
from	<from-concept>
to	<to-concept>
Roles	<scheduler ; schedules>
Transfer	<transfer -name>

Control link

This link is used in the Y-model for specifying the order of the components invocation. It is a sort of temporal connection

Name	< control-name>
sort of	<*temporal* association >
from	<from-concept>
to	<to-concepts>
Roles	<scheduler; schedules>
Transfer	<transfer name>

Ontology of transfer links

We can distinguish two types of transfers:

- *simple* defining a transfer of information between concepts (See fig 4).

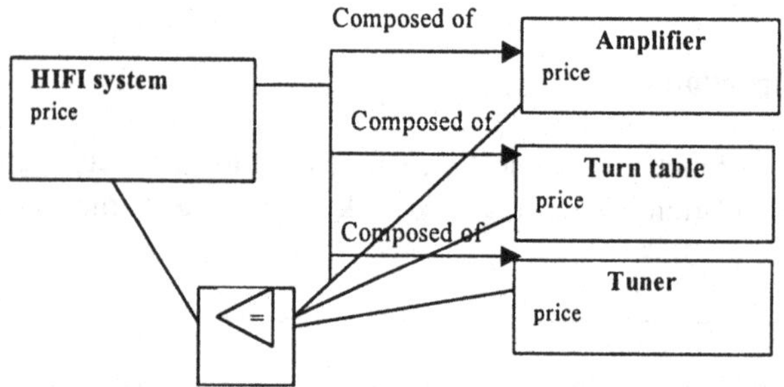

Fig. 6. Example of a complex transfer

- *complex* defining a transfer of information between many concepts related with the same sort of semantic. The semantic links having a complex transfer must have the same source concept or the same destination concept (See example Fig 6).

We can define some of transfer links according to their translator functions.

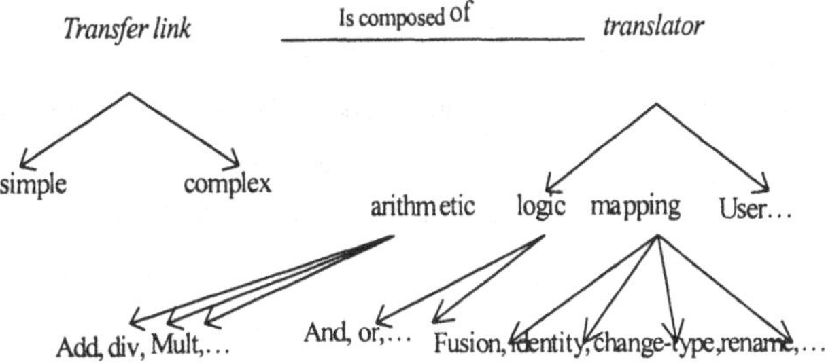

Fig. 7. Some transfer links

Rename translator link

An example of a mapping translator link might be renaming where the source attribute is renamed with the destination one:

Name <*translator-rename-name*> *Sort of* <translator> *input* <attribute-source> *output* <attribute-destination> *transfer function* <*rename(attribute-source, attribute- destination)*>

Example of the use of semantic and transfer ontologies

In this section we will see an example of our use of semantic- and transfer ontologies for defining components allowing the construction of applications (fig. 8).

Inter concept semantic and transfer links: we have identified the various sorts of inter-concept semantics links used in the Y-model as:

- PSM/domain: The association of a PSM with a domain, describes the knowledge that the former requires from the latter. In order to consider a PSM as a black box, its knowledge requirements are specified using *assumptions* [2]. We use *association links* to represent it. Data transfers between tasks and PSMs, as well as their inter-ontology mapping [23] are defined using mapping transfer links.
- Task/domain: This describes the association of a task with a domain in terms of knowledge requirements. We use an association link as defined below to specify it. The transfer can be defined using mapping transfer links.
- Task/PSM : A task can be *realised-by* either one or several PSMs and each PSM can *decompose* a task into sub-tasks. These in their turn, are *realised-by* sub-PSMs, until reaching terminal and non-decomposable PSMs. We define two *new* semantic links expressing the task/PSM relations. We refer to them as: *realised-*

by and *composed of.* The associated transfer link allows for example inter-ontology mapping.

- The *Realised-by link*: It allows relating a task to the different PSMs helping to solve it.

The *Composed of link*: It expresses the decomposition of the PSM into different sub-tasks.

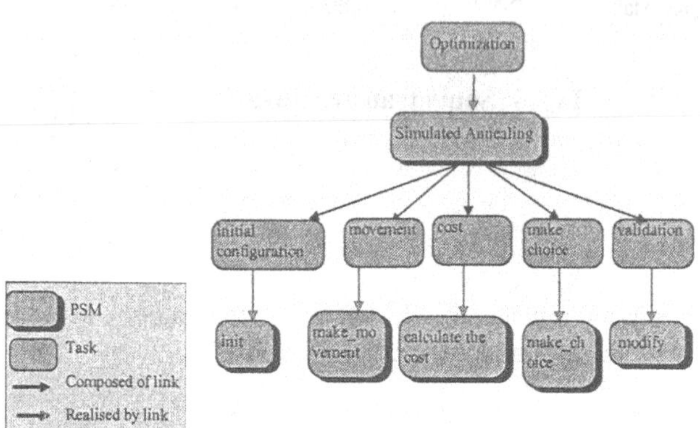

Fig. 8. The decomposition/realisation graph: Task/PSM decomposition of a specific and real case. It is about the optimisation of the concentrator location in an access network.

Intra-concept semantic links : we have identified the various sorts of intra-concept semantic links used in the Y-model as:

- *Specialisation* (*sort of* relation): it is used to specialise a task, a PSM, a domain or a link.
- *Instantiation link (is-a* link): Used to instantiated a concept. The concept can be task, PSM, or domain.
- *Control*: it specifies the order of invocation of tasks, PSMs, or the union of two domains.
- *Strategy*: this specifies the strategic knowledge used when there is a choice to be made among PSMs, tasks, or domains.
- *User intra-concept link*: as for inter-concept links, the user can extend the links we have identified by defining his/her own intra-concept link.

Intra-concept transfer link :Transfer links allow the propagation of information between related concepts: task/task and PSM/PSM used in the decomposition/realisation graph (figure 8); and domain/domain for a mapping between domains sharing common structures such as: electrical and telecommunication networks.

Application level

Our library is composed of: task, PSM, domain, semantic link and transfer link ontologies. Applications are collections of related task, PSM, domain knowledge components using inter-concept – like the *realised*-by - and intra-concept links - like the *specialisation* link. Figure 8 is an example of an application. The optimisation task is related to the simulated annealing PSM using a *realised by* semantic inter-concept link and a *mapping* transfer inter-concept link.

Implementation (object) level

The different hierarchies of the Y-model we have seen (fig. 2) are represented at the object level using meta-class, class and instance hierarchies. The meta-ontological level is represented at the meta-class level. The ontological (ontology library) level is defined by instantiating the meta- class (meta- ontology) level for representing the different ontologies. It is then represented at the class level. The inheritance graph enables us to maintain the levels of abstraction of the different sort of semantic and transfer links in the ontologies listed above. The instantiation graph permits to describe meta-classes, classes and instances.

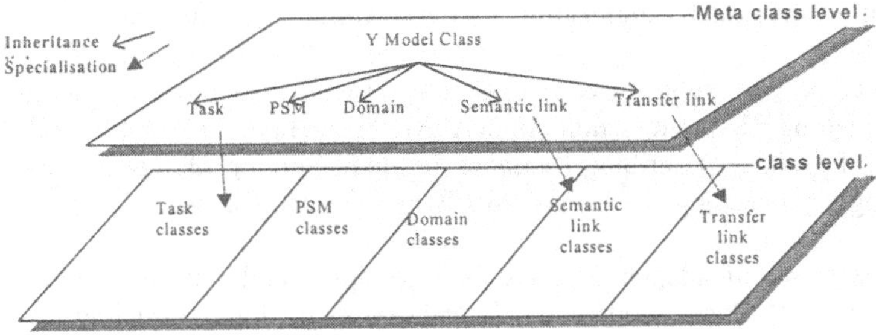

Fig. 9. Implementation level

Conclusion

We have described in this article the ontologies of semantic links and transfer links. We have seen the our use of them in the y-model for structuring a library of tasks, PSMs domains and links components. We have based our studies on work done in many fields: software engineering, object modelling, software engineering, and knowledge representation. The main contribution of our work is the explicit description it offers of the different concept interactions (relationships) by means of two new concepts, intra- and inter-concept links defined through different levels of description; secondarily, the use of object modelling concepts for representing them employing an ontology of semantic and transfer links. CGs have been used for knowledge representation in general and some work has been done specifically for ontology representation. We think that the theory underlying CGs can help to have a more formal description of the descriptive ontologies presented in this paper. This work is under development for an application concerning the re-use of tasks and PSMs in the telecommunications domain.

References

1. J. Angele, D. Fensel & R. Studer, Domain and Task Modelling in MIKE, Proc. IFIP WG. Joint Working Conference, Geneva. Chapman & Hall 1996.
2. R. Benjamins & C. Pierret-Golbreicht, Assumptions of Problem Solving Method. in 8th European Knowledge Acquisition workshop, EKAW-96, Springler-Verlag, 1996.
3. R.Benjamins & G. Van heijst, Modelling Multiple Models, In proceeding of CESA96, IMACS Multiconference, 1996.
4. P. Beys & R. Benjamins & G. van Heijst, Remedying the Reusability-Usability Tradeoff for Problem-solving Methods, proceedings of (KAW'96), Banff, Canada, Novembre 9-14, 1996.
5. B. Chandrasekaran & T.R, Josephson, The Ontology of Tasks and Methods. Symposium on ontological Engineering, AAAI Spring Symposium Series, Stanford, CA. 1997.
6. B. Chandrasekaran, T.R, Johnson & J.W Smith, Task-Structure Analysis for Knowledge Modelling, Communication of the ACM, 35(9) :124-137, 1992.
7. E. Coelho & G. Lapalme, Describing Reusable Problem- Solving Methods with a Method Ontology, proceedings of (KAW'96), Banff Alberta, Canada, Novembre 9-14, 1996.
8. D. Fensel : The tower of adapter Method for developing and reusing problem solving methods. In E. Plaza & al (eds.), *Knowledge acquisition, modeling and management*, (LNAI), 1319, Springler-Verlag, Berlin, 1997.
9. D. Fensel, H. Erikson, M. Musen & R. Studer, Conceptual and Formal Specifications of Problem-Solving Methods. In International Journal of Expert Systems, 9(4), 1996.
10. T. R. Gruber: A Translation Approach to Portable Ontology Specifications, Knowledge Acquisition, 5(2), 1993.
11. G. Klinker & Al, Usable and Reusable Programming Constructs, Knowledge Acquisition, 3:117-136, 1991.
12. M. Magnan, Objets complexes, In Oussalah & Al (Eds): Ingénierie objet concept et techniques, InterEditions, Masson, (in French) 1997.
13. J. McDermott, Preliminary steps towards a taxonomy of problem solving methods, in S. Marcus edt, Kluwer academic publisher, Boston 88.
14. M. Molina, Y. Shahar, J.Cuena, & M. Musen, A Structure of problem-Solving Methods for Real-time Decision Support : Modelling Approaches Using

PROTEGE-II and KSM. proceedings of (KAW'96), Banff Alberta, Canada, Novembre 9-14, 1996.

15. E. Motta, Trends in knowledge modelling: Report on 7th KEML Workshop. The Knowledge Engineering Review, Volume 12/Number 2/June 1997.
16. K. Messaadia, M. Oussalah using semantic links for reuse in KBS, in proceeding of DEXA'98.
17. A. Newel, The knowledge level, artificial Intelligence 18, 1982, 87-127.
18. C. Pierret-Golbreich, TASK MODEL: A Framework for the Design of Models of Expertise. and Their Operationalisation. KAW'96. Banff, Alberta, Canada 1994.
19. C. Pierret-Golbreich, TASK un environnement pour le developpement de systèmes à base de connaissances flexibles, habilitation à diriger les recherches, (in French) LRI, Orsay, 1996.
20. C. Oussalah & al, A framework for modeling the structure and behavior of a system including multi level simulation, IASTED INT. Symp. On Applied Simulation an Modeling, ASM, Grindelwald, Switzerland, February 1988.
21. M. Oussalah & K. Messaadia, The ontologies of semantic and transfer links to appear in proceeding of EKAW 99.
22. Uta Priss, Relational Concept Analysis: Semantic Structures in Dictionaries and Lexical Databases. Dissertation, TH-Darmstadt, october 96.
23. R. Studer, H. Eriksson, J.H. Gennari, S.W. Tu, D. Fensel & M. Musen, Ontologies and the Configuration of Problem-Solving Methods, proceedings of (KAW'96), 9-14, 1996.
24. A. Tchounikine, Activité dans les bases de données objets : le concept de Schéma Actif, thèse , University of paul sabatier, (in French)Toulouse 1993.
25. A. Valente, J. Breuker & B. Bredeweg, Integrating Modelling Approaches in the CommonKADS Library, In proceedings of the AISB93, 121-130, IOS Press, Amterdam, 1993.
26. G. Van Heijst, A.Th Shreiber & B.J Wielinga, Using explicit ontologies in KBS development, International Journal of Human-Computer Studies, 46(2/3):128-292, 1997.
27. B. Wielinga, A. Schreiber & J. Breuker, KADS: A Modelling Approach to Knowledge Engineering. Knowledge Acquisition. 4, 5-53. 1992.
28. M.E Winston, R. Chaffin, D.J, Herrmann, A Taxonomy of Part – Whole Relations, Cognitive Science 11, 417-444, (1987).

A Peircean Framework of Syntactic Structure

József Farkas and Janos Sarbo

University of Nijmegen, The Netherlands
janos@cs.kun.nl

Abstract. A semiotic framework for the syntactic structure of language is introduced. From properties of syntactic signs a parsing algorithm is derived. Using English as an example it is shown that, by means of its syntactic structures, the English language implements signs, analogous to those of Peirce's semiotic triads.

1 Introduction

Traditional models of natural language take as their starting point that hierarchical structure is somehow given. However, this assumption is sometimes too rigid, and this may partly explain the limited success of such models in natural language processing. The approach proposed in this paper is based on Peirce's semiotic ([5]) which provides us with a deeper foundation of language. In this approach, which is monostratal, hierarchical structure arises, via the interaction of language symbols, as a result of linguistic semiosis. Because interactions are events, language may be considered a set of symbol-events, a process. We apply the process view of language to syntactic (and morphologic) structure, and by using English as an example we illustrate the potential of the Peircean framework in parsing ([6]). A comparison with the 'generative' approach, from a philosophical point of view, can be found in [3].

2 The Peircean view

Peirce's semiotic is strongly related to his categories. In his doctrine of categories Peirce states that all phenomena present three aspects which, though irreducible to one another, have a different degree of dependency. The aspect of firstness is the aspect by virtue of which each phenomenon has an absolutely novel *quality*, unrelated to anything whatever. The aspect of secondness is the aspect by virtue of which each phenomenon involves an *interaction*. The aspect of thirdness is the aspect by virtue of which each phenomenon involves some *habit* (lawfulness, meaning etc.).

Though secondness cannot be reduced to firstness, it presupposes firstness, and, similarly, though thirdness cannot be reduced to either firstness or secondness, it presupposes both firstness (through secondness) and secondness. This dependency of the categories is formalised by the ordering firstness<secondness<thirdness, where "<" is a total order on categories.

2.1 Peirce's semiotic

Peirce's early papers suggested a convergence of his theory of the three categories and his presentation of the various semiotic triads. The most important of these triads is the triad of sign, object, and interpretant, which is a kind of *ontological* triad telling us what there is in the world. Based on the ontological triad, Peirce defined three triads of sign: the triad of icon, index, and symbol, dealing with how signs refer to their objects; the qualisign, sinsign and legisign triad, referring to the sign itself, prior to any relational possibilities and actualities; and the triad of rheme, dicent sign and argument, characterising the formal rules that associate signs and objects. These triads, which are called the *relational*, *material* and *formal* triad ([7]), are shown in fig. 1.

		Ontological type		
		1 Material	2 Relational	3 Formal
Phenomenological type	1 Quality	Qualisign	Icon	Rheme
	2 Indexicality	Sinsign	Index	Dicent sign
	3 Mediation	Legisign	Symbol	Argument

Fig. 1. *Peirce's classification of signs*

The above order relation on categories can be applied to Peirce's signs, according to their category exhibited. For example, Icon<Index<Symbol is an expression of degeneracy with respect to the realisation of the sign's object and interpretant; Qualisign<Icon<Rheme is an expression of degeneracy with respect to the sign's ontological type.

2.2 Language as a process

Language appears only as a form of interaction, whether we speak it, write it, or read it. In terms of Peirce's categories, interactions, or events, represent the category of secondness. An event involves the fact *that* something happens, but says nothing whatever about *what* happens. The latter aspect is the aspect of thirdness. *What* happens in an event requires that the event be embedded in a context of events which are related to each other. Such web of related events is what is called a *process* ([2], [3]).

Language consists of symbols which are signs. Because signs are generated from signs, and in turn generate other signs, every sign must be related to an event. From this it follows that language symbols are sign-events which, by virtue of their intepretants, are embedded within a process.

3 Language and ontological perspective

Language is a process involving symbol-events which are themselves generated according to rules which, in Peircean terms, are habits evolving from interaction with other symbol-events. Language processes involve both syntactic and semantic rules or habit.

Linguistic symbols may be considered as gesture-events within a process of interactive responses. Syntactic symbols may be considered from two angles: the messenger and the receiver. From the point of view of the messenger, a syntactic symbol is a gesture announcing other gestures to be generated in view of the interpretant of the entire unit of meaning (e.g. a sentence). From the point of view of the receiver, a syntactic symbol-event elicits an abduction regarding a range of possible subsequent symbol-events. Thus, in English, the symbol-event 'the' elicits an indefinitely large field of possible subsequent symbol-events, but excludes, for instance, the possibility of the next symbol-event being 'is'.

From a syntactic point of view, symbol-events have a specific function, regardless of their semantic function. The syntactic value of the language symbols making up the unit of meaning may be seen in the function of the value which they have in forming such a unit.

By virtue of their secondness, events are marked by a binary relation. Therefore, linguistic symbol-events must be also binary. This is why, strictly speaking, one lexical item by itself has no meaning. The syntactic value of the symbol-events will therefore depend upon the *sort* of relation that obtains between two language symbols. If one of the symbols has by itself no information content and therefore is a mere quality (a phoneme or a visible character), it will need another symbol to actualise its 'potential' content. Such nexus of two symbols, one of which is self-sufficient, but the other has mere potential content, may be called a *proto-symbol* (P) which corresponds to the category of firstness. An example of this is the symbol-nexus of free morpheme and affix.

Similarly, when the nexus is constituted by an asymmetrical relation between one language symbol which derives its full content from its association with another language symbol which is in principle self-sufficient, it may be called a *deutero-symbol* (D) which corresponds to the category of secondness. An example of this is the symbol-nexus of adjective and noun, or the one of determiner and noun.

Finally, when the nexus consists of two language symbols which are self-sufficient but together generate the interpretant of the unit formed by the string, e.g. a sentence, it will be called a *trito-symbol* (T) which, by its aspect of thirdness, mediates between the language symbols constituting a unit of meaning, or a thought, in the Fregean tradition. An example of this is the symbol-nexus between verb and subject.

To complete the picture, it is necessary to say a word about the *triadic relation* characterising each of these signs, because without such relation, they would not be signs, let alone syntactic signs. But precisely what makes them *syntactic* signs is the very fact that they stand for specific *rules* or habits. Thus, the object of syntactic signs is the rule for which they stand. Their interpretant on the other hand is the generation of the selection of the next symbol-event. The interpretant of the entire string of language symbols is, from a syntactic point of view, the establishment of the correctness of the string, regardless of its semantic content.

Inasmuch as linguistic symbols are also syntactic symbols, proto-, deutero-, and trito-symbols constitute a Peircean triad of linguistic symbols. By virtue of their category exhibited, these signs define the ordering $P<D<T$ which in turn defines the *levels* of syntactic signs.

In the remaining, we will denote by X a level of syntactic signs, and by X' the level subsequent to X. A sign (or symbol-event) of some level X will be called an X-level sign (or symbol-event).

3.1 Classes of syntactic sign levels

Language implements syntactic signs basically by lexical items and their relations. These are called *syntactic structures* or, equivalently, *language units*, depending on whether we want to emphasise their structural or linguistic properties. In the mapping of syntactic signs to syntactic structures (*syntactic mapping*), the notion of argument and functor, an abstraction from the combinatorial properties of lexical items, plays a crucial role. This combinatorial property can be characterised as relational or argumental need. A lexical item has *relational need* if it can be a functor, and *argumental need* if it can be an argument in some relation.

By analysing the structure of the three types of syntactic sign, we can recognise an argument and a functor symbol in each of them. In the case of trito symbols, the functor is that symbol which has the most relational need in the determination of the interpretant. We tacitly assume that such a distinction can always be made.

We denote the constituents of an X-level symbol-event, the argument and the functor symbol, and the syntactic symbol itself as X_1, X_2 and X_3. By virtue of the category and dependency which different signs respectively exhibit, syntactic signs may be said to define the ordering $X_1 < X_2 < X_3$ which in turn defines the *classes* of level X. The total order on levels and classes can be extended, by flattening, to a total order on syntactic signs.

The syntactic sign emerging from a symbol interaction is called its *descendant*. A syntactic sign that has no combinatorial need is a *completed* or well-formed sign. Two symbols are said *incompatible* if they cannot establish a relation syntactically, and *compatible*, otherwise. A completed sign is incompatible with any symbol. A sign of class X_i (i=1,2,3) of some level X is denoted an X_i sign.

3.2 The emerging syntactic sign

From a receiver's point of view, input symbols have merely potential content according to the receiver's (parser's) hypothesis. The set of such hypotheses is called the parser's dictionary. Input symbols appear one after the other, interact, and syntactic signs emerge by *symbol relation*. This might be called the 'automatic' type of sign generation. Because the descendant sign contains, besides the meaning of its constituents, the additional meaning of the relation itself, the signs generated by symbol relation are monotonously increasing.

But there are also cases of a degenerate symbol interaction. One of them is the interaction between incompatible symbols. In such a case, one of the interacting symbols, which is a sign generated in one symbol-event, is coerced to an argument or a functor, but *not* both, in another symbol-event. This type of sign generation is called *symbol coercion*.

Symbol coercion affects the interpretant of one of the symbols, but leaves the other symbol unchanged. Accordingly, we will say that the symbol coerced increases its meaning, and enters a higher class and/or level of syntactic signs. By virtue of their more developed interpretants, the signs generated by symbol coercion are monotonously increasing.

From the monotonicity property of syntactic signs, it follows that a lower level combinatorial need must have priority over a higher level one. The handling of one combinatorial need may require the elaboration of another, recursively.

Syntactic signs are composite signs which meet certain criteria. Thus, if a syntactic sign consists of related signs, the signs involved must in principle be *contiguous* to one another. This requirement is based upon the triadic structure of a syntactic sign the object of which is always a rule expressive of the expectation that a certain type of language unit must be followed by another type of language unit. The contiguity property can be defined as a covering relation on syntactic signs (classes which are empty are transitively closed).

The contiguity property is the driving force behind symbol coercion. Let us assume that two symbols S_1 and, subsequently, S_2 are recognised, which are incompatible. Assume, furthermore, that S_2 has combinatorial need on some higher level. Then S_1 is forced to enter a higher class without symbol relation. This follows from the assumption that the entire input, e.g. a sentence, is a well-formed syntactic sign. Due to the monotonicity property, if S_2 (or its descendant) enters a higher level, then the contiguous signs S_1 and S_2 (or its descendant) will only be able to establish a symbol relation on a higher level. Therefore S_1 must enter a higher class, as well, conform to its combinatorial need.

Syntactic sign generation respects, besides the combinatorial need of the symbols, also their syntactic properties. For example, an interaction between an X-level sign, and a sign entering the X-level subsequently, is syntactically different from the one in which the two signs are interchanged.

3.3 Towards an algorithm for syntactic signs

The properties of symbol coercion are formalised as follows ($X_i \rightarrow Y_j$ denotes that a sign of class i of level X may enter class j of level Y, and $X_i \rightarrow Y_j \vee Y_k$ is a shorthand for $X_i \rightarrow Y_j \vee X_i \rightarrow Y_k$):

(α_1) $X_1 \rightarrow X_3$; (α_2) $X_1 \rightarrow X_1' \vee X_2'$; ($\alpha_3$) $X_3 \rightarrow X_1'$.

In sum, X_1 and X_3 signs can increase their meaning without symbol relation, but an X_2 sign, due to its indexicality (aspect of secondness), presupposes an X_1 sign, and must relate with it. This meets our expectation that a relational need must be fulfilled always, though an argumental need can be optional.

Because language possesses a finite number of lexical items only, some syntactic signs must be generated *incrementally*, via degenerate symbol relations. In such a relation, the mediation aspect is incomplete, and the descendant of the X-level symbol interaction will become an X_1 or X_2 sign on the same level. Accordingly, the rules of symbol relation are formalised as follows (the symbol relation of X_1 and X_2 is denoted as X_1-X_2):

(β_1) $X_1\text{-}X_2 \rightarrow X_1 \vee X_2$; ($\beta_2$) $X_1\text{-}X_2 \rightarrow X_3$; ($\beta_3$) $X_1\text{-}X_2 \rightarrow X_1' \vee X_2'$.

3.4 Cumulative signs

The third type of sign generation is related to the incremental nature of syntactic signs, according to which, there may be encountered simultaneously more than one sign of the same class. By virtue of its aspect of firstness, an X_1 class may contain a number of signs which are *unrelated*, but the collection of which is a sign. By virtue of its aspect of secondness, an X_2 class may contain a number of signs which are *unrelated*, but which share a *common referent*. By virtue of its aspect of thirdness, an X_3 class may contain a completed sign which must be a single sign. In each case, the signs belonging to a class must have the same combinatorial need, and must be consistent with each other, with respect to their meaning (i.e. with the interactions involved in the signs).

Based upon the above properties of sign classes we refine our definition of incompatibility as follows. Symbols will be said *incompatible* if (i) they are of different classes of a level and cannot establish a relation syntactically, or (ii) they share the same class, but have different combinatorial need or meaning, or (iii) one of the symbols is a completed sign.

Besides symbol coercion, another case of a degenerate symbol interaction is the one between compatible symbols of the same class. In such a case, the symbols involved are accumulated on a stack. This type of sign generation is called *stacking*. The need for a stack is explained as follows. Assume that the subsequent symbols, S_a and S_b, enter the same class of some level X. Let S_a and S_b be such that they cannot enter a higher level, because their combinatorial need on level X is not yet fulfilled. As a consequence, both symbols have to be stored, temporarily. Let another symbol, S_c, enter the other class of level X. S_c and S_a are *not* contiguous symbols and, therefore, should not interact, however, S_c and S_b must interact. These requirements can be satisfied by storing S_a and S_b on a stack. A stack of signs is considered a single sign, represented via the topmost item of the stack. Symbols generated by stacking are, trivially, monotonously increasing.

The stacking of symbols is subject to restrictions implied by the property of contiguity, the binary nature of interactions, and the properties of symbol classes. Let us denote the state of some level X by the pair of stacks of X_1 and X_2 (the X_3 stack has no effect on the stacking of symbols and therefore omitted). Let s_i (i=1,2) denote a stack, $\perp$ the empty stack, s the next symbol entering level X, and '/' a left-associative stack constructor. Then, symbol stacking must respect the following rules, specified as transitions on the state of level X:

$(s_1, \perp) \rightarrow (s_1/s, \perp)$; $(\perp, s_2) \rightarrow (s, s_2)$; $(s_1, s_2) \rightarrow (s_1, s_2/s)$.

In sum, an X_1 stack can grow if the X_2 stack is empty, or otherwise, by a single item only (the common referent); and an X_2 stack can grow if the X_1 stack is constant.

3.5 Primary signs

We assume that the input symbols enter a lowest class (prm) as a sequence of primary signs, e.g. phonemes or characters. Prm, which has the aspect of firstness, is by definition a class of syntactic signs. The input primary signs, which have no combinatorial need, are collected in prm, as long as their sequence forms a morphological symbol which is a dictionary entry. When this happens, the symbol receives its combinatorial need from the dictionary, and enters the lowest level, in particular, P_1 if it has no P-level relational need; and P_2, otherwise:

(γ) prm$\rightarrow P_1 \vee P_2$.

3.6 Mediating evaluation

Syntactic signs are yielded by symbol interaction. But the decision as to *when* the mediation takes place depends upon the type of evaluation, which can be lazy or greedy. In general, we will assume lazy evaluation of relations, because it can be more economic in some cases.

The lazy evaluation of syntactic sign-events affects the modelling of the terminator symbol (e.g. the point symbol) which, therefore, will be treated as an incompatible argument and a nullary functor on each level, thereby forcing the realisation of pending relational needs.

In sum, syntactic signs arise in language due to (1) the quality of contiguity, (2) symbol interaction, and (3) mediating evaluation which, respectively, have the aspect of the categories, firstness, secondness and thirdness.

The emerging syntactic sign may become part of a cumulative sign, or change its aspect of correspondence with its object, or establish a relation with another sign. In sum, symbol interactions do emerge by (1) symbol stacking, (2) symbol coercion, and (3) symbol relation, which, again, exhibit the aspects of Peirce's categories.

4 Syntactic mapping

We illustrate the syntactic mapping of language by using English as an example. In this mapping we capitalise on the semiotic properties of syntactic signs and on the conceptual distinctions that may be expressed in English ([1],[4]).

Trito-symbols correspond to the symmetric relation between two constituents which are both self-sufficient and require the presence of the other, e.g. the relation between noun and verb (subject and predicate). Such a relation is called *predication*(p).

Deutero-symbols correspond to the asymmetric relation between an action/state or participant, and its properties: both are self-sufficient, and the latter requires the presence of the former, but the reverse does not hold. In English, two instantiations of this type of relation can be identified: *modification*(m), e.g. the relation between adjective and noun; and *qualification*(q), e.g. the relation between determiner and noun.

The third type of symbols, proto-symbols, correspond to the morphological process of *affixation*(a). An affixation relation distinguishes between a root (or base), e.g. a free morpheme, and an affix: the root is self-sufficient; the affix has only potential meaning actualised by the root.

From the semiotic point of view, q- and m-signs form subsets of deutero-symbols. Using the analogy of the relational triad, q- and m-signs represent iconic and indexical meaning, respectively, and, therefore, these signs may be said to define the ordering $q<m$. Eventually, this yields the ordering $a<q<m<p$ which in turn defines the *levels* of the English syntactic signs.

Though the above characterisation of the English syntactic mapping is based upon simple considerations, it may not be language model independent. In section 6 we will show, however, that this mapping is certainly not accidental. It will be argued that the English language implements signs by means of syntactic structures analogous to those of Peirce's semiotic triads.

4.1 Syntactic relations

Syntactic relations emerge due to the combinatorial need of syntactic symbols. In general, a syntactic symbol can have argumental need, optionally, but its relational need is a function of that of its constituents, or, in the case of a lexical item, it is some constant value. Lexical items can contribute to the relational need of syntactic signs, on each level.

A lexical item has a potential combinatorial need, which is a finite set. The combinatorial need of a syntactic symbol generated by a symbol relation is the disjoint *union* of the combinatorial need of its constituents, possibly modified (i.e. restricted) by the interaction itself. The potential relational need of the types of lexical items is exemplified in fig. 2 (respectively, a '+' or '-' represents the presence or absence of a relational need on the level indicated by the column). The relational need of a particular lexical item is the subset of that of its type.

For example, the q-level relational need of adjectives and adverbs allows symbol-events like *keep awake*, or *walk by*; and their m-level relational need the modification relation like *happy girl*, or *walk quickly*. In the case of a preposition, the q-level relational need contributes to the relation with the obligatory argument (qualification); and the m-level one to the modification of the optional argument by the qualification yielded.

Verb–complement relation is classified as modification, typically. Indeed, such a symbol-nexus fits the definition of a deutero sign: verb and complement are both self-sufficient, but the verb derives its full content from the complement. Because the descendant symbol of a verb–complement relation has an indexical character (the verb points in the direction of its complement it is acting on), this type of relation must be identical with modification (we admit that the terminology might be confusing for the linguist).

In sum, a verb relates with its complement(s) due to its m-level relational need (which is fulfilled when all necessary complements are found), and with the subject, due to its p-level one. A copula or an auxiliary relates with its complement due to its q-level relational need, but the copula relates with the subject due to its p-level relational need.

The development of the relational need of syntactic signs can be illustrated as follows. The potential m-level relational need of a preposition will be actual if the q-level relation it is involved in does not disallow that. For example, there will be such need in the case of *in London*, and there won't be, in the case of *drive in*.

	a	q	m	p
primary	-	-	-	-
affix	+	-	-	-
noun	-	-	-	-
determiner	-	+	-	-

	a	q	m	p
preposition	-	+	+	-
adjective	-	+	+	-
adverb	-	+	+	-
verb	-	+	+	+

Fig. 2. *Potential relational need*

5 Example

Below we give the analysis of the sample example: *Mary drank some wine yesterday*. We leave out the morphological analysis, and assume that the input signs leave the a-level and enter q_1 or q_2 conform to their combinatorial need.

The potential relational need of the lexical items is as follows: *Mary*={}, *drank*={m,p}, *some*={q}, *wine*={}, *yesterday*={m}. Symbols having a non-empty relational need are written in capitals. In the table below, an item represents the content of the storage of a class (column) prior to the evaluation of the next input symbol (row).

step	next input	q_1	q_2	q_3	m_1	m_2	m_3	p_1	p_2	p_3
0	*Mary*(m)									
1	*drank*(D)	m								
2	*some*(S)	D		m						
3	*wine*(w)		S			D	m			
4	*yest*(Y)	w	S			D	m			
5	.	Y		s-w		D	m			
6	.	.		.	D-s-w	Y	m			
7	.	.		.	D-s-w-y		m			
8	.	.		.	.		.	m	D-s-w-y	
9	.	.		.	.		.	.		m-d-s-w-y

In step_1, the symbol *Mary* enters q_1. Next, the symbol *drank* enters q_1 (it has no q-level relational need, but it has argumental need, optionally). Because *Mary* and *drank* are incompatible on this level, therefore, q_1 (*Mary*) must increase its class by α_1. The next input, *some*, forces q_1 (*drank*) to increase its level by α_2 which, in turn, forces q_3 (*Mary*) to do the same by α_3 and α_1.

The rest of the evaluation is as follows (we specify for each step which rule is applied to which sign(s), but we omit the treatment of the point symbol):

step_4 $\beta_2(q_1,q_2)$; step_5 $\alpha_2(q_1)$, $\alpha_3(q_3)$, $\beta_1(m_1,m_2)$; step_6 $\beta_1(m_1,m_2)$; step_7 $\alpha_2(m_1)$, $\alpha_3(m_3)$; step_8 $\beta_2(p_1,p_2)$.

6 English syntactic symbols and Peirce's signs

The ordering of the classes of a level is depicted in fig. 3a (edges represent the "<" relation). In the case that degenerate signs are allowed, this graph can be paraphrased as a two-level scheme consisting of a finite automaton (FA) and a number of stacks. A state of the FA corresponds to a sign class, and a transition to an application of an α or β rule, represented by solid and dashed lines, respectively (but in later graphs we will use solid lines for both types of transition). The resulting graph is depicted in fig. 3b (an edge which is a cycle is omitted).

In virtue of the syntactic mapping, the classes of English syntactic signs define a total ordering as shown in fig. 3c. By using the interpretation of fig. 3b to fig. 3c, we get a two-level system (this is not illustrated). We map the X_3 and X_1' classes, e.g. a_3 and q_1, to same states (same signs, different interpretants). The initial state is prm, all others are final states; each state has a stack. The output language of the system is the set of signs in the different stacks, upon termination.

We argue that the above system is terminating. This follows from the properties of the α and β rules. Indeed, an α rule increases the class of the sign it is applied to, but there are finitely many classes; and a β rule fulfils a combinatorial need, but the combinatorial need of syntactic signs is finite.

The priority of a lower level combinatorial need over a higher level one implies a stack-like behaviour of the levels of syntactic signs; the properties of symbol stacking imply a stack-like behaviour within a level of syntactic signs. It follows from this that the individual stacks of the classes can be simulated by a single stack. Eventually we get that our two-level system is equivalent to a context free (CF) grammar, and therefore, it has the same complexity as CF parsing.

Below we describe an equivalency transformation of our two-level system with respect to its input and output languages. In step1 (see fig. 4), we remove $m_2 \rightarrow p_1$. If the descendant sign m_1-m_2 has no m-level combinatorial need and enters the p-level, then, equivalently, we can push it to the m_1-stack, and let the m-level sign generated subsequently force it to enter the p-level by α_2.

In step2 (cf. fig. 5), we close the $q_1 \rightarrow m_1 \rightarrow p_1$ transitions by $q_1 \rightarrow p_1$. Because these transitions only apply when the m-level stacks are empty, the storages are not affected by this transformation. Furthermore, we merge q_2 and m_1, because of the orthogonality of their signs (respectively, functor and argument signs), and the incompatibility of the q_1 and m_1 signs (both are argument signs). The q_2 stack is 'pushed' upon the m_1 stack, thereby providing that the property, that q_2 signs represent lower level combinatorial need, and therefore, have priority over m_1 signs, holds invariantly after the transformation.

Last, in step3 (see fig. 6a), we remove $a_2 \to q_1$ (cf. step1 above). Furthermore, we close $a_2 \to q_2 m_1 \to m_2$ by $a_2 \to m_2$ (the stacks are not affected by this transformation), and remove the redundant transition $q_1 \to m_2$. The resulting system is depicted in fig. 6b (edge directions are omitted).

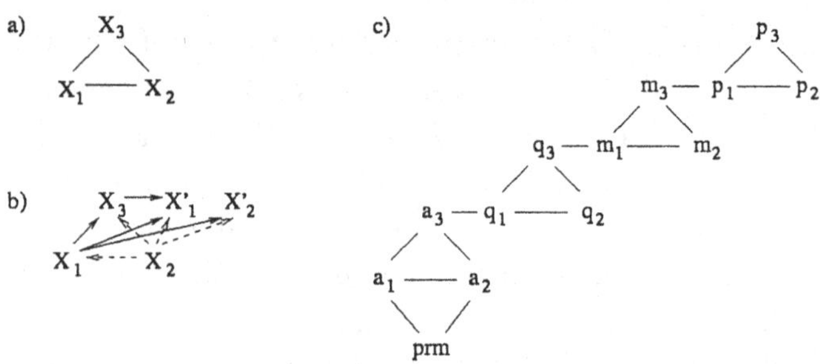

Fig. 3. *English syntactic signs*

Fig. 4. *Transformation step1*

The system of fig. 6b can be interpreted, the other way round, as a specification of a parsing algorithm, and also as a classification of syntactic signs (except for degenerate ones): states can be mapped to sign classes, the ordering of which is derived from the transitions, as shown in fig. 6b (cf. fig. 3a). A comparison of this classification with Peirce's triads shows their isomorphism (the reader might turn fig. 1 in 135° to help noticing this). The analogy between the corresponding signs is justified as follows:

Fig. 5. *Transformation step2*

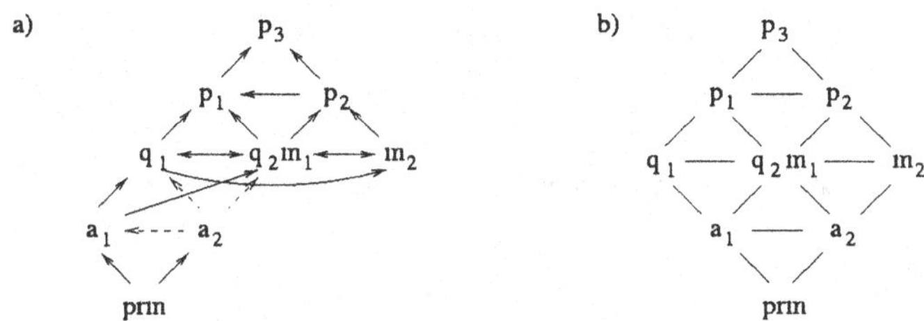

Fig. 6. *Transformation step3*

prm a pure quality, unrelated to anything else; a *primary.*

a_2 a particular quality, referring to an actually existing argument; an *affix.*

m_2 a sign involving the convention that arguments have certain properties which are general types; a *modifier.*

a_1 an image, a name of some 'thing', e.g. a free morpheme; a *root.*

q_2m_1 a morphological sign (a_3), or a qualifier (q_2), or a developing m-level sign (cf. section 3.3), each involving a reference to the argument; we call these signs collectively, a *relative.*

p_2 a sign involving the convention that arguments have some more basic properties; a *predicate.*

q_1 a sign representing the possible existence of some 'thing'; a *noun.*

p_1 a sign used to assert the actual existence of something, for example, a clause; a *nominal.*

p_3 a sign expressing a lawlike relation between subject (p_1) and predicate (p_2), a 'thought'; a *sentence.*

This completes our English syntactic mapping. In fig. 7, it is illustrated how English implements the signs of the 'real' world, syntactically. The upward-right diagonals represent the material, relational and formal ontological types; and modifier-, predicate- and nominal-formation, on the left- and righthand side of fig. 7, respectively; and, similarly, the upward-left diagonals represent the quality, indexicality and mediation phenomenological types; and, respectively, word-, expression- and sentence-formation.

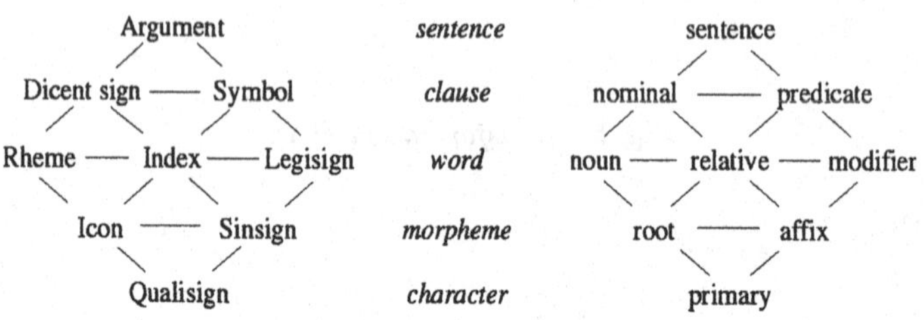

Fig. 7. *Peirce's signs and English syntactic signs*

Acknowledgements

We wish to thank Guy Debrock and Vera Kamphuis for the invaluable discussions during the course of the above work, and also the anonymous referee for the helpful remarks.

References

1. Aarts, F., Aarts, J.: English syntatic structures. Pergamon Press, Oxford (1982)
2. Debrock, G.J.Y., Sarbo, J.J.: Towards a Peircean model of language. Technical Report CSI-R9802, University of Nijmegen (1998)
3. Debrock, G.J.Y., Farkas, J., Sarbo, J.J.: Syntax from a Peircean perspective. In: Sandrini, P. (ed.): Proceedings of 5th International Congress on Terminology and Knowledge Engineering (in press). Innsbruck, Austria (1999)
4. Farkas, J., Kamphuis, V., Sarbo, J.J.: Natural Language Concept Analysis. Technical Report CSI-R9717, University of Nijmegen (1997)
5. Peirce, C.S.: Collected Papers of Charles Sanders Peirce. Harvard University Press, Cambridge (1931)
6. Sarbo,J.J.: Formal conceptual structure in language. In: D. M. Dubois (ed.): AIP Conference Proceedings of the American Institute of Physics (in press). Springer-Verlag (1999)
7. Sowa, J.F: Knowledge Representation: Logical, Philosophical, and Computational Foundations. PWS Publishing Company, Boston (1998)

A CG-Based Behavior Extraction System

Jeff Hess and Walling R. Cyre

Automatic Design Research Group
The Bradley Department of Electrical and Computer Engineering
Virginia Tech, Blacksburg, VA 24061-0111
cyre@vt.edu

Abstract. This paper defines "behavior extraction" as the act of analyzing natural language sources on digital devices, such as specifications and patents, in order to find the behaviors that these documents describe and represent those behaviors in a formal manner. These formal representations may then be used for simulation or to aid in the automatic or manual creation of models for these devices. The system described here uses conceptual graphs for these formal representations, in the semantic analysis of natural language documents, and for its word-knowledge database. This paper explores the viability of such a conceptual-graph-based system for extracting behaviors from a set of patents. The semantic analyzer is found to be a viable system for behavior extraction, now requiring the extension of its dictionary and grammar rules to make it useful in creating models.

1 Introduction

When designing, modeling, or creating a digital device, the engineer has many tools at his or her disposal that automatically translate between different levels of abstraction, from algorithmic model to gate-level design to silicon layout, and many tools that simulate the digital devices being designed. These translations and simulations require formal representations, such as VHDL or StateCharts, in order to work, but the engineer is often faced with the task of modeling and simulating a device for which he or she has only a natural language description, such as a specification or patent. While many modeling tools exist, very few automate the translation of descriptions from natural language into a formal model.

To aid in automating this translation, a system has been created to extract behaviors from natural language documents describing digital devices and systems. *Behavior extraction* is the act of analyzing the semantics of the natural language document to determine the behaviors and interactions of the devices and components described therein. Once information about these behaviors has been represented in a formal manner, it may be used as a modeling aid, for direct simulation, or for automatic synthesis.

The behavior extraction system described here, called the *semantic analyzer*, uses conceptual graphs [10] as the formal system of knowledge representation. Conceptual graphs are directed, connected, bipartite graphs that represent concepts, such as devices, actions, and attributes, and the relationships between them. The semantic analyzer uses conceptual graphs in the semantic analysis of the natural language document and to represent the meanings of words in its lexicon. The issue is whether the semantic analyzer is a viable method for helping to automate the translation of natural language specifications into behavioral models. To answer this question, the size of the lexicon is considered as sentences from a corpus of patents on direct memory access devices are analyzed.

2 Related Work

Conceptual graphs were introduced by John Sowa [10]. With Eileen Way, he described a semantic interpreter that uses conceptual graphs for the word-knowledge database, the actual analysis of parsed sentences, and the representation of results of this analysis [9]. The behavior extraction system described here is based on their approach, and an earlier semantic analyzer coded in Prolog [5]. Our lexicon of word definitions uses *word-sense graphs* to represent the meanings of words, similar to the canonical graphs used in Sowa and Way's interpreter. These word-sense graphs are modified abstractions that include valence information that indicates how the formal parameters can be identified. Although rejected by Graham Mann in his natural language analyzer [7], this valence information is very useful in this application for determining the roles that digital devices and signals play in the extracted behaviors. The word-sense graph have some of the characteristics of Fillmore's case frames for verbs [4], which define the relationships (conceptual relations) between a verb and nominal case fillers from a sentence. The sense graphs include *semantic markers* that help identify nominals in a sentence that fill case roles in the graph.

In earlier studies, it has been demonstrated that conceptual graphs representing individual sentences in device specifications can be integrated [6,2], how the integrated conceptual graphs can be simulated directly [1], and how engineering models for simulation and hardware synthesis can be derived from them [3].

3 Behavior Extraction Approach

The component that extracts behavior as conceptual graphs from sentences is called the *semantic analyzer*. It analyzes sentences that have been parsed by a phrase-structure grammar and generates *parse trees*. These parse trees contain all the information about the way that the sentence was parsed. Each internal node of a parse tree represents a phrase structure and was generated by a specific grammar rule; the leaves of the tree represent the terminals, the actual words of the sentence. Fig. 1 shows a parse tree for Sentence 1.

The processor receives the count from the register. *(1)*

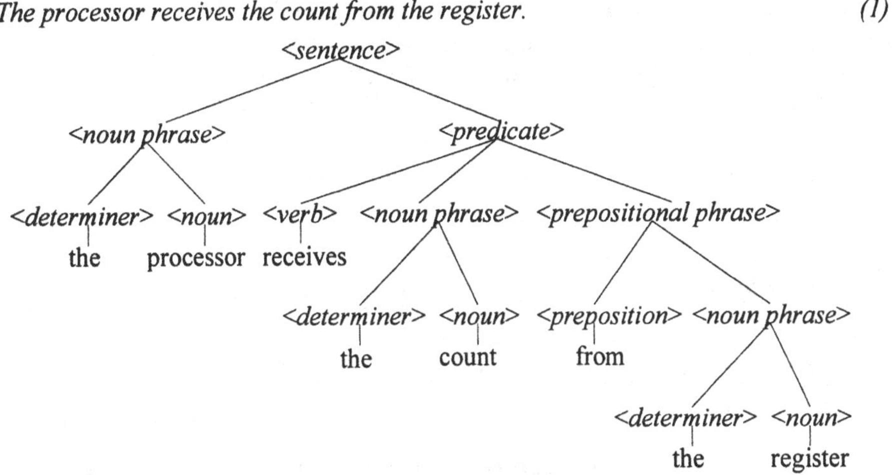

Fig. 1. Parse tree for Sentence 1.

The semantic analyzer uses two sources of knowledge in analyzing a parse tree: a set of *analysis rules* and the *dictionary* (lexicon). Each grammar rule used to create the parse tree has associated with it one or more analysis rules, which tell the semantic analyzer how to construct a conceptual graph, called a *meaning graph*, representing the meaning of the syntactic structure generated by the grammar rule. This meaning graph is constructed from the meaning graphs of the node's descendents, which are word-sense graphs for the leaf nodes. If more than one analysis rule exists for a particular grammar rule, each rule is applied, creating multiple meaning graphs for that node. The dictionary contains definitions for words in the form of conceptual word-sense graphs. The dictionary may have multiple definitions for a word, again creating multiple analysis possibilities.

Although analysis starts with the root node of a parse, all analysis rules require that all non-leaf children of a node (i.e., all non-terminal constituents of a grammar rule) be analyzed. Thus, the analysis may be thought of as following a depth-first traversal of the parse tree, or a bottom-up analysis.

For example, in the parse tree shown in Fig. 1 above, the semantic analyzer would start with the analysis rule for the *<sentence>* node, which requires that the children *<noun phrase>* and *<predicate>* be analyzed. The analysis rule for *<predicate>* requires that its children *<noun phrase>* and *<prepositional phrase>* be analyzed. Finally, the analysis rule for *<prepositional phrase>* requires that its child *<noun phrase>* be analyzed. Thus, the first phrase structure to be analyzed is this *<noun phrase>*, "the register". The analysis rule for this type of *<noun phrase>* requires that the *<noun>* be looked up in the dictionary and that the *<determiner>* be attached appropriately (the result of this will depend on the type of determiner). The noun "register" has one word-sense graph definition in the dictionary: [register < memory]. (This modified conceptual graph notation indicates that the concept type *register* is a subtype of the *memory* concept type in the concept type hierarchy.) The semantic analysis rule for this *<noun phrase>* structure recognizes the definite article "the" in the *<determiner>* position, and changes the referent of the concept to an individual marker. The meaning graph resulting from this step is shown by a distinctive type font in Fig 2.

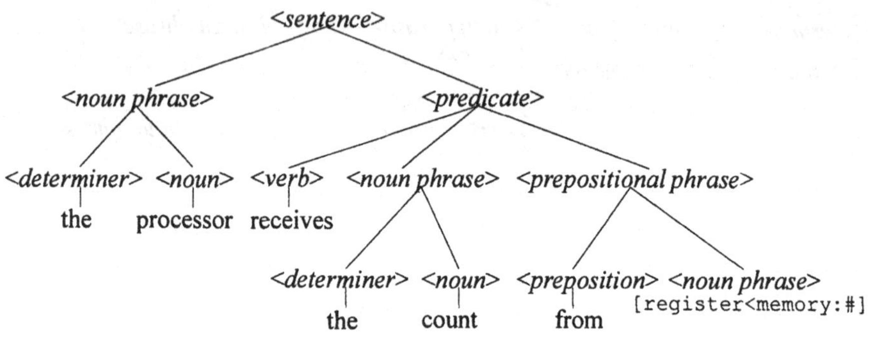

Fig. 2. First step in example analysis.

Now that this *<noun phrase>* has been analyzed, its parent *<prepositional phrase>* can be analyzed. The analysis rule for the *<prepositional phrase>* requires that the *<preposition>* be used to as the *syntactic marker* for attaching the meaning graph of the *<noun phrase>* to the structure it modifies. The results of this step are shown in Fig. 3.

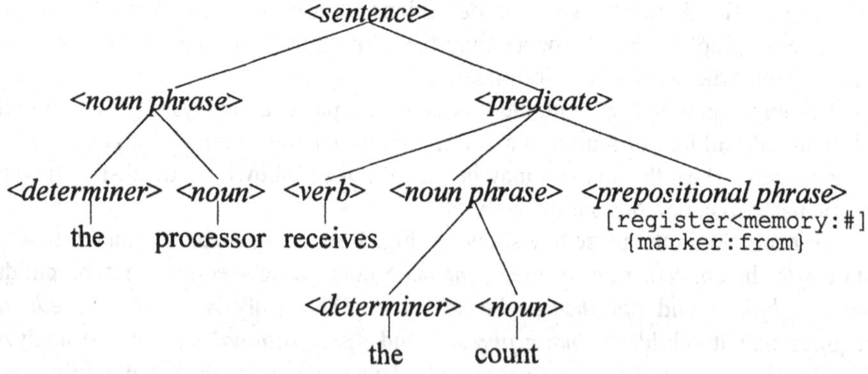

Fig. 3. Second step in example analysis.

For the *<predicate>* to be analyzed, the other descendent *<noun phrase>* must be analyzed. Its analysis is exactly the same as the previous *<noun phrase>*. Again, the notation [count < value] indicates that the concept type count is a subtype of the value concept type. The results of this step are shown in Fig. 4.

To synthesize a meaning graph for the *<predicate>* the word-sense graph for its verb, "receives", must be retrieved from the dictionary. this verb's word-sense graph is shown in Fig. 5. The concepts marked with question marks are called *unidentified.* These unidentified concepts may become joined with the meaning graphs of other phrases in the sentence, and act as case roles [4] to be filled by the meaning graphs of the other phrases. That is, the unidentified roles are slots in a template associated with the verb. These slots may be filled by nominals or adverbials of the sentence. Adverbials include prepositional phrases, subordinate clauses, modals (can, must, should) and simple adverbs (synchronously, however, quickly, not).

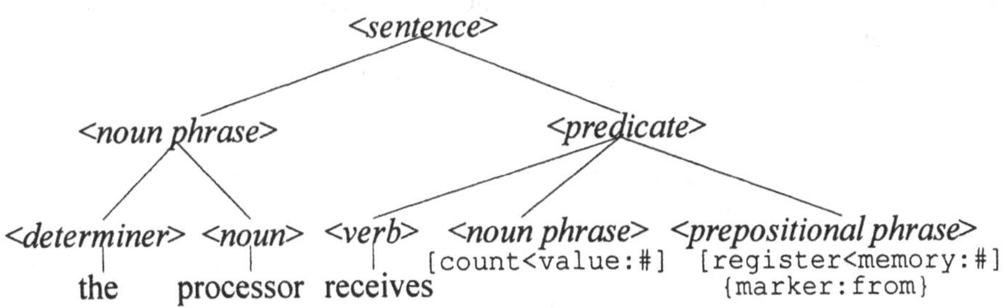

Fig. 4. Third step in example analysis.

In the sense graph shown in Figure 5, the syntactic markers listed after the question marks indicate the syntactic markers of the sentence that can signal or constrain potential role fillers. For example, the grammatical subject of the verb may identify the device playing the agent role, and the direct object of the verb may identify the value playing the operand role. The prepositions "via," "through," "on," "over," or "along" indicate that the meaning graph of the noun phrase which is the object of that prepositional phrase is a candidate filler for the instrument role. The "\" preceding a marker (subject, object) indicates it is marked by sentence word order rather than from a preposition or subordinating conjunction.

In the earlier version of the semantic analyzer, sense graphs were limited to trees rooted at the verb concept node and having leaves representing unary relations (modals, negatives, adverbs) or concepts related by binary relations. In the current version, sense graphs can be general rooted graphs. This is necessary in some cases, such as for the verb allows. In the sentence, "The device allows a signal to terminate processing." the subject of *allow*, is actually the agent of the purpose of *allow*, namely *terminate*, and *signal*, the object of *allow* is the instrument of *terminate*. The verb *allow* itself, simply affixes a unary *possible* relation onto *terminate*. This clearly requires a more general sense graph.

```
[receive]-
   (agent)→[device:?\subject]
   (operand)→[value:?\object]
   (source)→[device:?from]
   (instrument)→[device:?via,through,on,over,along],.
```

Fig. 5. Word-sense graph for the verb "receives".

To finish the analysis of the *<predicate>*, the meaning graphs for the *<noun phrase>* and the *<prepositional phrase>* are joined with unidentified concepts in the word-sense graph for the *<verb>*. The semantic analysis rule requires that the *<noun phrase>* join a concept with the special syntactic marker "\object," and that the *<prepositional phrase>* join a concept with the syntactic marker "from." The results of this step are shown in Fig. 6.

Before the complete *<sentence>* can be analyzed, the first *<noun phrase>* must be analyzed. Its analysis is exactly like that of the other two noun phrases. The results of this step are shown in Fig. 7.

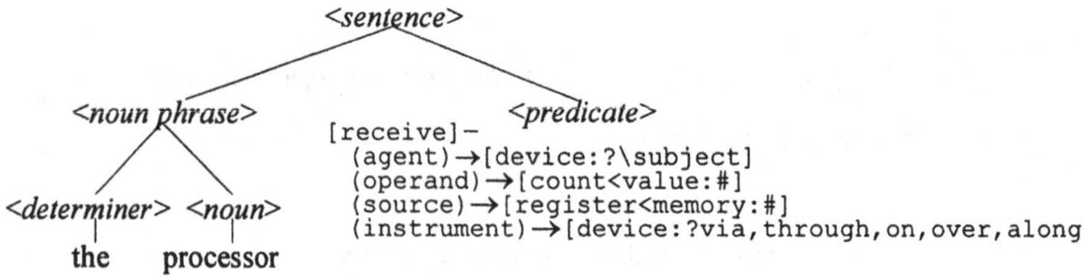

Fig. 6. Fourth step in example analysis.

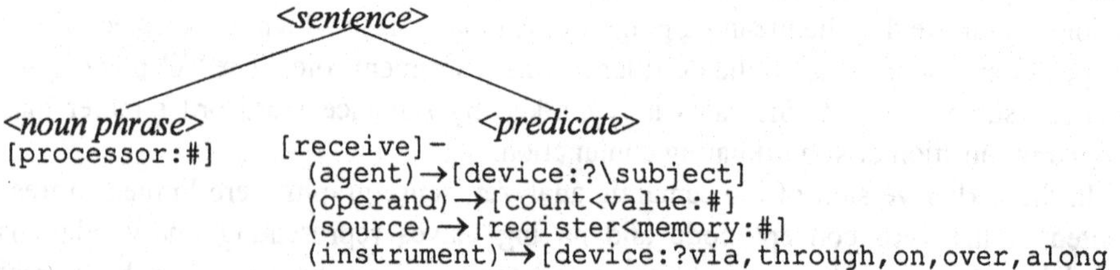

Fig. 7. Fifth step in example analysis.

The final step is the analysis of *<sentence>*, which requires that the *<noun phrase>* be used to identify a concept in the *<predicate>* using the special syntactic marker "\subject." The results of this step are shown in Fig. 8.

<sentence>

[receive] -

(agent) → [processor:#]

(operand) → [count<value:#]

(source) → [register<memory:#]

(instrument) → [device:?via,through,on,over,along]

Fig. 8. Final step in example analysis.

The "receive" behavior has now been extracted, and all information on related devices and values has been included in the resulting graph. The remaining unidentified *device* concept indicates that there may be a device playing this instrument role. This may be identified later either by information from other sentences or could be added manually by the user.

While this example covers only relatively simple words and grammatical structures, all are handled in a similar manner: analysis rules tell the semantic analyzer how to assemble word-sense graphs from the dictionary, taking advantage of the syntactic markers to guide the filling of case roles in the verb word-sense graphs.

4 Viability Analysis

To evaluate the viability of this system, a set of patents related to direct memory access (DMA) controller devices was analyzed. This corpus uses a rather limited vocabulary, under 4000 words. As definitions for these words are added to the dictionary and grammar analysis rules are added to the grammar, the *coverage* (number of sentences successfully analyzed out of the total number of sentences) should improve. If each new sentence requires a new definition for each word, then the system is not very practical.

To expand the coverage, likely target sentences are identified. These sentences may be ones with relatively few words lacking word-sense graphs. Sentences that describe on common behaviors like send and receive were of particular interest. The test sentences were then parsed to determine their parse trees. The parse trees were used to determine what dictionary entries and analysis rules are necessary to create a meaning graph for the sentence.

For example, consider that the dictionary and analysis rules are present to handle the example sentence analyzed in the section above. Now consider the parse tree shown in Fig. 9 for Sentence 2.

The controller sends the value to the processor through the bus. *(2)*

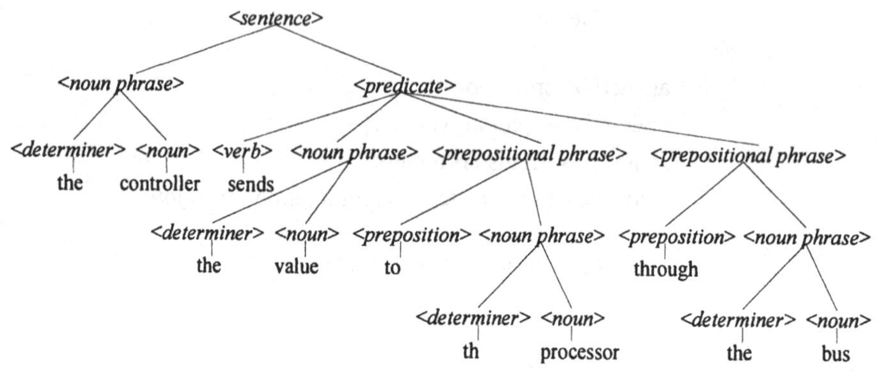

Fig. 9. Parse tree for Sentence 2.

Sentence 2 requires word-sense graphs for the nouns "controller," "value," "bus" and for the verb "send." The word-sense graphs for the three nouns will be simple, but the word-sense graph for "send" will require several unidentified concepts. Fig. 10 shows a possible word-sense graph for "send". Note that syntactic markers are included for a device in the instrument relation that is not required for the analysis of this sentence. In designing word-sense graphs, it is desirable to provide roles to cover a variety of instances. It is not desirable or necessary to include all possible roles. For example, a conditional or enablement role can be attached to any action or event type verb with the syntactic markers *if.* Because this role is so pervasive, it is handled by word sense graphs for these subordinating conjunctions. Temporal relation roles, such as marked by *before*, *after*, *during* and *while* are similarly introduced by word sense graphs for these words in the dictionary. Coordinating conjunctions also have their own word sense graphs. For example, *and* invokes a graph hypothesizing a concept whose type is the least upper bound of the conjoined phrase heads, and relates the phrase head meaning graphs by *part* relations.

```
[send]-
        (agent)→[device:?\subject]
        (operand)→[value:?\object]
        (destination)→[device:?to]
        (instrument)→[device:?through,via,on,along],.
```

Fig. 10. New word-sense graph for the verb "send".

Along with these new word-sense graphs, an analysis rule must be added to handle a predicate that consists of a verb, a noun phrase, and two prepositional phrases. With this in place, the semantic analyzer may now handle this sentence, producing the meaning graph shown in Fig. 11. The meaning graphs are easily converted to a standard conceptual graph by deleting the supertypes in the concept type fields as indicated in Fig. 12.

```
[send]-
        (agent)→[controller<device:#]
        (operand)→[value:#]
        (destination)→[processor:#]
        (instrument)→[bus<carrier:#],.
```

Fig. 11. Meaning graph for Sentence 3.

```
[send]-
        (agent)→[controller:#]
        (operand)→[value:#]
        (destination)→[processor:#]
        (instrument)→[bus:#],.
```

Fig. 12. Conceptual graph for sentence 3.

Performing this process on a few sentences often allows other sentences, similar in structure and/or vocabulary, to be analyzed. In this way, the dictionary and analysis rules are built and coverage is expanded. The graph in Fig. 13 shows the results after coverage was expanded to about 200 sentences from the corpus.

The coverage growth is approximately linear with respect to the dictionary size, requiring on average 1.3 new word-sense graphs per additional sentence. This is acceptable for the relatively small number of sentences being analyzed the relatively large number of words per sentence, and demonstrates that the system can handle a variety of vocabulary and grammatical structures. Since the vocabulary in a restricted domain such as digital devices is rather limited, the number of word sense graphs that must be added to analyze each new sentence should drop as the dictionary reaches the limit. It can also be expected that the number of graphs per sentence should drop toward the end of each document since authors tend to use a repetitive style. The figure below contains examples of more interesting sentences from the patent corpus together with their analyses.

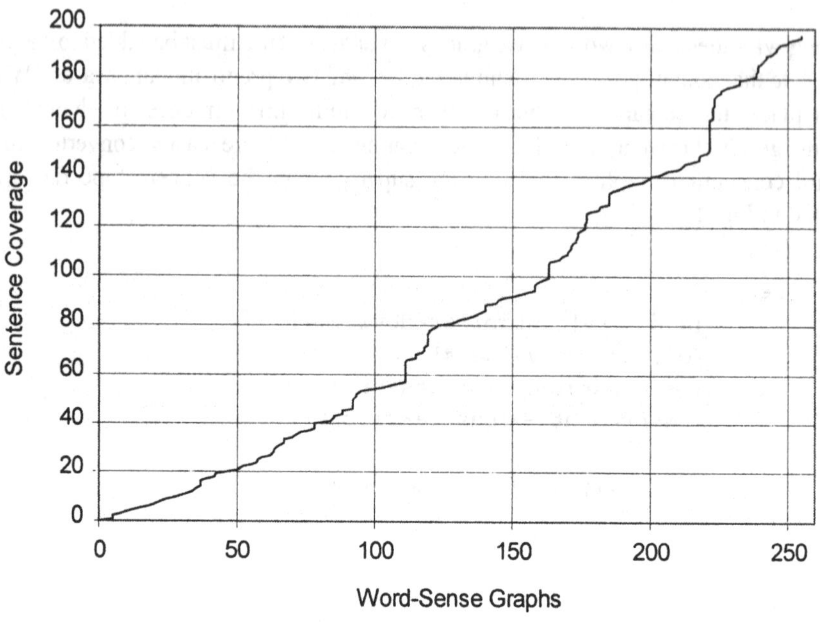

Fig. 13. Coverage results.

5 Analyses of Complex Sentences

This section provides additional examples of complex sentences successfully analyzed by the semantic analyzer. The first sentence includes a subordinating conjunction (when) that adds the *cause* role to the sense graph for *load.* Also, note the generic *value* concept and its part relations introduced by the coordinating conjunction *and.* Sentence 5 provides an example of a temporal relation, starts-after-finishes, introduced by the subordinating conjunction *after.* Both Sentences 5 and 6 illustrate the semantic analyzer's capability for handling long noun phrases. When the elements of a simple noun phrase appear to be attributes (indicated by adjective lexical category) they are reported as such, but when they seem to form a name, indicated by being part of a nominal compound, then they are reported in tact so the user can better compare them for conferences. The full analysis of noun phrases is developed by the parser for later use in coreference detection by the methods reported at an earlier conference[8].

When the DMA controller receives a controller address signal from the microprocessor, the microprocessor loads an initial memory address and a word count into the DMA controller. *(3)*

```
[load<write]-
   (agent)→[microprocessor<processor:#]
   (destination:into)→[controller<device:#dma controller]
   (operand)→[value]-
      (and)←[address:memory address]→
                              (attribute)→[attribute:initial]
      (and)←[count<value:word count],
   (cause)→[receive:\subordinate]-
      (agent)→[controller<device:#dma controller]
      (operand)→[signal<value:controller address signal]
      (source)→[microprocessor<processor:#],,.
```

Fig. 14. Sentence 3 and its analysis.

Generator 56 outputs an update signal S6 when it receives ready signal S8 from cache memory 16. *(4)*

```
[output<send]-
   (agent)→[generator<device:56]
   (operand)→[value:update signal s6]
   (cause:when)→[receive]-
      (agent)→[device:it]
      (operand)→[value:ready signal s8]
      (source)→[memory:cache memory 16],,.
```

Fig. 15. Sentence 4 and its analysis.

The input/output controller 309 provides a direct memory access request signal DRQ after receiving the first data to be transferred. *(5)*

```
[provide<send]-
   (agent)→[controller<device:#309]→
                           (attribute)→[attribute:input/output]
   (operand)→[value:direct memory access request signal drq]
   (starts after finishes)→[receive]→
                     (operand)→[6:data<value:#]-
                                   (operand)←[transfer]
                                   (ordinal)→[ordinal:first],,.
```

Fig. 16. Sentence 5 and its analysis.

When the direct memory access request signal DRQ is provided from the input/output controller 309 to the direct memory access controller 303, the direct memory access controller 303 makes the DRQH signal low. *(6)*

```
[make<write]-
   (agent)→[controller<device:
                              #direct memory access controller 303]
   (operand)→[signal<value:#drqh signal]
   (result)→[value:low]
   (cause)→[provide<send]-
              (operand)→[value:
                    #direct memory access request signal drq]
              (destination)→[controller<device:
                    #direct memory access controller 303]
              (source)→[controller<device:#309]-
                    (attribute)→[attribute:input/output],,,.
```

Fig. 17. Sentence 6 and its analysis.

6 Conclusions and Future Work

The semantic analyzer is successful at extracting behaviors from the patent texts, which includes identification of the devices and values that participate in the behavior, and the relations between the behaviors and participants. Expanding coverage of the knowledge base requires only moderate additions of new dictionary entries for each new sentence that must be analzed.

However, the corpus of 200 sentences that have been analyzed is relatively small and only represents a few of the many sentences in the patent documents. In order to create useful models from the extracted behaviors, much higher coverage will be required. Thus, the dictionary, grammar, and analysis rules must be expanded to cover more sentences. Automated methods of knowledge acquisition are being considered to accelerate dictionary growth. Some relevant work has been reported that exploits statistical correlation of terms for classification.

Currently, behaviors are extracted sentence by sentence. However, descriptions of behaviors often span multiple sentences. Consider the Sentences 5 and 6 above. The devices to which they refer are clearly the same, and the DRQ signal provided in (5) is the same as the one requested in (6). The conceptual meaning graphs for these sentences can be joined on their common references (coreference links) to create a graph representing the meaning of both sentences. This requires *coreference detection.* Some work on this has already been done [8], but has yet to be integrated into the semantic analyzer. Coreference detection will also resolve anaphora, identifying the device called "it" in sentence 4 as "generator 56." These identifications are necessary for building useful behavioral models from the meaning graphs of sentences.

The semantic analysis method reported here requires that the grammar produce syntactic analysis (parse) that corresponds to the meaning of the sentence and that the semantic analyzer has adequate rules and word knowledge to construct a meaning from that parse. Unfortunately no information is reported if any failure occurs. A back-up (or back-off) method is being developed to provide at least some information when a full analysis fails.

7 Acknowledgments

This work was funded in part by the National Science foundation, Grant MIP-9707317.

References

1. Cyre, W.R.: Executing Conceptual Graphs, Proc. Sixth International Conference on Conceptual Structures, Montpellier, France, 51-64, Aug. 10-12, 1998.
2. Cyre, W.R.: Capture, Integration and Analysis of Digital System Requirements with Conceptual Graphs, IEEE Transactions on Knowledge and Data Engineering, 9 (1), 8-23, Jan/Feb 1997.
3. Cyre, W.R., Armstrong J.R., and Honcharik, A.J.: Generating Simulation Models from Natural Language Specifications, Simulation 65 (1995): 239-251.
4. Fillmore, C.J.: The Case for Case, Universals in Linguistic Theory, in E. Bach and R.T. Harms eds.: Holt, Rinehart and Winston, New York, 1968.
5. Greenwood, R. and Cyre, W.R.: Conceptual Modeling of Digital Systems from Informal Descriptions, Proc. MASCOTS'93, San Diego, CA, Jan 1993.
6. Kamath, R. and Cyre, W.R.: Automatic Integration of Digital Systems Requirements using Schemata, Proc. Third International Conference on Conceptual Structures, Santa Cruz, CA, 44-58, Aug 14-18, 1995.
7. Mann, G.A.: Conceptual Graphs for Natural Language Representation. Technical Report 9311, University of New South Wales, Sidney, Australia, 1993.
8. Shankaranarayanan, S. and Cyre, W. R.: Coreference Detection in Automatic Analysis of Specifications, Proc. 1994 International Conference on Conceptual Structures, College Park, MD, 16-20, August 1994.
9. Sowa, J.F., and Way, E.C.: Implementing a semantic interpreter using conceptual graphs, IBM Journal of Research and Development, 30 (1986), 57-69.
10. Sowa, J.F.: Conceptual Structures. Reading, MA: Addison-Wesley Publishing Company, 1984.

Extending the Conceptual Graph Approach to Represent Evaluative Attitudes in Discourse

Bernard Moulin, Hengameh Irandoust

Laval University, Computer Science Department and Research Center on Geomatics
Pouliot Building, Ste Foy,
Quebec G1K 7P4, Canada
ph: (418) 656 5580, Fax: (418) 656 2324
{moulin, hengameh@ift.ulaval.ca}
http://www.ift.ulaval.ca/~moulin/

Abstract. In this paper we present a three-tiered framework extending the conceptual graph theory in order to represent evaluative attitudes expressed by locutors in a discourse. The descriptive plan models the locutors' (including the narrator's) speech acts and mental attitudes. The evaluative plan contains locutors' evaluative attitudes which provide an account of several linguistic modalities in discourse. The argumentative plan gathers locutors' discursive goals and models the discourse's argumentative structure.

1 Introduction

In his seminal work on Conceptual Graphs (CGs) Sowa [17] proposed to use monadic relations to represent modal operators such as "Obligation", "Interdiction", "Possibility" which apply on propositions. Peirce's writings [16] were Sowa's inspiration for modal reasoning using CGs, but this topic was not detailed in Sowa's work. Other researchers proposed more detailed accounts to use Peirce's existential graphs to reason in a modal logic S5 [18] [2] or in a tense logic framework [14]. These approaches correspond to the standard way of representing modalities in a logical framework and they are useful for classical modal reasoning. However, logical modal frameworks are not sufficient to deal with modalities in natural language [7]. To illustrate this point, let us examine few examples taken from a corpus of various texts written in French (literature, journalism, etc.).

(S1) Denise paraît très naïve.
Denise seems to be very naive.

(S2) Denise me semble faussement naïve.
Denise seems to me to be falsely naive.

(S3) Denise a l'air naïf, mais elle ne l'est pas du tout.
Denise looks naive, but she is not.

(S4) Heureusement, Denise est une femme pragmatique.
Fortunately, Denise is a pragmatic woman.

(S5) Selon Paul, Denise est une femme pragmatique.
According to Paul, Denise is a pragmatic woman.

These examples cannot be represented in a classical logical modal framework (deontic, epistemic, temporal, etc.), and we find no satisfactory representation for them in a classical CG framework. In linguistics, such phenomena come under the domain of modality because they express the locutor's attitude with respect to the propositional content of her utterance. From a logician's point of view, modalities are impersonal (ex. "it is necessary to ...", "it is forbidden to ...", etc.), whereas most modalities found in a discourse are related to locutors and indeed personalized. Adopting a linguist's point of view, Le Querler [7] defines three kinds of modalities. A *subjective modality* expresses a relation between the locutor and the propositional content of her utterance. An *intersubjective modality* involves a relation between the locutor and the addressee with respect to a propositional content. An *objective modality* expresses a subordination relation between two propositions and depends neither on the locutor's judgement, nor on her appreciation or will.
Subjective modalities are divided into two sub-categories: epistemic and appreciative modalities. *Epistemic modalities* denote the locutor's degree of certainty about the proposition she asserts (ex.: "Paul is *certainly* on his way to school").
Appreciative modalities are those by which the locutor evaluates the propositional content of her utterance (ex.: "I *appreciate* your coming").
Following Le Querler [7], we consider that a modality is the expression of the locutor's attitude with respect to the propositional content of her speech act. When performing a speech act, a locutor not only transfers certain information to the addressee, but most often expresses an attitude with respect to that information. For example, adverbs such as "falsely" (sentence S2) or "fortunately" (sentence S6) are lexical items that denote such attitudes. Hence, it is necessary to explicitly enhance the representation of such speech acts with a structure that denotes the locutor's attitude with respect to the propositional content. In this paper, we propose a formal framework within the conceptual graph theory in order to represent various kinds of modalities which are found in discourses and result from locutors' evaluations of situations. We propose to use a framework composed of a descriptive plan and an evaluative plan. We present the basic structures characterizing the descriptive plan in Section 2 and the evaluative plan in Section 3. In Section 4 we show how this framework can be extended to model argumentation structures.

2 A Two-Tiered Framework

Le Querler's definition of modality is compatible with the approach that Moulin [9] [10] [12] proposed to represent the temporal structure of discourse along with locutors' speech acts. Both approaches acknowledge the importance of explicitly introducing the locutor's interaction with the discourse utterances in order to represent the context of utterance of speech acts [9][10] [12] or to represent modality [7]. Hence, we propose a representational framework for discourses that gives a central importance to locutors by explicitly introducing structures that represent the context of utterance and the context of evaluation of the locutors' speech acts. We differentiate two representational plans for any discourse: the descriptive plan and the evaluative plan.

The ***descriptive plan*** is based on a conceptual framework which extends Sowa's CG approach [17] by introducing knowledge structures that can be used to represent temporal knowledge, to specify the context of utterance of locutors' speech acts and to model pragmatic phenomena such as indexicals and verb tenses [9][10] [11][12]. We consider that a discourse is composed of a succession of speech acts performed by the discourse's narrator. In a discourse, a narrator may describe certain situations that characterize her environment, report on her mental attitudes, report on speech acts and/or mental attitudes of other persons. We consider that when producing a discourse, a locutor considers her own spatio-temporal localization as the main coordinate system ("here and now"), which sets a reference for localizing the temporal situations that she evokes, as well as other person's speech acts that she reports. Hence, each locutor acts physically, mentally or illocutionarily from her own perspective (we can speak of speech acts as illocutionary acts or communicative acts).

In [9] we claimed that much of temporal linguistic phenomena found in discourse can be explained by explicitly introducing time coordinate systems in the semantic representation of discourses and by specifying the temporal relations which link temporal situations to these time coordinate systems. When we analyse discourses, we notice that locutors use several time coordinate systems, such as:

- The *official time coordinate system* which provides dates, hours, years, that are used as time references for localizing time intervals on an absolute time scale.
- The *narrator's perspective* which is the narrator's time coordinate system.
- The narrator can describe temporal situations taking place in a temporal localization which is different from her own temporal location. A set of temporal situations can be localized according to a time coordinate system which is called a *temporal localization*. We can also represent temporal localizations embedded within another temporal localization.
- When a narrator reports someone's words in a discourse, the time coordinate system changes and is localized at the time when the person spoke. We call an *agent's perspective*, the time coordinate system attached to a person whose words are reported in a discourse using a direct style.

The main structures found in the descriptive plan are situations, narrator's perspectives, agent's perspectives, agent's attitudes, temporal localizations and temporal relations that relate the time intervals associated with these structures. Let us take an example (Figure 1):

(S6) Yesterday Paul said to Mary that Denise is naive and she knows it.

A ***situation*** is a triple <SD, SPC, STI> where:

- The situation description SD is a pair [SITUATION-TYPE, situation-descriptor]. The situation type is used to semantically distinguish different kinds of temporal situations: events, states, processes, etc. The situation descriptor identifies an instance of a situation and is used for referential purposes.

- The situation propositional content SPC is a non-temporal conceptual graph which makes explicit the semantic characteristics of the situation;
- STI is the time interval associated with the temporal situation.

Graphically, a temporal situation is represented using a rectangle composed of two parts. The rectangle represents the situation's time interval. In the upper part of the rectangle we indicate the situation description SD, as well as the relevant parameters of the situation time interval STI. In the lower part of the rectangle we represent the situation propositional content SPC. In Figure 1, the sentence "Denise is naive" is represented by situation SIT [state, Sit1], one of the last embedded rectangles.

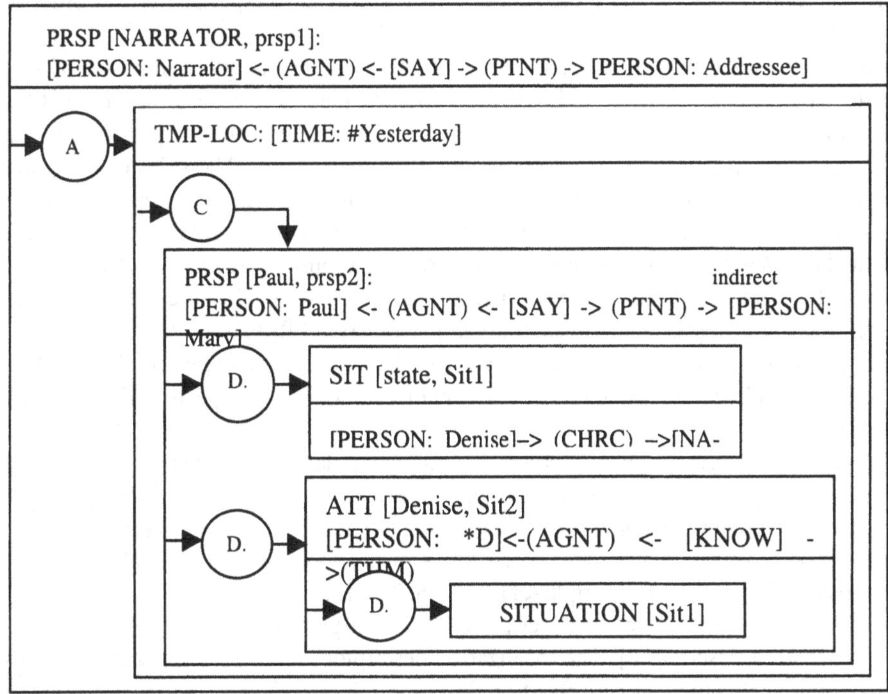

Fig.1. Representation of "Yesterday, Paul said to Mary that Denise is naive and she knows it". Rectangles represent the various temporal contexts (PRSP: perspectives; TMP-LOC; temporal localization; STATE: a situation of type 'state') and circles represent temporal relations (A.: AFTER; D.: DURING; C.: CONTAIN)

A ***narrator's perspective*** is a triple <NPD, NPS, NPTI>

- The narrator's perspective description NPD is a pair [perspective-descriptor, perspective description]. The perspective-descriptor identifies the perspective instance in a discourse. The perspective-description is specified by a conceptual graph which describes the narrator's action (usually implicit in the text) when creating the discourse:

  [PERSON: narrator]<- (AGNT) <- [SAY] -> (PAT) ->[PERSON: addressee].
- The narrator's perspective set NPS of a given narrator's perspective NPx is the set of all temporal situations, localizations and agents' perspectives which are in the

temporal scope of NPx (i.e. which are evoked by the narrator's speech acts described in the perspective description).

- NPTI specifies the characteristics of the time interval associated with the time coordinate system origin corresponding to the narrator's perspective. Usually it is implicit.

A narrator's perspective NPx sets the time coordinate system origin for a given discourse. The time intervals of the temporal situations, localizations and perspectives within NPx's temporal scope are localized with respect to the narrator's perspective time interval using temporal relations. The narrator's perspective is used in relation to the temporal situations and localizations which are in its perspective set, in order to determine verb tenses [11].

We graphically represent a narrator's perspective as a rectangle composed of two parts. In the upper part, we indicate the narrator's perspective description and time interval parameters. In the lower part of the rectangle we indicate the perspective set: this part corresponds to the temporal scope of the perspective. The sides of the lower part of the rectangle symbolize the perspective's time interval which represents the time coordinate system origin. The temporal relations link the side of the rectangle with the rectangles representing the temporal situations, localizations and perspectives which are contained within the narrator's perspective set. In Figure 1 the embedding rectangle represents the narrator's perspective PRSP[NARRATOR, prsp1]. Within the perspective set (lower part of the rectangle) there is a temporal localization TMP-LOC: [TIME: #Yesterday] which contains an agent's perspective PRSP [Paul, prsp2].

Speakers often refer to specific points in time, using dates or indexicals such as "last year", "yesterday" or "tomorrow". Doing so, they refer to secondary time coordinate systems. A temporal localization sets a secondary time coordinate system which is positioned relatively to a temporal perspective or to another temporal localization.

A ***temporal localization*** is a triple <TLD, TLS, TLTI>.

- The temporal localization description TLD is a pair [localization-descriptor, localization-description]. The localization-descriptor identifies the instance of temporal localization. The localization-description is specified by a conceptual graph such as [TIME: #yesterday] in our example (Figure 1).
- The temporal localization set TLS of a temporal localization TLx is the set of all temporal situations, localizations and agent's perspectives which are in the temporal scope of TLx: their time intervals are included in the time interval of TLS.
- The temporal localization time interval TLTI specifies the characteristics of the time interval associated with the temporal localization.

We graphically represent a temporal localization as a rectangle composed of two parts. In the upper part, we indicate the temporal localization description and time interval parameters. In the lower part of the rectangle we indicate the temporal localization set. In Figure 1 the second embedded rectangle represents a temporal localization which corresponds to "Yesterday" in sentence S6. Indexicals such as "yesterday" are resolved with respect to the perspective in which the localization is embedded [9].

An ***agent's perspective*** has the same characteristics as a narrator's perspective. It is specified by a triple <APD, APS, APTI> where APD identifies the agent's perspective description, APS represents the agent's perspective set and APTI is the time interval associated with the origin of the time coordinate system represented by the agent's perspective. The graphical conventions used to represent an agent's perspective are the same as those used for the narrator's perspective. In Figure 1 the third embedded rectangle depicts the agent's perspective PRSP [Paul, prsp2] which corresponds to "Paul said to Mary" in sentence S6 and is expressed by the CG:

[PERSON: Paul] <- (AGNT) <- [SAY] -> (PTNT) -> [PERSON: Mary]

An ***agent's attitude*** is a structure similar to an agent's perspective, but it expresses an agent's mental attitude (belief, wish, fear, hope, goal, etc.) instead of representing a speech act. An example is found in Sentence S6; "she knows it". This is represented in Figure 1 by the second rectangle embedded in the rectangle of PRSP [Paul, prsp2]. This is a situation Sit2 which describes Denise's attitude that "she knows ...", represented by the conceptual graph [PERSON: *D]<-(AGNT) <- [KNOW] ->(THM). The theme of this mental attitude is the coreferred situation Sit1.

All the rectangles symbolically represent the time intervals associated with the corresponding structures. Hence, we represent the temporal relations holding between these structures using circles whose arrows link the corresponding rectangles. For example in Figure 1, the rectangle associated with the temporal localization TMP-LOC is linked to the rectangle associated with the agent's perspective PRSP[Paul, prsp2] by the relation marked by C. (an abbreviation of the CONTAIN relation [1]).

The narrator's perspective corresponds to the context of utterance of the speech acts performed by the narrator when uttering the discourse. The agent's perspectives represent the context of utterance of the speech acts performed by the locutors whose words are reported by the narrator. Hence, the descriptive plan puts a special emphasis on the agents' interaction with the situations that they report in their speech acts.

3. The Evaluative Plan

The *evaluative plan* is used to model subjective modalities [7]. In this plan we find the locutor's subjective attitudes towards the propositional content of her discourse and more specifically her appreciation of the particular situations that she is depicting in the descriptive plan. This corresponds to that part of the locutor's discourse which does not assert facts but comments on them. These subjective attitudes can be evoked in a discourse in different ways. The locutor may use an adverb such as "falsely", "fortunately" (epistemic modalities), "luckily", "by chance", etc. The locutor may also express a judgment about the depicted situation by introducing it with a clause such as "I am pleased that ...", "I appreciate that ...", "It is fun that ...", "It is so difficult to ...".

The evaluative plan contains structures that are parallel to the structures found in the descriptive plan. A locutor's subjective attitude relative to a situation is specified in a structure called *agent's orientation* which is associated with the same time interval as the agent's perspective in which the situation description appears.

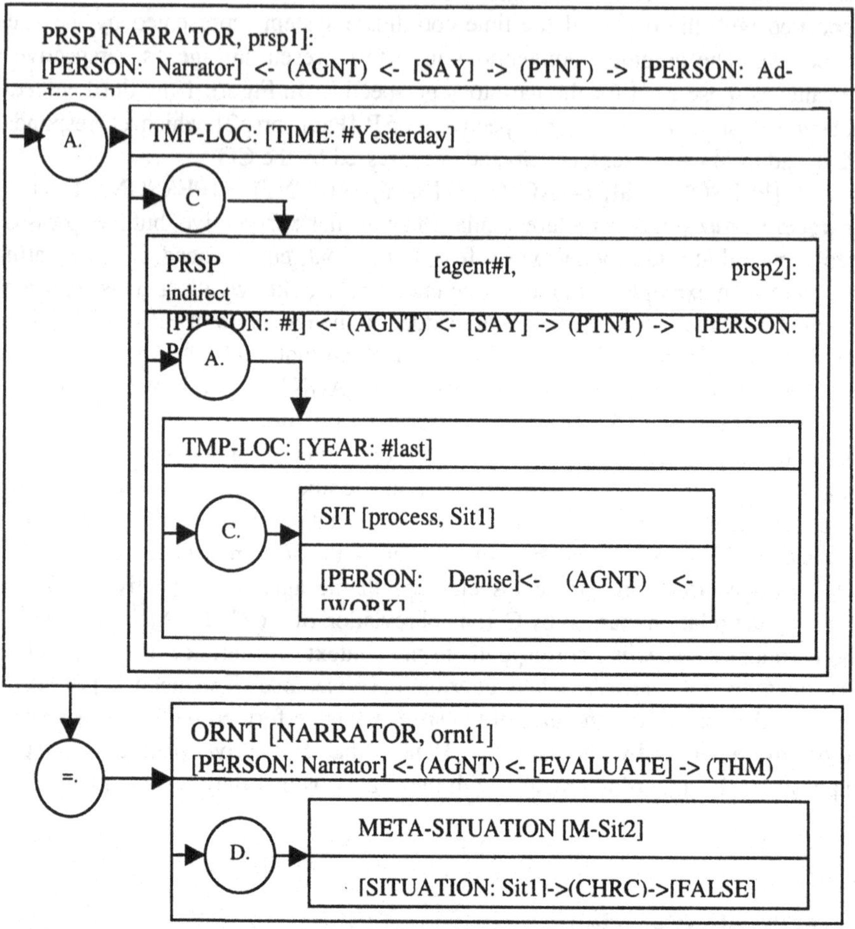

Fig.2. Representation of "Mistakenly, I said to Peter yesterday that Denise worked in Toronto last year". The sign '=.' denotes the temporal relation EQUAL. #I is an indexical corresponding to I in the text: it is solved with respect to [PERSON: Narrator] in PRSP [NARRATOR,prsp1].

As an agent's perspective, an ***agent's orientation*** is a triple <AOD, AOS, AOTI>

- The agent's orientation description AOD is a pair [orientation-descriptor, orientation description]. The orientation-descriptor identifies the agent's orientation instance in a discourse. The orientation-description is specified by a conceptual graph which describes the agent's action stating her evaluation. By default, when the appreciative act is not explicitly expressed in the text, we use the following conceptual graph:
  [PERSON: agent-name] <- (AGNT) <- [EVALUATE] -> (THM) -> .

- The agent's orientation set AOS contains meta-situations which denote the properties that the agent assigns to the situation she evaluates. These meta-situations are the theme (relation THM) of the CG specifying the orientation description
- AOTI is the time interval of the agent's orientation. It is equal to the time interval of the agent's perspective which is associated with the agent's orientation.

We graphically represent an agent's orientation as a rectangle composed of two parts. In the upper part, we indicate the agent's orientation description and the parameters of AOTI, when necessary. In the lower part of the rectangle we indicate the orientation set. The sides of the rectangle symbolize the orientation's time interval.

Let us examine the following example (S7) which is represented in Figure 2:

Hier par erreur, j'ai dit à Peter que Denise travaillait à Toronto l'année dernière.

Mistakenly, I said to Peter yesterday that Denise worked in Toronto last year .

The narrator asserts two things: 1) Yesterday he said to Peter that Denise worked in Toronto last year (which is represented by the narrator's perspective: PRSP [NARRATOR, prsp1]); 2) he was mistaken (which is represented by the agent's orientation: ORNT [NARRATOR, ornt1]).

The rectangles of the perspective and orientation are related by a temporal relation EQUAL ('=.' in Figure 2) which denotes that their time intervals are equal. Within the orientation set (lower rectangle of the orientation structure), there is a meta-situation M-Sit2 which indicates that situation Sit1 is false.

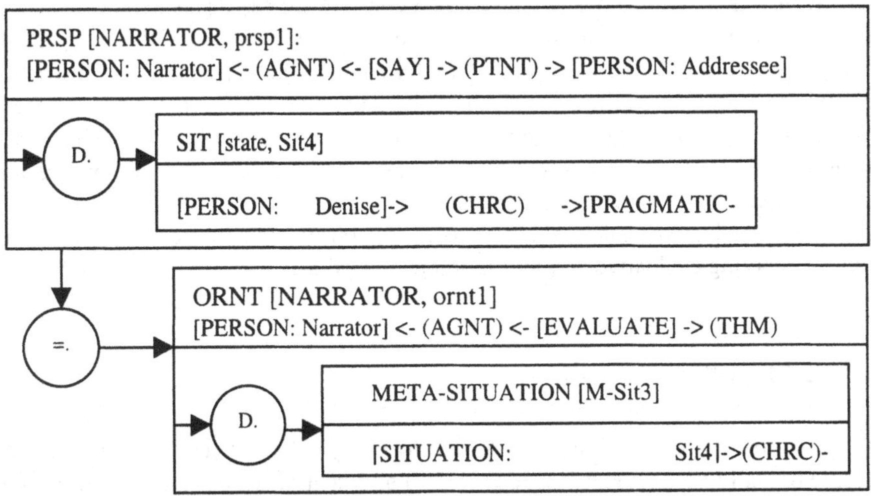

Fig.3. Representation of " Fortunately, Denise is a pragmatic woman"

The markers of appreciative modality (verbs, noun phrases, adjectives) generally establish a relation of syntactic dependence with the propositional content of the phrase and lead to a *de dicto* interpretation. That is the modality is external to the *dictum* and its scope extends on the subject-predicate relation as a whole. The adverbs used in sentences (S2) and (S5) indicate respectively the negative and the positive appreciation of the narrator with respect to the content of her discourse. Although the syntactic construction is different in these sentences, in both cases, the modality characterises the speaker's attitude towards the entire situation. Figure 3 is the representation of sentence S4.

The narrator may not only express her own judgements but also comment on the evaluations and intentions of the locutors she mentions in her discourse as in (S8):

Durant l'intermède, on aura compris qu'à ses yeux [Dominique Strauss-Kahn], les tentations narcissiques du pouvoir sont un danger permanent.

During the break, we understood that in his opinion, the narcissic temptations of power are a constant danger.

Let us comment upon the structures (Figure 4) which represent sentence S8. Let us recall that in the CG theory any concept appearing in a graph can be referred to through anaphoric references. However, in sentence S8 the concept referent of "his opinion" is not found in the graph. We need to introduce a discourse referent [PERSON: Strauss-Kahn *Agent1] in order to solve the anaphor "to his opinion". The notion of discourse reference is borrowed from [6]. Hence, in the Discourse Referent structure (top of Figure 4) we put any concept that needs to be referred to in the discourse in addition to the concepts that appear in the descriptive and evaluative plans.

In Figure 4 the set of the narrator's perspective [NARRATOR, prsp1] represents the sentence portion "During the break, we understood that". The expression "in his opinion" indicates that the narrator provides his interpretation of Strauss-Kahn's belief that "the narcissic temptations of power are a constant danger". This is represented in the evaluative plan by the narrator's orientation structure [NARRATOR, ornt2] which contains an attitude Sit2 of *Agent1 (corefering to Strauss Kahn in the discourse referent structure). The theme of this attitude is situation Sit3 corresponding to sentence portion "the narcissic temptations of power are a constant danger". Notice that the attitude Sit1 in the descriptive plan corresponds to the sentence portion "we understood" and that its theme is situation Sit2.

Indeed, the evaluative plan enables us to model fairly sophisticated linguistic phenomena denoting evaluative attitudes found in a discourse.

4. The Argumentative Plan

In a discourse a competent speaker is able to easily differentiate connotated words (people casually speak about "loaded words") and neutral words. Locutors often use connotated words on purpose, in order to trigger specific reactions from their interlocutors (ex. "anger", "pride", "shame", etc.). For instance in sentence S1 (in Section 1), the narrator expresses a judgement on an individual named Denise and uses the connotated adjective "naive" which reflects her subjective view. This is a simple assertion, yet, the use of the adjective 'naive' appears as a personal judgement. In our model we mark the corresponding concept by a symbol ψ which indicates that the corresponding word is the result of a value judgement expressed by the locutor. In Figure 5 representing sentence S1, the concept NAIVE is marked by the ψ symbol. The value judgements that a narrator expresses in a discourse usually aim at influencing the addressee's opinion. We say that the narrator aims at changing the addressee's mental model by trying to make the addressee adopt certain mental attitudes conveyed by the discourse. In fact, the narrator works toward an end when speaking with the addressee: she tries to reach certain discursive goals which usually involve the addressee's adoption of certain mental attitudes. This is very clear in so called "argumentative discourses" such as political discourses, debates between politicians, lawyers' pleas, sales talks, etc.

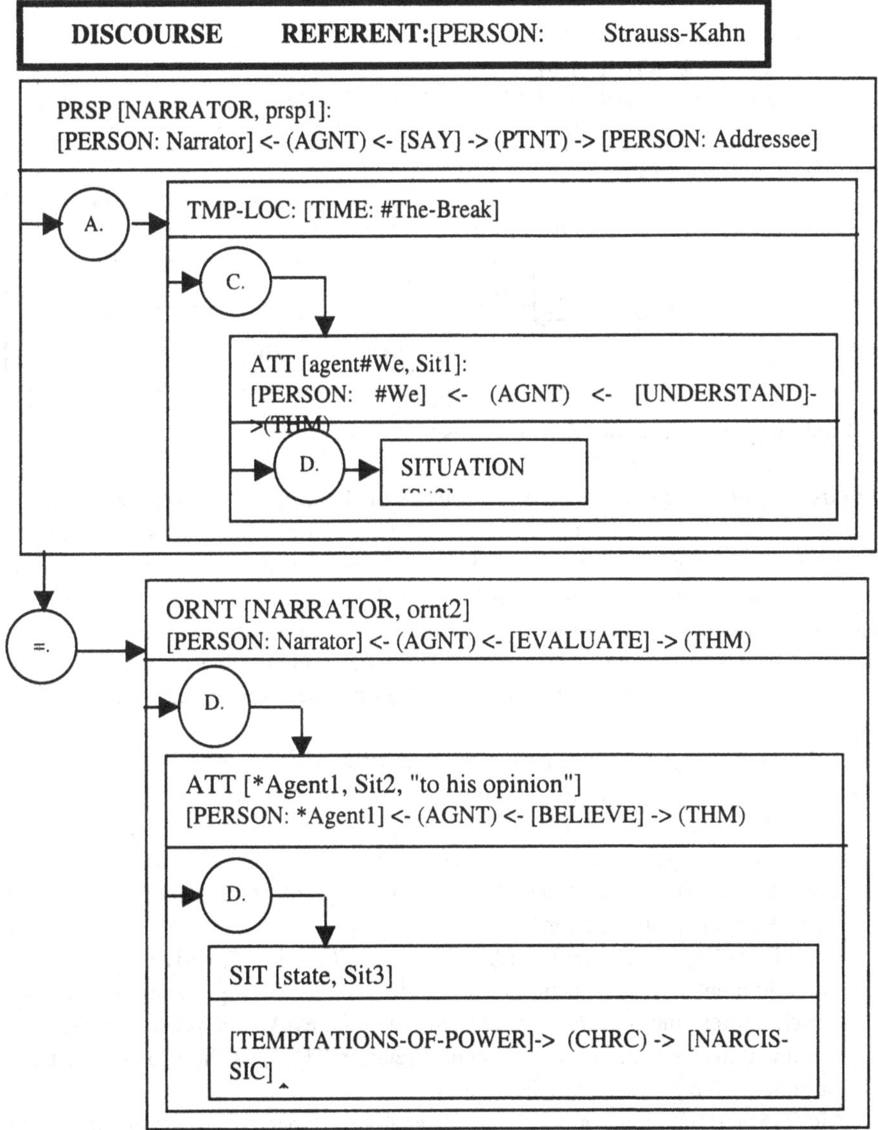

Fig.4. Representation of "During the break, we understood that in his opinion, the narcissic temptations of power are a constant danger."

As several other researchers (see for instance [5]), we think that in any discourse the narrator aims at reaching certain goals and that it is important to model them in order to fully understand a discourse. In addition, when a narrator reports on a dialog between locutors, understanding the locutor's discursive goals may help the addressee to understand the discourse content.

The narrator's and locutors' discursive goals are captured in a third plan that we call the *argumentative plan* which is parallel to the descriptive and evaluative plans.

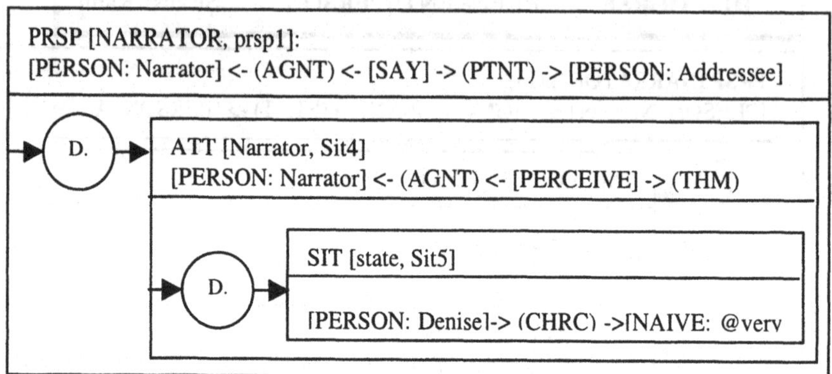

Fig.5. Representation of "Denise seems to be very naive"

The ***argumentative plan*** models the structure of the argument developed by the locutors in a discourse. It contains two types of structures: the agent's argumentative perspectives and the argumentative situations. Those structures are obtained from the analysis of argumentative expressions found in the discourse.

The *agent's argumentative perspective* is parallel to the structure of the agent's perspective (found in the descriptive plan). The latter contains the agent's speech acts from which discursive goals are extracted. Both perspectives are associated with the same time interval.

The ***argumentative perspective*** is is a triple <ARPD, ARPS, ARPTI>

- The argumentative perspective description ARPD is a pair [arg-perspective-descriptor, arg-perspective description]. The arg-perspective-descriptor identifies the agent's argumentative perspective instance in a discourse. The arg-perspective-description is specified by a conceptual graph which describes the agent's argumentative action:

  [PERSON: agent-name] <- (AGNT) <- [INTEND] -> (THM) ->.
- The argumentative perspective's set ARPS contains the argumentative situations which express the discursive goals that the agent evokes in her discourse. These argumentative situations are the theme (relation THM) of the CG specifying the argumentative perspective's description
- ARPTI is the time interval of the argumentative perspective. It is equal to the time interval of the corresponding agent's perspective.

An *argumentative situation* is a structure which relates the situations found in the descriptive or evaluative plans and the discursive actions that the narrator (or locutor)

intends to perform in the discourse. We think of locutors' interactions as exchanges of conversational objects (COs). A *conversational object* is a mental attitude (belief, goal, wish, etc.) along with a positioning which a narrator transfers to another locutor during a conversation [13]. The locutor positions herself relative to a mental attitude by performing actions like "proposing", "accepting", "rejecting": this is called the locutor's positioning relative to that mental attitude.
The *argumentative situation* shows how certain situations of the agent's perspective or orientation are intended to trigger certain conversational objects on the addressee's part. The syntax of an ***argumentative situation*** is:

Arg-Sit := *Sit-Set* => *COSet*
CO-Set := [*CObj*] | [*CObj* OR *CObj*] | [*CObj* AND *CObj*]
Sit-Set := [*SIT* {Sit_1, Sit_2, ... Sit_n}]
CObj := *POSITIONING-TYPE (Addressee, Mental-Attitude)*
POSITIONING-TYPE := *ACCEPT* | *CONFIRM* | *REJECT* | *INQUIRE*

Where Sit_i is any situation appearing in the agent's perspective or orientation; *POSITIONING-TYPE* characterizes a positioning that the narrator intends to trigger in the addressee's mental model relative to a given *Mental-Attitude*.

As an illustration, let us examine sentence (S3): "Denise looks naive, but she is not", where the connector "but" clearly separates the facts (what is perceived) from the narrator's evaluation. As Ducrot [3] puts it : "The speaker, after having uttered the first proposition *p*, expects the addressee to draw a conclusion *r*. The second proposition *q*, preceded by a *but*, tends to avoid this conclusion by signalling a new fact that contradicts it. The whole movement would be : "*p* ; you are thinking of concluding *r* ; do not do so, because *q*". Analysing sentence S3, this can be paraphrased as: p : Denise looks naive (descriptive plan)
r : You might think she is really naive (implicit)
q : p is false (evaluative plan)

In Figure 6 we display the three plans of sentence S3's representation: the descriptive plan (PRSP [NARRATOR, prsp1]), the evaluative plan (ORNT [NARRATOR, ornt1]) and the argumentative plan represented by the argumentative perspective ARGMT [NARRATOR, argmt1]. We notice the addition of the argumentative relation BUT which relates situation Sit1 representing "Denise looks naive" in the perspective prsp1 and situation M-Sit3 representing "she is not" in the orientation ornt1.
In the argumentative perspective argmt1 the narrator's discursive goals are represented by two argumentative situations Asit1 and Asit2 which are represented by the two rectangles embedded in the rectangle representing argmt1's perspective set. These two rectangles are related by a "pointed arrow" (•→) indicating that Asit2 temporally follows and replaces Asit1. The argumentative situation Asit1 denotes that situations Sit1 and Sit2 (corresponding to "Denise looks naïve") trigger that the addressee accepts or confims the mental attitude BEL(Addressee, Sit2) (meaning that the "addressee believes that "Denise is naïve"). The argumentative situation Asit2 denotes that situation M-Sit3") triggers that the addressee accepts the mental attitude ¬BEL(Adressee, Sit2) (meaning that the "addressee does not believe that "Denise is

naive") which replaces the mental attitudes triggered by Asit1. This precisely reflects the interpretation steps of a *But-statement* described by Ducrot [3].

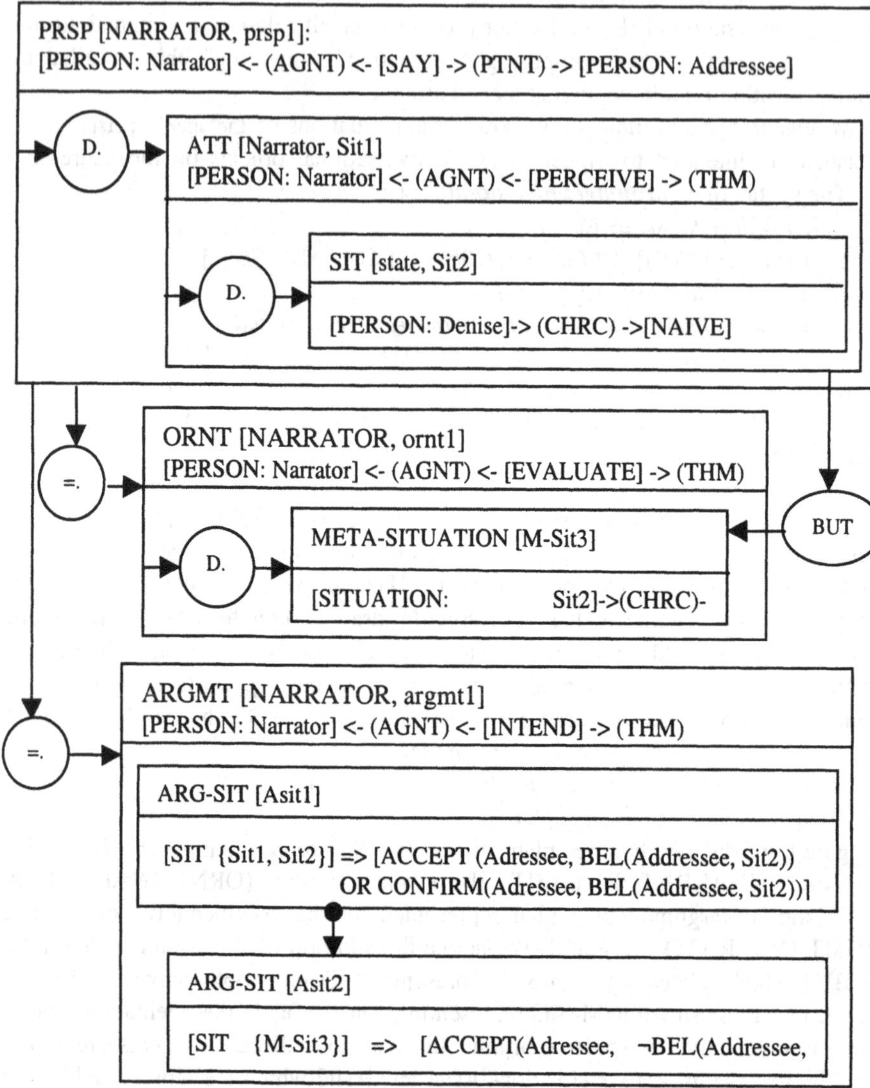

Fig.6. Representation of "Denise looks naive, but she is not"

5 Conclusion

In this paper we proposed a three-tiered framework in order to represent the knowledge contained in discourses. The descriptive plan models the locutors' (including the narrator's) speech acts and mental attitudes. The evaluative plan contains locutors' evaluative attitudes which provide an account of several linguistic modalities in discourse. The argumentative plan gathers the locutors' discursive goals and provides a model of the argument structure in the form of conversational objects. All these structures allow various kinds of reasoning mechanisms that we will describe in a forthcoming paper. More specifically we are currently investigating how the extension of the CG theory can be applied to model arguments as defined in argumentation theory [4] [15].

References

1. Allen J. F. (1983) Maintaining Knowledge about Temporal Intervals. *Communications of the Association for Computing Machinery*, vol 26 n11.
2. Braüner T. (1998) Peircean graphs for the modal logic S5, In M. L. Mugnier, M. Chein (edts.), *Conceptual Structures: Theory, Tools and Applications*, Springer Verlag, LNAI n. 1453, 255-269.
3. Ducrot O. (1991) *Dire ou ne pas Dire, Principes de sémantique linguistique*, Herman, Collection Savoir, Paris.
4. Eemeren F. van, Grootendorst R., Kruiger T. (1987) *Hanbook of Argumentation Theory*, Dordrecht: Foris.
5. Grosz B.J., Sidner. C.L. (1986), Attention, intentions and the structure of discourse, *Computational Linguistics*, n. 2 (3), July September, 175-204.
6. Kamp H., Reyle U. (1993) *From Discourse to Logic*, Dordrecht: Academic Press.
7. Le Querler N. (1996), *Typologies des modalités*, Presses Universitaires de Caen, Fr.
8. Mineau G., Moulin B., Sowa J.F. (edts.), *Conceptual Graphs for Knowledge representation*, Springer Verlag, LNAI n.69,
9. Moulin B. (1992) A conceptual graph approach for representing temporal information in discourse, *Knowledge-Based Systems*, vol5 n3, 183-192.
10. Moulin B. (1993) The representation of linguistic information in an approach used for modelling temporal knowledge in discourses. In [8], 182-204.
11. Moulin B., Dumas S. (1994) The temporal structure of a discourse and verb tense determination, In W. M. Tepfenhart, J. P. Dick, J. F. Sowa, (edts.) (1994), *Conceptual Structures: Current Practices*, LNAI n. 835, Springer Verlag, 45-68.
12. Moulin B. (1997) Temporal contexts for discourse representation: an extension of the conceptual graph approach, *the Journal of Applied Intelligence*, vol 7 n3, 227-255.
13. Moulin B., Rousseau D., Lapalme G. (1994), A Multi-Agent Approach for Modelling Conversations, Proceedings of the International Conference on Artificial Intelligence and Natural Language, Paris, 35-50.
14. Øhrstrøm P. (1996) Existential graphs and tense logic, In P. W. Eklund, G. Ellis, G. Mann (edts.), Conceptual Structures: *Knowledge representation as Interlingua* Springer Verlag, LNAI n. 1115, 203-217.
15. Plantin C. (1990) *Essais sur l'argumentation*, Paris: Editions Kimé.
16. Roberts D.D. (1973) *The Existential Graphs of Charles S. Peirce*, The Hague: Mouton.
17. Sowa, J. F. (1984) *Conceptual Structures*, Reading Mass: Addison Wesley.
18. Van den Berg H. (1993) Modal logics for conceptual graphs, In [8], 411-429.

Implementing a Semantic Lexicon

Sait Dogru[1] and James R. Slagle[1]

Computer Science Dept.
University of Minnesota
Minneapolis, MN 55454
{dogru, slagle}@cs.umn.edu
http://www.cs.umn.edu

Abstract. The human lexicon is emerging as the single most important aspect of modern NLP applications. Current research focus ranges from the organization and structure of lexicons to fleshing it out with lexical entries with their full semantics. We have implemented a model of the human lexicon that addresses the automatic acquisition of the lexicon as well as the representation of the semantic content of individual lexical entries. We have used Conceptual Structures as the representation scheme. The system analyzes paragraphs of English sentences, maps the extracted lexical and semantic knowledge to CS graphs, and reasons about them. The system augments its vocabulary by using a persistency mechanism that restores the previously defined lexical items in later sessions.

1 Introduction

Early NLP systems had small vocabulary, rigid grammar rules with limited coverage, and they could not be used in realistic environments [1], [2], [3], [4], [5], [6], [7], [8]. Research on modern theories of grammar, in contrast, has focused on the lexical semantics as the most important aspect of a successful NLP system. One of the latest and most successful modern linguistic frameworks, Head–Driven Phrase Structure Grammar (HPSG) [9], [10], for instance, has only a handful of grammar rules, but depends almost entirely on the structure of its lexical entries through *signs*. This lexical approach to NLP has produced the new generation of successful NLP applications. Consequently, the acquisition and representation of lexical information have become fundamental issues in modern computational linguistics [11], [12], [13], [14], [15], [16], [17], [18], [19]. There are two major problems with the current research into these very important issues:

- *Acquisition Problems:* Unfortunately, many lexicons are still being created manually [19], [20], [21], [22], [23], [24]. Manual creation of lexicons is extraordinarily hard, and it involves difficult and sometimes arbitrary decisions on the part of the lexical enterer. It demands skills of highly trained linguists or knowledge engineers. The manual approach also gets in the way of reusability of the lexicons, requiring lexicons to be duplicated for different purposes. Alternative approaches include statistical methods and automatic identification of related words in machine readable dictionaries (MRDs), again with similar shortcomings.

- *Semantic Representation:* The recent attention on the lexicon has not resulted in new advancements in the semantic content of the lexical entries. The traditional Montegian ***semantic compositional*** approach through the lambda expressions is still being used widely. This view fails to capture the semantics of words and at best is limited to capturing structural relations among a set of related words.

We have implemented a two–level theory of the human lexicon (see section 3). Our theory addresses both the acquisition of the lexicon as well as the representation of the semantic content of lexical entries. The human lexicon has been traditionally viewed as a simple lookup table, ignoring the underlying semantic layer. We show how both of the levels can be naturally captured within the framework of our theory by analyzing definitions of words of all major categories of the English grammar, as found, for instance, in paper or online dictionaries. This process starts by accepting sequences of words in paragraphs and tagging these words and identifying the major syntactic constituents. Each sentence in the paragraph is then analyzed, both semantically and within the contextual constraints of the previously analyzed sentences, to produce an internal semantic representation in terms of Conceptual Graphs, which are evaluated via our implementation of the formula translator. These graphs capture the denotational semantics of individual lexical words. We show how this two–level view of the lexicon enables one of the widest–coverage grammars of English and how it can be applied to novel problem domains.

The rest of this paper is organized as follows: After reviewing previous work on lexical acquisition in section 2, we give a brief description of the theory behind our system in section 3. We then describe our implementation framework called *NATIVE* in section 4. To help in understanding the examples in a sample session with our system in section 6, we also present our notation for CS linear form in section 5. Finally, we summarize our experiences with CS in the conclusions.

2 Previous Work

2.1 Cyc

Improved NL understanding (NLU) was one of the big promises of Cyc [16], [25], [26], [27], [28] right from the start. This promise was crucial especially since NLU would be used to automate knowledge entry into Cyc and help keep Cyc on schedule. Thus, NLU in Cyc is defined as a the process of mapping from an English expression to one in CycL. Luke is Cyc's knowledge based (KB) lexical acquisition tool to help build a very large robust lexicon [28]. While this term usually refers to finding and/or building relationships among words, Luke tries to find relationships between words and KB entities. Thus, the word ***acquisition*** is a misnomer, since most of the acquired knowledge is entered by a database expert and/or a knowledge engineer familiar with Cyc and CycL.

Unfortunately, the acquisition system in Luke is simply a long question-answering session. Among these questions are: all tensed forms of verbs, whether regular or irregular, plural forms of nouns, its synonyms and related nouns, varying degrees of adjectives that can be used to describe a particular noun, and most importantly, the semantic representation of the entry in terms of one or more CycL expressions. The user is assumed to be extremely familiar with the Cyc ontology also, as he/she is assumed to use internal Cyc entities such as ***$%GeoPoliticalEntity***, ***$%urbArea*** and ***$%numInhabitants*** while describing the word ***city***.

2.2 Other Manual Approaches

There are a number of projects aimed at creating lexicons manually. Some of these include:

- **COMLEX** (an acronym for Common Lexicon) at New York University, where a large, relatively theory–neutral lexicon of English is being implemented [19], [20], [22], [29]. Lexical entries in COMLEX resemble LISP–style s–expressions: a part–of–speech (POS) followed by a zero or more keyword–value pairs enclosed in a pair of parentheses. A word with multiple POS will have one entry for each of its POS. Keywords identify features such as orthography, inflected forms, features, subcategorization information, and such.
- **WordNet** at Princeton University, where a large network of English words are linked together by semantic relations based on the similarity of their individual senses [23], [30],
- the **Penn Tree Bank** at the University of Pennsylvania, where large corpora including the Brown corpus and other financial news text is annotated with syntactic structure, and finally
- the **Japanese Electronic Dictionary Research program (EDR)** where lexical information including syntactic and semantic knowledge is manually integrated into a unified structure (***concept dictionary***). It aims to produce a collection of interrelated dictionaries, including a Japanese and an English dictionary. Entries in individual dictionaries are linked to the concept dictionary [24].

Such manual efforts at creating large lexicons in general, and COMLEX in particular, have serious drawbacks. First of all, human errors are inevitable. In COMLEX, for instance, typographical and judgment errors are inevitable. Also, certain features may be omitted. For example, verbs like run, send, walk, and jump take a long list of directional prepositions. COMLEX can assign a long and comprehensive list of directional prepositions but this results in over-generation. Finally, they usually ignore the semantic content of lexical entries. Current manual projects, including COMLEX, have realized this significance and are now in the process of incorporating some semantics into their lexicons.

2.3 Automated Construction of Semantic Lexicons

Automated strategies [31], [32], [13], [11], [33], [34] exploit MRDs (machine-readable dictionary) to construct another lexical knowledge base to use in NLP applications. This new knowledge base resembles a massive interconnected semantic net in traditional sense. Nodes, or words, are connected via arcs labeled with semantic relations such as Hypernym, Part_of, Location, and Purpose. As in any semantic network, simple semantic associations can be captured. In a sense, the information that was hidden in the MRD is now explicit and distributed over the entire knowledge base. This means that, at least in a restricted sense, *word meanings* are available to NLP tools that use such a lexicon, which in turn can make further inferencing.

A successful application of this type is MORELS [13], that tries to automatically identify morphological relations using the lexicon described above. In one application, MORELS automatically learns the meaning of the denominal *-er* morpheme. It is able to link such nouns as *geography* and *geographer* (as [[*geography*]+*er*]), and *cartography* and *cartographer* (as [[*cartography*]+*er*]). MORELS is still in development and its performance is far from perfect. For example, it has problems in automatically identifying suppletive paradigms such as king/royal, dog/canine, etc.

Even though much work still remains in this line of research, it shows some very promising signs as to its ultimate success. Even at this early stage of the research, such a lexicon is able to address, better than Cyc (section 2.1), certain NLP problems that *appear* to require *common-sense* knowledge. In resolving prepositional phrase attachment ambiguities, a simple one–level hypernym relation expansion suffices in many cases [31].

2.4 Statistical Approaches

With the availability of large tagged and/or untagged corpora, statistical approaches to NLP are gaining popularity [35], [12], [21], [36], [37], [38], [39], [40], [41], [17], [18], [42]. Popular statistical techniques include *n*–grams [35], [17], Hidden Markov models [38], [35], decision trees [41], and decision lists.

Statistical measures analyze co–occurrence patterns between words to produce a statistical measure of semantic relatedness between them. Thus, as would be expected, they are not sufficient in detail, and reveal only that two words are statistically correlated with one another, but can provide no information in a real sense about the semantic nature of this relationship. Statistical techniques also fail to capture the wider context information. The performance of such techniques degrades dramatically when the data is sparse [18], [40].

3 Modeling the Human Lexicon

We model the human lexicon as two–level, highly interconnected structure. We refer to the first level as the *index level* and to the second as the *semantic level.* Figure 1 shows the part of the lexicon defining the adjectives *positive* and *negative.*

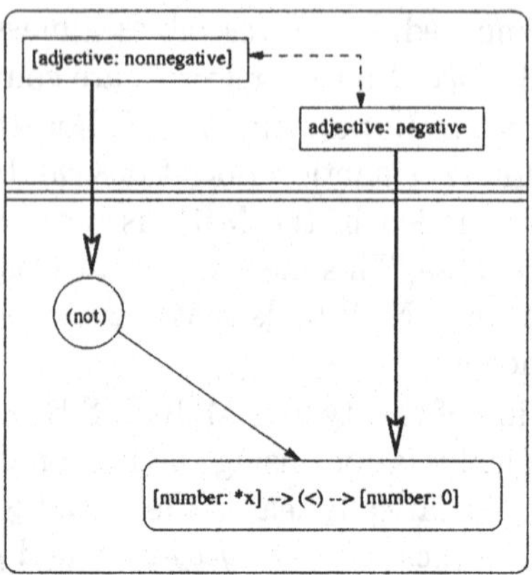

Fig. 1. A cross–section of the lexicon.

3.1 The Index Level

The index level contains structures that correspond to words of a language, such as verbs, nouns, and adjectives. We envision that such an index level is used for each distinct human language, while the semantic layer is shared (and augmented) among them. For each entry, the index level also keeps information about its grammatical variations, including inflections, word endings, its irregular forms, and so on. An entry for a verb, for example, will contain information about its category, subcategorization lists, various tense forms, and usage examples. A word may also be connected to several other related entries at this level, such as its synonyms or antonyms. Finally, each entry has one or more anchors to the structures in the second level. There will be one anchor for each distinct *sense* of the word. This level is the familiar view of the lexicon in the traditional sense given in a dictionary such as LDOCE[43]. As we discussed in section 2, current research has focused on creating lexicons at this level.

3.2 The Semantic Level

The second and more important level contains the practical, *semantic* knowledge that the corresponding index level entries denote within a particular language. The semantic level may contain definitions for both simple words that constitute single structures in the memory, and larger, more complex words that are defined in terms of several other structures, in which case common sub–structures are reused. Each structure defines the semantics of a lexical entry. These two levels are obviously highly inter–connected. One of the reasons is economy: words similar in meaning have separate, albeit related, entries at the index level, but share some of their structures in the semantic memory. Some definitions may be completely contained within other entries. For example, while a pair of synonyms may share the same semantics and thus the same structures at the second level, in case of a synonym–antonym pair, the definition of one will be contained within the definition of the other, as in Figure 1.

We should emphasize that this layer is not required for parsing or, to a certain degree, for interpreting sentences. The semantic level is required, though, if any real understanding is desired, or if a psychologically plausible model of semantics and pragmatics is to be considered in processing sentence fragments. Many types of inferences that seemingly require common sense reasoning, and hence explicit hand–coded rules are easily inferred through the most basic levels of understanding at the lexicon level as we saw when we compared MORELS with Cyc in section 2.3.

4 Main Components of the System

Figure 2 shows the block diagram of the *NATIVE* framework. As can be seen, *NATIVE* consists of modules represented by squares and two structures represented by the two large rectangles: the lexicon and the world model. The lexicon initially contains only a few primitive, built–in lexical entries. It grows as the user defines more lexical words. It serves as an input source to *Parser* in parsing the initial input string, and to *interpreter* in its semantic analysis. The lexicon is persistent across *NATIVE* sessions. The world model, in contrast, acts like a short–term memory for *NATIVE* and contains the specific information about a model of the world that the user has described during a single session of interaction with *NATIVE* . The world model serves as a repository of all the information about a particular domain of discourse being described to *NATIVE* through English sentences. It contains a set of CS graphs that is model–theoretically equivalent to the English description. The world model also serves as an input source to *Interpreter* in its semantic analysis and to *PhiLog* in answering queries about the current state of affairs of the world. Both the lexicon and the world model is created and updated through *PhiLog*.

In discussing the individual modules, it is helpful to work through an example sentence and see the functionality of each module. As we go through the modules, we will give a complete step–by–step construction of the internal meaning for a simple sentence. Our sentence is:

a big cat is on a small mat.

We will see more detailed examples in section 6.

4.1 Parser

The *Parser* is the first module that acts on the input from the user and acts more like a chunker. It tags the input words and combines them into syntactically correct larger *chunks* whenever possible. This module does not try to build anything larger than a noun phrase.

In our example, the *Parser* generates the following structure. It has constructed two noun phrases, complete with the adjectives, corresponding to *a big cat* and *a small mat.* The main verb is also expanded to include other useful information.

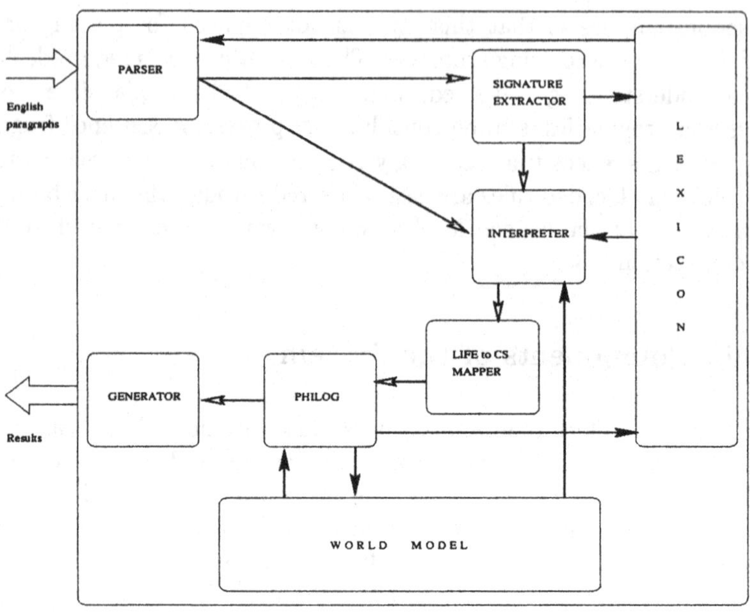

Fig. 2. The *NATIVE* block diagram.

```
S = [[np,
      [det,a,[[number,singular]]],
      [adjList,[adj,[unspecified],[big,[base],[subcat]]]],
      [noun,_A: cat,[_A,cats,count,entity,[orig_def]]]],
     [verb,
      [is,
       [tense,v1],
       [forms,is,was,was,n_o_n_e],
       [subjCase,[[number,singular]]],
       [subcat,[cat,np],[cat,adj],[]],
       [defin,put,def,here]]],
     [prep,on],
     [np,
      [det,a,[[number,singular]]],
      [adjList,[adj,[unspecified],[small,[base],[subcat]]]],
      [noun,_B: mat,[_B,mats,count,entity,[orig_def]]]],
     [punctuation,"."]].
```

4.2 Signature Extractor

The signature extractor works only on input strings that are *definitions*, i.e., English sentences that define a new word, or a new sense for an existing word, such as a noun or an adjective. It accepts the output of the parser and *extracts* such syntactic information as the category, the complement list, if any, that the word licenses, and the different inflectional forms of the word such as tense endings and singular and plural forms.

This module does not get invoked on our example sentence as the sentence is not a definition but a description, i.e., a particular cat is on a particular mat.

4.3 Interpreter

The output of the chunker is still raw, syntax–oriented, and may contain many kinds of ambiguities and inconsistencies. The *Interpreter* comes into play at this stage and resolves different kinds of ambiguities to come up with an internal semantic representation.

Returning to our example, the *Interpreter* produces a complex internal structure that we have not included here due to space limitations. The analysis has discovered that the input is an affirmative sentence. It has a subject (*the cat*), and the main verb has no subcategorization elements that it has licensed. The sentential preposition list includes the interpreted prepositional phrase *on a small mat*. We also note that there is much more information at this level than was available at the chunked input level. For example, a unique internal name, *gen0*, has been assigned to denote the particular cat, corresponding to the indefinite reading of the noun phrase *a ... cat*.

4.4 CS Mapper

The interpreted input has too much information than we would care to have or examine. Thus, we wish to convert it to the internal linearized CS graph form. Note that this translation results in a more intuitive and better representation of the intended meaning of the input sentence. Many people even unfamiliar with the CS theory can correctly guess what the graph means.

```
[cat: *gen0] -
       --> (in1) --> [adj: big]
       --> (on) --> [mat: *gen1] --> (in1) --> [adj: small],   .
```

4.5 PhiLog

The CS output is simple and intuitive but unfortunately it is not in a format that facilitates automated deduction. We thus need to obtain an equivalent set of first–order formulae for the graph and augment our model description through our formula operator, *PhiLog* [44]. This final step in our example generates the following output:

```
on(Gen0, Gen1) :-
        cat(in1=> Gen0),
        big(in1=> Gen0),
        small(in1=> Gen1),
        mat(in1=> Gen1).
```

4.6 Generator

This module interacts with the user. In the current implementation, this module has been kept very simple: it simply responds to the user with an affirmative or a negative answer, along with a list of variable–value bindings, to convey the results. (For a more fully developed English language generator module, see [45]).

5 CS Linear Notation

We explain the CS linear notation used in *PhiLog* (section 4.5) before presenting an actual session with the *NATIVE* system. This parser is part of PhiLog and is the foundation of the persistency mechanism in our system: It is first used to bootstrap the lexicon by loading primitive lexical entries expressed in CS graphs. In later sessions, it is used to save and load newly defined lexical entries.

We use a fairly complete coverage of the standard linear notation. Only λ–expressions and concepts with nested graphs in the type field are omitted. For convenience, proposition concepts are written without the type (i.e., proposition) and the referent graph(s) are put in a pair of braces ({ and }). Similar shortcuts are also defined for *if, then,* and *else.* A CS graph denotes either an assertion or a query. An assertion graph ends with a period. A query graph ends with a question mark. A comment can be put in graphs if bracketed between /* and */. Variable names are preceded with the star sign (*), and constants (*individual markers*, to be more accurate) with the pound sign (#). Numbers may be written without the pound sign. For a concept with links to more than two relations, the concept is ended with a dash sign (-) and the remaining relations are continued on separate lines. A comma (,) ends the last continuation line.

6 Example Session with NATIVE

In this section, we show what a typical interaction looks like with our system. We define several words by entering their English descriptions. We also query the system in English sentences using the newly defined words. We have listed the actual output of the system verbatim, with no editing.

1. **Definition:** We start by defining the meaning of the word *positive.*

   Input: *a number is positive if it is bigger than zero.*

   CS Graph:

```
adj positive(in1 => *gen0) is
 { [number: *gen0]
   [number: *gen0] --> (in1) --> [adj: bigger] --> (than) --> [number: 0]
 } .
```

2. **Definition:** Together with the next example, we define the factorial numbers. The system recognizes factorial as a function, and thus represents it as a graph of actor type.

   Input: *the factorial of zero is one.*

   CS Graph:

```
actor factorial(out => *gen3, of => 0) is
 {   [number: *gen3]
     [number: 0]
     [number: *gen3] --> (is) --> [number: 1]
 } .
```

3. **Definition:** The next definition for the factorial deals with the more general case. The order in which these definitions are given is not important. In particular, the word *positive*, as defined above, ensures that our system does not suffer from the clause ordering problem associated with most logic programming languages, such as Prolog.

   Input: *the factorial of a positive number is the product of the number and the factorial of the difference between the number and 1.*

   CS Graph:

```
actor factorial(out => *gen6, of => *gen5) is
{ [number: *gen6]
  [number: *gen5] --> (in1) --> [adj: positive]
  [number: *gen6] --> (is) --> [number: *gen8] <-- (out) <-- [fn: product] -
                                  --> (of) --> [number: *gen5]
                                  --> (and) --> [number: *gen10]
  <-- (out) <-- [fn: factorial] --> (of)
  --> [number: *gen12] <-- (out) <-- [fn: difference] -
                                  --> (between) --> [number: *gen5]
                                  --> (and) --> [number: 1], ,
} .
```

4. **Query:** We ask the system for the factorial of 5.

   Input: *what is the factorial of 5?*

   CS Graph:

```
[number: *gen20] -
          <-- (out) <-- [fn: factorial] --> (of) --> [number: 5]
          --> (is) --> [number: *gen18],  ?
```

   Solution *Gen18 = 120, Gen20 = Gen18.*

5. **Query:** This query shows that the CS graphs generated for the English definitions are flexible as their logic-language counterparts, i.e., there is no distinction between input and output arguments.

   Input: *what is 120 the factorial of?*

CS Graph:

```
[number: 120] --> (is) --> [number: *gen24] <-- (out)
<-- [fn: factorial] --> (of) --> [number: *gen22] ?
```

Solution *Gen22 = 5, Gen24 = 120.*

6. **Query:** This query shows how adjectives are used within the context of their use. Even though we have not (explicitly) defined the factorial for negative numbers, the system correctly handles them.

   Input: *what is the factorial of -5?*

   CS Graph:

   ```
   [number: *gen37] -
               <-- (out) <-- [fn: factorial] --> (of) --> [number: -5]
               --> (is) --> [number: *gen35],  ?
   ```

   Solution **** No*

7. **Definition:** To represent *prime* numbers, the system generates a quantified concept as an intermediary step in the semantic representation.

   Input: *a positive number is prime if it is divisible by two numbers that are between 1 and itself.*

   CS Graph:

   ```
   adj prime(in1 => *gen82) is
    {   [number: *gen82] --> (in1) --> [adj: positive]
        [number: *gen82] --> (in1) --> [adj: divisible]
         --> (by) --> [number: *gen83{*}@2]
        [number: *gen83] --> (in1) --> [adj: between] -
                                 --> (in2) --> [number: 1]
                                 --> (and) --> [number: *gen82],
    } .
   ```

8. **Query:** Instead of asking the system whether a given number is prime or not, we can ask the system to generate all prime numbers between a given range.

   Input: *what is a prime number between 20 and 25?*

   CS Graph:

   ```
   [number: *gen106] -
             --> (in1) --> [adj: prime]
             --> (is) --> [number: *gen105],  ?
   [number: *gen106] --> (in1) --> [adj: between] -
                           --> (in2) --> [number: 20]
                           --> (and) --> [number: 25],  ?
   ```

   Solution *Gen105 = 23, Gen106 = Gen105.*

7 Conclusions

We have implemented a model of the human lexicon that addresses the automatic acquisition of the lexicon as well as the representation of the semantic content of individual lexical entries. The system analyzes paragraphs of English sentences, maps the extracted lexical and semantic knowledge to CS graphs, and reasons about them.

We have used CS as the representation scheme in a system to develop a model of the human lexicon. We have chosen CS to represent the lexicon for its close relationship with natural languages, for its support for natural language constructs, and for its almost direct representation of natural language sentences. We also use CS to bootstrap the system. Since the lexicon is stored and maintained as a set of CS graphs, one can supply definitions for lexical entries directly in terms of CS graphs. This technique has the additional benefit that we can replace the CS engine by a more powerful and complete one, without affecting the most important components of the system, such as the chunker and the interpreter.

Our implementation of the CS engine (see *PhiLog* in section 4.5) is partial with respect to CS formalism. In particular, it cannot handle events and nested contexts. It also has difficulty in dealing with plural concepts that are in multiple relationships with each other. Although CS as a formalism addresses these issues in theory, there is no standard implementation of these features in practice.

References

1. Michael P. Barnett. *Computer Programming in English.* Harcourt, Brace and World, 1969.
2. D. Barstow. *A Knowledge Based System for Automatic Program Construction*, volume 5. 1977.
3. A. W. Biermann. Approaches to automatic programming. In M. Rubinoff and M. C. Yovits, editors, *Advances in Computers*, volume 15, pages 1–63. Academic Press, 1976.
4. D. G. Bobrow. *Natural Language Input for a Computer Problem-Solving System*, pages 146–226. MIT Press, 1968.
5. J. G. Carbonell and G. H. Hendrix. A tutorial on natural language processing. *Communications of the ACM*, 24(1):4–8, January 1981.
6. C. Green. A summary of the psi programming synthesis system. In *Proc. of the 2nd International Conference on Software Engineering*, pages 4–18, 1976.
7. G. E. Heidorn. Automatic programming through natural language: A survey. *IBM Journal of Res. and Dev.*, (4):302–313, 1976.
8. G. E. Heidorn. English as a very high level language for simulation programming. *SIGPLAN Notices*, (9):91–100, 1974.
9. Carl Pollard and Ivan Sag. *Information-Based Syntax and Semantics, Vol. 1: Fundamentals.* CSLI Publications", 1987.
10. Carl Pollard and Ivan Sag. *Head-Driven Phrase Structure Grammar.* CSLI Publications", 1994.
11. L. Vanderwende. Ambiguity in the acquisition of lexical information. In *Proceedings of the AAAI 1995 Spring Symposium Series*, pages 174–179. 1995.

12. N. Calzolari and A. Zampolli. *Methods and Tools for Lexical Acquisition*, pages 4–24. Springer–Verlag, 1990.
13. J. Pentheroudakis and L. Vanderwende. Automatically identifying morphological relations in machine–readable dictionaries. In *Proceedings of the Ninth Annual Conference of the UW Centre for the New OED and Text Rersearch*, pages 114–131. 1993.
14. Beth Levin. Lexical semantics: Tutorial. In *Proceedings of 33rd Annual Meeting of the Association for Computational Linguistics*. 1995.
15. G. Serasset. Interlingual lexical organisation for multilingual lexical databases in nadia. In *COLING*, pages 278–282, 1994.
16. D. Lenat, R. Guha, K. Pittman, D. Pratt, and M. Shepherd. Cyc: Towards programs with common sense. *Communications of the ACM*, 33(8):30–49", August" 1990.
17. P. F. Brown, V. Della Pietra, P. V. deSouza, J. C. Lai, and R. L. Mercer. Class–based *n*–gram models of natural language. *Computational Linguistics*, 18(4):467–479, 1992.
18. Walter Daelemans. *Memory–Based Lexical Acquisition and Processing*. Springer–Verlag, 1994.
19. R. Grishman, C. Macleod, and A. Meyers. Comlex syntax: Building a computational lexicon. In *Coling*, 1994.
20. C. Macleod, R. Grishman, and A. Meyers. The comlex syntax project: The first year. In *ARPA Human Language Technology Workshop*, 1994.
21. A. Sanfilippo. *LKB Encoding of Lexical Knowledge*. Cambridge University Press, 1992.
22. C. Macleod, R. Grishman, and A. Meyers. Creating a common syntactic dictionary of english. In *SNLR: International Workshop on Sharable Natural Language Resources*, 1994.
23. G. Miller. Wordnet: An on–line lexical database. *International Journal of Lexicography*, 3(4):235–312, 1990.
24. Ltd. Japan Electronic Dictionary Research Institute. Edr electronic dictionary technical guide. *Project Report*, (TR–042), August 1993.
25. D. Lenat, R. Guha, K. Pittman, D. Pratt, and M. Shepherd. Cyc: A midterm report. *AI Magazine*, pages 32–58, Fall 1990.
26. R. Guha and D. Lenat. Enabling agents to work together. *Communications of the ACM*, 37(7):127–142, July 1994.
27. D. Lenat and R. Guha. *Building Large Knowledge Based Systems*. Addison Wesley, 1990.
28. J. Barnett, K. Knight, I. Mani, E. Rich, and M. Shepherd. Knowledge and natural language processing. *Communications of the ACM*, 33(8):50–71, August 1990.
29. C. Macleod, R. Grishman, and A. Meyers. Developing multiply tagged corpora for lexical research. In *International Workshop on Directions of Lexical Research*, 1994.
30. G. Miller, C. Leacock, Tengi R., and R. Bunker. A semantic concordance. In *Proceedings of the Human Language Technology Workshop*, pages 303–308. Morgan Kauffmann, 1993.
31. W. Dolan, L. Vanderwende, and S. D. Richardson. Automatically deriving structured knowledge bases from on–line dictionaries. In *Proceedings of the First Conference of the Pacific Association for Computational Linguistics*, pages 5–14. 1993.
32. W. Dolan. Word sense ambiguation: Clustering related senses. In *Proceedings of COLING*, pages 712–716. 1994.

33. H. Alshawi. *Analysing the Dictionary Definitions*, pages 153–170. London: Longman Group, 1989.
34. Y. Wilks, D. Fass, C. Guo, J. McDonald, T. Plate, and B. Slator. *Parsing in Functional Unification Grammar*, pages 193–228. London: Longman Group, 1989.
35. Eugene Charniak. *Statistical Language Learning*. MIT Press, 1993.
36. M. Meteer, R. Schwartz, and R. Weischedel. *POST: Using Probabilities in Language Processing*, pages 960–965. Morgan Kauffmann, 1991.
37. K. A. Church. *Stochastic Parts Program and Noun Phrase Parser for Unrestricted Text*, pages 136–143. ACL, 1988.
38. J. Kupiec. *Augmenting a Hidden Markov Model for Phrase–Dependent Word Tagging*, pages 92–98. 1989.
39. Mark Lauer. How much is enough?: Data requirements for statistical nlp. In *Second Conference of the Pacific Association for Computational Linguistics*, 1995.
40. Mark Lauer. Conserving fuel in statistical language learning: Predicting data requirements. In *Eighth Australian Joint Conference on Artificial Intelligence*, 1995.
41. D. M. Magerman. Statistical decision–tree models for parsing. In *Proceedings of the 33rd Annual Meeting of the ACL*, 1995.
42. K. Chen and H. Chen. Extracting noun phrases from large–scale texts: A hybrid approach and its automatic evaluation. In *ACL*, 1994.
43. *Longman Dictionary of Contemporary English.*
44. S. Dogru and J. Slagle. Philog: An implementation of formula translator in a sorted logic programming language. Number 95–027. April 1995.
45. S. Dogru and J. Slagle. A system that translates conceptual structures into english. In H. D. Pfeiffer and T. Nagle, editors, *Conceptual Structures: Theory and Implementation*, pages 283–292. Springer–Verlag, July 1992.

Analysis of Task-Oriented Conversations into Conceptual Graph Structures

Bernard Moulin[1], Mohamed Gouiaa[1] and Sylvain Delisle[2]

[1] Laval University, Computer Science Department, Pouliot Building
Ste-Foy, Quebec, G1K7P4, Canada
{Moulin, Gouiaa}@ift.ulaval.ca
[2] Département de mathématiques et d'informatique
Université du Québec à Trois Rivières
C.P. 500, Trois-Rivières, Québec, Canada, G9A 5H7
E-mail : Sylvain_Delisle@uqtr.uquebec.ca

Abstract. In this paper we present the main characteristics of the MAREDI system which is used to generate a conceptual representation of written transcription of task-oriented dialogs. We show how several techniques of the CG theory (concept type lattice, concept schematas, case relations, anaphoras) can be put to use in a practical system integrating other technologies (robust parsing, neural nets).

1 Introduction

In this paper we report on the main results that we obtained in the MAREDI Project which aims at developing a natural language processing (NLP) system for the analysis of written transcriptions of task-oriented conversations. In Section 1 we describe the context and main assumptions of this research as well as the system's general architecture. In Section 2 we present the general characteristics of two of the system's main components: a robust syntactic parser for oral utterances and a neural network used to determine the types of locutors' speech acts. In Section 3 we present the semantic component which puts into work several notions borrowed from the conceptual graph (CG) theory [16]. The originality of this project stems from the integration of several techniques complementing classical notions of the CG theory in order to build a practical NLP system for oral dialogs.
Several natural language processing systems have been developed in the CG community for various kinds of applications such as knowledge-based machine-aided translation [1], the analysis of medical texts [15] and text generation [12], to quote only a few. All these systems, as traditional NLP systems, are based on the assumption that the texts to be analyzed follow the ideal grammar that is implemented in the system. Indeed, this is correct for analyzing most written texts. However, we cannot rely on such an assumption when it comes to oral dialogs which are full of anomalies (repeti-

tions, ellipses, missing or unnecessary terms). In the MAREDI project (MArkers and the REpresentation of DIscourses) we explore ways of taking advantage of the various markers found in oral discourses in order to determine the semantic content of task-oriented dialogs. In a task-oriented dialog locutors interact in order to jointly perform a task. We worked on a corpus of transcriptions of dialogs [13] [5] in which an instructor gives directives to a manipulator in order to draw a geometric figure. The instructor and manipulator were located in different rooms, each being in front of a terminal. They interacted using a microphone system through which the dialogs were recorded. The instructor gave directives to the manipulator who drew a picture on her terminal and simultaneously, the instructor could view it on her own terminal. Here is an example of a dialog in which we can observe various anomalies (Inst: instructor; Man: manipulator; i_x and m_y are the utterance's identifiers)

Inst (i1) euh main'nant tu vas prendre un petit un p'tit carré	*/ Uh now you take a small a small square*
Inst (i2) et le mettre en bas à gauche	*/ and put it in the left bottom corner*
Inst (i3) pour faire une maison pour euh	*/ to draw a house uh*
Man (m4) geste (exécution)	*/ gesture (drawing)*
Inst (i5) maintenant tu en prends un autre	*/ now you take another one*
Man (m6) geste (prise et déplacement)	*/ gesture (taking and dragging)*
Man (m7) avec un toit aussi ?	*/ with a roof, right?*
Inst (i8) non non	*/ No no*

The objective of the MAREDI Project is to develop a system for analysing transcriptions of oral dialogs in order to generate a conceptual representation of the conversation (Moulin et al. 1994). Locutors' interactions are modeled as exchanges of conversational objects (COs). A *conversational object* is a mental attitude (belief, goal, wish, etc.) along with a positioning that a speaker transfers to the addressee during a conversation [11]. The speaker positions herself relative to a mental attitude by performing actions like "proposing", "accepting", "rejecting": this is called the speaker's positioning relative to that mental attitude. Figure 1 is a graphical representation of the model of conversation that we want to generate after analysing dialog transcriptions such as our example. Ovals represent mental attitudes and labelled arrows represent locutors's positionnings.

Our approach is based on the assumption that various markers found in a discourse can be used to orient the analysis [7] and more specifically to identify the categories of locutors' speech acts [6]. The main steps of the analysis are: 1) Divide the original text into utterances; 2) Delete noisy terms such as "Uh"; 3) Identify markers in utterances (such as "now", "right", "so") and extract them from the text; 4) Proceed with the syntactic analysis of utterances (including ill-formed or fragmentary utterances); 5) For each utterance, identify the category of speech act performed by the locutor (this is a pragmatic analysis which takes into account the numerous indirect speech acts found in oral dialogs); 6) Based on the category of speech act, identify the locutor's positioning and the type of mental state on which it applies (this is done using a correspondence table between categories of speech acts and couples of (positioning, mental state type)); 7) Proceed with a semantic analysis of the speech act's propositional content based on the results of the syntactic analysis; 8) Create the conversation's conceptual model.

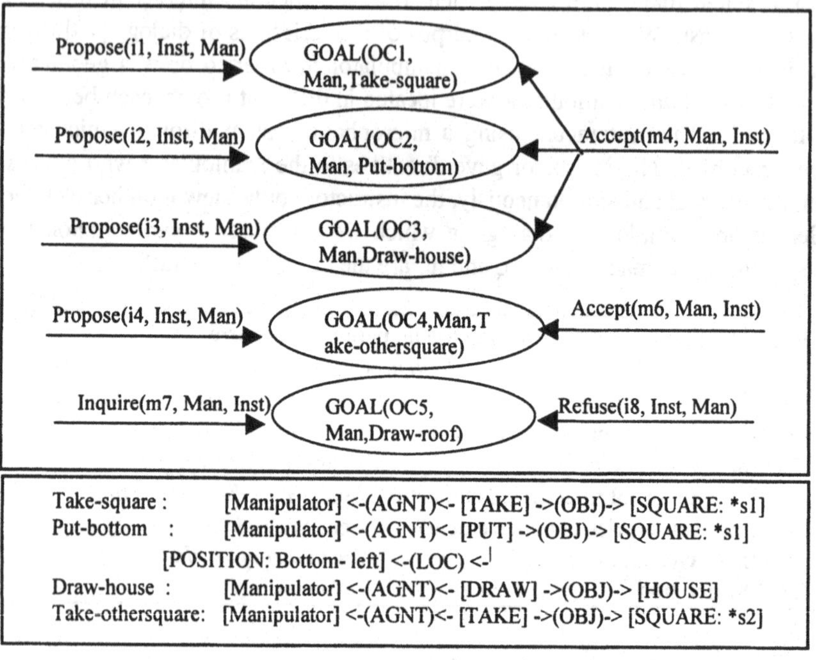

Fig.1. Diagram illustrating the conversational model

2. The Syntactic Parser and Neural Analyser

Figure 2 presents the architecture of the MAREDI System. Written transcriptions of oral dialogs are processed by a *Pre-syntactic Filter* which detects lexical markers (such as "Uh", "right", "OK"). The filtered text (which may be fragmentary) is processed by the *Robust-Syntactic Parser* that generates a syntactic tree and sends it to a *Post-syntactic Filter*. This filter identifies syntactic markers (such as imperative or interrogative constructions) and sends them to the *Neural Analyzer* which processes them along with the lexical markers provided by the *Pre-Syntactic Filter*. The *Neural Analyzer* identifies the category of speech act that was performed in this utterance and sends it to the Integrator. The syntactic tree is also sent to the *Semantic Analyzer* which determines the semantic interpretation of the speech act's propositional content. The *Semantic Analyzer* is presented in more details in Section 3. The Integrator integrates the information provided by the *Neural Analyzer* and the *Semantic Analyzer* in order to generate a conceptual representation of the utterance which is integrated in the *Conceptual Model of the Dialog* in a form equivalent to the example of Figure 1.

The *Pre-syntactic* and *Post-syntactic Filters* are pattern matching programs which identify relevant markers in the text or in the syntactic tree, using tables containing the various kinds of markers that can be found in our corpuses.

The *Syntactic Parser* is composed of four modules [2, 3]: 1) the *Supervisor* coodinates the interactions of the 3 other modules; 2) the *Kernel-Analyzer* uses a standard grammar of written French in order to process well-formed utterances; 3) the *Recovery Module* detects anomalies (repetitions, interruptions, noise, ellipses) in ill-formed utterances and corrects them in order to transform the syntactic tree into a canonical form which will be used by the Semantic Analyzer; 4) the Lexical Ambiguity Resolver validates the syntactic category of each word in the utterance. The system uses a specification of the French grammar based on Chomsky's theory of Government and Binding [4]. According to this theory syntactic rules take into account the argument structure of verbs, nouns and adjectives: this greatly facilitates the resolution of lexical and syntactic ambiguities. Chomsky's theory also provides transformation mechanisms which are used by the Recovery Module. Anomalies are detected, thanks to heuristics developed during the analysis of our corpuses. The proper transformations are applied in order to restore a syntactically correct structure for the utterance. Hence, the Syntactic Analyzer is able to analyze well-formed French sentences as well as a number of ill-formed structures and to provide at least fragments of syntactic trees.

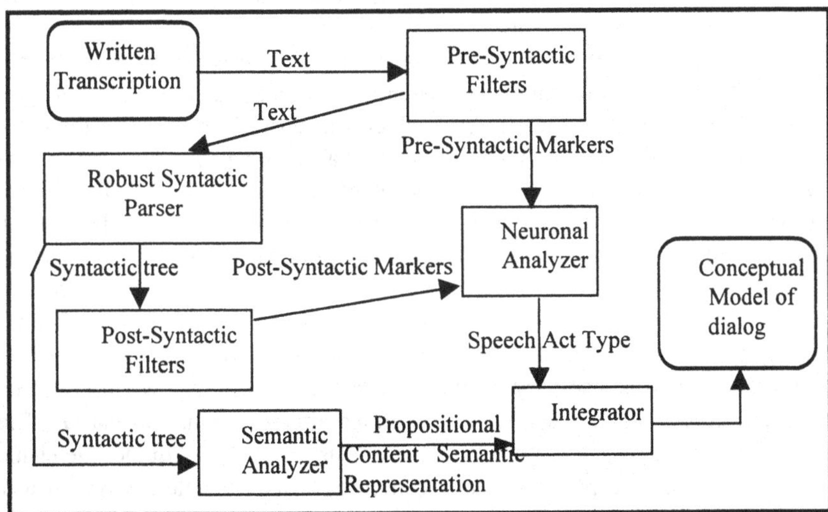

Fig.2. Architecture of the MAREDI System

The Neural Analyzer [6, 7] is a localist neural network (without learning capabilities) which takes the text markers as entries and proposes corresponding speech act categories. The core of the Neural Analyzer is a sub-network that matches combinations of markers with categories of speech acts. However, in many cases, it is not possible to uniquely identify the speech act category corresponding to a given combination of markers. For instance, "comme ça" ("like that") is a marker that may denote either a question or a confirmation, or even an assertion about the manner to do something. So, the Neural Analyzer contains 2 other sub-networks that take into account the role played by the locutors (i.e. "manipulator" and "instructor") and the context of the utterance (the characteristics of the preceding utterance). Using these sub-networks the system is able to limit the number of possible continuations for a given utterance in terms of the markers that can be found in the following utterance. Colineau [6] experimentally showed that the Neural Analyzer is able to analyze almost 85% of the utterances of the drawing task corpus and 88% of the utterances of a different task-oriented corpus. She showed how to improve these results.

3. The semantic Analyzer

The Semantic Analyzer is based on a case-based analysis [8] [9] and on a template-based and inference-driven mapping approach [14]. Such an approach allows us to take advantage of the correspondence existing between syntactic patterns and semantic cases. An utterance is composed of a modal part (tense, mode, modality) and a proposition. The proposition is composed of a main predicate (the verb) and a collection of nominal components (nouns and adjectives) which are related to the verb by various grammatical relations. The objective of the analysis is to identify in an utterance the concepts corresponding to the verb, nouns and adjectives and the case relations associating those concepts.

Here is the general idea of our analysis strategy which is centered on the semantic characterization of verbs found in utterances. We use a *dictionary of concept schemata* [16] corresponding to the verbs which can possibly be used in the corpus. To each verb corresponds one or several schemata (also called templates) in which the "verb concept" is associated by conceptual relations (corresponding to semantic cases) to acceptable concept types. In concept schemata cases may be mandatory or optional: this enables the system to analyze similar sentences where some cases are omitted. We also use a *concept type lattice* [16] which contains the various kinds of concepts corresponding to words found in our corpus. It is used to validate the conformity of concepts found in utterances with the semantic cases associated with the verb predicates. It is also very useful for anaphoric resolution. We also use a collection of *syntactic patterns* that we have defined after analysing our corpus. There are two kinds of patterns. The "modal patterns" are used to determine the elements of the utterance's modal part. The "proposition patterns" enable the system to find in the syntactic tree the parts that correspond to the semantic cases associated to the verb concept. Hence, each syntactic part is associated with a semantic role whose validity is checked using selection constraints based on the acceptable concept types for a given schema.

Figure 3 presents the general architecture of the semantic analyzer [10] where the main modules are identified by numbers Pi. The utterance's syntactic tree is received by the *Interface with Syntactic Parser* (P1) which generates *Intermediate Syntactic Structures* that will orient the analysis. If the utterance is an idiomatic expression, the

Intermediate Structures contain markers that are passed to the *Processor of Idiomatic Expressions* (P2) which uses marker tables to determine the corresponding category of speech act and transfers the information to module P9 for the creation of the semantic structure. If the utterance is not an idiomatic expression, the *Intermediate Syntactic Structures* contain the decomposition of the utterance's syntactic constituents. If there is no verb, the *Processor of Elliptic Utterances* (P3) is activated: it tries to find the missing verbal context using *Concept schematas* and *Verbal contexts of preceding utterances*.

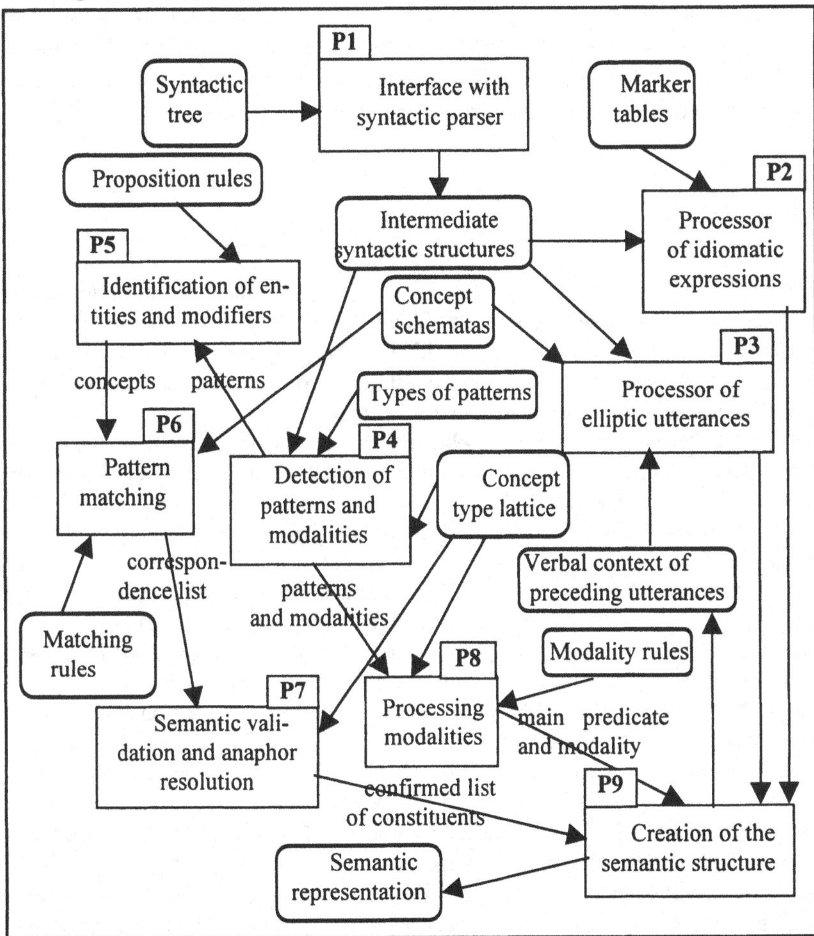

Fig.3. Architecture of the semantic analyzer

Module P4 uses a repository of *Types of patterns* and the *Concept type lattice* in order to detect the *Syntactic patterns* in the *Intermediate Syntactic Structures* and to identify modal elements. Those elements are passed to Module P8, *Processing modalities*, which uses *Modality rules* to identify the relevant modal parameters and to transfer them to Module P9. The patterns detected by Module P4 are used by module P5 to identify the concepts corresponding to the utterance's syntactic components with their

modifiers (adjectives), using *Proposition rules*. The process of *Pattern Matching* (P6) determines the semantic roles of those concepts, using the *Concept schematas* and *Matching rules*. The output is a *Correpondence list* that is checked by the process of *Semantic Validation and Anaphor Resolution* (P7), using *the Concept Type Lattice*. The result is a *Confirmed list of constituents* which is passed to module P9. Module P9 creates the resulting *Semantic Representation* of the utterance which has a form equivalent to the example given in the lower rectangle of Figure 2.
In this paper we presented the main characteristics of the MAREDI system which is used to generate a conceptual representation of written transcription of oral task-oriented dialogs. We showed how several techniques of the CG theory (concept type lattice, concept schematas, case relations, anaphoras) can be put to use in a practical system integrating other technologies (robust parsing, neural nets).

References

1. Angelova G, Bontchera K. (1996) DB-MAT: Knowledge acquisition, processing and NL generation using conceptual graphs, In Eklund et al. (edts,) *Conceptual Structures: Knowledge Representation as Interlingua*, LNAI n. 1115, Springer Verlag, 115-129.
2. Boufaden N.(1998) Analyse syntaxique robuste des textes de dialogues oraux, Master's Thesis, Département d'informatique, Université Laval.
3. Boufaden N., Delisle S., Moulin B. (1998) Analyse syntaxique robuste de dialogues transcrits: peut-on vraiment traiter l'oral à partir de l'écrit?, Proceedings of the International Conference TALN'98, Paris, 112-121.
4. Chomsky N. (1981), *Lectures on Government and Binding*, Dordrecht : Foris, Collection : Studies in generative grammar.
5. Colineau N. (1994), Vers une compréhension des actes de discours, rapport de DEA en Sciences Cognitives, Institut National Polytechnique de Grenoble (France).
6. Colineau N. (1997), *Etude des marques discursives dans un dialogue finalisé*, Thèse de doctorat ès Sciences Cognitives, Université Joseph Fourier, Grenoble (France).
7. Colineau N., Moulin B. (1996) Un modèle connexionniste pour la reconnaissance d'actes de dialogue, Conf. Informatique et Langue Naturelle, Nantes (France), 157-174.
8. Delisle S., Barker K., Copeck T. & Szpakowicz S. (1996), Interactive Semantic Analysis of Technical Texts, *Computational Intelligence*, vol. 12, n. 2, 273-306.
9. Fillmore C. (1968), The Case for Case, In E. Bach et R.T. Harms (edts.), *Universals in Linguistic Theory*, Holt, Rinehart and Winston.
10. Gouiaa M. (1998) Modélisation conceptuelle des conversations à partir de l'analyse de textes de dialogues transcrits: analyseur sémantique et générateur de la structure conversationnelle, Master's Thesis, Département d'informatique, Université Laval.
11. Moulin B., Rousseau D., Lapalme G. (1994), A Multi-Agent Approach for Modelling Conversations, Proc. of the Intl. Conf. on AI and Natural Language, Paris, 35-50.
12. Nicolov N., Mellish C., Ritchie G. (1995) Sentence generation from conceptual graphs, In Ellis G., Levinson R., Rich W., Sowa J.F. (edts*.), Conceptual Structures: Application, implementation and theory*, LNAI n. 954, Springer Verlag, 74-88.
13. Ozkan N. (1994) *Vers un modèle dynamique du dialogue : analyse de dialogues finalisés dans une perspective communicationnelle*, thèse de Doctorat en Sciences Cognitives, Institut National Polytechnique de Grenoble (France).

14. Palmer M. S. (1990) *Semantic processing for finite domains*. Cambridge Univ. Press.
15. Rassinoux A-M, Baud R.H., Scherrer J.R. (1994) A multilingual analyser of medical texts, in Tepfenhart W.M., Dick J.P., Sowa J.F., *Conceptual Structures: Current Practices*, LNAI n. 835, Springer Verlag, 84-96.
16. Sowa, J. F. (1984) *Conceptual Structures*, Reading Mass: Addison Wesley.

Using Conceptual Graphs as a Common Representation for Data and Configuration in an Active Image Processing System

J. Racky, M. Pandit

University of Kaiserslautern
Department of Electrical Engineering
Institute for Control and Signal Theory
67653 Kaiserslautern – Germany
Jens.Racky@e-technik.uni-kl.de

Abstract. Active image processing is an attempt to overcome some limitations of the classical, intuition based approach for developing image processing solutions. It's goals are to allow for intuitive specification of image processing tasks and to take advantage of the ability to influence the scene. The general structure is that of interconnected, self-configuring stages. To formalize communication between these stages conceptual graphs have been found useful to serve for both, as a representation for the scene model as well as for the stage configuration.

1 Introduction

Many industrial image processing tasks are visual inspection tasks. Most of them can be divided into *object recognition* followed by *measurements*.
The recognition process is usually (e.g. [1]) viewed as a sequential process called the *image processing pipeline*.
The major task in designing an image processing solution consists of the selection of suitable operators, their parameterization and interconnection. Usually the design process is highly intuitive and solutions are very task specific, as task description is not formalized.
Thus the knowledge and experience of the application engineer concerning the properties of the environment (location of camera and light sources, dust, ...), the objects (reflection properties, relevant views) and the available image processing operators still play the most important roles in the development process.
Another disadvantage of conventionally designed image processing systems is the lack of feedback within the system. By using a static configuration they are often unable to react to changes in the environment which require a re-parameterization or even a reconfiguration.
Some task may even just become solvable if there is feedback not only within the 'software part' of the system, but also on the scene itself (active vision), e.g. a variation of the illumination setup.

2 Active Image Processing

The paradigm of Active Image Processing (AIP) is an attempt to overcome some of the limitations of conventionally developed image processing solutions.

In the following we give an overview over the concept of AIP with respect to our implementation.

AIP systems shall

- allow a non image processing expert to teach the system to solve particular segmentation tasks[1] : The task is entered by presenting and explaining an example to the system. Explanation consists of drawing a sketch of the scene over the example image, marking target regions and generalizing parameter ranges of image features (e.g. 'region is always bright').
- also configure the scene setup, e.g. the illumination
- be able to reconfigure themselves dynamically to react/adapt to changes in the environment or to even solve the task, e.g. by using illumination sequences.

They have the general structure show in fig. 1.

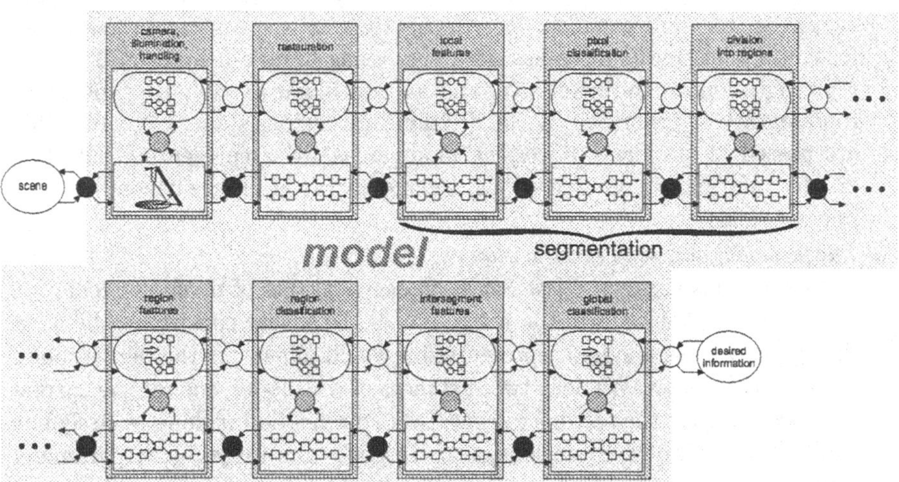

Fig. 1. General structure of an active image processing system.

[1] Tasks are considered to be repetitive: While the general structure of the scene remains the same, certain parameters vary, e.g. position, orientation, interferences.

AIP systems consist of three major components:

1. *Extended image processing pipeline.*
   A sequence of stages which reflects a division into subtasks which appear (in this sequence) for almost every image processing task. Each stage either calculates new informations (feature stages) or groups elements by similar features (classification stages). In contrary to conventional systems AIP seamlessly integrates a stage to influence the scene. In general some or all stages will be invoked several times during a recognition process, if subtasks have to be solved before the main task can be solved or in case a new scene requires a reconfiguration.
2. *Local configuration.*
   To perform it's task each stage is divided into a *data processing unit* (lower box) which performs the image transformations (e.g. filtering, thresholding, feature calculation, etc.) and a *local configuration unit* (upper box) which creates the local configuration. The latter uses a specific view of the underlying image model as it's reference and (informational) input from the next stage as it's task specification. It is able to send (sub)task specifications to the previous stage, if necessary. Configurations are generated by several numerical and non-numerical algorithms and are always represented as conceptual graphs (CGs). Every execution unit translates this CG representation into a program for execution by an (embedded) interpreter.
3. *Systematic connection.*
   The division into stages and the general demand for feedback requires each stage to be connected with it's neighbors. Configuration modules are connected on the informational level using CGs as a specification 'language' for communicating (sub)tasks. Data processing units are connected using a representation on the pixel level and an additional (optional) graph-based representations generated by the division-into-regions stage. Configurations are passed to the data processing modules as CGs representing data-flow graphs, information on images (e.g. the mean gray level) are passed back via referent fields.
4. *Image model and task specification.*
   Each stage has it's own 'view' on an image (e.g. sets of pixels, regions, ...). Together these views form the image model. To ensure consistency between the stages we use a single representation which allows for the derivation of these specific views. We need two instances of the model: one for the current scene and one for the reference scene.Task specification is done by assigning additional attributes to regions, e.g. 'target', 'non-target' and parameter ranges.

2.1 Image Model

We use two image models in addition to the representation on the pixel level. They correspond to different levels of abstraction.

Numerical Model The numerical model (NM) represents images as hierarchical region graphs. Each region is either a group of other regions or an elementary region. Each elementary region is characterized by it's *(exo)skeleton* represented by B-splines segments and the *face* of the region characterized by various texture parameters (max, min, mean values, ...).
The NM is obtained in two ways: (1) It is entered by the user as an explanation of an example image by drawing a sketch of the scene and specifying the desired (segmentation) result. (2) It is generated by the ***division-into-regions*** stage.

Symbolical Model The symbolical model (SM) represents images as conceptual graphs which are basically derived from the NM. It allows the configuration algorithms to query specific information (e.g. inclusion, neighborhood, texture, etc.), as well as specific views from this representation in a formalized way.
For shortness we desist from giving a formal description of all concept and relation types in this paper. Instead we illustrate the idea using an example.
Fig. 2 shows an image as it would have been produced by the pixel classification stage or been drawn by the user.

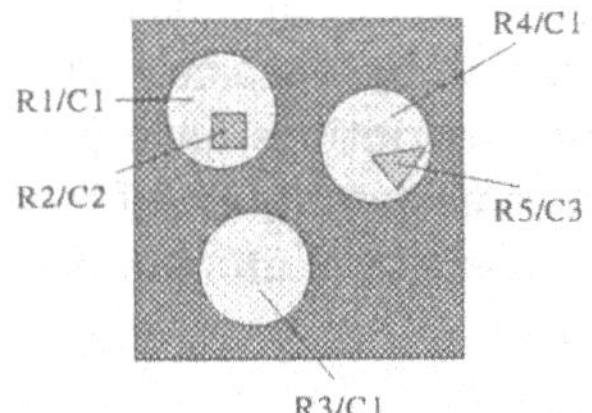

Fig. 2. Example of the region oriented image representation showing on the left an image containing 5 regions (R1...R5) from three classes/shapes (C1..C3). The right image shows a possible target (R3) as a segmentation task specification.

This example can be translated to the following CG (regarding only shape class and inclusion properties):

```
[image:example]-
    ->(incl)->[region:#1]-
                        ->(chrc)->[class:#1]
                        ->(incl)->[region:#2]->(chrc)->[class:#2]
    ->(incl)->[region:#3]->(chrc)->[class:#1]
    ->(incl)->[region:#4]-
                        ->(chrc)->[class:#1]
                        ->(incl)->[region:#5]->(chrc)->[class:#3]
```

By defining a transitivity rule it can be (explicitly) stated that a nested region (e.g. `[region:#2]`) is also included in the image. The

problem is, however, that adding another region to the image (e.g. `[region:#3]->(incl)->[region:#6]->(chrc)->[class:#2]`) makes the new image a specialization (in CG terms) of the original image. A solution which works for the original image, however, can *not* generally be specialized to a solution for the new image. This problem has already been mentioned by Ho [4] whose solution was to break the graph into a conjunctive set of graphs ('strands'). Our approach breaks up the graph into graphs in which only one class concept or one relational (e.g. '`incl`') type appears.
For the example this leads to

```
[class:#1]-
          <-(chrc)<-[region:#1]
          <-(chrc)<-[region:#3]
          <-(chrc)<-[region:#4]
[class:#2]<-(chrc)<-[region:#2]
[class:#3]<-(chrc)<-[region:#5]
```

```
[region:#1]->(incl)->[region:#2]
[region:#4]->(incl)->[region:#5]
```

As we consider only translation invariant image transformations *individual* regions (of the same class) can not be accessed, but just the set of all regions of one class.
Regions can be characterized in various ways, i.e. using an exact numerical model, a simplified numerical description, a description by a geometric primitive or by specifying a group of other regions. These descriptions have just a few properties in common, like position, rotation and some description of it's shape. It is thus not advisable to include all possible parameters into the definition. Instead only the common properties are included:

type region(x) **is**

```
[region: *x]-
          ->(location)->[2D_Measure]
          ->(rotation)->[Angle]
          ->(boundary)->[Shape]
```

All other (possible) properties are stored as schemata to be used in constructing a specific image model for a specific configuration algorithm. This applies also to descriptions of spatial relations (i.e. inclusion and neighborship).

2.2 Configuration

Our system has been designed to solve segmentation tasks, like the simple example shown in the previous section. At this point we should mention the difference between a self-configuring system and an automatic image interpretation system like the one described in [4]: Whereas the latter employs a fixed (not to be mistaken for simple) segmentation algorithm and tries to classify it's results, self-configuring systems additionally try to find such a segmentation algorithm. The general goal of the configuration is to enable the system to filter out as many of the non-target areas as early as possible: a region which can be identified just by classifying it's shape should not be identified using it's relations to other regions. A stage may, however, *request* non-target regions from preceding stages (if they are needed to solve it's task).

Our system employs *local* configuration units, which has several advantages: (1) The division into a sequence of stages *implicitly* reflects the sequence of subtasks which always have to be solved. (2) Subtask specification is *formalized* as a prerequisite for (3) using different/alternative configuration approaches. The goal is further to find simple solutions, by which we mean transformations on the pixel level[2], because then the computation of the NM and SM can be omitted.

Representation The configuration on each stage is represented as a CG forming a data flow graph: Concepts correspond to *data* (i.e. variables), relations represent *operations*. Thus we need only one representation for all configurations computed by different configuration algorithms. As the representation is the same as for the image/scene description, task specification and the configuration, solutions can be stored in a database, enabling the configuration algorithms to query the database not only for image properties but also for known solutions of partial tasks. Communication between the stages is done using CGs. They can 'dock' to the current configurations to modify it according to the request (e.g. decrease a threshold used for pixel-classification if bigger regions are requested). Translation to visual languages (which have been become quite popular in recent years) is simple.

Fig. 3 shows the algorithm (i.e. the configuration of the inter-segment-features stage) to extract region R3 from the image in Fig. 2.

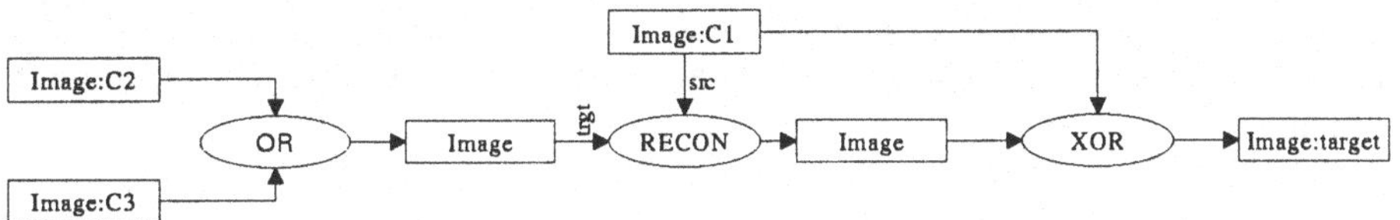

Fig. 3. Algorithm to extract target from example in Fig. 2.

It works in the following way, using just transformations on binary images at the pixel level: As there exist three shape classes in this image, the segment-classification stage separates the regions R1...R5 into three images: {R1, R3, R4} ∈ C1, {R2} ∈ C2 and {R5} ∈ C3. The inter-segment-feature stage now has to combine these images in a way that finally an image just containing R3 results.

[2] To avoid confusion: The NM and SM are used only for the configuration process, as their calculation is very expensive. If possible (for the current scene) they are not calculated during normal (repetitive) operation.

This is possible because R3 can be identified from the fact that it is the only region (of class 1) which does not include another region.
Separation is done in several steps: (1) R2 and R5 are merged into a (temporary) image using a pixel-wise OR. (2) The morphological reconstruction operator (see e.g. [3]) puts those regions of the `src` image into a new image that are (at least partially) covered by some regions of the `trgt` image (i.e. R1 and R4). (3) Finally the pixel-wise XOR of that image with the original one gives the desired result.

Bootstrap First the user enters his description of the scene by drawing a sketch of it as an overlay over an example image and specifying some region and relation properties (fixed/non-fixed region, variability of shape, size, etc.). The system then computes additional parameters, i.e. texture descriptors, shape parameters and finally inclusion and neighborhood relations. The pixel and shape classification stages are configured to use general classifiers (e.g. Maximum-Likelihood) which are now trained. Finally shape abstractions (elementary shapes, e.g. ellipses, rectangles) and spatial relations are computed, before the numerical model is translated to the CG representations, implemented by the CoGITaNT system ([2]): (1) Every region becomes a `region` typed (individual) concept. (2) Every class of texture and shape becomes an (individual) concept, to which the class features are linked via (subtypes of) `attr` relations. (3) The region concepts are linked to the corresponding class concepts. (4) Spatial relations (`inclusion`, `neighborship`) each introduce another set of graphs.
Now the system has a running configuration as far as no spatial relations have to be employed. Thus configuration starts at the inter-segment classification level, before trying to simplify the configurations of the other stages (e.g. replacing a general classifier by a threshold).

Configurating Algorithms Often a suitable configuration is not expressible as a chain of transformation operators (what many existing self-configuring systems assume).
As there exists no general algorithm to find a configuration for a stage, the system uses several alternative algorithms which range from very specific solutions (e.g. [5]) to general search techniques like genetic algorithms and genetic programming, as well as artificial neural networks for learning image transformations.
In either case feasible configurations are represented as conceptual graphs and stored together with a description of the task. By this the system will be able to solve known partial task without having to use expensive search algorithms again, even if they might have been used to find the solution for the first time. In this paper we can just give an example of how this can be done.

Example: Inter-segment Classification As in every (rule based) planning process the system has to be able to describe the result of the application of an action and to determine the set of possible actions.

In our example we consider just inclusion relations, the real system additionally covers spatial relations (neighborhoods).
The graph rule

```
[region:*r1]-
            ->(chrc)->[class:*c1]
            ->(inc)->[region:*r2]->(chrc)->[class:*c2]
```

$\Longrightarrow$

```
[class:*c1]<-(src)<-[Recon]-
                           ->(marker)->[class:*c2]
                           ->(result)->[class:*c3]<-(chrc)<-[region:*r1]
```

describes a morphological reconstruction operator which puts a region `r1` of class[3] `c1` which includes a region of class `c2` into a new class `c3`. Rules of this kind provide a link between image *features* and image *transformations*. By reasoning over features uniquely characterizing a region algorithms to isolate regions can be obtained. The problem is, however, to find proper pairs of features and algorithms.

3 Summary

To overcome some of the limitations of conventional image processing systems we introduced the paradigm of active image processing. An important feature is the division into interacting stages and their self-configuration. This carried out the need to formalize the representation of images at different levels of abstraction as well as the representation of the system configuration.. We showed how our implementation employs conceptual graphs to represent images, task specifications and configurations. Use of a common representation not only simplifies the formal description of the system but also provides a natural basis for the learning of solutions.
We thank the DFG (Deutsche Forschungsgemeinschaft) for supporting this project and LIRMM, Montpelier, for providing the CoGITaNT platform.

References

1. Dana H. Ballard and Christopher M. Brown. *Computer Vision.* Prentice-Hall, 1982.
2. David Genest and Eric Salvat. A Platform Allowing Typed Nested Graphs: How CoGITo Became CoGITaNT, Proceedings. In *Proceedings of ICCS'98*, volume 1453 of *Lecture Notes in Computer Science*, pages 154–161. Springer, 1998.
3. H.J.A.M. Heijmans. Mathematical morphology: A modern approach in image processing. *SIAM Review*, 37(1):1–36, 1995.
4. Kenneth H. L. Ho. Learning fuzzy concepts by examples with fuzzy conceptual graphs. In *Proceedings of the 1st Australian Conceptual Structures Workshop*, 1994.
5. J. Racky and M. Pandit. Automatic generation of morphological opening–closing sequences for texture segmentation. In *Proceedings of NSIP99*, to appear.

[3] we use 'class' synonymical with 'image', as the contents (regions) of every are put into separate images.

A Software System for Learning Peircean Graphs

Torben Bräuner[1], Claus Donner[2], and Peter Øhrstrøm[3]

[1] InterMedia
Aalborg University
Fredrik Bajers Vej 7 C, 9220 Aalborg East, Denmark
torbenb@intermedia.auc.dk

[2] Department of Communication
Aalborg University
Langagervej 8, 9220 Aalborg East, Denmark
cd@hum.auc.dk

[3] Centre for Philosophy and Science-Theory
Aalborg University
Langagervej 6, 9220 Aalborg East, Denmark
poe@hum.auc.dk

Abstract. In this paper, we describe a software system for learning graph-based reasoning. The system is designed for learning Peircean graphs in an interactive fashion that allows each student to proceed in his or her own pace. The student may use the built-in exercise library working from easy exercises towards harder ones, or may work at exercises of his or her own choice. The system has been successfully tested by a number of first-year students. Our software system comes in two versions: One for reasoning within propositional logic and one for reasoning within the modal logic S5.

1 Introduction

It was a major event in the history of diagrammatic reasoning when Charles Sanders Peirce (1839 - 1914) developed graphical methods for reasoning within what we now call propositional logic, first-order predicate logic, and modal logic. See [16] and ([9], Chapter 3.6). Peirce's line of work was taken up again in 1984 where conceptual graphs, which are generalisations of Peircean graphs, were introduced by John Sowa, [13]. Since then, conceptual graphs have gained widespread use within natural language processing, knowledge based systems, knowledge engineering and database design, among others. Also notable is the recent book [1] which witnesses a general interest in reasoning with diagrams.

Peirce completed his work on graphical systems for reasoning within propositional and predicate logic but left unfinished similar systems for various modal logics - see the account given in [8]. In the paper [2], a formal system of Peircean graphs for reasoning within the modal logic S5 was given, and it is likely that other graphs systems can be found using the Peircean idea of experimentation in logic. In the present paper, we describe a software system for learning graph-based reasoning within S5. The system is designed for learning Peircean graphs in an interactive fashion that allows each student to proceed in his or her own pace and it can be used by an individual student as well as by a class under an instructors supervision. The system has been successfully tested by a number of first-year students. Our software system comes in two versions: One for reasoning within propositional logic and one for reasoning within the modal logic S5. The version for the modal logic S5 is based on the formal system of Peircean graphs given in [2]. The software system is written in Visual Prolog 5.02 for Windows

When the system is activated, a window appears on the screen. At the top of the window there is a toolbar with buttons which are used to manipulate a graph represented in the window, cf. the inference rules. Thus, the window corresponds to what Peirce called "the sheet of assertions". Every attempt at manipulating the graph represented in the window is checked by the system. If an attempt at manipulating the graph is consistent with the inference rules, the graph is manipulated accordingly. If the attempt is illegal, a dialog box with an error message is generated. The student may use the built-in exercise library working from easy exercises towards harder ones, or may work at exercises of his or her own choice. When the student is working at an exercise graph from the built-in exercise library, the system checks after each manipulation whether the exercise graph in question has been derived. The system indicates successful derivation of the exercise graph by generating a dialog box with a message. The system then continues to the next exercise in the library.

Besides the software system described in the present paper, we only know of one other software system for learning Peircean graphs where a description is publically available. This is a software system developed by K. L. Ketner and described in the booklet [6]. The booklet has a disk with the software system attached and it also provides an introduction to Peircean graphs. Ketner's ideas as well as his system are certainly interesting. However, we consider our software system as a considerable improvement compared to Ketner's: His system runs under DOS 2.0 and the interaction between the system and the student is not fully graphical. For example, graphs are entered in linear style using the keyboard. Furthermore, Ketner's system is only for reasoning within propositional logic. The availability of another software system for learning Peircean graphs has recently been announced at the conceptual graphs mailing list (message from M. Goldstein at January 6th, 1999). Also this system runs under DOS, but it is more user friendly than Ketner's. However, the interaction is not fully graphical, and furthermore, the system is only for reasoning within propositional logic.

The paper is structured as follows: In the second section, we give the philosophical background of our software system. Graph-based systems for reasoning within propositional logic and the modal logic S5 are given in the third section. Then, in section four, we present our software system for learning Peircean graphs for the modal logic S5. Also, we describe how an example derivation can be carried out. In section five we describe a test of the system carried out by first-year students. Possible further work is discussed in section six.

2 Philosophical Background

According to Peirce a diagram should mainly be understood as "an Icon of intelligible relations", ([10], 4.531). He pointed out that his diagrams of what he called existential graphs represent "propositions, on a single sheet, and arguments on a succession of sheets, presented in temporal succession", ([11], p. 662), and that the system of existential graphs may "be characterized with great truth as presenting before our eyes a moving picture of thought.", (unpublished manuscript, MS 291; 1905). In a modern context these Peircean ideas certainly suggest the use of graphical computer interfaces for the representation of logical inference. In particular, the use of such tools would be in good accordance with Peircean thought for the purpose of logic teaching. In 1908 Peirce stated the following:

> The aid that the system of graphs thus affords to the process of logical analysis, by virtue of its own analytical purity, is surprisingly great, and reaches further than one would dream. Taught to boys and girls before grammar, to the point of thorough familiarization, it would aid them through all their lives. For there are few important questions that the analysis of ideas does not help to answer. The theoretical value of the graphs, too, depends on this. ([10], 4.619)

Following this line of thinking, logic should be taught by means of graphs. In fact, our personal experiences in logic teaching within the humanities also suggest that one may benefit a lot from the use of graphs in the teaching of such logic courses. As Morgan Forbes, ([4], p. 398), has pointed out, existential graphs have the advantage of looking like nothing our students have seen before. This substantiates the hope that various phobias related to mathematics may not be triggered when the graphs are presented. But it is much more important that the use of graphs helps the student to understand how the process of reasoning about the real world may in fact be modelled. In this way the use of graphs may help the students to focus on the very nature of rationality. This is very important since one purpose of logic teaching is a better understanding of the notion of rationality. The use of graphs strongly suggests that rationality should be viewed as a process.

As Robert W. Burch, [3] has pointed out Peirce's diagrammatical reasoning may be seen as some sort of game. He regarded the system of existential graphs as inseparable from a rather game-like activity which is carried out by two fictitious persons, the Graphist and the Grapheus. The two persons are very different. The Graphist is an ordinary logician, and Grapheus is the creator of the universe of discourse, who also makes additions to it from time to time, ([10], 4.431). The idea in the logical game is that the Graphist wants to transform the graph on the sheet of assertion using certain rules of transformation. Each time a tranformation is proposed by the Graphist, the Grapheus either accepts or rejects the proposal. In a modern context, the acceptance of the idea of logic as a game has obvious consequences for the way a system for logic teaching should be designed. This makes it likely that logic teaching may benefit from the wide distribution of computer games.

According to Peirce the use of diagrams in logic can also be compared with the use of experiments in chemistry. Just as experimentation in chemistry can be described as "the putting of questions to Nature", the experiments upon diagrams may be understood as "questions put to the Nature of the relations concerned", ([10], 4.530). As Peirce saw it, the ultimate aim of logic is to understand the nature of reasoning - in particular the nature of inference. According to Peirce, ([10], 7.276), there are three chief steps in the process of inference: 1) the putting together of facts 2) experimentation, observation, and experimental analysis 3) the generalization of experimental results. The first step is to state the basis of the inference to be carried out. The facts are all represented on the sheet of assertions. The second step is the same kind of process as the chemist uses in his laboratory. The chemist uses his physical apparatus to explore how the given materials can be transformed. In a similar way, the logician may employ the apparatus of diagrams to see what follows from the facts represented on the sheet of assertion. The third step is to create a series of transformations leading from the premisses to the conclusion which the logician wants to derive.

Peirce emphasized that in addition to this way of performing experiments upon the diagram using a given apparatus, there may be some other or higher kind of experimental work. The point is that sometimes further progress in science is blocked until a new method of experimentation has been invented. In logic the invention of such a new method corresponds to the construction of new kinds of representation and the introduction of new rules of transformation.

If Peirce's conception of philosophical logic is accepted, it seems reasonable to further investigate the possibilities of using computer-based tools not only for learning purposes but also for the purpose of Peircean experimentation in diagrammatical reasoning. Once a graphical system for multi-modal logic has been developed it can probably provide a platform to perform the kind of logical experiments which Peirce had in mind. In that case, the result will not only be a tool with which students cover more material and gain deeper understanding of logic and formal proof, but also an important tool for research in reasoning and philosophical logic.

3 Graphs for Propositional and Modal Logic

Before considering graphs, we give a short introduction to propositional and modal logic along the lines of [12]. We make use of the traditional Hilbert-Frege style. See also [5]. Formulae for propositional logic are defined by the grammar

$$s \ ::= \ p \mid s \wedge \ldots \wedge s \mid \neg(s)$$

where p is a propositional letter. Parentheses are left out when appropriate. The connectives $\wedge$ and $\neg$ respectively symbolises "and" and "it is not the case that". Given formulae ϕ and ψ, we abbreviate $\neg(\neg\phi \wedge \neg\psi)$ and $\neg(\phi \wedge \neg\psi)$ as respectively $\phi \vee \psi$ and $\phi \Rightarrow \psi$. It follows that $\vee$ and $\Rightarrow$ symbolises respectively "or" and "implies". We assume that a set of axioms and proof-rules for propositional logic has been given.

Formulae for modal logic are defined by extending the grammar for propositional logic with the additional clause $\Box(s)$. The connective $\Box$ symbolises "it is necessary that". Given a formula ϕ, we abbreviate $\neg\Box\neg\phi$ as $\Diamond\phi$. It follows that $\Diamond$ symbolises "it is possible that".

Definition 1. *The axioms and proof-rules for S5 are constituted by the axioms and proof-rules for propositional logic together with the following:*

K $\vdash \Box(\phi \Rightarrow \psi) \Rightarrow (\Box\phi \Rightarrow \Box\psi)$.
T $\vdash \Box\phi \Rightarrow \phi$.
S5 $\vdash \Diamond\phi \Rightarrow \Box\Diamond\phi$.
Necessitation *If* $\vdash \phi$ *then* $\vdash \Box\phi$.

It may be worth pointing out the relation between S5 and other well known modal logics. We get the modal logic S4 if axiom S5 is replaced by the axiom $\vdash \Box\phi \Rightarrow \Box\Box\phi$, called S4. It is straightforward to show that the logic S5 is stronger than the logic S4 in the sense that any formula provable in S4 is provable in S5 also. We get the modal logic T if axiom S5 is left out, and the modal logic K is obtained by leaving out S5 as well as T. The various modal logics described correspond to different notions of necessity. We shall here concentrate on the modal logic S5 which plays an important role within philosophy as well as within Artificial Intelligence and computer science in general.

In the following we give an account of propositional and modal logic using Peircean graphs. Graphs for propositional logic are defined by the grammar

$$s \ ::= \ p \mid s...s \mid \neg(s)$$

where p is a propositional letter. The number of occurrences of s in a string $s...s$ might be zero in which case we call the resulting graph *empty*. A graph can be rewritten into a formula of propositional logic by adding conjunction symbols as appropriate. The $\neg(...)$ part of a graph $\neg(\phi)$ is called a *negation context*. We say that a graph is *positively enclosed* in an enclosing graph if and only if it occurs within an even number of negation contexts. Negative enclosure is defined in an analogous fashion. A graph can be written in non-linear style by writing

$$\boxed{\phi}$$

instead of $\neg(\phi)$. In non-linear style, there is no order on graphs in an enclosing graph. Neither is the shape of a context of significance. A notion of derivation for propositional graphs is introduced in the definition below.

Definition 2. *A list of graphs ψ_1, ..., ψ_n constitutes a derivation of ψ_n from ψ_1 if and only if each ψ_{i+1} can be obtained from ψ_i by using one of the following inference rules:*

Insertion *Any graph may be drawn anywhere which is negatively enclosed.*

Erasure *Any positively enclosed graph may be erased.*

Iteration *A copy of any graph ϕ may be drawn anywhere which is not within ϕ provided that the only contexts crossed are negation contexts which do not enclose ϕ.*

Deiteration *Any graph which could be the result of iteration may be erased.*

Double Negation *A double negation context may be drawn around any graph and a double negation context around any graph may be erased.*

It is well known that Hilbert-Frege propositional logic is equivalent to the graph-based formulation in the sense that the same graphs/formulae are derivable, see for example the proof in [2].

Graphs for modal logic are defined by extending the grammar for defining graphs of propositional logic with the additional clause $\Box(s)$. The $\Box(...)$ part of a graph $\Box(\phi)$ is called a *modal context*[1]. Note that modal contexts do not matter for whether a graph is positively or negatively enclosed. We say that a graph is *modally enclosed* if it occurs within a modal context. In non-linear style we write

$$\boxed{\phi}$$

instead of $\Box(\phi)$. A notion of derivation for modal graphs is introduced in the definition below. The definition is taken from the paper [2].

Definition 3. *A list of graphs ψ_1, ..., ψ_n constitutes a derivation of ψ_n from ψ_1 if and only if each ψ_{i+1} can be obtained from ψ_i by using either one of the inference rules for propositional logic or one of the following:*

Negative Modality Introduction *A modal context may be drawn around any negatively enclosed graph.*

Positive Modality Introduction *A modal context may be drawn around any positively enclosed graph ψ which is not modally enclosed provided that each propositional letter which is within a negation context enclosing ψ, but which is not within ψ, is modally enclosed.*

Note that the condition regarding propositional letters in the rule Positive $\Box$-Introduction is vacuous if the graph ψ is not enclosed by any (negation) contexts at all. Also, note that in the rule for iteration only negation contexts can be crossed when copying a graph (modal contexts cannot be crossed). In [2] it is proved that the graph-based formulation of S5 is equivalent to a Hilbert-Frege formulation in the sense that the same graphs/formulae are derivable.

---

[1] A historical remark should be made here: Peirce's modal contexts correspond to $\neg(\Box(...))$ in our system. Our choice of definition for modal contexts deviates from Peirce's because we want to keep the notions of negation and necessity distinct. This is in accordance with Sowa's definition of modal contexts for conceptual graphs, [13]. It is straightforward to restate our graph-rules in terms of Peirce's modal contexts by adding negation contexts as appropriate and by defining positive and negative enclosure such that negation contexts as well as modal contexts are taken into account.

4 The Software System

In this section we present our software system for learning Peircean graphs. The software system comes in two versions: One for reasoning within propositional logic and one for reasoning within the modal logic S5. The latter software system is an extension of the former. We consider here the latter.

The system is activated by double-clicking the icon titled 'Graph'. After a moment, the window displayed in Figure 1 appears on the screen. A graph is

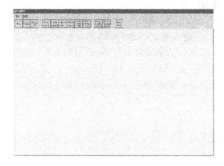

Fig. 1.

represented in the window. The window is empty when it appears; therefore the empty graph is represented. At the top of the window, there is a tool bar with a button for each inference rule (note that double negation involves two inference rules; one for insertion and one for erasure of double negation). See Figure 2. The buttons of the tool bar are used to manipulate the graph represented in the

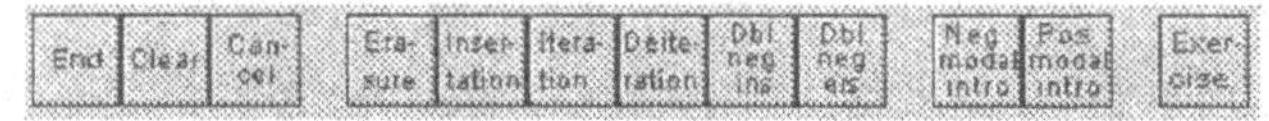

Fig. 2.

window, cf. the inference rules. Thus, the window corresponds to what Peirce called "the sheet of assertions". At the tool bar, there are also buttons titled 'Cancel', 'Clear', 'End' and 'Exercise'. The 'Clear' button clears the window, that is, the sheet of assertions. We shall return to the 'Cancel', 'End' and 'Exercise' buttons.

To manipulate the graph in the window according to an inference rule, the button corresponding to the inference rule in question is clicked and one or two arguments are entered. The number of arguments depends on the choice of inference rule. If two arguments are needed, they are entered one by one. A dialog box with an error message is generated by the system if the arguments entered are illegal. By clicking the 'Cancel' button, one can cancel the arguments entered and make the system ready for another click at a button corresponding to an inference rule.

When contexts and graphs in the window are needed as arguments, they are entered by marking them according to the following conventions: A context is marked by clicking it followed by clicking the 'End' button. An empty graph is marked by clicking its position while pressing the 'Shift' button at the keyboard followed by clicking the 'End' button. A non-empty graph is marked by clicking each of its outermost contexts and propositional letters followed by clicking the 'End' button. Thus, the 'End' button is used to end a sequence of clicks. The clicked contexts and propositional letters are highlighted by the system.

Below we describe each of the buttons corresponding to inference rules. We describe first the buttons corresponding to the rules for propositional logic.

Insertion takes as arguments a graph to be inserted and a context into which the insertion is to take place. The first argument is entered using a dialog box generated by the system. The second argument is entered by marking.

Erasure takes as argument a graph to be erased. The argument is entered by marking.

Iteration takes as argument a graph to be iterated and a context into which the iteration is to take place. The arguments are both entered by marking.

Deiteration takes as arguments a graph justifying the deiteration and a graph to be deiterated. The arguments are both entered by marking.

Double Negation Insertion takes as argument a graph to be double negated. The argument is entered by marking.

Double Negation Erasure takes as argument the outermost context of the double negation to be erased. The argument is entered by marking.

We describe next the buttons corresponding to the additional rules for the modal logic S5.

Negative Modality Introduction takes as argument a graph to be modalised. The argument is entered by marking.

Positive Modality Introduction takes as argument a graph to be modalised. The argument is entered by marking.

When carrying out the inference rule Insertion, a graph is entered using a dialog box generated by the system. This is the left hand side dialog box displayed in Figure 3. A graph is represented in the dialog box. The dialog box is empty immediately after generation, so the graph thus represented is empty. A negation context is entered at a certain position in the graph in the dialog box by clicking the 'Neg context' button followed by clicking the position in question. A modal context is entered in an analogous fashion using the 'Mod context' button. A propositional letter is entered at a certain position in the graph in the dialog box as follows: The 'Prop letter' button is clicked. This makes the system generate a second dialog box where the propositional letter is typed. This second dialog box is the right hand side dialog box displayed in Figure 3. The second

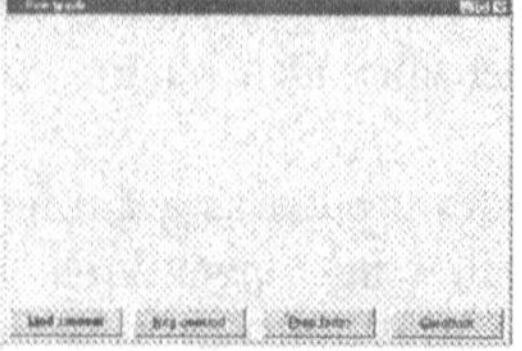

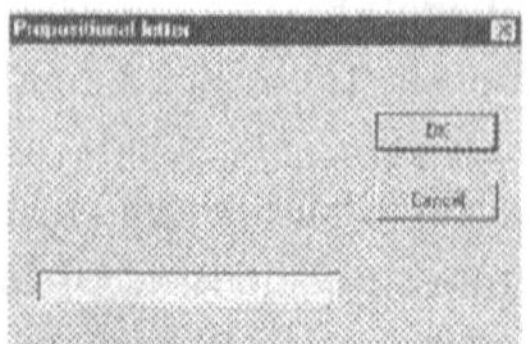

Fig. 3.

dialog box is closed by clicking the 'OK' button. Finally, the position in question is clicked. The dialog box for entering graphs is closed by clicking the 'Continue' button.

4.1 The Exercise Library

Clicking the 'Exercise' button activates the built-in exercise library. The system then generates a window where an exercise graph is represented. During activation of the exercise library, the student is expected to try to derive the graph in the window. Immediately after the exercise library has been activated, exercise graph number one is represented in the window. This is the easiest exercise. The exercises are ordered by hardness. When the student is working at an exercise graph from the library, the system checks after each manipulation whether the exercise graph in question has been derived. The system indicates successful derivation of the exercise graph by generating a dialog box with a message. The system then continues to the next exercise in the library. The built-in exercise library is deactivated by clicking the 'Exercise' button again.

4.2 An Example Derivation

In what follows, we will describe how an example derivation can be carried out in our system. We want to carry out a derivation of the graph corresponding to the formula $\neg(\Box p \wedge \neg p)$, that is, to $\Box p \Rightarrow p$. This is the axiom T. The derivation takes four steps. The starting point of the first step is the empty graph. This is the graph in the window displayed in Figure 1. We begin by clicking the 'Dbl neg ins' button. The system now expects the user to enter as the argument a graph to be double-negated. We enter the empty graph by marking it. The result is the graph in the left hand side window displayed in Figure 4.

In the second step, we begin by clicking the 'Insertion' button. This makes the system generate a dialog box. The system now expects the user to enter as the first argument a graph to be inserted. Using the dialog box, we enter the graph constituted by an occurrence of the propositional letter p. Following this, the system expects the user to enter as the second argument a context into which the insertion is to take place. We enter the outermost context by marking it. The result is the graph in the right hand side window displayed in Figure 4.

In the third step, we begin by clicking the 'Iteration' button. The system now expects the user to enter as the first argument a graph to be iterated. We enter

Fig. 4.

the graph constituted by the one and only occurrence of the propositional letter p by marking it, that is, by clicking the occurrence of p followed by clicking the 'End' button. Following this, the system expects the user to enter as the second argument a context into which the iteration is to take place. We enter the innermost contexts by marking it. The result is the graph in the left hand side window displayed in Figure 5.

Fig. 5.

In the fourth step, we begin by clicking the 'Neg modal intro' button. The system now expects the user to enter as the argument a graph to be modalised. We enter the graph constituted by the left hand side occurrence of the propositional letter p by marking it. The result is the graph in the right hand side window displayed in Figure 5 which is the one we wanted to derive.

5 Test of the System

A first testing of the propositional version of the software system has been carried out at Aalborg University. 22 first-year communication students and 10 first-year philosophy students participated in the test. These students had all earlier been introduced to basic propositional logic (3-4 hours), but they had never studied any proof theory. As a preparation to the test, the students were given an introductory lecture (3 hours) on the fundamentals of Peircean graphs. These lectures took the form of presentation of 1) how to translate between the standard formalism in propositional logic and the formalism of Peircean graphs, and 2) how propositional graphs can be manipulated with the rules of Erasure, Insertion, Iteration, Deiteration, and Double Negation. A few days after this introductory lecture the students were given a short introduction to the system in the computer lab. Each student was placed in front of a computer and asked to translate the following 5 tautologies into Peircean graphs and to prove them using the system:

(Problem 1) $p \Rightarrow p$
(Problem 2) $p \vee \neg p$
(Problem 3) $p \Rightarrow (p \vee q)$
(Problem 4) $((p \Rightarrow q) \Rightarrow p) \Rightarrow p$
(Problem 5) $((p \Rightarrow q) \wedge (q \Rightarrow r)) \Rightarrow (p \Rightarrow r)$

The students could try to apply any of the rules as often as wanted. All students were able to solve the problems within two hours. All computer operations of all students were logged during the test, and after the test each student was asked to answer a number of questions regarding the qualities of the system.

The material from the test has been analysed by us. Although the sample of students is small and the amount of data is rather limited, the material does seem to indicate that the following statements are likely to hold, or at least should be regarded as reasonable hypotheses to be tested with new and more detailed empirical investigations:

- The students mainly fall into two groups, A and B, of almost the same size. The A-students used the system as some kind of computer game, and they attempted to solve the problems by a trial and error method. The B-students, on the other hand, accepted the system as a tool for learning about graphical proofs in propositional logic. For the A-students the frequency of errors was almost the constant during the test. For the B-students the frequency of errors was clearly decreasing during the test time.
- The typical errors were: 1) attempts to erase sub-graphs from negatively enclosed areas, 2) attemps to insert sub-graphs in positively enclosed areas, 3) attempts to iterate sub-graphs over a "valley of contexts".
- Almost nobody could use deiteration correctly from the beginning. This rule seems to be rather difficult to understand and to apply.

It is very interesting that some of the students (mainly from the B-group) were able to find a very nice solution of (Problem 5) without the use of Deiteration. This was a surprise for us, since we had assumed that "the natural solution" of this problem would involve the use of deiteration. The graphs involved in crucial steps of the derivation solving (Problem 5) are displayed in Figure 6. The very fact that the several students were able to find this nice solution without making

Fig. 6.

errors in the proofs, indicate that they had in fact gained a remarkable level of familarity with proof techniques during the very short time they had worked with the Peircean graphs. On the other hand, it seems that the A-students learned next to nothing from their use of the system. They were simply able to get to the graphs to be proved by trial and error.

The tentative conclusion from the test we have carried out seems to be that it is in fact possible to teach students elementary techniques in proof theory by means of a system like the one we are developing. On the other hand, it is obvious that all or at least most of the students should belong to the B-students, if the system should be counted as an effective tool for logic teaching in the relevant context. So if one wants this kind of computer based logic teaching to be effective for all students, one has to make the strategy of the A-students impossible or at least very difficult. This would involve forcing students to reflect on their errors. This may be done using a series of error messages. It may even be worth considering making a special kind of error message in the communication with the user if she or he performs the same kind of error for the second or the third time. It is, however, an open question whether it is possible by such methods to convert all or most of the A-students to B-students.

If the system should provide a good understanding of the process of reasoning, it may not be sufficient that the system checks for logical errors in each move. In addition, the system should offer some training in the strategy of navigating toward a particular outcome of interest. It is not clear how this kind of training should be designed. Perhaps the system should be constructed such that if it could "suggest" a move at any stage in the process - like some chess programs can.

Like Stenning, Cox, and Oberlander, ([14], p.16), we strongly disagree with those who have described the use of graphical visualizations for teaching purposes as "an obvious case of curriculum infantilization". It is certainly very likely that most logic students can benefit from the use of graphical methods in logic teaching. On the other hand, it is also clear some student can benefit more from it than others. But it is likely that the use of graphical methods in logic teaching may guide the 'visual' thinker to a deeper understanding of logic and reasoning which he could only obtain in other ways with great difficulties.

6 Further Work

In the present paper a software system for learning Peircean graphs has been described. Following the paper [2], we have concentrated on a formal system for reasoning within the modal logic S5. We plan to consider systems of Peircean graphs for other modal logics as well. Here we in particular have in mind multi-modal logics such as tense logics. In a multi-modal logic, there are more than one modality. In the case with tense logics, there are usually two modalities, namely the past tense modality, "it was the case that" and the future tense modality, "it will be the case that". The past and future tense modalities are respectively symbolised as P and F. There might also be a third modality, namely the so-called Ockhamistic necessity modality, "it is now unpreventable that".

The founding father of tense logic was Arthur N. Prior (1914–1969). See the account given in [9]. The leading idea motivating Prior's work was that linguistic items like 'Socrates is sitting' are already complete sentences expressing propositions. This has to be compared to the traditional idea that such items are predicates expressing properties of instants. Putting a sentence in past or future tense was symbolised by Prior using the modal operators P and F. Thus, if p symbolises 'Socrates is sitting', then Pp and Fp respectively symbolises 'Socrates was sitting' and 'Socrates will be sitting'. So the development of tense logic was originally motivated by philosophical and linguistic considerations. But nowadays tense logic also plays an important role within Artificial Intelligence and computer science.

We would like to design software systems for learning tense-logical graphs analogous to the software system for learning S5 graphs which we have described in the present paper. However, to design a software system for learning graphs for a given logic, it is obviously a prerequisite to have a formal system of Peircean graphs for the logic in question. The software system described in the present paper is based on the proof-theoretically motivated formal system for S5 given in [2]. In the mentioned paper it is shown how the inference rules for the given graph-based formulation of the modal logic S5 are analogous to Gentzen style rules as appropriate for S5 (Gentzen style is one way of formulating a logic which is characterised by particularly appealing proof-theoretic properties). Alternatively, one can use graph-based analogues of Hilbert-Frege axioms and rules for a logic. This is what is behind the graph-based formulations of the modal logics K, T and S4 given in [15]. Also, this is what is behind the graph-based formulations of various tense logics given in [7]. We shall leave this highly interesting issue to further work.

References

1. G. Allwein and J. Barwise, editors. *Logical Reasoning with Diagrams.* Oxford University Press, 1996.
2. T. Braüner. Peircean graphs for the modal logic S5. In M.-L. Mugnier and M. Chein, editors, *Proceedings of Sixth International Conference on Conceptual Structures*, volume 1453 of *LNAI*, pages 255–269. Springer-Verlag, 1998.
3. R. W. Burch. Game-theoretical semantics for Peirce's existential graphs. *Synthese*, 99:361–375, 1994.
4. M. Forbes. Peirce's existential graphs. A practical alternative to truth tables for critical thinkers. *Teaching Philosophy*, 20:387–400, 1997.
5. G. E. Hughes and M. J. Cresswell. *An Introduction to Modal Logic.* Methuen, 1968.
6. K. L. Ketner. *Elements of Logic: An Introduction to Peirce's Existential Graphs.* Arisbe Associates, 1996.
7. P. Øhrstrøm. Existential graphs and tense logic. In *Proceedings of Fourth International Conference on Conceptual Structures*, volume 1115 of *LNAI.* Springer-Verlag, 1996.
8. P. Øhrstrøm. C. S. Peirce and the quest for gamma graphs. In *Proceedings of Fifth International Conference on Conceptual Structures*, volume 1257 of *LNAI.* Springer-Verlag, 1997.
9. P. Øhrstrøm and P. Hasle. *Temporal Logic: from Ancient Ideas to Artificial Intelligence.* Kluwer Academic Publishers, 1995.
10. C. S. Peirce. In C. Hartshorne, P. Weiss, and A. Burke, editors, *Collected Papers of Charles Sanders Peirce*, volume I-VIII. Harvard University Press, 1931-58.
11. D. D. Roberts. The existential graphs. *Computers Math Applic.*, 23:639–663, 1992.
12. D. Scott, editor. *Notes on the Formalisation of Logic.* Sub-faculty of Philosophy, University of Oxford, 1981.
13. J. F. Sowa. *Conceptual Structures: Information Processing in Mind and Machine.* Addison-Wesley, Reading, 1984.
14. K. Stenning, R. Cox, and J. Oberlander. Contrasting the cognitive effects of graphical and sentential logic teaching: Reasoning, representation and individual differences. *Language and Cognitive Processes*, 10:333–354, 1995.
15. H. van den Berg. Modal logics for conceptual graphs. In *Proceedings of First International Conference on Conceptual Structures*, volume 699 of *LNAI.* Springer-Verlag, 1993.
16. J. Zeman. Peirce's graphs. In *Proceedings of Fifth International Conference on Conceptual Structures*, volume 1257 of *LNCS.* Springer-Verlag, 1997.

Synergy : A Conceptual Graph Activation-Based Language

Adil KABBAJ

INSEA, Rabat, Morocco, B.P. 6217
Fax : (212) 7 77 94 57
akabbaj@insea.ac.ma

Abstract. This paper presents the core of *Synergy*; an implemented visual multi-paradigm programming language based on *executable Conceptual Graph* (CG). Execution is based on a *CG-activation mechanism* for which *concept lifecycle*, *relation propagation rules* and *referent instantiation* constitute the key elements. In this paper we define the activation mechanism and the CG structure (concept, relation, context, co-reference) used in Synergy as well as the concept type definition, the encapsulation mechanism and the knowledge base of Synergy. Examples are given to illustrate some aspects of the language. Hybrid object-oriented and concurrent object-oriented use of Synergy are presented in other papers [9, 10].

1 Introduction

Conceptual Graph (CG) theory is proposed by Sowa [18, 19] as a graphic system of logic and recently as a CG Interchange Format (CGIF) [21]. In [19], Sowa notes that many popular diagrams (e.g., type hierarchies, dataflow diagrams, state-transition diagrams, Petri-Nets, etc.) can be mapped to CG and in [20] he proposes CG as a logical foundation for object-oriented systems. Thus, CG is presented by Sowa as a formalism for the representation of knowledge (both declarative and procedural) with a logical interpretation; execution should be done with proof techniques of logic using rules of inference. Other CG execution mechanisms have been proposed in literature [1, 3, 4, 7, 13, 14, 15, 16].

In [7, 8] *a CG activation-based mechanism* is proposed as a computation model for *executable conceptual graphs*. Unlike the other proposed mechanisms, our CG activation mechanism has been used to produce an *implemented* CG general-purpose programming language.

Activation-based computation is an approach used in visual programming, simulation and system analysis where graphs are used to describe and simulate sequential and/or parallel tasks of different kinds : procedural, process, functional, event-driven, logical and object oriented tasks [2, 11, 12, 17, 23, 24].

Activation-based interpretation of CG [7, 8] is based on *concept lifecycle*, *relation propagation rules* and *referent instantiation*. A concept has *a state* (which replaces and extends the notion of control mark used by Sowa [18]) and the concept lifecycle is defined on the possible states of a concept. Concept lifecycle is similar to process lifecycle (in process programming) and to active-object lifecycle (in concurrent object oriented programming), while relation propagation rules are similar to propagation or firing rules of procedural graphs, dataflow graphs and Petri Nets [8].

Since 1995, we have designed and implemented a visual language, called *Synergy* that is based on such an activation mechanism. Synergy is a visual multi-paradigm programming language; it integrates functional, procedural, process, reactive, object-oriented and concurrent object-oriented paradigms. The integration is done using CG as the basis knowledge structure, without actors or other external notation.
Synergy is designed for visual programming, modeling and simulation. It can be used for different purposes and in many fields : programming languages, simulation, database, information systems, software engineering, knowledge acquisition, knowledge base systems, intelligent tutoring systems and multi-agent systems. Synergy has been used for the development of a concurrent-object oriented application (e.g., a visual agent-oriented modeling of the Intensive Care Unit) and in modeling some components of an intelligent tutoring system [9]. It has been used also for the specification of a multi-agent system [25]. A hybrid object-oriented use of Synergy is illustrated in [10].

The Synergy language constitutes the main part of the SYNERGY environment [8] ; a CG tool composed of a CG graphic editor, the Synergy language, an engine for dynamic formation of knowledge bases and an information retrieval process [8]. The Synergy language and its environment have been implemented with Microsoft Visual C++ and a Visual J++ implementation will be available in summer 1999. The web SYNERGY site (www.insea.ac.ma/InseaCGTools/Synergy.html) is under construction and a SYNERGY group is being formed to enhance, extend and use SYNERGY in many ways.

This paper presents the core of the current version of Synergy. Section 2 defines the notion of Synergy application (e.g., a Synergy program) and the CG structure used in Synergy. It introduces also and briefly the primitive operations of the language. Section 3 introduces type definitions and encapsulation that enables the definition of a modular knowledge base. The use of co-references in Synergy is introduced in section 4. Section 5 describes the CG activation mechanism of Synergy and its interpreter; how CGs are executed. Concept lifecycle and relation propagation rules are introduced in this section. As an illustrative example, section 6 presents an application in project management. Finally section 7 gives a conclusion with an outline of some current and future works.

2 The basic elements of Synergy

Definition 1. An *application* (e.g., a program) in Synergy is represented as a CG with two concepts/contexts : [LongTermMemory] and [WorkingMemory]. The description of the first context represents the knowledge base (KB) of the application and the description of the second represents the working space where the user specifies his requests. A Request is represented by a CG and it is interpreted or executed according to the content of the KB and to the primitive types (built-in) hierarchy.

Definition 2. The context [LongTermMemory] contains the *knowledge base (KB) of a Synergy application* which is represented by a CG. The KB corresponds to concept types hierarchy augmented by type definitions, individual specification, types schemas and types synonyms.

In this paper, we consider a KB composed of concept types hierarchy with only type definitions and individual specifications. The KB of a new application will contain only the concept [Universal]. Using the CG graphic editor of SYNERGY, the user can add new concept types with their definitions eventually. Fig. 1 shows the new application "Sample1.syn" with the content of its [LongTermMemory] (e.g., its KB) after the addition of some concept types with definitions of some of them (like [BscProc :_BasicProcedures] and [Factorial :_Fct]).
To have a concise KB, a Synergy KB (for instance, the KB in Fig. 1) contains only the information given by the user, the primitive types are kept apart in the primitive types (built_in) hierarchy.

Definition 3. A *CG* in Synergy is a set of concepts related eventually by dyadic relations.

Fig. 1 illustrates different cases of CG : a) the top-level description of an application (like "Sample.syn") is a CG composed of two concepts without relations, b) the contain of the context [LongTermMemory] is a connected CG (which represents the KB) and c) the definition of "Factorial" contains a connected CG with three unconnected concepts.

Definition 4. A *concept* is described by four fields :
[Type :Referent =Description #State]. The referent, if specified, can be an identifier of an instance, a variable identifier (that can be bound to a referent) or a co-referent. A description, if specified, can be a simple data (for example, a real, a boolean or a string) or a composed data (a CG). The state field is defined on the following set : {steady, trigger, check-preconditions, wait-preconditions, wait-value, in-activation, wait-affectation, wait-end-affectation}.

Remarks: Nothing is really specific to Synergy in definition 4 ; the description field of a concept is already used by Sowa [19, 20], Esch [5], Ellis [4] and others. The state field can be viewed as a generalization of the control mark introduced by Sowa in [18].

Remark. As illustrated by Fig. 1, the CG graphic editor of SYNERGY represents a concept by a rectangle. To have a concise description of a CG, a relation is not encircled with a circle and the description and the state of a concept are hidden. They are shown as required by the user. An iconic indication is given however : if the concept has a description, a small triangle is shown at the left-bottom of the rectangle. If the description is a CG then the surface of the triangle is red and it is blue if the description is a simple data (unfortunately the colors are not printed !). Also, each state of a concept has a visual representation. For instance, the rectangle is green if the concept state is "trigger" and it is red if the state is "in_activation".

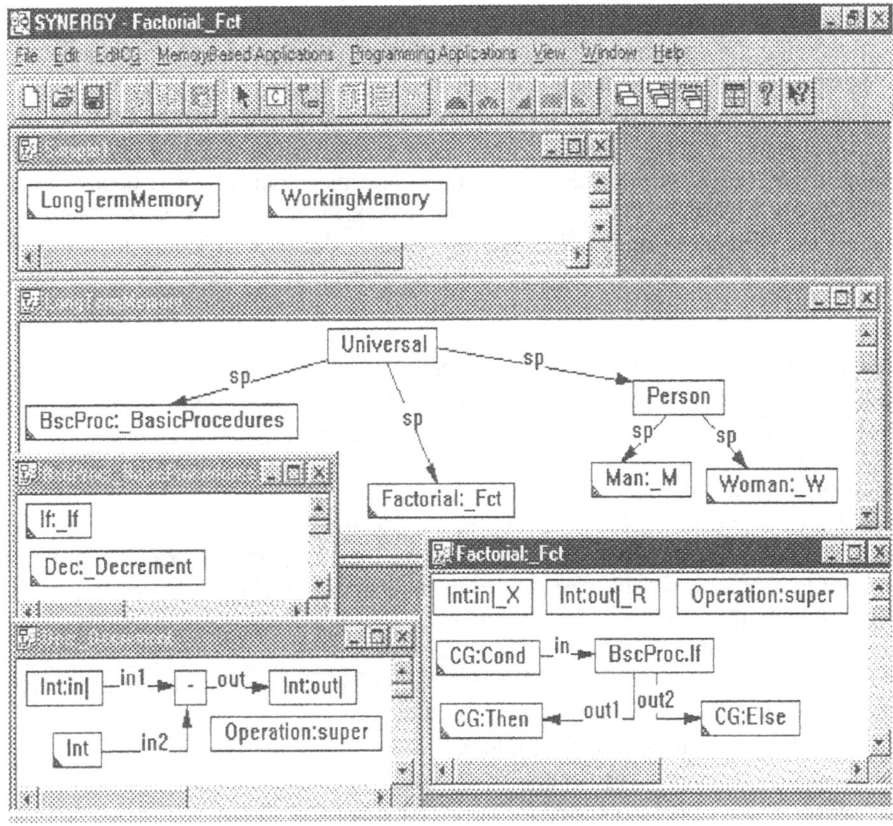

Fig. 1. The "Sample1.syn" application.

The type of a concept can be a primitive type, specified in the Synergy primitive types hierarchy, or it can be a defined type, specified by the user in the KB of the application. A concept type definition is considered in section 3.

Definition 5. The *primitive concept types hierarchy* is composed of primitive data types ("Int", "Real", "String", "Bool", "CG", "Image", "Sound", "Video", "Text", "Window") and primitive operation types (assignation, I/O, arithmetic, relational, boolean, list, CG, type_hierarchy, concept, relation and Multi-Media) :

- A *list* is represented as a CG and *list operations* manipulate the CG as a list of elements.
- *CG operations* correspond to "ExpandConcept", "ContractDef" and to the generic operation "MatchCG" and the derived ones (Project, Subsume, Unify,

MaximalJoin, Specialize, Generalize and other variants). CG operations are defined on simple and compound CG with co-referents.

- *Type_hierarchy operations* are "IsSuperTypeOf", "MinimalCommonSuperType" and "MaximalCommonSubType".
- *Concept operations* are subdivided in two categories : operations that have an access to a specific field of a concept (especially the type, the referent or the state fields like the operations "GetType", "SetType", "GetState" and "SetState") and operations that delete or add a new concept to a given CG.
- *Relation operations* correspond to delete or to add a new relation between two concepts of a given CG.
- *Multi-Media operations* : the Visual C++ implementation of these operations is based on a subset of the Microsoft device-independent "Media Control Interface" (MCI). With these operations, Synergy enables the user to create multi-media applications. We are considering the implementation of these operations with Visual J++.

Beside the primitive data and operation types, Synergy provides a set of "primitive" relations, called the Synergy-Relation-Set (SRS) = {"in", "out", "grd", "next", "sp", "inst", "sta", "synm"}.

Definition 6. A *relation* between two concepts can be either an element of the Synergy-Relation-Set (SRS) or it can be an identifier provided by the user. The primitive relations "in", "out" "grd" (guard) and "next" are data/control relations and the primitive relations "sp" (specializationOf/generalizationOf), "inst" (instanceOf), "sta" (situationFor) and "synm" (synonymOf) are "knowledge base relations"; they are used to organize a KB of a Synergy application.
Note : in each of the following statements, C and C' represent respectively the source and the target concepts of the relation in question.
C —in→ C' : C is an input argument for C'.
C —out→ C' : C has C' as an output argument.
C —grd→ C' : C is a precondition for the execution of C'.
C —next→ C' : After the execution of C, execute C'.
C —sp→ C' : the type of C' is a sub-type of the type of C and the description of C' is a specialization of the description of C. The two descriptions represent the definition bodies of the types of the two concepts.
C —inst→ C' : C' represents a specific referent of the type of C.
C —sta→ C' : C' represents a situation (e.g., a schema) for the type of C.
C —synm→ C' : C' represents the list of synonyms for the type of C.
C —R[,α]*→ C' : R represents a data/control relation ("in", "out" or "grd") and α is either "f" the functional attribute, "/" the cut_forward_propagation attribute or, "\" the cut_backward_propagation attribute. R[,α]* means that the relation R can have zero-to-many (three) of the above attributes (for instance, C —in,f,/→ C' means that C' will consume the description of C –according to the semantic of the functional attribute– and that the computation of a new value for C will not trigger C' –due to the presence of the cut_forward_propagation "/" attribute–).

Multiple-assignment and *procedural interpretation* of "in", "out" and "grd" relations are considered by default in Synergy: for instance, with C1—in→C2 or C3—grd→C4, the concept C2 (or C4) will consult the description of C1 (or of C3 for C4) without consuming it. Also, with C5—out→C6, C5 can assign a new description (value) to C6 even if this later has already one.
Functional interpretation of the above three relations will be adopted however if the user post-fixes the relation name with the optional functional attribute "f" : with C1—in,f→C2 or C3—grd,f→C4, the concept C2 (or C4) will consult the description of C1 (or of C3 for C4) and consume it. And with C5—out,f→C6, C5 can assign a new description to C6 only if C6 has not one, otherwise C5 will wait for the consumption of C6 description.

Fig. 1 illustrates the use of the relation "sp" to describe the KB and the use of data relations ("in" and "out") to relate input and output arguments to the corresponding operations. For instance, the definition of the operation [Dec :_Decrement] contains a call to the primitive operation [-] which has two inputs and one output arguments. Data and control relations are illustrated more fully in section 6.

3 Concept type definition and encapsulation

Definition 7. *Concept type definition* "**type** Type(_x) **is** CG" is represented in Synergy as a concept : [Type :_x = CG]. The concept type with its definition should be added to the KB of the application and it should be related to its super-type(s) with the relation "sp".

Fig. 1 gives the definition of three types : "BscProc", "Dec" and "Factorial".
If the defined type represents a treatment, then the user should specify the concept [Operation :super] or [Process :super] in its definition to indicate how the type should be interpreted. For instance, the types "Factorial" and "Dec" in Fig. 1 are defined as a kind of operation (e.g., once the operation terminates its execution, its description is destroyed as the case for the "record activation" of a procedure in procedural programming. This is not the case however if the treatment is defined as a process).
Also, if a concept type represents a treatment with arguments, then the corresponding parameters should be specified in the definition of the type. The type "Factorial" for instance (Fig. 1) has one input parameter ([Int: in|_X]; its referent is prefixed with "in|") and one output parameter ([Int :out|_R]; its referent is prefixed with "out|").

Synergy enables the user to define new concept types and moreover, *the type can be defined in the context of another type definition*. Fig. 1 shows an example: the operations [If] and [Dec] are defined in the context [BscProc].

Definition 8. *Encapsulation/Modularity principle* is incorporated in Synergy by allowing a *concept type to be defined in the context of another concept type definition*.

Such a contextual definition of a concept type has been proposed and used by Sowa [20] to model object-oriented encapsulation principle. The same use of context is done in Synergy, for both object-based and object-oriented programming. The encapsulation principle is very relevant to the efficient organization of a KB; very often "auxiliary" concept types are defined only to enable an abstract definition of other concept types. Those auxiliary types should be encapsulated to allow for a more concise KB. With definition 8, Synergy provides such a possibility.
Definition 8 above leads to the following extension in the formulation of a concept type :

Definition 9. The *type of a concept* can be *simple*, of the form "Id" where "Id" is the identifier of a primitive or defined type, or it can be *composed*, of the form "Id1.Id2" where "Id1" is either a type identifier or an instance identifier, and "Id2" a type identifier. If "Id1" is a type identifier then "Id2" is defined in the context of "Id1" definition. If "Id1" is an instance identifier then "Id2" is defined *in the class of* the instance "Id1".

In the definition of "Factorial" (Fig. 1), the concept [BscProc.If] has a composed type, e.g., the type "If" is defined in the context [BscProc]. Composed type "Id1.Id2" where "Id1" is an instance identifier concerns message or method call and it is considered in [10].

4 The use of co-reference in Synergy

In the CG theory, the pair (*x, ?x) shows a co-reference link between co-referent concepts : the concept with referent "*x" indicates the first occurrence or the *defining node* of the variable x while a concept with referent "?x" is a *bound node* and it indicates a subsequent reference to the concept where the variable is defined [19, 20]. In [5] Esch notes that "the basic thing to remember about contexts and co-reference is that it closely models scope of variables in block structured languages". This remark is especially true for Synergy and as discussed below, it is extended to other situations.
In Synergy, the pair (x, x.) is used instead of (*x, ?x). For instance, the concept [Employee : Employee2] is the *defining concept* for the referent "Employee2" while the concept [Employee : Employee2.] is a *reference concept.*
Synergy provides an extended form of co-reference called *composed co-reference* : the concept [String : Employee2.Address.Zip] (e.g., a reference to the "Zip of the Address of Employee2") is a reference to the defining concept of the referent "Zip" which is (or should be) contained in the description of the defining concept of the referent "Address" which is (or should be) itself contained in the description of the defining concept of the referent "Employee2".
A composed co-referent, like "Employee2.Address.Zip" is similar to the composed identifier in Pascal for example (e.g., Employee2.Address.Zip : Zip could be a field of the record Address which is a field of the record Employee2).

Definition 10. A *co-referent* can be *simple*, of the form "Id." where "Id" is the referent of the refereed concept, or it can be *composed*, of the form "Id1.Id2. ... .IdN-1.IdN" which specifies the path to follow, through the embedded contexts refereed by the associated referents "Id1", "Id2", ..., to reach the concept with the referent "IdN" which should be contained in the context refereed by "IdN-1".

The example of section 6 illustrates the use of composed co-referents. A more detailed illustration is given in [7, 8, 9, 10] where composed co-referents are used in the context of object-oriented and concurrent object-oriented applications.

Composed co-referent is an important feature specific to Synergy. It enhances the expressive power of the language and it enables the specification of a reference to a concept that is not present at the specification time. A simple or composed co-referent in Synergy specifies how to determine at the execution time the defining concept. Co-referent resolution procedure interprets the co-referent identifier to establish, at the execution time, the co-reference link. Co-referent resolution is described in [8, 10].

5 The CG activation mechanism and the interpreter of Synergy

Interpretation in Synergy is based on *CG or context activation*. Indeed, initially the Synergy interpreter starts with the interpretation of the working memory context [WorkingMemory]; the first active context. This latter contains the user requests.
Context activation or interpretation corresponds to the interpretation of its description ; the activation of a CG (which could be set of CGs). CG activation begins with *the parallel* interpretation of some concepts of the CG, then data and control relations (if they are present) spread or propagate the activation through the graph. Interpretation of a concept is done according to its *lifecycle* while the spreading activation is conform to the *relations propagation rules.*

We explain briefly the notion of parallelism in Synergy and then we define the notions of concept lifecycle and relations propagation rules.

Parallelism and the Synergy interpreter
As noted above, the activation of a context involves *parallel activation* of some of its concepts, each concept is activated according to its lifecycle. Thus, *the interpreter has to simulate a parallel interpretation of concept lifecycles.*
More specifically, a concept can represent a primitive operation or a context. Hence, the Synergy interpreter should manage *the parallel* activation of several contexts and/or the execution of several primitive operations.
The association of a state to a concept is the main reason that makes such a parallelism possible. In effect, each concept is interpreted (e.g., evolves, changes) according to its lifecycle, the activation (or interpretation) of the context is thus decentralized.
As a parallel language, Synergy enables the user to write parallel programs, its interpreter is able to execute those programs and its graphical environment allows the

user *to see and debug* the parallel activation of those programs. This "conceptual parallelism" is at the design/language level. It can be implemented on a parallel machine (e.g., "physical parallelism") or it can be interpreted and simulated on a sequential machine. In our current implementation of Synergy, we have considered the second option. Of course, some constraints have been assumed, especially for parallel execution of primitive operations. For instance, we assumed that during the execution of a primitive, no interaction could occur between the primitive and the other parts of the Synergy "program".

Concept lifecycle

Concept lifecycle is a state transition diagram where states correspond to the possible states of a concept and transitions to the conditions/actions on the concept and on the data/control relations linked to the concept.

The concept C in the "trigger" state is asked to "determine its description and to execute it". The concept C can have already a description, or (if not) its type is defined and in this case the description will be determined by instantiation, or its type is a primitive operation type. If none of these possibilities hold, the interpreter will attempt to compute the description by backward propagation (through the "out" relation) and the concept C becomes at state "wait-value".

Once C has a description or its type is a primitive operation type, its state changes in general from "trigger" to "check-preconditions"; the state where the concept C checks its preconditions (e.g., C' is a precondition for C if : C' —in→ C or C' —grd→ C). A concept C' that represents a precondition for the concept C can be triggered (by backward propagation through the "in" or "grd" relation) and C changes its state to "wait-preconditions".

If all the pre-conditions are satisfied, C becomes at state "in-activation" and its description is executed. If the type of C is a primitive operation then the primitive is executed, if the description of C is a CG then the CG is executed (*in parallel* to other active CGs), if the description is a primitive data (like an integer or a string) then its execution is null.

Once the concept C terminates its execution, its state changes to "wait-affectation" if it has results to affect to the corresponding output arguments. After the affectation step, the concept C returns to the "steady" state.

A more detailed description of concept lifecycle is given in [8].

Data/control relations propagation rules

Relation propagation rules apply only to data and control relations as well as co-reference link (viewed as a "virtual" data relation between a bound node C1 and its defining node C2 : C1 –coref→ C2).

A data/control relation can propagate a state forward (from the source to the target concept of the relation) and/or backward (from the target to the source of the relation).

Forward propagation rule. *If* an active context contains a branch C1 –Rel→ C2 where the state of C1 is "trigger" and Rel ∈ {"in", "out", "next", "coref"} *then* the relation Rel will propagate forward the state "trigger" to the concept C2.If Rel = "grd" then the propagation will be done only if the description of C1 is different from "false".

Backward propagation rule. *If* an active context contains a branch C1 –Rel→ C2 where the state of C2 is "trigger" and Rel ∈ {"in", "out", "grd", "coref"} *then* the relation Rel will propagate backward the state "trigger" to C1 if one of the following conditions is satisfied :

- Rel = "in" and the type of C2 is a strict activity (e.g., C2 description can be executed only if all its input arguments have a description) and C1 is at state "steady" and it hasn't a description.
- Rel = "out" or "coref" and C2 hasn't a description, its type isn't a primitive operation and it isn't defined (e.g., its description cannot be determined by instantiation).
- Rel = "grd" and C1 is at state "steady" and it hasn't a description.

Forward and backward propagation role of a relation Rel can be inhibited by the optional attribute "cut_forward_propagation; /" ("Rel,/") or "cut_backward_propagation, \" ("Rel,\"). For instance, if an active context contains C1 –in,/→ C2 and the concept C1 is at state "trigger" then the relation "in,/" will not propagate the state "trigger" to the concept C2.
With the optional attributes "/" and "\", the user can control the propagation of concepts activation in a CG. This is similar to the use of the "cut" to control the backtracking in the PROLOG language.
A more detailed description of CG activation and the Synergy interpreter is given in [8].

6 An example : a method for project schedule

To improve the skills and techniques for project management, Spinner [22] presents a method for developing the project schedule. We describe briefly this method and then its implementation with Synergy. Several features of the core of Synergy are illustrated by this example : data and control relations, co-references, operation treatment, process treatment and lazy treatment, forward and backward activation and other features.
Project schedule is based on the specification of the project network which describes the dependence between several activities. Each activity is associated to an arc and it has two attributes : the "time estimate" for the activity (which is given) and the "total float" which is computed. Each node of the network is described by two attributes too: the "earliest start time" and the "latest finish time" which are also computed, except for the earliest start time of the first node.
The first step of the method is to calculate the timing of each activity : compute for each node the earliest start and the latest finish times. The computation of the earliest start time (identified later by Erlst_st) for each node N is defined as follows :
If the node N is the target of only one arc : M –ActivityM→ N *Then*
N.Erlst_st := M.Erlst_st + ActivityM.TimeEstimate; /* TimeEstimate of ActivityM */

Else N.Erlst_st := Maximum((M1.Erlst_st + ActivityM1.TimeEstimate), ...,
(Mj.Erlst_st + ActivityMj.TimeEstimate)); /* we
consider all the predecessor nodes of N */

The value of the earliest time (Erlst_st) attribute of a node depends on the values of the Erlst_st attributes of the predecessor nodes. Hence, with a kind of forward propagation starting from the first node, the Erlst_st attribute of each node can be computed. The computation of the latest finish time (Ltst_ft) attribute of each node is defined as follows :

If the node N is the end node of the network *Then* N.Ltst_ft := N.Erlst_st
Else If the node N is the source of only one arc : N –ActivityM→ M *Then*
N.Ltst_ft := M.Ltst_ft - ActivityM.TimeEstimate;
Else N.Ltst_ft := Minimum((M1.Ltst_ft - ActivityM1.TimeEstimate), ...,
(Mj.Ltst_ft - ActivityMj.TimeEstimate)); /* we consider all the
successor nodes of N */

The value of the latest finish time (Ltst_ft) attribute of a node depends on the values of the Ltst_ft attributes of the successor nodes. With a kind of forward propagation starting from the end node, the Ltst_ft attribute of each node can be computed. Note that the value of the Ltst_ft of the end node corresponds to the value of its Erlst_st attribute.

Thus, the first forward propagation should be done first, to compute the Erlst_st of each node and then the second forward propagation should follow to compute the Ltst_ft attribute as well as the "total float" of each activity M –Activity→ N : N.Ltst_ft - M.Erlst_st.

Our approach in the implementation of this method is to incorporate this latter in the project network so that the network could perform itself the steps of the method.

To perform the two forward activation of the project network, we introduce a process P as indicated in Fig. 2.

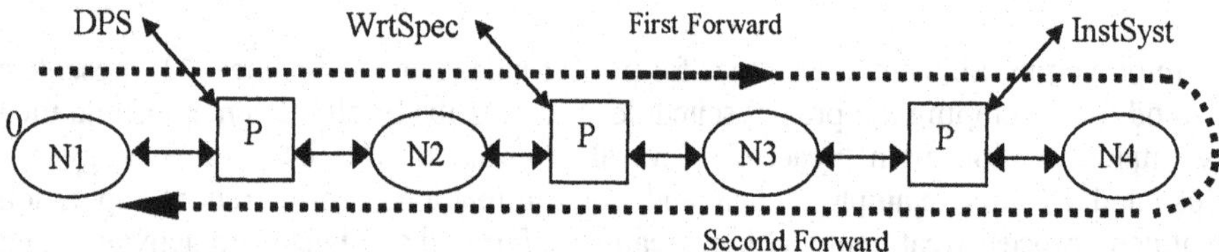

Fig. 2. A simple project network that incorporates directly, thanks to the process P, the Spinner method. In this Fig., activity DPS stands for "Design Pump System", WrtSpec for "Write Specifications" and InstSyst for "Install System".

During the first forward activation of the network, the process P will take as inputs the Erlst_st attribute of a node Nj (Nj ↔ P ↔ Nj+1) and the TimeEstimate of the corresponding activity to compute the Erlst_st attribute of Nj+1. And during the second forward activation of the network, P will take as inputs the Ltst_ft attribute of Nj+1 and the TimeEstimate of the corresponding activity to compute the Ltst_ft attribute of Nj and then to compute the TotalFloat attribute of the activity.

The first forward activation is triggered by the value of the Erlst_st attribute of the first node N1 and the second forward activation is triggered by the value of the Ltst_ft attribute of the end node which is equal to the value of its Erlst_st attribute.

Fig. 3 shows the KB (knowledge base) of the Synergy application that implements the above approach. The type "Node" is defined as an abstract entity with two attributes (the concept [Int: Erst] which represents the Erlst_st attribute and the concept [Int: Lft] which represents the Ltst_ft attribute). A similar definition is given for the type "Job" with its two attributes (the attribute TimeEstimate represented by the concept [Int: Etm] and the attribute TotalFloat represented by the concept [Int: TFlt]). Fig. 3 shows also the description of the simple project network of Fig. 2 ([Project_Network: ProjNetwork1] in Fig. 3). The network is represented as a CG and the abstract bi-directional link used in Fig. 2 (↔) is replaced by the detail of the connection; connections between the process P and its arguments.

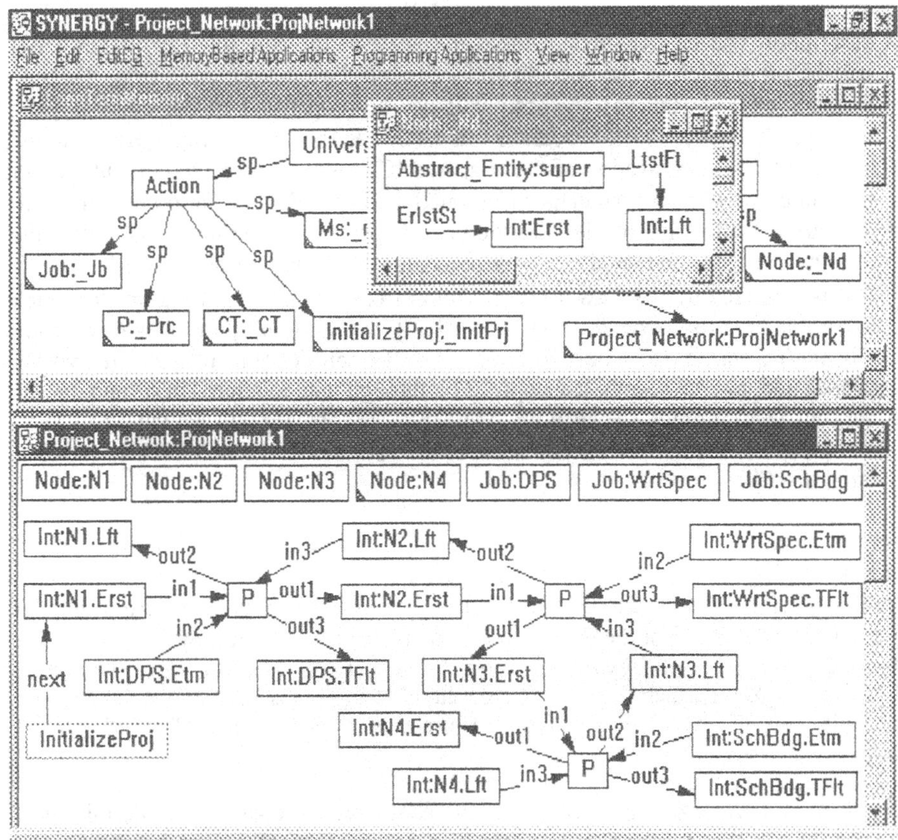

Fig. 3. Formulation in Synergy of the Project Network of Fig. 2

Let us consider now the definition of the process P (Fig. 4). P should be defined as a *lazy process* :

- *P should be a process* : the same description of [P] could be executed twice; in the first and the second forward steps. Thus, the description of [P] should not be destroyed after its first execution. This *process interpretation* will be adopted for a concept [P] since the definition of its type contains the concept [Process: super] (Fig. 4).
- *P should be a lazy process* : P does not need all its input arguments and should not wait for all their values to start its execution. In fact, only some input arguments (the two first) will be used in the first execution while others (the second and the third input arguments) will be used in the second execution. Since the definition of the type "P" contains the concept [Lazy: super], the Synergy interpreter will adopt a *lazy evaluation* for a concept [P]; once triggered by an argument, [P] can begin its execution without asking and waiting for the values of its input arguments.

The process P plays two roles (see its definition in Fig. 4) :
- *If* it is triggered by the forward propagation from its first input argument (which corresponds to its parameter [Int: in1|_N1Erst]), *then* only the left part of its description will be executed ! Indeed, the state of the concept [Int: in1|_N1Erst] will be propagated to the concept [CT] through the relation "in3" (Fig. 4). Note that the state of [Int: in1|_N1Erst] will not be propagated to the concept [Ms] through the relation "in2,/" because this latter has the "cut_forward_propagation" attribute "/". The activation of the triggered [CT] will involve the computation of the value of the output parameter [Int: out1|_N2Erst] (Fig. 4). No other activity or animation will exist in the description of [P] in this case. Thus, after the termination of the activated [CT], the activation of [P] will terminate too.
- *If* it is triggered by the forward propagation from its third input argument (which corresponds to its parameter [Int: in3|_N2Lft]), *then* the left part of its description will not be considered ! Indeed, the state of the concept [Int: in3|_N2Lft] will be propagated to the second concept [CT] through the relation "in3" and to the concept [Ms] through the relation "in1" (Fig. 4). The parallel activation of [CT] and [Ms] will involve the computation of the values of [Int: out2|_N1Lft] and [Int: out3|_JbTFlt].

Let us consider now the type "CT" (Fig. 4). "CT" is called twice in the body of the process P (Fig. 4): CT(">", "+", _N1Erst, _JbEtm, _N2Erst) and CT("<", "-", _N2Lft, _JbEtm, _N1Lft). The operation/type CT has two *operations/types as arguments*. A pseudo-code definition of the operation CT can be expressed as follows :

```
define CT(in Type:_SupInf, Type: _PlusSubst, Int:_X, Int: _Y; inout Int: _Z) is begin
    Int Tp := _X _PlusSubst _Y;  /** the value of _PlusSubst is either "+" or "-" **/
    If (Tp _SupInf _Z) Then _Z := Tp;  /** the value of _SupInf is either ">" or "<" **/
end;
```

Fig. 4 gives a Synergy reformulation of the above pseudo-code for the definition of the type "CT". We use the Synergy primitive operation "SetType" to implement the possibility of modifying the type of a concept at the execution time (for instance, to change the type of [Action] with the type "+" or "-").

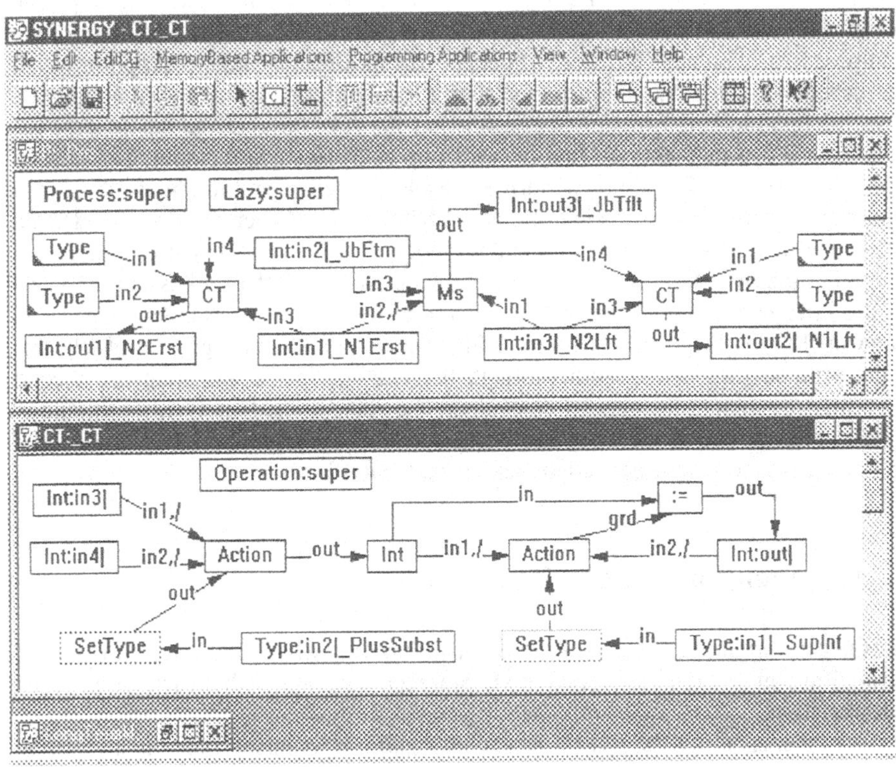

Fig. 4. The Project Schedule application (suite)

Now that the specification in Synergy of the project schedule application was given, let us follow its use (e.g., its execution). To initiate the activation of the project network, the request composed of the concept [Project_Network :ProjNetwork1. #?] should be added to the context [WorkingMemory].
The concept [Project_Network :ProjNetwork1. #?] is a reference to the description of the individual "ProjNetwork1" (Fig. 3). Being at state "trigger; ?", this concept will be considered by the Synergy interpreter (once activated). The first action concerning the interpretation of the concept [Project_Network:ProjNetwork1. #?] is to resolve the co-reference "ProjNetwork1.". Since no concept with the referent "ProjNetwork1" exists in the current context ([WorkingMemory]), the search will continue in the "father" context (in this case [LongTermMemory]). This context contains the required concept (Fig. 3).
Once the description is located, the next action is to interpret it. At this time, the state of [Project_Network :ProjNetwork1.] becomes "in-activation; !@". At start, the concept [InitializeProj #?] in [Project_Network :ProjNetwork1] is the only concept to be interpreted; the other concepts are in state "steady" (Fig. 3). The function of [InitializeProj] is basically to initialize the Erlst_st of the first node N1 and the TimeEstimate of each job.

Once the execution of [InitializeProj] is terminated, the relation "next" will propagate forward the state "trigger" to the concept [Int: N1.Erst] (Fig. 3). The resolution of the composed co-reference "N1.Erst" will enable the localization of its description which corresponds to the value 0. Since this latter is a simple data, its execution is null and then the relation "in1" will propagate forward the state "trigger" to the concept [P] (Fig. 3). At this moment, the first forward step of the method (compute the Erlst_st attribute of each node) will be performed, thanks to the lazy definition of the process P and to the forward propagation role of the relation "in". When the Erlst_st attribute of the end node N4 is computed, the description of the node N4 will be interpreted involving the execution of the assignment operation. This latter will assign the value of the Erlst_st attribute to the Ltst_ft attribute of the same node N4. As a result, the concept [Int: N4.Lft] will be triggered (Fig. 3). Being an input argument of [P], this latter will be triggered in turn (through the relation "in3"), initiating the second forward activation of the network to compute the value of the Ltst_ft (Lft) attribute of all the nodes as well as the "total float" attribute of all the jobs.
It is more easy to see the activation then to comment it !

7 Conclusion, current and future works

This paper introduced the core of the language Synergy which illustrates how the CG formalism can be used as a visual multi-paradigm language with an activation-based interpretation.
Synergy makes use of the CG formalism as a uniform foundation for the integration of sequential and/or parallel procedural, process, functional, event-driven, object oriented and concurrent-object oriented paradigms. Indeed, programs according to these paradigms can be represented and integrated in Synergy using a minimum of basic notions common to them all. Those notions correspond to the basic elements of the CG theory : CG structure, context, co-reference, type hierarchy, conceptual structures and knowledge base.

Several current and future works are underway or planned to enhance, extend and use different components of the SYNERGY CG tool. One of the current projects is to re-implement SYNERGY with Java. Other projects concern the use of SYNERGY in the development of several applications : object-oriented applications, concurrent object-oriented and multi-agents applications, multi-media applications, intelligent tutoring and training applications, case-based reasoning, etc.

References

1. Bos C., B. Botella, and P. Vanheeghe, Modeling and Simulating Human Behaviors with Conceptual Graphs, in Proc. Of the Fifth International Conference on Conceptual Structures, ICCS'97, Springer (1997)
2. Brauer W., W. Reisig and G. Rozenberg (eds.), Petri Nets: Applications and Relationships to Other Models of Concurrency, Springer-Verlag (1986)

3. Cyre W. R., Executing Conceptual Graphs, in Proc. Of the 6th International Conference on Conceptual Structures, ICCS'98, Springer (1998)
4. Ellis G., Object-Oriented Conceptual Graphs, in Proc. of the Third Intern. Conf. on Conceptual Structures, ICCS'95, Santa Cruz, CA, USA (1995)
5. Esch J., Contexts, Canons and Coreferent Types, Proc. Second International Conference on Conceptual Structures, ICCS'94, College Park, Maryland (1994)
6. Hee K. M., P. M. P. Rambags and P. A. C. Verkoulen, Specification and Simulation with ExSpect, in Lauer (Ed), Functional Programming, Concurrency, Simulation and Automated Reasoning, Springer-Verlag (1993)
7. Kabbaj A. and C. Frasson, Dynamic CG: Toward a General Model of Computation, Proc. Third International Conference on Conceptual Structures, ICCS'95, Santa Cruz, CA (1995)
8. Kabbaj A., Un système multi-paradigme pour la manipulation des connaissances utilisant la théorie des graphes conceptuels, Ph.D Thesis, DIRO, Université de Montréal, June (1996)
9. Kabbaj A., Rouane K. and Frasson C., The use of a semantic network activation language in an ITS project, Third International Conference on Intelligent Tutoring Systems, ITS'96, Springer-Verlag (1996)
10. Kabbaj A., Synergy as an Hybrid Object-Oriented Conceptual Graph Language, Seventh International Conference on Conceptual Structures, ICCS'99 this volume.
11. Lakos C., From Coloured Petri Nets to Object Petri Nets, in Michelis G. and M. Diaz (Eds.), Application and Theory of Petri Nets, Springer (1995)
12. Liddle S. W., D. W. Embley and S. N. Woodfield, A Seamless Model for Object-Oriented System development, Bertino E. and S. Urban (Eds.), Object-Oriented Methodologies and Systems, Springer-Verlag (1994)
13. Lukose D., Executable conceptual structures, Proc. First International Conference on Conceptual Structures, ICCS'93, Quebec City, Canada (1993)
14. Lukose D., Complex Modelling Constructs in MODEL-ECS, in Proc. Of the Fifth International Conference on Conceptual Structures, ICCS'97, Springer (1997)
15. Mineau G. W., From Actors to Processes : The Representation of Dynamic Knowledge Using Conceptual Graphs, in Proc. Of the 6th International Conference on Conceptual Structures, ICCS'98, Springer (1998)
16. Raban and Delugach, Animating Conceptual Graphs, in Proc. Of the Fifth International Conference on Conceptual Structures, ICCS'97, Springer (1997)
17. Shlaer S. and S. J. Mellor, Object Lifecycles - Modeling the World in States, Prentice-Hall (1992)
18. Sowa J. F., Conceptual Structures : Information Processing in Mind and Machine, Addison-Wesley (1984)
19. Sowa J. F., Relating Diagrams to Logic, Proc. First International Conference on Conceptual Structures, ICCS'93, Quebec City, Canada (1993a)
20. Sowa J. F., Logical foundations for representing object-oriented systems, J. of Experimental and Theoretical AI, 5 (1993b)
21. Sowa J. F., Conceptual Graph Standard and Extensions, in Proc. Of the 6th International Conference on Conceptual Structures, ICCS'98, Springer (1998)
22. Spinner M. P., Improving Project Management Skills and Techniques, Prentice-Hall (1989)
23. Thakkar S. S. (ed.), Selected Reprints on Dataflow and Reduction Architectures, IEEE Computer Society Press (1987)
24. Törn A., Systems Modelling and Analysis Using Simulation Nets, in C. A. Kulikowski and al. (eds.), AI and ES Languages in Modelling and Simulation, North-Holland (1988)
25. Zouaq A., Modélisation en Synergy d'un système multi-agents, Mémoire de fin d'étude, INSEA, Rabat, Morocco (1998)

On Developing Case-Based Tutorial Systems with Conceptual Graphs

Pak-Wah Fung and Ray H. Kemp

Institute of Information Sciences and Technology
Massey University
Palmerston North, New Zealand
{P.W.Fung, R.Kemp}@massey.ac.nz

Abstract. In this paper, a case-based intelligent tutoring system (CBITS) using conceptual graphs to represent the cases is described. Among others, two core issues to be addressed by CBITS developers are indexing and how learning activities can be constructed from the cases. Addressing the former, the authors discovered that the minimum common generalization of a set of graphs is a powerful means of embedding different tutorial primitives into the cases. This approach provides an extremely rich indexing vocabulary for the cases and relieves the developers from speculative assignment of indexes. For constructing learning activities an operational semantics for the case graphs is defined and with this semantics the instructor can reason about the graph structure and provide intelligent guidance to the students in exploring the case contents. Newtonian mechanics is the testing domain of the proposal but its underlying methodology should be equally applicable to other subject domains.

1 Introduction

Intelligent tutoring systems (ITS) is an exciting area and the development of this field will no doubt have a lot of impact to our educational systems. One of our long term ambitions is to develop an open-ended campus-wide learning resource which may include online exercises, worked examples, past students' projects or real-world case studies, etc. From anywhere and at any time, the students can call upon a domain specific tutorial agent to provide individualized guidance on solving problems or explaining the case contents in terms of fundamental principles. This paper reports our progress in representing the collected cases and the mechanism of making the case contents usable in tutorial settings.

We adopted the position of case-based tutoring which is chiefly based on the experimental discovery (e.g. [1], [10], [11], [17] and [21]) that experts excel mainly in their own domains, and there is very little evidence that a person highly skilled in one domain can transfer that skill to another. The excellence of competent problem solvers is due to their possession of a good deal of domain-specific knowledge, an achievement that requires many years of intensive experience over many cases. This phenomenon applies to learning music, to science and to chess playing. No one can reach genius levels of performance without at least ten years of practice. Therefore it is well-grounded to draw the conclusion that the expertise of competent problem solvers is developed through their exposure to very many cases pertaining to their own particular subject domain. One of the key differences in domain knowledge between a more competent problem-solver and a less competent one is the number and range of cases to which they have been exposed. The remaining questions are: what structure best represents the cases and how can we help the students acquire the domain knowledge embedded in the cases by developing an ITS? Cognitive psychologists (e.g. [14] and [20]) suggest that expert knowledge is in schemata form and that schemata are induced or abstracted from many specific examples [15], [9]. To reflect these finding, the case representation language must offer an abstraction mechanism for generalizing different cases. Conceptual graphs [19] provides such a mechanism.

2 On Indexing Case Features

In case-based tutoring environments, the learning activities are designed around a specific situation, i.e. by tackling problems specified within particular contexts. It is expected that the students acquire their generic knowledge structure through generalizing the cases they have experienced and the knowledge can be transferred to a larger class of similar and yet novel problems. It, therefore, follows that case representation is critical and operations on the representation must support indexing, retrieval and comparison or adaptation.

Superficially, encoding the problem context and the solution methods seems to constitute the necessary case components adequately but the system may be held back when facing an unusual user request which is beyond the scope of the indexes chosen by the system developer. In our subject domain of Newtonian mechanics, a typical mechanics case such as "finding the acceleration of a block resting on an inclined plane" may be indexed under the surface feature - a block resting on an inclined plane, as well as under the deep feature - Newton's second law of motion. Indexing cases in this way is also in accord with the 'mainstream' approach (i.e. problem description, solution methods and outcomes) delineated by [12].

The cases can be used in two ways within a tutorial setting: (1) as worked examples by providing detailed solutions or (2) initially hiding the answers and helping the students to discover it via a guidance procedure. However, supposing a student is attempting to solve a problem that is about a skier rushing down a ski slope. In actual fact the worked example just mentioned is entirely relevant to the present problem but its tutorial utilities will be lost if the system is unable to recognize the similarity between the stored case (i.e. a block resting on an inclined plane) and the student's current problem (i.e. a skier rushing down a ski slope). Better still, the system should be smart enough to classify the case as relevant when facing 'unexpected' requests such as "a boy plays on a playground slide" or "resolution of vector into two perpendicular components", etc.

Addressing this problem demands the system to be capable of manipulating the cases at different levels of abstraction and relieving the developers from 'guessing' what sort of future situations the case can be considered as relevant. A case not only represents the application of general knowledge (say Newton's second law of motion) in a specific context (finding the acceleration of a block resting on an inclined plane), it should also allow the detection of cross-contextual similarities across difference cases.

3 Benefits of Representing Cases with Conceptual Graphs

We are not the pioneers (e.g. [16]) of exploiting the advantages of representing cases with graphs. In [16], the case details are encoded using directed graphs but directed graphs can only denote binary relationships and are therefore inadequate or cumbersome in representing relationship among three or more objects. Conceptual graphs (CG) was chosen as the knowledge representation scheme not only because of its linguistic and logical basis, but also because of its separate consideration of the relation node allows us to represent n-ary relationship among objects. Fig.2 below shows an example 3-ary relations between three concept nodes in our subject domain.

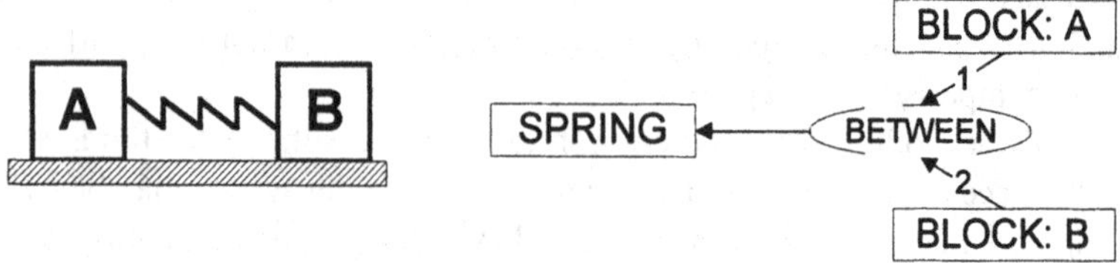

Fig. 1 A 3-ary relationship among 3 concept nodes

The graph denotes the physical configuration of 3 objects: a spring being placed between two blocks with referent labels 'A' and 'B' respectively. In linear format, it is represented as:

[SPRING] ← (BETWEEN) -
1 ← [BLOCK: A]
2 ← [BLOCK: B]

Another advantage of using CG is the global concept type hierarchy and attaching concept nodes with referent have elegantly represents concept subsumption and concept instantiation. The explicit use of 'inst-of' and 'isa' links in directed graphs becomes unnecessary.

4 Cases: Instantiation of Knowledge Schemata

One central theme of case-based learning is learning by induction, i.e. the generic knowledge structures, namely schemata in psychological terms, are induced from multiple cases. This assertion implies a plausible correspondence between the problem representation of the cases. If we intend to teach the students a physical principle via studying cases or solving problems, the respective cases must be derivable from the schema representing the law in generic terms. Any computer emulation of case-based tutoring must thus contain the following features:

1. The domain knowledge involved in the cases can be represented as semantic schemata
2. There are algorithms to find correspondences between abstract structures

If we are representing cases with graph structures, the first feature requires a specialization and generalization mechanism on the structures and the second one builds on the fact that the structures are computational. These notions have corresponding elements in the CG model.

4.1 Specialization and Generalization of Conceptual Graphs

What does it mean to assert that two problems, say P1 and P2, are instantiation of a physical principle? Generally, physics principles like Newton's laws of motion or conservation of energy are abstract descriptions of how the physical world behaves. When a problem solver successfully applies a particular physics principle to solve a problem, this implies that by instantiating the physics principle he/she can make sense

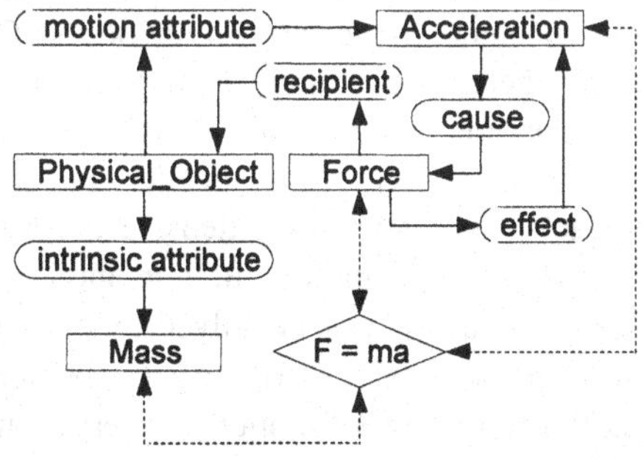

Fig. 2 The canonical graph for Newton's 2nd law

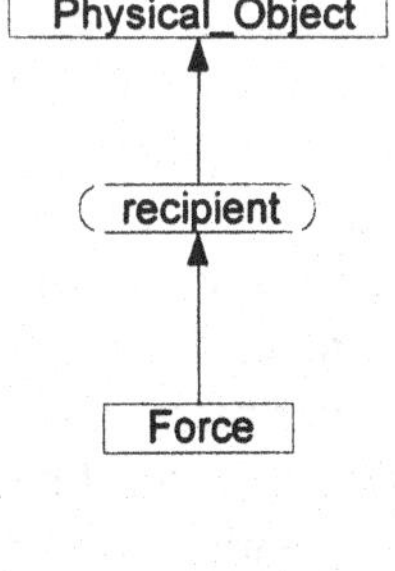

Fig. 3 A typical schema

of the problem situation and is able to draw inferences. In the CG perspective, the case graph is derivable from the well-formed graphs, namely canonical graphs, that representing the modelled world in an abstract sense. The canonical graph for Newton's second law of motion (i.e. whenever a force is exerted on a physical object the object accelerates) is depicted in Fig.2. Fig.3 illustrates the schema for the general situation of a force acing upon an object.

For the cases that were encoded using CG, as, for example, the one shown in Fig. 4, one can see that it is a specialization of the generic principle denoted by the graph in Fig. 3. In other words, Fig. 3 is both an abstract description of and a generalization of Fig. 4. Supposing we want to retrieve some cases about "a force acting on an object" for the students to practise. This task has been transformed to one of matching the target graph, i.e. Fig. 3, with the case graphs stored in the case library. By all means the case of Fig. 4 is matchable in this circumstance and can be retrieved for further processing. Of particular importance in this approach is that the explicit effort of indexing the case becomes unnecessary because the case features have been implicitly embedded.

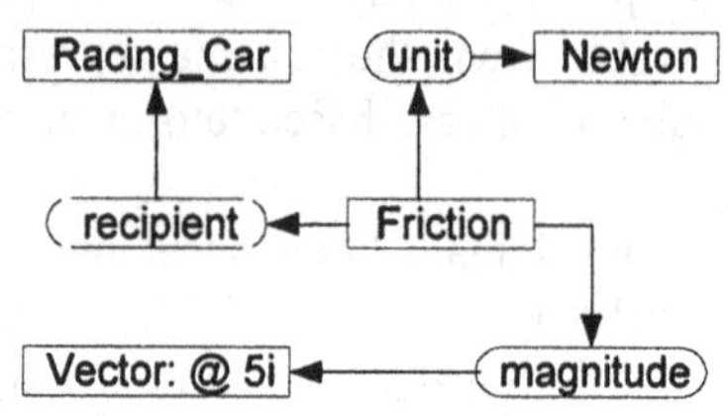

Fig.4 A typical case

4.2 Schema Induction: Finding Common Generalization across Cases

In theory, generalization can occur with just one case but this seldom happens in reality. Teaching experience informs us that most students need to be exposed to multiple cases before a generic knowledge structure can be induced. So the question is: what constitutes a correspondence between two (or more) problem representations? It is based on the notion of common generalization of the CGs representing the corresponding problems. As there can be more than one possible generalization, all these generalizations can be ordered in a generalization hierarchy.

The minimum common generalization (MCG) in the generalization hierarchy represents both situations in the least abstract way. The more extensive the MCG between two cases, the more features they have in common. In our application domain, mechanics, wherever we say that two problems can be solved by a similar physics principle, the MCG of the problems must consist of one of the canonical graphs representing the classic laws such as Newton's laws of motion, conservation of energy, conservation of momentum, and so forth.

Besides having a thorough understanding of domain specific principles, expert problem-solvers are often endowed with some powerful domain independent problem-solving strategies such as the principle of decomposition or the principle of invariance. Canonical graphs representing these principles are also potential candidates for the MCG graphs. All of these are being considered as tutorial primitives. For real cases which often comprise more than one tutoring primitive, the attractiveness of using CG lies with the integration of all primitives into one single coherent unit (see Fig. 5) and obviously the more extensive a case graph is, the more tutoring primitives it may contain.

It seems that everything discussed is relating to the domain of physics, but in fact some primitives are domain independent and can be used in many other areas. Consider the following example: our current learning objective concerns the exponential increase of some entities after a certain period of time. The system can retrieve cases ranging from the domain of banking (principal plus compound interest) to ecology (growth of fish population). That is why we claim that CG-based case representation allows cross-contextual matching and retrieval. The case developer does not need to speculate on what future scenarios the cases can be utilized as the indexing vocabulary had implicitly been encoded in the graph. Essentially speaking, each node is an index by itself and all other [concept]→(relation)→[concept] tuples form another index class.

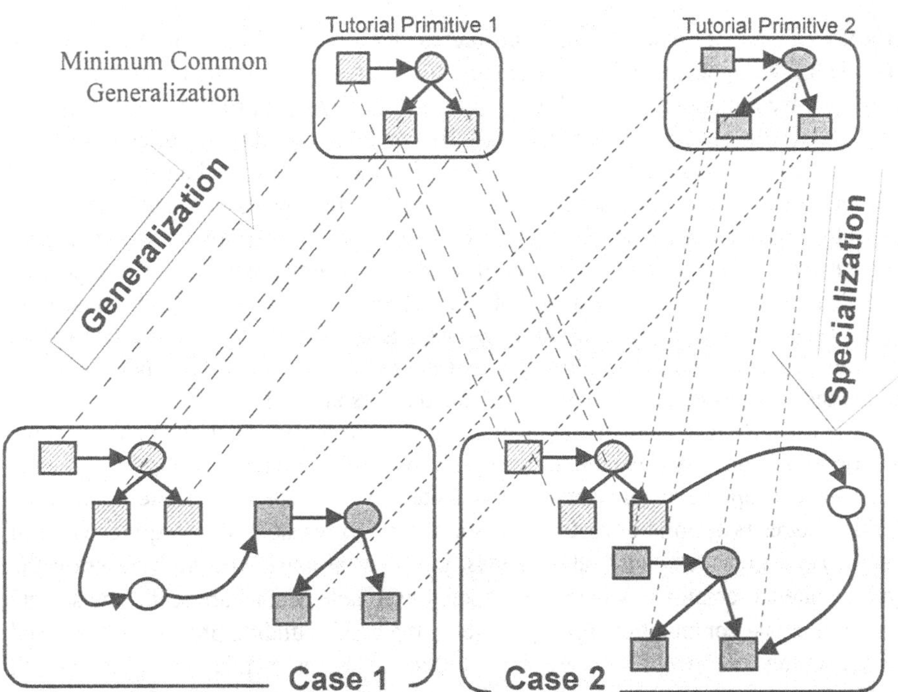

Fig. 5 A generic framework for embedding primitives into cases

5 An Operational Semantics for Conceptual Graphs

In the previous sections, we addressed the issue of retrieving cases based on the notion of matching case graphs with the tutorial primitive graph. The next question is: "Now we have a case, but so what?" The pedagogical utilities of a case are very limited if the system is not endowed with the ability to convey the knowledge embedded in the case to the users. A typical physics case is generally set in the context of problem-solving with some initially given physical quantities. During the process, the knowledge base of the problem-solvers is invoked and inference procedures are executed to derive the unknown physical quantities sought. To be useful in a tutoring context, a case should contain the knowledge components which reveal such a process explicitly to the users. In our testing domain of mechanics, mathematical computation plays a significant role so merely consider symbolic relationships among objects is inadequate. Besides the usual concept and relation nodes, we also need to consider the third class of nodes, the *actors*, described in [19]. Although our problem-solving cases can be faithfully represented by CG, the construction of the graph remains in a 'black-box' so far and the world model is represented declaratively. That is to say the user has no way of inspecting the internal processes to see how the graph is constructed. To be useful in a tutorial context, the knowledge components of a case should be made available to the students. By available, we mean the system should be able to justify each problem-solving step to be shown to the students. In other words, the procedural mechanism of achieving the solution must be made explicit but the issue is how can this mechanism be defined and executed within a graph-structured framework. In Sowa's original formulation, the actor nodes are an attachment of a CG, and the group of actors constitute a *data flow graph*. The execution mechanism of the data flow graph is like the firing process in a Petri Net [13]. In the semantics to be defined here, the CG and the data flow graph are synthesized into one single global graph, instead of treating them as two separate graphs. A case has three types of nodes: *concept* nodes, *symbolic relation* nodes and *mathematical relation* nodes (called actors in [19]).

Two assumptions were made in making the synthesis. First, human cognitive functions in studying a concrete case are viewed as a graph construction process. Relevant concept nodes are created and linked to each other via some appropriate relation nodes (symbolic or mathematical). A case represented by a graph consists of sets of concept nodes and relation nodes, but to what extent the students grasp the graph contents remains unknown until some observable actions are seen. The second assumption is a computational perspective on the graph building process and is based on the notion of concept node *marking*. Initially, the sets of nodes in a case are all transparent to the users because they are not yet marked. The set of nodes representing the initially given physical quantities are marked first. Each problem-solving step is viewed as generation of new graph nodes, but they are implemented as the nodes states change from unmarked to marked. To mark a set of nodes, the mathematical relation nodes (simply called operators) which link the marked and the unmarked nodes have to be *fired*. The procedures of solving the problem are defined as the firing sequence for marking the target concept nodes. The subgraph associated with a particular fired node represents the semantics of the knowledge behind its firing.

The semantics is derived from Sowa's original formulation and from a class of high-level Petri Nets called Predicate Transition Networks (PTN) [8]. In PTN, each input place can carry a number of pre-defined typed object called *tokens*, which are said to be marked whenever the tokens reside in the place. When all the input places of a transition are marked (a condition which *enables* the transition), the transition can be fired at any time and the token will be deposited to its corresponding output place(s) after the firing. This token-fire notion is borrowed for our model as there are a lot of similarities between a net and a graph. Whenever a concept node is instantiated to a specific individual, it is said to be marked. Firing a mathematical relation node (i.e. executing the built-in mathematical operations) will put a token (a referent for a concept node) into the concept node which is being pointed at by the head of the directed arc. The formulation may be considered as a remnant of the classic general problem-solver GPS [4] which considers problem-solving as a process of searching the problem space. The initial state of the problem is described by a set of marked concept nodes, and the goal is to mark the sought concept nodes, while the set of situations on the way to the goal are the intermediate states. Every time an operator is fired and marks new concept nodes, it is like invoking a problem-solving operator to transform the problem state from one state (an old marking) to another (a new marking).

5.1 Formal Definition of Case Contents

Definition 5.1.1: A Case $\mathcal{C}$ consists of:

1. A finite, directed graph, defined by the tuple $<C, R, R_m, E, E_m>$;
2. *C*, *R* and R_m are three disjoined sets of vertices (i.e. *C* ⬆ *R* = *ĕ* ; *C* ⬆ R_m = *ĕ* ; *R* ⬆ R_m = *ĕ* and *C* ⬆ *R* ⬆ R_m = *ĕ*) where *C* represents the set of concept nodes, *R* represents the set of symbolic relation nodes, and R_m represents the set of mathematical relation nodes (or simply operators);
3. *E* is a set of directed arcs, each arc connecting a concept *c ї C* to a symbolic relation *r ї R* or vice versa, i.e. *E í (C à R) ğ (R à C)* ; and
4. E_m is another set of directed arcs, each arc connecting a concept *c ї C* to a mathematical relation r_m *ї* R_m or vice versa, i.e. E_m *í (C à* R_m *) ğ (* R_m *à C)*.

Definition 5.1.2: Input/Output Sets, Arcs and Concepts.

1. There are four functions: two *input* functions I_c , I_r and two *output* functions O_c and O_r.
2. I_c : *C* ® R_m and O_c : *C* ® R_m are mappings from concepts to sets of mathematical relations.

3. $I_r : R_m \rightarrow C$ and $O_r : R_m \rightarrow C$ are mappings from mathematical relations to sets of concepts.
4. For each concept $c \in C$, an input set $I_c(c)$ and an output set $O_c(c)$ are defined as:
   - $I_c(c) = \{ r_m \in R_m \mid (r_m , c) \in E_m \}$; (r_m , c) is called the input arc of c, and r_m is called the input mathematical relation of c.
   - $O_c(c) = \{ r_m \in R_m \mid (c , r_m) \in E_m \}$; (c , r_m) is called the output arc of c, and r_m is called the output mathematical relation of c.
5. For each mathematical relation $r_m \in R_m$, an input set $I_r(r_m)$ and an output set $O_r(r_m)$ are defined as:
   - $I_r(r_m) = \{ c \in C \mid (c , r_m) \in E_m \}$; (c , r_m) is called the input arc of r_m, and c is called the input concept of r_m.
   - $O_r(r_m) = \{ c \in C \mid (r_m , c) \in E_m \}$; (r_m , c) is called the output arc of r_m, and c is called the output concept of r_m.

Definition 5.2.3: Marking of Concept Nodes

1. For every $c \in C$, if *referent(c)* $\neq$ * and *referent(c)* $\in I$, i.e. c is being instantiated to a specific individual by having an individual marker, then c is *marked*.
2. A marking m of a graph $G = < C , R , R_m , E , E_m >$ is a function mapping from the set of concepts C to the set of Boolean variables. i.e. $m : C \rightarrow Boolean$.
3. The marking m can be represented by a *n-vector*: $m = (m_1, m_2, \ldots , m_n)$, where $n = |C|$ and each $m_i \in \{ T , F \}$, $i = 1, 2, \ldots , n$.
4. A marked graph $M = < G , m >$ is a graph with marking m.

Definition 5.2.4: Enabling of Mathematical Relations

A mathematical relation node $r_m \in R_m$ is *enabled* whenever each concept $c \in I_r(r_m)$ is marked. For the case graph in Fig.9, only r_{m1} is enabled at that marking.

Definition 5.2.5 Firing of Mathematical Relations

1. When a mathematical relation node is enabled, it can be fired at any time.
2. Every time a mathematical relation is fired, every $c \in O_r(r_m)$ will be marked.
3. For every $c \in O_r(r'_m)$, where r'_m is the fired mathematical relation, the referent of c is evaluated according to the formula(e) inscripted in the respective $r'_m \in I_c(c)$.
4. The definition of firing is different from the original PTN formalism in that no token is removed from the input concepts, because it makes little sense in our application of problem solving that a specific concept will become a general concept again after the firing.

An example case graph is shown in Fig. 6 in which only C_1 and C_5 are marked as initial conditions. At this marking, only r_{m1} can be fired to mark C_2 but either r_{m2} or r_{m3} can be fired subsequently.

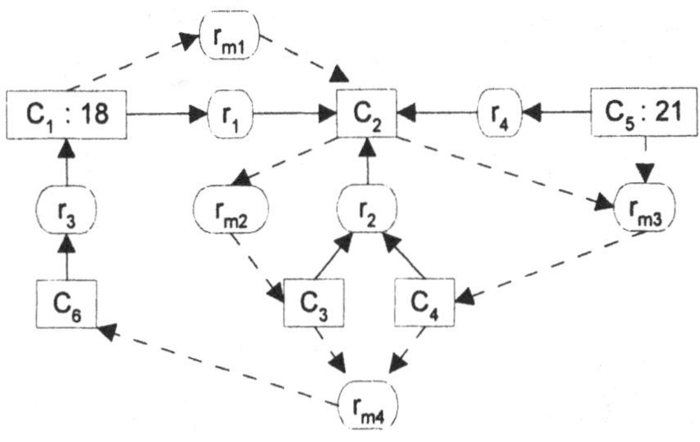

Fig. 6 A marked case graph with marking $m = (T, F, F, F, T, F)$.

5.2 Modelling Data-driven Tutoring

Having defined the essential constructs in representing procedural knowledge, we are in a position to show how they can be used in modelling different styles of reasoning. According to the finding reported in [18], novices tend to adopt a working back strategy, whereas more competent problem-solvers usually opt for a working forward strategy. A working forward approach operates from the initial set of data given by the problem statement, successively invoking equations which can be solved with the givens until the goal of finding values of the target physical quantities can be achieved. Let us attempt to capture such a mode of reasoning within our framework by referring to Fig. 6. In this case, we have six concepts (i.e. C_1, C_2, ..., C_6) which relate to each other symbolically via r_1 to r_4 and mathematically via r_{m1} to r_{m4}. The values of C_1 and C_5 are given at the beginning. The goal of the problem is to find out the value of the concept C_6. From the initial marking, we can only fire r_{m1} [giving (T, T, F, F, T, F)]. After firing, r_{m1}, r_{m2} and r_{m3} become enabled and therefore there are two alternatives available [i.e. firing r_{m2} gives (T, T, T, F, T, F) or firing r_{m3} gives (T, T, F, T, T, F)]. This process can be continued until a marked C_6 is obtained (i.e. the goal of the problem-solving activities). Fig. 7 below depicts the whole process of creating successive markings. Even this simple example indicates the student can gain access to quite a large solution space for him/her to explore but in the mean time the tutor can keep track of what can/cannot be done.

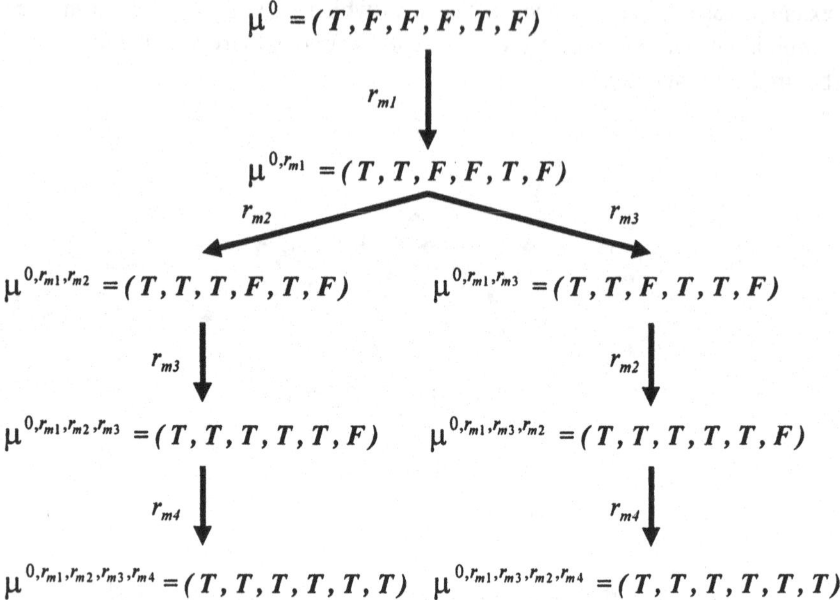

Fig. 7 A complete marking tree showing the data-driven reasoning process

This section only shows how forward reasoning can be modelled but in fact it is much more powerful and flexible. Readers can refer to [5] to see how other problem-solving approaches can be modelled as well. These include goal-driven (or backward reasoning), bi-directional reasoning and hierarchical reasoning.

6 CLASP: A Case-based Learning Assistant System in Physics

At this stage, a prototype called CLASP, has been developed to test the idea. Thus far only two types of activities associated with examples have been identified: providing solutions for studying, and exercises with answers; hence the modes of interaction in the CLASP prototype are also designed around these two themes. When the users issue a request (in terms of the problem description of their own problems) the system will search through its whole case library and provide them cases which match their request. The style of presenting the case will follow the user's wishes, but only two modes of interaction (solution studying and guided-problem-solving) are available. This is to reflect the common way of using examples in physics textbooks.

In the study mode, the system presents the whole case (i.e. both the problem and solution statements) for the user to study. This looks like an electronic reference book and the student may browse through the relevant cases. However, a special agent called *the case-questioner* was developed to question the users on the contents of the examples. The motivation of questioning is to promote self-explanation [2] by the students on the example solutions. In the guided-problem-solving mode (the term 'interactive mode' was used), the system only presents the problem situation to the users, but appropriate system guidance will be provided in solving the problems. Schematically, the overall architecture of the current edition of the CLASP prototype is summarized in Fig. 8. The students interact with the CLASP system with the support of the back-end knowledge base.

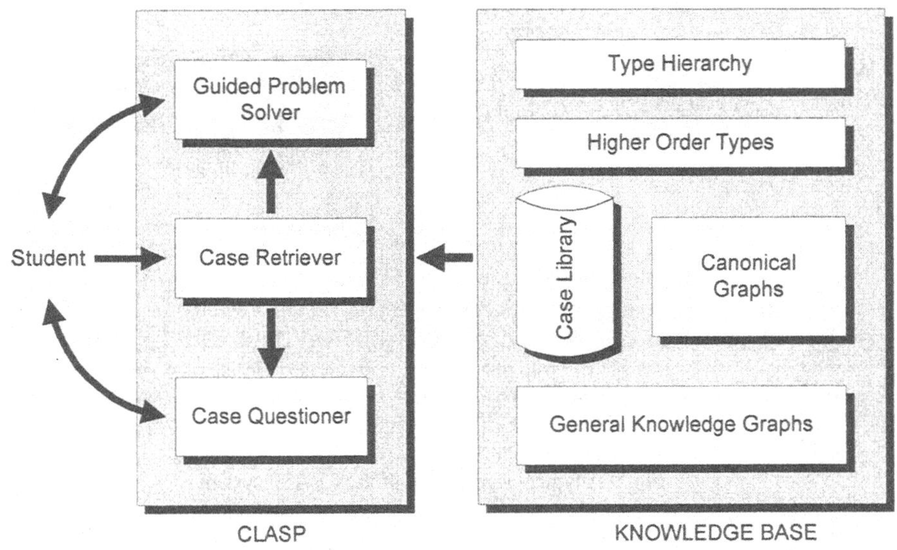

Fig. 8 Schematic description of the CLASP architecture

The deliberate separation of the knowledge base from the main CLASP system enables the flexibility and convenience of adding new knowledge, new cases, new type definitions and so forth without needing to know how the system functions. Different inference engines are developed separately and embedded in the guided-problem-solver, the case-retriever and the case-questioner respectively. These modules have independent rights of access to the knowledge base.

Once a particular case is selected, the case's problem statement will be displayed (see Fig. 9 for an example). At this stage of development, the user has three choices: studying the complete solution (i.e. study mode); asking the system to guide him/her to solve the problem (i.e. interactive mode); or to leave the system.

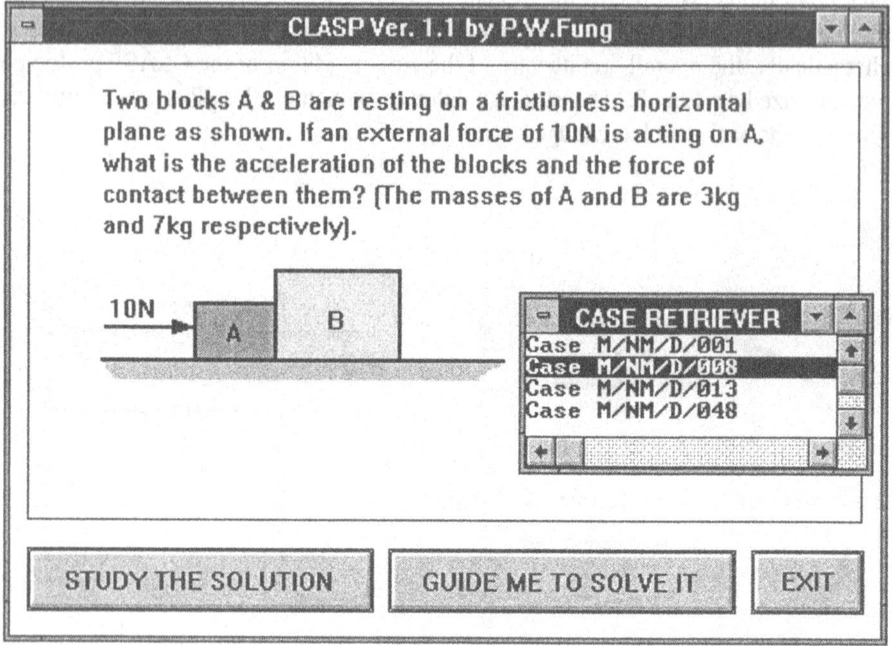

Fig. 9 A typical screen shot of the CLASP system

7 Guided Problem Solving

In CLASP, problem-solving is modelled as a graph search. When a problem situation, such as the one in Fig. 9, is encountered, the initial data are represented as concept nodes being instantiated to specific values and they are displayed to students on the working pad (Fig. 10). Now the problem-solver can start tackling the problem by searching through the graph and seeing what additional information can be inferred from the initial given data. For the system to perform the tasks, the expertise has already been encoded in the conceptual graphs, therefore the next step to be taken is modelled by searching the graph to find out which operators can be fired. The inferred steps may be unfolded or kept hidden for a while as a hint to advise the student. The intelligence of the system's problem-solving ability comes from its inference engine, being implemented as different graph search methods.

The explanatory capability of the system comes from the matching of the input-operator-output nodes with the consequences of the general knowledge graphs. Whenever an operator is fired, the associated nodes will be matched against the consequences of the general knowledge graphs. If one is found, and it should be, then that particular graph will be tagged. If the student requests a justification of the step taken, the system can explain the graph in general terms. For example, the firing of an algebraic summation operator on the values of masses of two physical objects will match the consequence of the general knowledge graph in Fig. 8 so the whole graph can be retrieved for explanation. The working pad, showing the problem space, and the explanation combinations supply the integration of what and why the step happened and the whole process becomes transparent to the student.

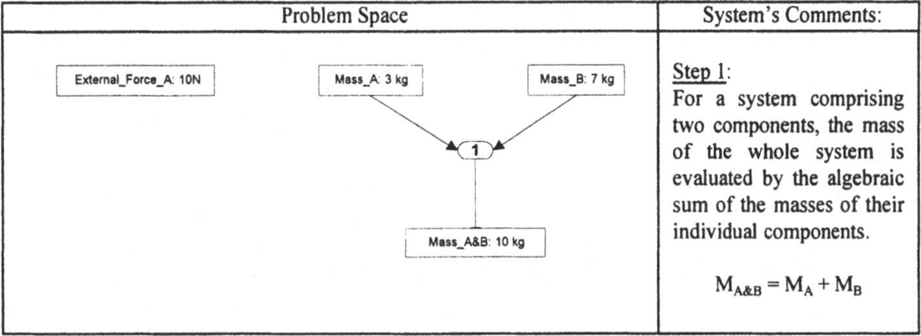

Fig. 10 The problem space and the system's comments

8 Concluding Summary

CGs have been widely used in other knowledge-based applications but very few in building instructional systems. The work reported here may serves to strengthen the utility of CG in this area. The contribution of the project can be roughly divided into two categories: theoretical and application. On the theoretical side, we have developed an alternative perspective on matching cases in case-based reasoning. What constitutes the case contents was also formally defined and an operational semantics was defined on a combined structure between a CG and a data flow graph. Regarding the application achievements, the operational semantics was used in implementing four tutoring methodologies. The methodologies are very powerful in guiding students solving problems. Some other features of the system were not mentioned due to space limitation. They include generating different categories of questions from a case graph [7] to promote self-explanation from the students. The model proposed in this paper can also perform qualitative reasoning [6] and causal order between system variables can be represented succinctly. In the short term, we plan to test run the system with evaluation from students. The issue of complexity was temporary sidelined in developing the graph matching algorithm because of the initial objectives of the project were mainly educational. Once the students are satisfied with the tutoring capability of the system, we will look into the issue of efficiency before deploying it in the campus-wide network.

References

1. Chase, W.G. & Simon, H.A. (1973). Perception in Chess. *Cognitive Psychology*, Vol. 4, pp. 55-81.
2. Chi, M.T.H., Bassok, M., Lewis, M.W., Reimann, P. & Glaser, R. (1989). Self-Explanation: How Students Study and Use Examples in Learning to Solve Problems, *Cognitive Science*, Vol. 13, pp.145-182.
3. Collins, A., Brown, J.S. & Newman, S. (1989). Cognitive Apprenticeship: Teaching the crafts of Reading, Writing and Mathematics. In L.B. Resnick (Eds.). *Knowing, Learning and Instruction: Essays in honours of Robert Glaser*. Hillsdale, NJ: Lawrence Erlbaum Associates.
4. Ernst, G.W. & Newell, A. (1969). *GPS: A Case Study in Generality and Problem Solving.* NY: Academic Press.
5. Fung, P.W. (1996). *Designing an Intelligent Case-based Learning Environment with Conceptual Graphs*. PhD Thesis, Dept. of Computation, UMIST, England.
6. Fung, P.W. (1997). Generating Qualitative Predictive Problems in a Case-based Physics Tutor. In B. du Boulary & R. Mizoguchi (Eds.) *Artificial Intelligence in Education; Frontiers in Artificial Intelligence and Applications Vol. 39*. Amsterdam: IOS Press.
7. Fung, P.W. & Adam, A. (1996). Questioning Students on the Contents of Example Solutions. In *Proceedings of European Conference on Artificial Intelligence in Education*, Lisbon.
8. Genrich, H.J. & Lautenbach, K. (1981). System Modelling with High-Level Petri Nets. *Theoretical Computer Science*, Vol. 13, pp. 109-136.
9. Gick, M. & Holyoak, K.J. (1983). Schema Induction and Analogical Transfer. *Cognitive Psychology*, Vol. 15, pp.1-38.
10. Glaser, R. & Chi, M.T.H. (1988). Overview. In M.T.H. Chi, R. Glaser & M.J. Farr (Eds.) *The Nature of Expertise*. Hillsdale, NJ: Lawrence Erlbaum Associates.
11. Hayes, J.R. (1985) Three Problems in Teaching General Skills. In J. Segal, S. Chipman & R. Glaser (Eds.), *Thinking & Learning, Vol.2*. Hillsdale, NJ: Lawrence Erlbaum Associates.
12. Kolodner, J. (1993). *Case-Based Reasoning.* San Mateo, CA: Morgan Kaufmann Publishers.
13. Peterson, J.L. (1981). *Petri Net Theory and the Modeling of Systems.* Englewood Cliffs. NJ: Prentice-Hall Inc.
14. Rumelhart, D.E. (1980). Schemata: The Basic Building Blocks of Cognition. In R. Spiro, B. Bruce & W. Brewer (Eds.), *Theoretical Issues in Reading Comprehension.* Hillsdale, NJ: Lawrence Erlbaum Associates.
15. Rumelhart, D.E. & Norman, D.A. (1981). Analogical Processes in Learning. In J.R. Anderson (Ed.), *Cognitive Skills and their Acquisition.* Hillsdale, NJ: Lawrence Erlbaum Associates.
16. Sanders, K.E., Kettler, B.P. & Hendler, J.A. (1997). The Case for Graph-Structured Representations. In D.B. Leake & E. Plaza (Eds.) *Case-based Reasoning Research and Development. Lecture Notes in Artificial Intelligence, Vol. 1266*, pp. 245-54. Springer-Verlag.
17. Schank, R.C. & Cleary, C. (1995). *Engines for Education.* Hillsdale, NJ: Lawrence Erlbaum Associates.
18. Simon, D.P. & Simon, H.A. (1978). Individual Differences in Solving Physics Problems. In R. Siegler (Ed.), *Children's Thinking: What Develops?* Hillsdale, NJ: Lawrence Erlbaum Associates.
19. Sowa, J. (1984). *Conceptual Structures: Information Processing in Mind and Machine.* Reading, MA: Addison-Wesley.

20. Thorndyke, P.W. (1984). Applications of Schema Theory in Cognitive Research. In J.R. Anderson & S.M. Kosslyn (Eds.), *Tutorials in Learning and Memory.* San Francisco: Freeman.
21. Voss, J.F. & Poss, T.A. (1988). On the Solving of Ill-Structured Problems. In M.T.H. Chi, R. Glaser & M.J. Farr (Eds.), *The Nature of Expertise.* Hillsdale, NJ: Lawrence Erlbaum Associates.

Embedding Knowledge in Web Documents: CGs versus XML-based Metadata Languages

Philippe Martin and Peter Eklund

Griffith University, School of Information Technology,
PMB 50 Gold Coast MC, QLD 9726 Australia
{p.eklund,philippe.martin}@gu.edu.au

Abstract. The paper argues for the use of general and intuitive knowledge representation languages for indexing the content of Web documents and representing knowledge within them. We believe these languages have advantages over metadata languages based on the Extensible Mark-up Language (XML). Indeed, the representation and retrieval of precise information is better supported by languages designed to represent semantic content and support logical inference, and the readability of such a language eases its exploitation, presentation and direct insertion within a document.To further ease the representation process, we propose techniques allowing users to leave some knowledge terms undeclared. We illustrate these ideas with WebKB[1], a precision-oriented information retrieval/annotation tool, and show how lexical, structural and knowledge-based techniques may be combined to retrieve or generate knowledge or Web documents. Finally, to overcome the scalability problems of storing knowledge within Web documents, we propose some ideas for scalable and cooperatively built knowledge repositories.

1 Introduction

Large-scale search engines for the WWW retrieve entire documents effectively. However, they can be considered imprecise because they do not exploit and hence retrieve the semantic content of Web documents. Such content cannot yet be automatically extracted from general documents. Manually structuring Web documents, e.g. via mark-up languages such as XML[2], allows more precise information to be retrieved using string-matching and structure-matching tools, e.g. Web robots such as Harvest[3], WebSQL[4] and WebLog[5]. However, this approach is not scalable because *fine-grained* information is only retrieved if the documents are thinly structured and the querier knows the structures, their exact names and forms. More flexible and precise knowledge representation and retrieval can be achieved with knowledge representation languages that support

[1] http://meganesia.int.gu.edu.au/~phmartin/WebKB/
[2] http://www.w3.org/XML/
[3] http://harvest.transarc.com/
[4] http://www.cs.toronto.edu/~websql/
[5] http://www.cs.concordia.ca/~special/bibdb/weblog.html

logic inference. Many "metadata" languages are currently being developed to allow people to index Web information resources by knowledge representations (logical statements) and store them in Web documents. However, these metadata languages are insufficient to satisfy several requirements necessary to allow *precise, flexible and scalable* information retrieval.

A first requirement for that is that the metadata language is sufficiently *intuitive and concise to be* **easy** *to use* by people (after a short period of training). Most current knowledge-oriented metadata languages are built above XML, e.g. RDF[6] and OML[7]. The choice of XML as an underlying format ensures that standard XML tools will be usable to exchange and parse these metadata languages. However, since XML is verbose, the metadata languages built above XML are also verbose and are difficult to use without specialized editors. Such editors do not eliminate the need for people to use a language for representing knowledge (except in application-dependent editors that only allow predefined "frames" to be filled). Consequently, as noted by the authors of Ontobroker[8] [1], with XML-based languages information has to be written in two versions, one for machines and another for humans. Additionally, standard XML tools are of little interest to manage these languages since specialized editors, analyzers and inference engines are required. To reduce information redundancy, Ontobroker provides a notation for embedding attribute-value pairs inside an HTML hyperlink tag. These tags may be used by the document's author to delimit an element to represent. Thus, each element may be implicitly referenced in the knowledge statement within the tag enclosing the element.

Along this same line, a document's author should be allowed to let some *knowledge statements be visible* to the reader. This is an obvious requirement when an especially intuitive notation can be used, e.g. when graphics can be made with a visual language[9] or sentences can be written using a "controlled language"[10], — a subset of natural language which eliminates sources of ambiguity. This visualization feature is also handy with any notation when the document provides explanations about the knowledge statements it stores. In this way, for example, a knowledge base and its associated documentation can be integrated within the same document and both accessed using classic searches (string-matching, navigation from the table of content, etc.) as well as knowledge-based searches. Though the Ontobroker metadata language was designed to reduce information redundancy, statements cannot be shown since they are within HTML tags. Futhermore, like RDF, the Ontobroker metadata language is essentially a notation for attribute-value pairs. Such a representation is general but basic and hard to read. Finally, since the indexation of document elements are made via HTML tags, only the document's author can index any of its parts. Others are limited to only those elements that are accessible via URLs.

---

[6] http://www.w3.org/RDF/

[7] http://wave.eecs.wsu.edu/CKRMI/OML.html

[8] http://www.aifb.uni-karlsruhe.de/WBS/broker/

[9] http://www.cpsc.ucalgary.ca/~kremer/home.html#visualLanguages

[10] http://www-uilots.let.uu.nl/Controlled-languages/

A metadata language should also be sufficiently *precise and general* to allow users to represent any Web-accessible information at the desired level of precision. This implies that the metadata language is based on an expressive formal model and that it has a notation allowing the user to exploit the expressivity of the formal model. Any formalism equivalent to first-order logic and permitting the use of contexts is an appropriate candidate, e.g. KIF[11] and Conceptual Graphs (CGs)[12] [10]. It is important not to restrict users but, for efficiency reasons, a search engine may ignore some features in knowledge statements. For example, a CG-based search engine may ignore references to sets within CGs and still exhibit adequate precision (the CGs with references to sets are also retrieved). The ontobroker metadata language and RDF are general but not precise in the sense that they are oriented towards the representation of entire documents (not arbitrary parts of them) and do not propose conventions to represent logic-based features, e.g. quantifiers and operators. This limits the capacity of the statements to be shared.

In summary, the three first requirements for *precise, flexible and scalable* information retrieval implies (i) several easy to use notations, some intuitive, some precise and expressive, and (ii) the possibilities to insert them anywhere in a Web document. We have satisfied these requirements by building a Web-accessible tool (CGI server[13]) named WebKB[14] [6][7] which interprets "chunks" of knowledge statements in Web documents. Each chunk, i.e. each group of statements, must be delimited by two special HTML marks ("`<KR>`" and "`</KR>`") or the strings "$(" and ")$". These chunks are visible unless the document's author hides them with HTML comment tags. The knowledge representation language used in each chunk must be specified at its beginning, e.g.: "`<KR language="CG">`" At present, WebKB can interpret the linear notation of CGs plus less expressive but simpler linear notations we have invented: a formalised English, a frame-like CG linear notation and structures that relate document elements by semantic relations (some of these structures come from HTML). These simpler notations are translated into CGs. This formalism has been chosen first because it has a graphical notation and a linear notation, both concise and easily comprehensible, and secondly because we can reuse two CG inference engines (CoGITo [3] and Peirce [2]) that exploit subsumption relations defined between formal terms for calculating specialization relations between graphs — and therefore between a query and facts in a knowledge base. Hence, statements and queries may be made at different levels of granularity. In the future, other notations may be accepted and other formalisms exploited.

Another requirement is that *not all the terms in the knowledge statements should have to be explicitly declared and organized by each user.* Indeed, declaring and organizing terms is a tedious and often complex work that detters most users, and probably one of the main reasons why so few hypertext systems have

---

11 http://logic.stanford.edu/kif/kif.html

12 http://meganesia.int.gu.edu.au/~phmartin/WebKB/doc/CGs.html

13 http://www.w3.org/CGI/

14 http://meganesia.int.gu.edu.au/~phmartin/WebKB/

been knowledge-based (MacWeb [8] is an exception). This requirement is a rationale for semi-formal knowledge representation languages such as concept maps[15], as opposed to logic-based formalisms such as KIF. The use of semi-formal statements is at the expense of knowledge precision and accessibility but allows rapid expression and incremental refinement of knowledge. When forewarned by a special command ("no decl"), WebKB accepts CGs that include some undeclared terms. We show below how the imprecision may partially be compensated by exploiting ontologies. Another informal feature accepted by WebKB are notations for sets within CGs: WebKB ignores them during searches but displays each retrieved CG in the form it was entered (thus, notations for sets are displayed).

HTML and XML do not allow a user to reference — and hence index — any part of a document that s/he has not created. An *indexation notation allowing a document element to be referred by its content or occurrence in a document* is required. WebKB provides such a notation.

Simply representing knowledge within documents is insufficient, *knowledge-based and string-based commands* are also necessary. It is handy to be able to use them *within documents* and — if desired — have the results automatically inserted in the place of the commands. In the hypertext literature, such a technique is known as *dynamic linking*, and the generated document is called a *dynamic document* or a *virtual document* [8]. This idea has many applications, e.g. adapting a document content to a user. A procedural or declarative language is necessary to combine the commands and their results. Web robots (e.g. Harvest, WebSQL, WebLog) perform some document generation in that way but current metadata languages only allow knowledge representation. WebKB permits the generation of virtual documents and combines lexical, structural and knowledge-based data management by proposing (i) commands for searching and joining CGs, (ii) Unix-like file management commands working on Web-accessible documents, (iii) a simple Unix shell-like script language to combine commands. These commands may be inserted in documents. They may also be directly sent to WebKB by programs or manually from form-based interfaces, e.g. the WebKB interfaces. Figure 1 shows the WebKB tool menu and the "Knowledge-based Information Retrieval/Handling Tool".

The four following sections respectively illustrate the ideas of the last four paragraphs. Though this document-based approach is handy, its scalability is limited. For example, before using knowledge query commands, the WebKB user must either directly assert some knowledge or use loading commands (such as "load URL") to specify Web documents that include the knowledge to exploit. Considering its features, WebKB may be seen as a knowledge-based *directed* Web robot and *private* annotation tool. To allow users to benefit from the knowledge of users they do not know, and therefore to enable WebKB to also be a knowledge-based *shared* annotation tool, we are extending it to handle *a cooperatively built knowledge repository.* We address this issue in section 6.

[15] http://www.cpsc.ucalgary.ca/~kremer/home.html#CM

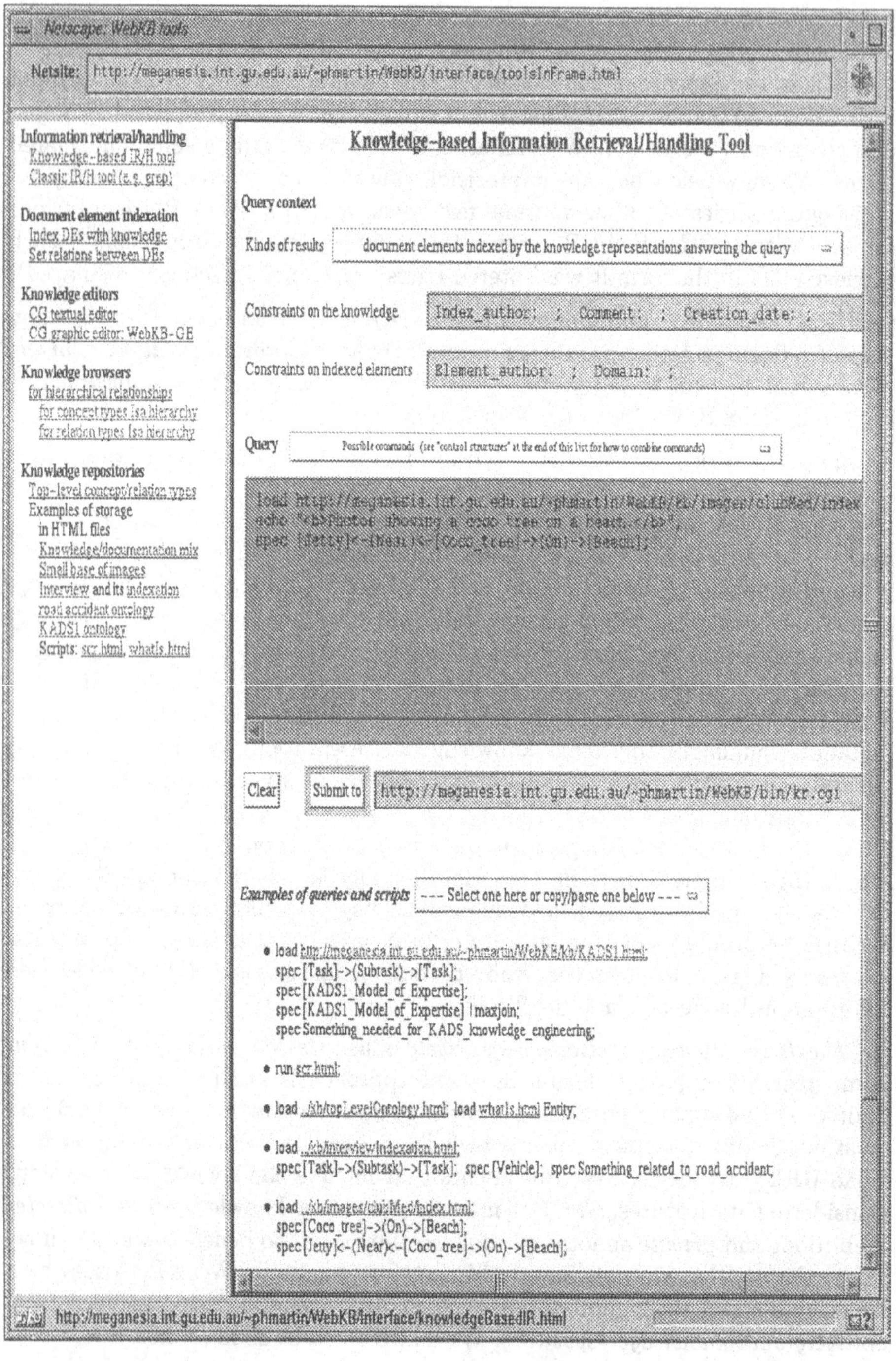

Netscape: WebKB tools

Netsite: http://meganesia.int.gu.edu.au/~phmartin/WebKB/interface/toolsInFrame.html

Information retrieval/handling
Knowledge-based IR/H tool
Classic IR/H tool (e.g. grep)

Document element indexation
Index DEs with knowledge
Set relations between DEs

Knowledge editors
CG textual editor
CG graphic editor: WebKB-GE

Knowledge browsers
for hierarchical relationships
for concept types Isa hierarchy
for relation types Isa hierarchy

Knowledge repositories
Top-level concept/relation types
Examples of storage
in HTML files
Knowledge/documentation mix
Small base of images
Interview and its indexation
road accident ontology
KADS1 ontology
Scripts: scr.html, whatIs.html

Knowledge-based Information Retrieval/Handling Tool

Query context

Kinds of results: document elements indexed by the knowledge representations answering the query

Constraints on the knowledge: Index_author: ; Comment: ; Creation_date: ;

Constraints on indexed elements: Element_author: ; Domain: ;

Query: Possible commands (see "control structures" at the end of this list for how to combine commands)

```
load http://meganesia.int.gu.edu.au/~phmartin/WebKB/kb/images/clubMed/index
echo "<b>Photos showing a coco tree on a beach</b>";
spec [Jetty]<-(Near)<-[Coco_tree]->(On)->[Beach];
```

Clear | Submit to: http://meganesia.int.gu.edu.au/~phmartin/WebKB/bin/kr.cgi

Examples of queries and scripts: --- Select one here or copy/paste one below ---

- load http://meganesia.int.gu.edu.au/~phmartin/WebKB/kb/KADS1.html;
  spec [Task]->(Subtask)->[Task];
  spec [KADS1_Model_of_Expertise];
  spec [KADS1_Model_of_Expertise] | maxjoin;
  spec Something_needed_for_KADS_knowledge_engineering;
- run scr.html;
- load ../kb/topLevelOntology.html; load whatIs.html Entity;
- load ../kb/interviewIndexation.html;
  spec [Task]->(Subtask)->[Task]; spec [Vehicle]; spec Something_related_to_road_accident;
- load ../kb/images/clubMed/index.html;
  spec [Coco_tree]->(On)->[Beach];
  spec [Jetty]<-(Near)<-[Coco_tree]->(On)->[Beach];

http://meganesia.int.gu.edu.au/~phmartin/WebKB/interface/knowledgeBasedIR.html

Fig. 1. The WebKB tool menu and knowledge-based Information Retrieval/Handling Tool. *This example shows how a document containing CGs is sent to the WebKB server and how the command "spec" is used to retrieve CGs and the images they index*

2 Representing Knowledge

To represent knowledge within documents, we advocate the use of knowledge representation languages over XML-based metadata Languages. To compare the alternatives, Figure 2 shows how a simple sentence may be represented with CGs in WebKB, with KIF and with RDF. The sentence is: "John believes that Mary has a cousin who has the same age as her".

The CG representation (top) seems simpler than the others. The semantic network structure of CGs (i.e. concepts connected by relations) has three advantages: (i) it restricts the formulation of knowledge without compromising expressivity and this tends to ease knowledge comparison from a computational viewpoint; (ii) it encourages the users to explicit relations between concepts (as opposed, for instances, to languages where "slots" of frames or objects can be used); (iii) it permits a better visualization of relations between concepts.

Even if CGs seem relatively intuitive, they are not readable by everyone. Often, simpler notations could be used. For instance, WebKB accepts alternative notations for CGs. We call two of them "Frame-CGs" and "Formalised English". In Frame-CGs, the above sentence could be represented in that way: `[Mary. Age: a. Cousin: [Person. Age: a] ].` Here are two possibilities in Formalised English: `John believes {Mary has for age A and has for cousin a person who has for age A}.` and `{Mary has for cousin a person who has for chrc an age chrc of Mary}(Believer:John).`

By default, WebKB accepts that statements expressed in Frame-CGs and Formalised English include undeclared terms. On the opposite, unless forewarned by the command "no decl", WebKB requires that terms used in CGs are declared.

3 Allowing Undeclared Terms in Knowledge Statements

The user may not want to take the time to declare and order most of the terms s/he uses when representing knowledge. This may, for example, be the case when a user indexes sentences from various documents for private knowledge organisation purposes.

To permit this, and still allow the system to perform some minimal semantic checks and knowledge organisation, we propose the casual user represent knowledge with basic declared relation types and leave undeclared the terms used as concept types. This method has the following rationales.

- If knowledge statements are made from concepts linked by basic relations, i.e. if the complexity is manifest within concept types rather than in relation types, only a limited set of relation types are necessary for an application. WebKB already proposes a top-level ontology of 200 basic relation types[16] [4] [5] collecting common thematic, mathematical, spatial, temporal, rhetorical and argumentative relations types.

[16] http://meganesia.int.gu.edu.au/~phmartin/WebKB/kb/topLevelOntology.html

```
<KR language="CG">
load "http://www.bar.com/topLevelOntology";    //Import this ontology
Age < Property;                     //Declare Age as a subtype of Property
Cousin(Person,Person) {Relation type Cousin};

[Person:"John"]<-(Believer)<-[Descr: [Person:"Mary"]-
                                        { (Chrc)->[Age: *a];
                                          (Cousin)->[Person]->(Chrc)->[*a];
                                        } ];
</KR>
```

```
<KR language="KIF">
load "http://www.bar.com/topLevelOntology";    //Import this ontology
(Define-Ontology Example (Slot-Constraint-Sugar topLevelOntology))
(Define-Class Age (?X)  :Def (Property ?X))
(Define-Relation Cousin(?s ?p) :Def (And (Person ?s) (Person ?p)))
(Exists ((?j Person))
   (And (Name ?j John) (Believer ?j '(Exists ((?m Person) (?p Person) (?a Age))
                                        (And (Name ?m Mary) (Chrc ?m ?a)
                                             (Cousin ?m ?p) (Chrc ?p ?a)
                                        ))
                  ))) </KR>
```

```
<RDF xmlns:rdf="http://www.w3.org/TR/WD-rdf-syntax#"
     xmlns:t="http://www.bar.com/topLevelOntology">
  <Class ID="Age"><subClassOf resource="t#Property"/></Class>
  <PropertyType ID="Cousin"><comment>Relation type Chrc</comment>
                            <range  resource="t#Person"/>
                            <domain resource="t#Person"/></PropertyType></RDF>
<RDF xmlns="http://www.w3.org/TR/WD-rdf-syntax#"
     xmlns:t="http://www.bar.com/topLevelOntology"
     xmlns:x="http://www.bar.com/example.html"> <!-- x refers to this file -->
  <Description aboutEach="#Statement_01">
     <t#Believer>John</t#Believer> </Description>
  <t#Person bagID="Statement_01">
     <t#Name>Mary</t#Name>
     <x#Chrc><x#Age ID="age"></x#Age></x#Chrc>
     <x#Cousin><t#Person><x#Chrc resource="#age"/></t#Cousin>
  </t#Person> </RDF>
```

Fig. 2. Comparing knowledge representation with CGs, KIF and RDF

- WebKB can use relation signatures to give suitable types to the undeclared terms used as concept types. For instance, in the top-level ontology proposed by WebKB, the relation types *Input, Output, Agent, Method, Sub-Process* and *Purpose* are all defined to have a concept of type *Process* as the first argument. Hence, in the previous example, WebKB can infer that *Knowledge_design* must be a subtype of *Process*.
- We have merged the natural language ontology WordNet[17] (120,000 words linked to 90,000 concept types) into our top-level ontology (cf. [4] [5]). When the WebKB shared repository is implemented and initialized with these ontologies, it will be possible for WebKB to semi-automatically relate the undeclared terms used as concept types to precise concept types in the repository, thanks to links between words and concept types and to constraints imposed by the relation signatures. Consider for example, the following CG where the terms *Cat* and *Table* have not been declared: `[Cat]->(On)->[Table]`. In WordNet, the word *cat* has 5 meanings (feline, gossiper, X-ray, beat and vomit) and the word *table*, 5 meanings (array, furniture, tableland, food and postpone). In the WebKB ontology, the relation type *On* connects a concept of type *Spatial_entity* to another concept of the same type. Thus, WebKB can infer that "beat" and "vomit" are not the intended meanings for *Cat*, and "array" and "postpone" are not the intended meanings for *Table*. To further identify the intended meanings, WebKB could prompt the following questions to the user: "does *Cat* refer to feline, gossiper, X-ray or something else?" and "does *Table* refer to furniture, tableland, food or something else?".
- Knowledge statements are more readily comparable if they follow the same conventions. The convention of using basic relations is thus important. (The opposite convention — using primitive concepts and complex relations — would be much harder to follow). Consider for example the sentence "Mary is 20 years old". Following our conventions it is better to use the concept type *Age*, e.g. `[Person:"Mary"]->(Chrc)->[Age:@20]`,
rather than the relation type *Age*, e.g. `[Person:"Mary"]->(Age)->[Integer:20]`, unless this relation type has been predefined by a user: [18]

```
relation Age (x,y) is  [Age]- { (Chrc)->[Living_entity:*x];
                                (Measure)->[Integer:*y];
                              }
```

The commands "decl" and "no decl" enable the user to overide the default modes for the acceptatance of undeclared terms. Furthermore, an exclamation mark before a type explicitly tells the system that the type was deliberately left undeclared. Quoted sentences may also be used: they are understood by WebKB as individual concepts of type "Description".

Another facility of the WebKB parser is that, like HTML browsers, it ignores HTML tags (except definition list tags) in knowledge statements. However, when

---

[17] http://www.cogsci.princeton.edu/~wn/

[18] This solution implies that the inference engine expands the relation type definition when comparing graphs. Few CG engines can perform type expansion.

these statements are displayed in response to a query, they are displayed using the exact form given by the user, including HTML tags. Thus, the user may combine HTML or XML features with knowledge statements, e.g. s/he may put some types in italics or make them the source of hypertext links.

4 Indexing Any Document Element Using Knowledge

4.1 General Cases

We call a Document Element (DE) any textual/HTML data, e.g. a sentence, a section, a reference to an image or to an entire document. This definition excludes binary data but includes textual knowledge statements. WebKB allows users to index any DE of a Web-accessible document (or later of our repository) by knowledge statements, or connect DEs by relations. Figure 3 shows an example of each case.

```
$(Indexation
   (Context: Language: CG;
             Ontology: http://www.bar.com/topLevelOntology.html;
             Repr_author: phmartin; Creation_date: Mon Sep 14 1998;
              Indexed_doc: http://www.bar.com/example.html; )
   (DE: {2nd occurence}  the red damaged vehicle )
   (Repr: [Color: red]<-(Color)<-[Vehicle]->(Attr)->[Damaged] )
)$

$(DEconnection
   (Context: Language: CG;
             Ontology: http://www.bar.com/topLevelOntology.html;
             Repr_author: phmartin; Creation_date: Mon Sep 14 1998;)
   (DE: {Document: http://www.bar.com/example.html} )
   (Relation: Summary)
   (DE: {Document: http//www.bar.com/example.html}
                   {section title: Abstract})
)$
```

Fig. 3. A language for indexing or connecting any Web-accessible document element

XML provides more ways to isolate and reference DEs than HTML. Since WebKB exploits the capacities of Web-browsers, the XML mechanisms may be used by the WebKB users. However, XML does not help users to annotate others' documents since DEs cannot be referenced if they have not been explicitly delimited by the documents' authors. Therefore, the WebKB facility of referring to a DE by specifying its content and its occurrence number will still be useful.

4.2 A Simple Example

The above indexation notations allow the statements and the indexed DEs to be in different documents. Thus, any user may index any element of a document on the Web. Figure 1 presents a general interface for knowledge-based queries and shows how a document containing knowledge must be loaded in the WebKB processor before being queried.

WebKB also allows *the author of a document* to index an image by a knowledge statement directly stored in the "alt" field of the HTML "img" tag used to specify the image. We use this special case of indexation to present a simple illustration of WebKB's features. This example, shown in Figure 5, is a good synthesis but in no way representative of the general use of WebKB — it is not representative because it mixes the indexed source data (in this case, a collection of images), their indexation, and a customized interface to query them, in a single document. Figure 4 shows a part of this document that illustrates the indexation. The result of the query shown in Figure 5 is displayed in Figure 6.

```
<h3><a name="WarmRegion">Warm areas</a></h3>

<h4>Islands and beaches</h4>

<p><KR language="CG">
<img src="http://meganesia.int.gu.edu.au/~phmartin/WebKB/kb/images/club
     alt="[Boat]-
            { (Near)->[Island];
              (On)->[Sea];
              (On)<-[Person];
            }">
<img src="http://meganesia.int.gu.edu.au/~phmartin/WebKB/kb/images/club
     alt="[Curved]<-(Attr)<-[Jetty]->(Near)->[Island]">
<img src="http://meganesia.int.gu.edu.au/~phmartin/WebKB/kb/images/club
     alt="[Straight]<-(Attr)<-[Jetty]->(Near)->[Island]">
<img src="http://meganesia.int.gu.edu.au/~phmartin/WebKB/kb/images/club
     alt="[Straight]<-(Attr)<-[Jetty]->(Near)->[Island]">

<p><img src="http://meganesia.int.gu.edu.au/~phmartin/WebKB/kb/images/c
     alt="[House]->(On)->[Island]">
<img src="http://meganesia.int.gu.edu.au/~phmartin/WebKB/kb/images/club
     alt="[Coco_tree]->(On)->[Beach]">
<img src="http://meganesia.int.gu.edu.au/~phmartin/WebKB/kb/images/club
     alt="[Coco_tree]->(On)->[Beach]">
<img src="http://meganesia.int.gu.edu.au/~phmartin/WebKB/kb/images/club
     alt="[Coco_tree]-
            { (On)->[Beach]->(Near)->[Jetty:*j]->(Attr)->[Straight];
              (Near)->[*j];
            }">
</KR>
```

Fig. 4. The HTML source of the indexation of the images shown in Figure 5

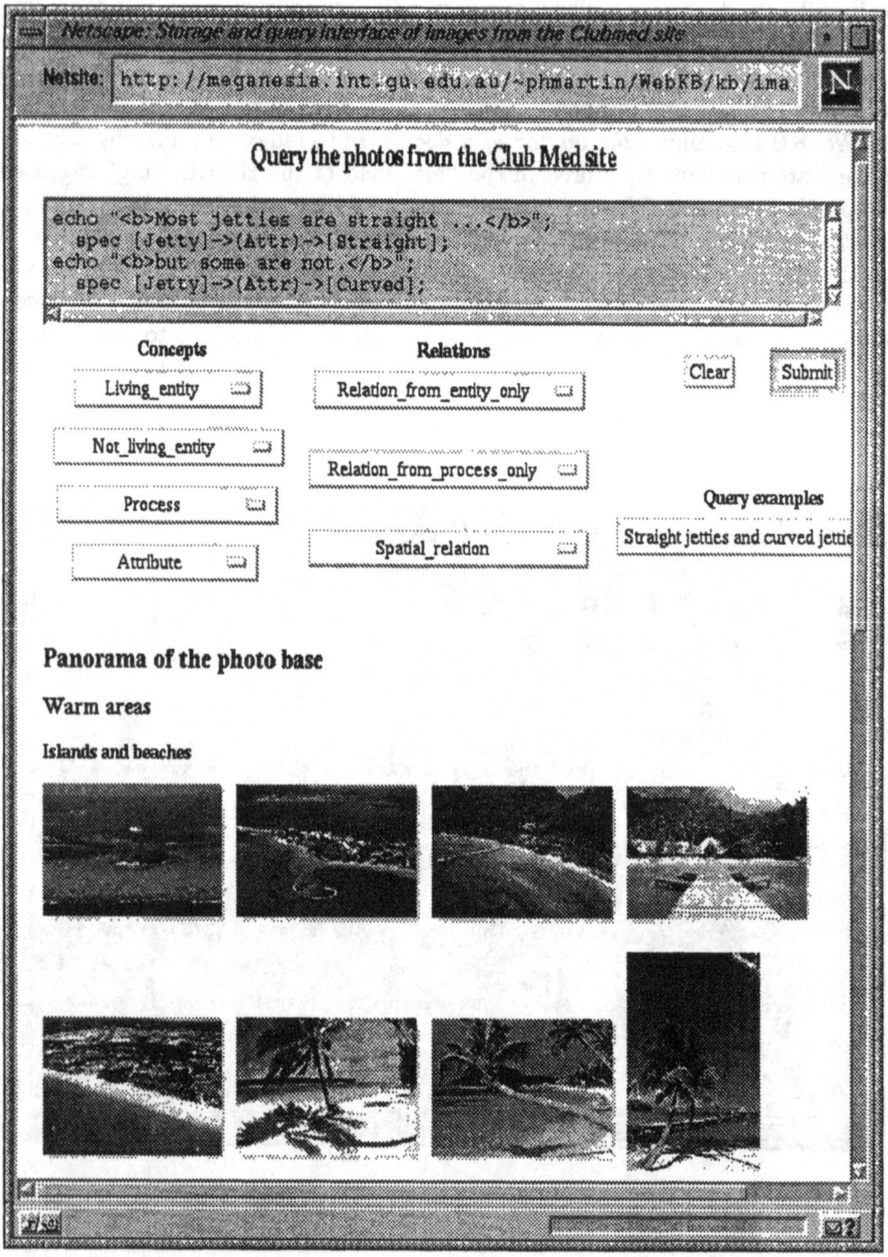

Fig. 5. Images, knowledge indexations and a customized query interface contained within a same document *(the example query shows how the command "spec" which looks for specializations of a CG can be used to retrieve images indexed by CGs. The results are shown in Figure 6)*

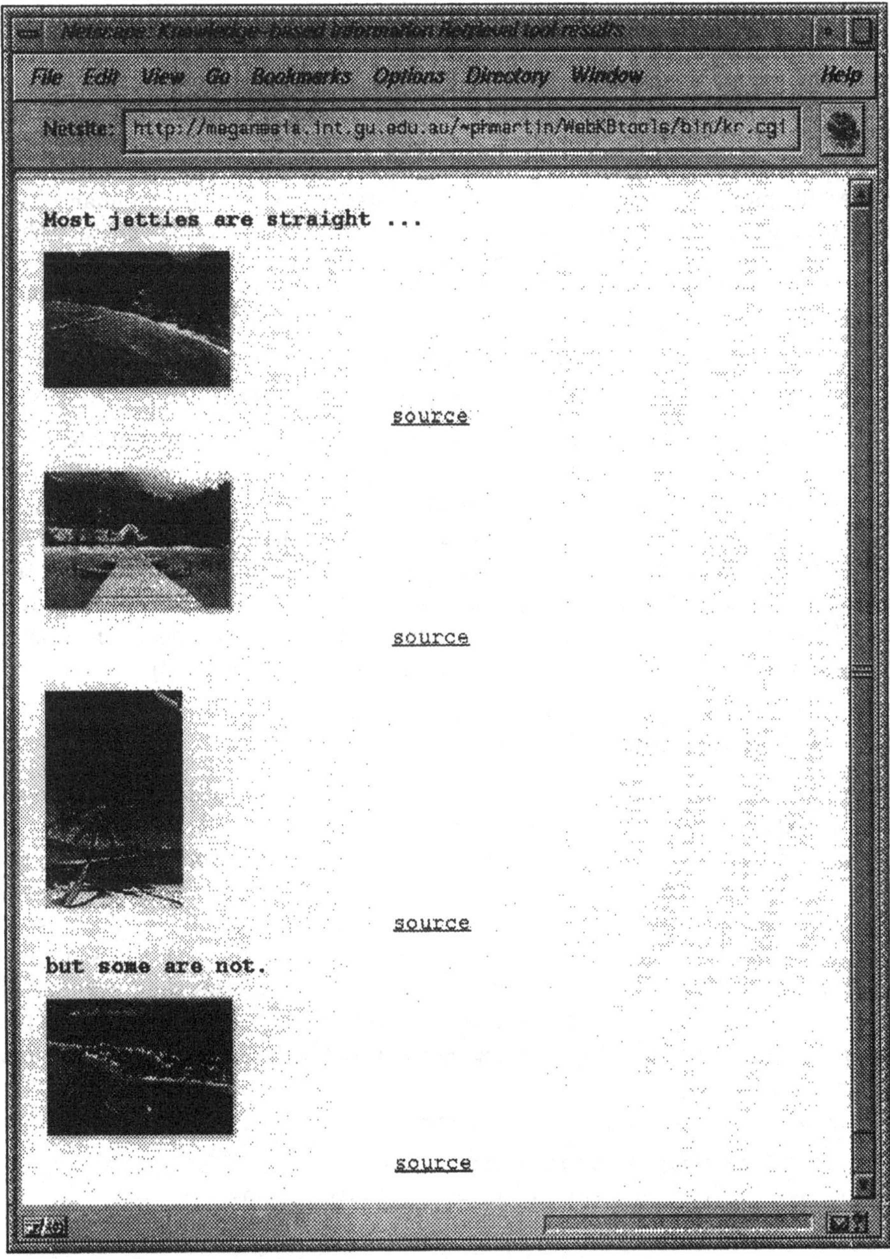

Fig. 6. The document generated in response to the query in Figure 5

5 Commands to Exploit Knowledge and Documents

5.1 Lexical and Structural Query Commands

Because WebKB proposes knowledge representation and query commands, and a script language, we have not felt the need to give it a lexical and structural query language as precise as those of Harvest, WebSQL and WebLog. Instead, we have implemented some Unix-like text processing commands for exploiting Web-accessible documents or databases and generating other documents, e.g. cat, grep, fgrep, diff, head, tail, awk, cd, pwd, wc and echo. We added the hyperlink path exploring command "accessibleDocFrom". This command lists the documents directly and indirectly accessible from given documents within a maximal number of hyperlinks. For example, the following command lists the HTML documents accessible from http://www.foo.bar/foo.html (maximum 2 levels) and that include the word "knowledge" in their HTML source code.

```
> accessibleDocFrom -maxlevel 2 -HTMLonly http://www.foo.bar/foo.html
    | grep -i knowledge     //''-i'' to search without regard to case
```

Lexical/structural queries – and knowledge queries – may be embedded inside documents so parts of these documents may be generated by WebKB using document elements or knowledge stored in other documents. Alternatively, with HTML documents, Javascript may be used for associating a query to an hypertext link in such a way that the query is sent to the WebKB processor when the link is activated (then, as for any other query, the WebKB processor generates an HTML document that includes the results; if the query has been sent from a Web-browser, this "virtual" document is automatically displayed).

5.2 Knowledge Query Commands

WebKB has commands for displaying specializations or generalizations of a concept or relation type or an entire CG in a knowledge base. At present, queries for CG specializations only retrieve connected CGs: the processor cannot retrieve paths between concepts specified in a query. If a retrieved CG indexes a document element, it may be presented instead of the CG (Figure 6 gives an example). In both cases, hypertext links are generated to reach the source of each answer presented in its original document. What follows is an example of such an interaction, assuming that http://www.bar.com/example.html is the file where the indexation in Figure 3 has been stored, and *Something* is the most general concept type.

```
> load http://www.bar.com/example.html

> spec [Something]->(Color)->[Color: red]
 [Color: red]<-(Color)<-[Vehicle]->(Attr)->[Damaged];
        Source

> use Repr   //display represented DEs

> spec [Something]->(Color)->[Color: red]
 the red damaged vehicle
        Source
```

Queries for specializations give the user some freedom in the way s/he expresses queries: searches may be done at a general level and subsequently refined according to the results. However, the exact names of types must be known. To improve this situation, WebKB allows the user to give only a substring of a type in a query CG if s/he prefixed this substring by the character %. WebKB generates the actual request(s) by replacing the substring by the manually/automatically declared types that include that substring. Replacements that violate the constraints imposed by relation signatures or individual types are discarded. Then, each remaining request is displayed and executed. For example, `spec [%thing]` will trigger the generation and execution of `spec [Something]`.

Knowledge query commands may be combined with the script language to generate complex documents, perform consistency tests on the knowledge base, or solve problems procedurally. The WebKB site provides many examples of queries and scripts. For example, one script solves the Sisyphus-I room allocation problem[19]. The reader is invited to test these examples[20].

Here is an example of a script that shows that the procedural language frees us to add some special operators to our query language, such as the modal operators "few" and "most", since they are easily definable by the user.

```
spec [Something] | nbArguments | set nbCGs;
spec [Cat] | nbArguments | set nbCGsAboutCat;
set nbCGsdiv2 'expr $nbCGs / 2';
if ($nbCGsAboutCat > $nbCGsdiv2)
{ echo "Most CGs of the base are about cats"; }
```

5.3 Knowledge Generation Commands

The only type of knowledge generation commands in WebKB are commands that join CGs. Various kinds of joins may be defined but WebKB only proposes joins which, given a set of CGs, create a new CG specializing each of the source CGs. Though the result is inserted in the CG base, it may not represent anything true for the user, but provides a device for accelerating knowledge representation. For instance, in WebKB, CGs related to a type may be collected and automatically merged via a command such as this one: `spec [TypeX] | maxjoin`. The result may then serve as a basis for the user to create a type definition for TypeX.

The following is a concrete example for the maximal join command.

```
> maxjoin  [Cat]->(On)->[Mat]  [Cat:Tom]->(Near)->[Table]
[Cat:Tom]- { (On)->[Mat];  (Near)->[Table]; }
```

[19] http://meganesia.int.gu.edu.au/~phmartin/WebKB/kb/sisyphus1.html

[20] http://meganesia.int.gu.edu.au/~phmartin/WebKB/
or if this server is down: http://www.int.gu.edu.au/~phmartin/WebKB/

6 Scalable Cooperatively Built Knowledge Repositories

Some servers, called ontology servers, support shared knowledge repositories, e.g. the Ontolingua ontology server[21] and Ontosaurus[22]. However, they are not usable for managing large quantities of knowledge and, apart from AI-Trader [9][23], they do not allow the indexation and retrieval of parts of documents. Finally, support of cooperation between the users is essentially limited to consistency enforcement, annotations and structured dialogues, as in APECKS[24], Co4[25] and Tadzebao[26].

We are currently extending WebKB to handle a knowledge repository. As this implementation has just begun, we do not detail this extension[27]. However, here are the five points through which we address scalability: (i) a scalable multi-user persistent object repository to support the storage and exploitation of knowledge structures (we have chosen Shore[28]); (ii) algorithms allowing the exploitation of large-scale dynamic taxonomies efficiently (we have chosen Fall's algorithms[29]; (iii) visualisation techniques (mainly the handling of aliases for terms and the generation of views) to avoid lexical conflicts and enable users to focus on certain kinds of knowledge; (iv) protocols to allow users to solve semantic conflicts via the insertion of new terms and relations in the common ontology and, in some cases, in the knowledge of other users; (v) conventions for representing knowledge to improve the automatic comparison of knowledge from different users and hence their consistency and retrieval.

Though these five points permit the exploitation of a large knowledge repository (that is essential for efficiency reasons and practical use), it is also clear that for efficiency and reliability reasons, a unique server cannot be used to handle a universal knowledge repository by all Web users. Knowledge has to be distributed and mirrored on various knowledge servers. However, since there is no static conceptual schemas in knowledge bases, the techniques of distributed database systems - such as AlephWeb[30], Hermes[31] and Infomaster[32], cannot all be reused since they exploit a fixed conceptual schema (ontology) associated to each database. In knowledge bases, the ontology is constantly modified by the users.

---

[21] http://WWW-KSL-SVC.stanford.edu:5915/
[22] http://www.isi.edu/isd/ontosaurus.html
[23] http://www.vsb.informatik.uni-frankfurt.de/projects/aitrader/intro.html
[24] http://www.psychology.nottingham.ac.uk/staff/Jenifer.Tennison/APECKS/
[25] http://ksi.cpsc.ucalgary.ca/KAW/KAW96/euzenat/euzenat96b.html
[26] http://ksi.cpsc.ucalgary.ca:80/KAW/KAW98/domingue/
[27] http://meganesia.int.gu.edu.au/~phmartin/WebKB/doc/coopKBbuilding.html
[28] http://www.cs.wisc.edu/shore/
[29] http://www.cs.sfu.ca/cs/people/GradStudents/fall/personal/index.html
[30] http://www.pangea.org/alephweb.aleph/paper.html
[31] http://www.cs.umd.edu/projects/hermes/
[32] http://infomaster.stanford.edu/infomaster-info.html

A first step to the distribution of a knowledge repository is to duplicate it on several servers, with updates made on a server automatically duplicated in other servers. Some servers may be dedicated to searches and others to updates.

A second step is to have general servers and specialized servers. General servers would probably have knowledge bases with a content similar to the CYC knowledge base[33]. A specialised server would store the same knowledge as general servers plus knowledge related to a well-defined set of objects, e.g. knowledge expressed with the subtypes of certain types. Since these sets of objects are well-defined (extensively or via definitions), a general server would store the URLs of these servers and, when answering a query, would delegate the query to the relevant servers if more precision is required. These sets of objects might be determined by the managers of specialized servers, or according to the frequency of accesses to objects in knowledge repositories. Whatever the specialised server a user updates, if the knowledge it enters is relevant to other servers (e.g. if the knowledge is expressed with general terms), it should be automatically duplicated in these servers The rationale of all these duplications is to speed searches and simplify the query mechanisms by avoiding, whenever possible, parallel searches in various servers and then the composition of the results.

Other steps may be necessary, but what should be avoided in this knowledge-based approach (hence precision-oriented) is to let the specialized servers developp independently of each others instead of being part of a unique consistent virtual knowledge repository. Otherwise, conceptual queries and cooperation across the repositories would no longer be possible; this is the case in current traders where the repository most relevant repository to answer a query is automatically "guessed".

Finally, knowledge servers should not be limited to the storage of knowledge statements but should also allow the storage and handling of knowledge-based and document-based commands similar to the storage and handling procedure we described for documents.

7 Conclusion

Current information retrieval techniques are not knowledge-enabled and hence cannot give precise answers to precise questions. To overcome this problem, a current trend on the Web is to allow users to annotate documents using metadata languages.

On the basis of ease and representational completeness, we have argued for the use of general and intuitive knowledge representation languages such as CGs rather than the direct use of XML-based languages. To allow users to represent knowledge at the level of detail they require, we have proposed simple notations for restricted knowledge representation cases and a technique allowing users to leave knowledge terms undeclared. Our knowledge representation/retrieval tool WebKB supports this approach and allows its users to combine lexical, structural and knowledge-based techniques to exploit or generate Web documents.

[33] http://www.cyc.com/

We have shown why the WebKB features are needed for precision and scalability. To overcome the scalabilitity limitations inherent to directed Web robots and private annotation tools, we are extending it to also handle a scalable cooperatively built knowledge repository.

Acknowledgments

This work is supported by a research grant from the Australian Defense, Science and Technology Organisation.

References

1. Decker, S., Fensel, D.: Ontobroker: Ontology Based Access to Distributed and Semi-Structured Information. In: Meersman, R. (eds.), Semantic Issues in Multimedia Systems, Kluwer Academic Publisher, Boston (1999)
2. Ellis, G.: Managing Complex Objects. Ph.D thesis, Queensland University, Australia (1995)
3. Haemmerlé, O.: *CoGITo: une plate-forme de développement de logiciels sur les graphes conceptuels.* Ph.D thesis, Montpellier II University, France (1995)
4. Martin, Ph.: Using the WordNet Concept Catalog and a Relation Hierarchy for Knowledge Acquisition. In: Peirce'95, 4th Peirce workshop, California (1995) http://www.inria.fr/acacia/Publications/1995/peirce95phm.ps.Z
5. Martin, Ph.: *Exploitation de graphes conceptuels et de documents structurés et hypertextes pour l'acquisition de connaissances et la recherche d'informations*, PhD Thesis, University of Nice - Sophia Antipolis, France (1996)
6. Martin, Ph., Eklund, P.: WWW Indexation and Document Navigation Using Conceptual Structures. In: ICIPS'98, 2nd IEEE International Conference on Intelligent Processing Systems, IEEE Press (1998) 217–221
7. Martin, Ph., Eklund, P.: Embedding Knowledge in Web Documents. In: WWW8, 8th International World Wide Web Conference, Toronto, Canada (1999)
8. Nanard, J., Nanard, M., Massotte, A-M., Djemaa, A., Joubert, A., Betaille, H., Chauch, J.: Integrating Knowledge-based Hypertext and Database for Task-oriented Access to Documents. In: DEXA'93, LNCS Vol. 720, Springer-Verlag, Prague (1993) 721–732
9. Puder, A., Romer, K.: Generic Trading Service in Telecommunication Platforms. In: ICCS'97, 5th International Conference on Conceptual Structures, LNAI 1257 Springer Verlag (1997) 551–565.
10. Sowa, J.F.: Conceptual Structures: Information Processing in Mind and Machine. Addison-Wesley, Reading, MA (1984)

Synergy as an Hybrid Object-Oriented Conceptual Graph Language

Adil KABBAJ

INSEA, Rabat, Morocco, B.P. 6217
Fax : (212) 7 77 94 57
akabbaj@insea.ac.ma

Abstract. This paper presents the use of Synergy as *an Hybrid Object-Oriented Conceptual Graph Language* (HOO-CGL). Synergy is an *implemented* visual multi-paradigm language based on *executable conceptual graphs* with an *activation interpretation*, instead of a logical one. This paper describes the formulation in Synergy of basic concepts of the hybrid object-oriented paradigm: encapsulation, definition of a class with methods and daemons, method and daemon definitions, class hierarchy, instance and instantiation mechanism, inheritance (both property and method inheritance), method call, method execution and daemon invocation due to accessing data. An example is used to illustrate the presentation of such an Hybrid Object-Oriented Conceptual Graph Language.

1 Introduction and previous works

Due to the relevance of the object-oriented paradigm (more abstraction, encapsulation and modularity), its relationship to conceptual graphs has been investigated by several authors [2, 20, 13, 18, 3, 1].
As noted early by [2], melding conceptual graphs (CG) with object-oriented principles (OOP) to produce object-oriented CG "presents the possibility of greater semantic capture than can be obtained with either technique alone".
A CG-language that aims to incorporate OOP should support (at least) the concepts of class, instance, method, inheritance, instantiation and sending messages. This paper presents a running implementation of such an OO-CG language. Moreover, the language is enhanced by the incorporation of daemons to enable reactive OO-CG programming.
To motivate the need for such a language and to put this research in context, this section gives an overview of related works.

In the mapping of OO concepts to CG, it is straight-forward to map a declarative part of a class definition to a type definition, an instance description to a concept with a specific referent and description, and OO inheritance to type hierarchy. Several approaches have been proposed however for method description and call : map a method to an actor graph [2], map a method to a schema with bound actors [20], map

a method to a process defined with a CG (or a set of CGs) [18], map a method to a defined goal in Prolog+CG language [3] and map a method to a dyadic relation defined as a single state-transition description [1].
According to the standard view of CG [16], CG is a purely declarative notation and to perform actions, one should use actors and functional dataflow graphs [16]. In their formulation of methods, Hines and al. [2] and Wuwongse and Ghosh [20] follow this standard view. Thus, they commit method descriptions to the limited expressive power of functional dataflow graphs as outlined by Sowa [16, 17, 19] (see also [6] for a more detailed discussion).

Sowa [18] and Ellis [1] use CG to represent both declarative and procedural knowledge. They propose CG as a logical foundation and as a specification language for object-oriented systems. Sowa proposal is based on two aspects of CG : a) using contexts to realize encapsulation and, b) using Peirce's rules of inference to define method execution as theorem proving. Two additional rules have been proposed by Sowa to enable the insertion of a message in the context of the object.
Ellis [1] notes that if the method is defined outside the type definition and it is defined as a dyadic relation with a simple state transition semantic and a modus-ponens interpretation, then method execution (by theorem proving) will be more efficient and the two additional rules proposed by Sowa will be of no use.
Both Sowa and Ellis focused on method execution only and this process is presented through a simple example only. The generality of their models has not been considered and the models should be developed further to assess their claims (e.g., to consider in detail all the concepts of OOP, and to develop an executable language and a running implementation). These remarks apply also to the approaches of [2] and [20].
Also, defining methods outside the class definition, as actors [2], as schemas [20] or as relations [1], violates the encapsulation principle which constitutes the basis of OOP. Like Sowa and as explained in this paper, we use contexts to realize encapsulation and to define methods inside class definition.

Lukose presents in [13, 14] "actor graphs" as the result of the merger of CG and OO principles : "Actor Graph = CG + Actor"; CG is used to define the declarative part of an action while "Actor" represents the procedural part. OO principles are used to define a domain-dependent notation for the definition of actors : on top of the Prolog language and the modules of the Deakin ToolSet, Lukose developed a Prolog-based data structure in order to define actors as action classes and to build and use actor hierarchy. Methods of an actor are defined as rules of Prolog and CG is used by actors as data structure, beside other Prolog data structures. Thus, CG is neither used to define actor classes nor to define the methods of an actor class.

All the above works have been concerned by the class-OO paradigm. In fact, the OO paradigm is subdivided in general in class-OO paradigm and frame-OO paradigm, leading to two OO language families. The first paradigm is message-driven and computation is based on method execution while the second is data-driven and computation is based on daemon invocation due to accessing data [15]. It has been recognized early that the two OO paradigms are complementary [12] and, to get more flexibility and to deal with complex and real-world applications, they should be integrated in an hybrid OO paradigm. And in fact, several hybrid OO languages have been proposed [15].

This paper presents the *Synergy* language as *an hybrid or reactive object-oriented CG language*. Synergy is a visual multi-paradigm language based on CG activation [5, 6, 7, 8, 9, 10]. Synergy is multi-paradigm since it integrates functional, procedural, process, reactive, object-oriented and concurrent object-oriented paradigms. The integration is done using CG as the basis knowledge structure, without actors or other external notation. In Synergy, CG is used as a knowledge representation formalism with activation-based interpretation instead of a logical one.

The Synergy language constitutes the main part of the SYNERGY environment [10]. SYNERGY is a CG tool implemented with Microsoft Visual C++ (a portable Visual J++ implementation will be available in summer 1999) and it is composed of a CG graphic editor, the Synergy language, an engine for dynamic formation of knowledge bases [4, 6, 11] and an information retrieval process. The SYNERGY tool is described in [6, 10], the SYNERGY site (www.insea.ac.ma/InseaCGTools/Synergy.html) is under construction and a SYNERGY group is being formed to enhance, extend and use SYNERGY in many ways.

This paper is organized as follows : sections 2 recalls some basic elements of Synergy. Section 3 presents the anatomy of hybrid object-oriented CG programming in Synergy: definition of a class with methods and daemons, method and daemon definitions, class hierarchy, instance and instantiation mechanism, method call (message), inheritance mechanism (both property and method inheritance) and method/daemon execution. An example is used in section 3 to illustrate the different concepts of such an hybrid object-oriented CG language. Using the same example, section 4 presents some detail of methods and daemons execution. Finally, section 5 gives a conclusion with an outline of some current and future works.

2 An overview of the Synergy language

This section recalls some basic elements of the Synergy language that are mandatory to understand the following sections. We refer the reader to [10] for a more detailed presentation of Synergy.

An *application* (e.g., a program) in Synergy is represented as a CG with two concepts: [LongTermMemory] and [WorkingMemory]. The top window in Fig. 1 represents the application "OOSample". The contain of the concept [LongTermMemory] describes the knowledge base (KB) of the application in terms of concept type hierarchy augmented by type definitions, individual descriptions, type schemas and type synonyms. The whole KB is represented by a CG.

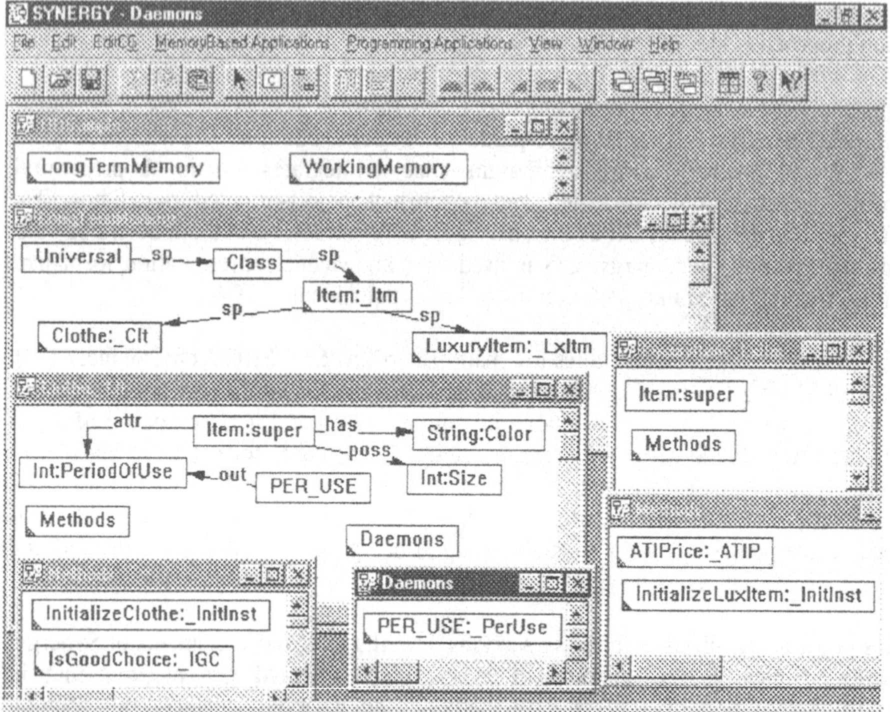

Fig. 1. An OO example - a class hierarchy. To have a concise drawing of CG, the graphic editor of SYNERGY does not encircle relations with circles and it shows only the type and the referent of a concept; the description and the state fields are shown only upon a demand from the user. However and as illustrated in this Fig. and in Fig. 2, the CG editor draws a small triangle at the left-bottom of a rectangle to indicate that the concept has a description. The triangle is blue if the description is simple and red if it is a CG. The state field has a visual representation too : the concept at state "trigger" has a green rectangle, the rectangle is "red" if the state is "in-activation", etc. For more information on the CG graphic editor of SYNERGY see [6, 10].

The second window in Fig. 1, named "LongTermMemory", shows the contain of the concept [LongTermMemory], which is the KB of the application "OOSample"; a simple class hierarchy.

In this paper, we consider KB composed of concept type hierarchy with type definitions only (without type schemas and type synonyms). Moreover, the types represent classes.

Relevant to this paper, please note that Synergy incorporates the encapsulation/modularity principle by allowing a concept type to be defined in the context of another concept type definition. As shown in the next section, this is useful for the encapsulation of methods (and daemons) in the class definition.

The concept [WorkingMemory] of an application describes the working space where the user specifies his requests, represented by CGs.

A *CG* in Synergy is a set of concepts related eventually by dyadic relations. A *relation* between two concepts can be either an element of the Synergy-Relation-Set (SRS) or it can be an identifier provided by the user. The SRS is composed of the primitive relations "in", "out", "next" and "grd" (guard) which are data/control relations and the primitive relations "sp" (specializationOf/generalizationOf), "inst" (instanceOf), "sta" (situationFor) and "synm" (synonymOf) which are "knowledge base relations"; they are used to organize the KB of a Synergy application.

A *concept* is described by four fields : [Type :Referent =Description #State]. The referent, if specified, can be an identifier of an instance, a variable (that can be bound to a referent) or a co-referent. A description, if specified, can be a simple data (for example, a real, a boolean or a string) or a composed data (a CG). The state field is defined on the following set : {steady, trigger, check-preconditions, wait-preconditions, wait-value, in-activation, wait-affectation, wait-end-affectation}. The state field represents the dynamic part of the concept. It is considered in detail in [6, 10].

3 The anatomy of Hybrid Object-Oriented Programming (HOOP) in Synergy

This section describes how Synergy supports the basic notions of HOOP : section 3.1 introduces class definitions, method and daemon definitions as well as class hierarchy, section 3.2 describes the instantiation mechanism and section 3.3 defines method call, inheritance and method/daemon execution.

3.1 Class hierarchy and class, method and daemon definitions

Definition 1. The root of any *class hierarchy* in Synergy is the "primitive" concept [Class]. Using the CG graphic editor of the environment SYNERGY, the user can define a new class and relate it to its direct super-class by the relation "sp" (specializationOf).

Fig. 1 shows a simple example of a class hierarchy, with a zoom-in on two classes ([Clothe] and [LuxuryItem]) including their "Methods" and "Daemons" contexts. This example will be used through the sub-sections of section 3. The definition of the class [Item] is given in Fig. 2.

Definition 2. A *class* is a concept type defined by its super-class, its declarative part and its procedural part.

- The *super-class* of the class is represented by the concept [SUPER_CLASS : super], with "SUPER_CLASS" that represents the direct super-type of the class type and "super" a key-word referent.
- The *declarative part* is composed of the *attributes* of the class. The attributes are represented by a set of concepts that could be connected to the super-class to form a connected CG.
- *Daemons* can be attached to attributes of the class. The set of daemons used in the class should be defined in the context [Daemons] contained in the class definition.
- The *procedural part* is composed of a set of *methods* defined in the context [Methods] which is contained in the class definition.
- Attributes, methods and daemons are optional.
- A *default value* for an attribute can be specified as a description for the corresponding concept.

In our example (Fig. 1 and 2), the class [Clothe] has [Item] as a direct super-class, it has three attributes ("Color", "Size", "PeriodOfUse"), two methods ("InitializeClothe" and "IsGoodChoice") defined in the context [Methods] and one daemon ("PER_USE") defined in the context [Daemons]. The daemon ("PER_USE") is used in the class definition to compute the value of the attribute "PeriodOfUse". Thus, if the value of this attribute is needed, the daemon [PER_USE] will be triggered.
The class [LuxuryItem] has also [Item] as a direct super-class, it has no attribute and has two methods ("ATIPrice" and "InitializeLuxItem"). The method "ATIPrice" is a redefinition of a method defined in the super-class [Item].

The class [Item] has [Class] as a direct super-class (Fig. 2), it has six attributes ("Reference", "Designation", "DateOfProduction", "NetPrice", "HTPrice", "Quantity"), five methods ("Retract", "Add", "InitializeItemPart", "ATIPrice", "TransPrice") and two daemons ("HTCPrice", "StockUnderFlow"). Each time the attribute "Quantity" gets a new value, the daemon [StockUnderFlow] will be triggered, since it has [Int: Quantity] as an input argument. Concerning the daemon [HTCPrice], it will be triggered if the value of the attribute [Real: HTPrice] is needed, since it has this attribute as an output.

Definition 3. The *definition of a method* or *a daemon* is represented by a CG (CG of Synergy, with the underlying computation model [6, 10]).

- A method and a daemon can be an operation (a procedure or a function) or a process. The user should specify, in the method/daemon definition, the concept [Operation : super] for the first case, or the concept [Process : super] for the second.
- A method and a daemon can have input argument(s)/parameter(s) and/or output argument(s)/parameter(s). A parameter is a reference to the corresponding argument.
- The definition of a method and of a daemon contains in general defining concepts (for local referents) and/or reference concepts; references to attributes of the class

where the method/daemon is defined, or to attributes of a super-class (direct or indirect), or even to defining concepts which should be contained, at time execution, in an outer context.

- The definition of a method and of a daemon can contain a concept that represents a call to : a) an operation defined locally in a context specified in the [LongTermMemory], b) an operation defined globally in the [LongTermMemory], or c) a primitive operation. In addition, the definition of a method can contain a concept that represents a call to a method of the same class or of a super-class (direct or indirect).

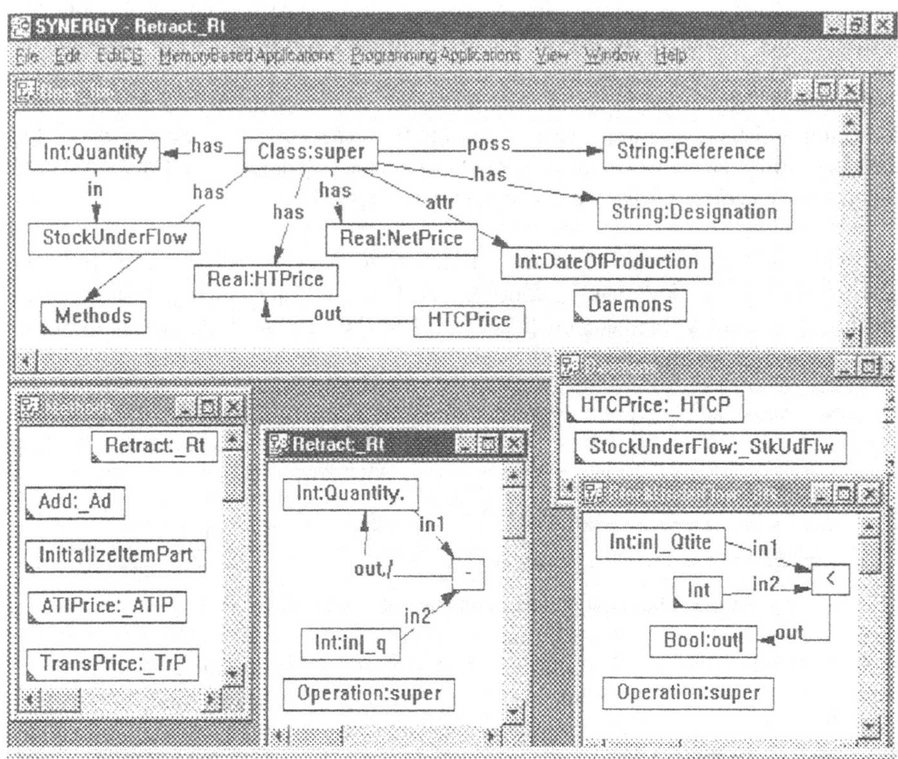

Fig. 2. Definitions of the class [Item] and of some of its methods and daemons

To illustrate the above definition, let us explain the definition of the method "Retract" and of the daemon "StockUnderFlow" (Fig. 2). The method [Retract] is defined as an operation with one input parameter [Int : in|_q] and no output parameter. "in|" and "out|" prefixes the referent of an input and an output parameter respectively. The definition of the type-method "Retract" contains a defining concept [Int : in|_q] for the referent-parameter "_q" and a reference concept [Int : Quantity.] to the attribute [Int : Quantity] of the class [Item]. In Synergy, the concept [Int : Quantity] represents

the defining node for the referent "Quantity" while the concept [Int : Quantity.] represents a bound or a referent node.
The daemon [StockUnderFlow] is defined as an operation with one input parameter [Int :in|_Qty] and one output parameter [Bool: out|_Res].

3.2 Instantiation mechanism

The Synergy working space, represented by the context [WorkingMemory], can be used by the user to specify entities (like instances of classes) and requests. For instance, [WorkingMemory = [Clothe: Clothe1]] contains the declaration of the class instance "Clothe1". At the specification time, the instance has no description, but during method execution the first access to the instance will trigger its instantiation.

Definition 4. The *creation of an instance of a class* consists in locating the class definition and then creating a description for the instance by a *partial copy* of the class definition. The copy is partial since the concepts [Methods] and [Daemons] (and of course, the content of the two concepts) are not copied.

Fig. 3 shows the result of "Clothe1" instantiation using the definition of the class [Clothe] (in Fig. 1). Note that the concepts [Methods] and [Daemons] are not copied but the attachment of the daemon [PER_USE] to the attribute [Int: PeriodOfUse] is copied.

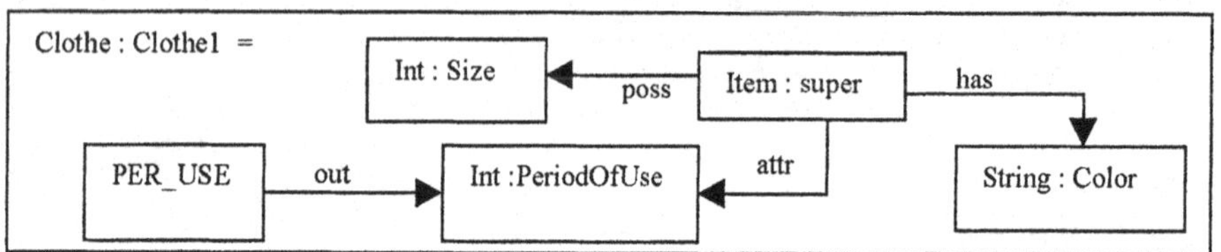

Fig. 3. The result of the instantiation of [Clothe: Clothe1]

In Synergy, the creation of an instance of a class is just a special case of *referent instantiation* ; if the referent is not an instance of a class, the instantiation consists in a *complete* copy of the referent type definition. The referent can be either an individual identifier, a variable identifier or an implicit variable (when the referent is not explicitly specified). As illustrated in the next section, instantiation of both referent and instance of a class is a basic operation in the Synergy interpreter.

3.3 Message, inheritance and method/daemon execution

While a method is defined in the context of a class definition, a message (or method call) should be interpreted in the context of an instance description.

Definition 5. *Sending a message* to an instance "Inst" for the execution of the method "Mtd" is represented by a concept with the composed type "Inst.Mtd" : [Inst.Mtd]. The method is interpreted in the context of the instance description.

A composed type in Synergy [10] has the form "Ident1.Type2" where "Ident1" is either a type identifier or an instance identifier. In the first case, the type "Ident1.Type2" is a reference to the type "Type2" which is defined in the context of "Ident1" definition. In the second case, the type "Type2" is defined in the class of the instance "Ident1".

Fig. 4 shows a request, formulated in the [WorkingMemory], that is composed of two messages : the first message [Clothe1.InitializeClothe] is sent to the instance "Clothe1" to execute the method "InitializeClothe", followed by a second message sent to the same instance to execute the method "ATIPrice".

Calling a method, from the body of another method or due to a message involves inheritance mechanism to locate the method. Beside this "method inheritance", method execution involves in general "property inheritance" to locate the defining concept for a property or attribute of the instance.

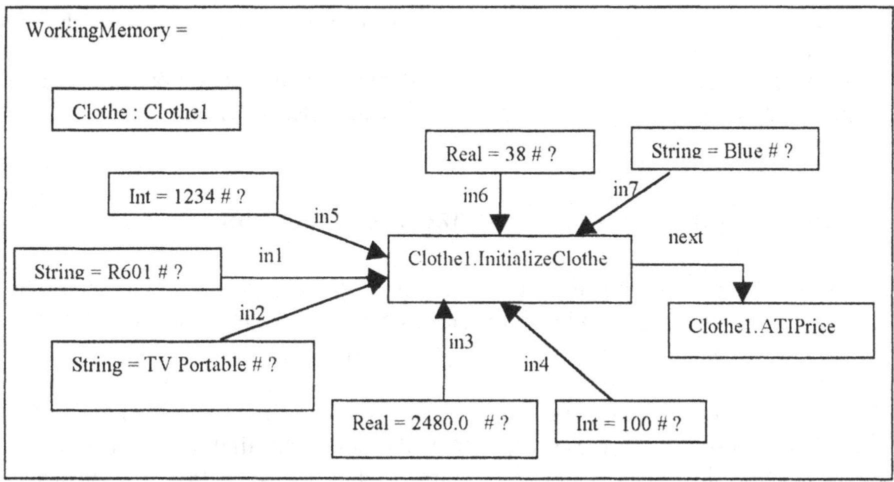

Fig. 4. Sending messages [Clothe1.InitializeClothe] and [Clothe1.ATIPrice]. Unlike the SYNERGY graphic editor, we show explicitly in this Fig. and in others the description and the state of a concept. "#?" stands for "state trigger" and "#!@" for "state in-activation".

Definition 6. *Inheritance* concerns the methods and the attributes of a class :

- *Method inheritance* : the definition of a method "Mtd" specified in a message [Inst.Mtd] is looked for in the definition of the class of the instance "Inst" and then upward in the class hierarchy until the method definition is located or the class [Class] is reached. In the same way, the definition of a method "Mtd" specified in the body of another method "Mtd1" is searched first in the definition of the class where the method "Mtd1" is defined and then upward in the class hierarchy, if necessary.

- *Property inheritance* : in the body of a method call [Inst.Mtd] a co-referent is resolved by looking for its defining concept in the description of the instance "Inst". If the defining concept is not found there, the search continues recursively in the "super" contexts embedded in the instance description until the defining concept is found or the context [Class: super] is reached in which case the co-referent is not a reference to a property (direct or inherited) of the instance, and the search of the defining concept will continue in the context that contains the instance description.

Definition 7. *Method and daemon executions* conform to *concept execution* or interpretation which constitutes the basis part of the Synergy interpreter.

Briefly, interpretation of a CG consists in the *parallel* interpretation of its concepts and a concept C is interpreted according to its lifecycle : a concept C at state "trigger ; ?" is requested to "determine and then interpret (i.e. evaluate or execute) the description". Once the description is determined (by instantiation for example, if the concept type is defined) and if the concept C has preconditions, it will enter a waiting phase until its preconditions are satisfied. At the end of this waiting phase, the concept' description is evaluated and then, the concept will enter another waiting phase until its post-conditions are satisfied. After that, the concept C returns to the state ``steady ; o``.
We refer the reader to [5, 6, 10] for a more detailed description of concept interpretation, CG interpretation and the Synergy interpreter in general.

4 An example of methods and daemons execution

As an example of methods and daemons execution, let us follow in some detail the execution of the message [Clothe1.InitializeClothe] (Fig. 4). Its input arguments are in "trigger" state. The "in" relation will propagate this state to [Clothe1.InitializeClothe]. Being at state "trigger" and since the concept referent does not have a description and its type is a composed type, the first action is to locate the definition of this latter : the interpreter will attempt to find the definition of "InitializeClothe" in the definition of the class "Clothe" (since Clothe1 is an instance of Clothe). Once located, the definition is then instantiated.

Note that the type of the message [Clothe1.InitializeClothe] becomes "Clothe@Clothe1.InitializeClothe". Indeed, once determined, the class of the instance prefixes the instance identifier to avoid the search of the class (several times) during message execution. The prefix will be eliminated at the end of the message execution, e.g., to become again "Clothe1.InitializeClothe".

Let us consider now the interpretation of [Clothe@Clothe1.InitializeClothe] (Fig. 5). The states of the arguments are transmitted to the corresponding parameters which become in “trigger” state. Then, by forward propagation from the first five parameters, the concept [InitializeItemPart] that is contained in the description of [Clothe@Clothe1.InitializeClothe] becomes in turn at state “trigger”. The same effect happens in *parallel* to the two primitive operations (:=); triggered by the six and the seven parameter respectively.

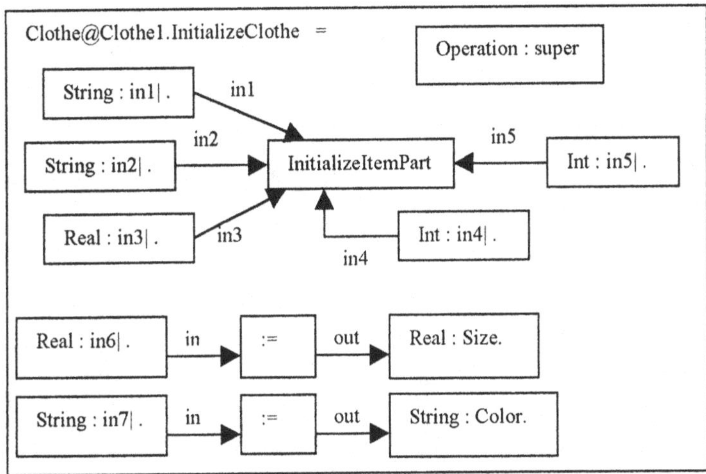

Fig. 5. Instantiation of [Clothe1.InitializeClothe]

Let us follow now the interpretation of the concept [InitializeItemPart #?] : the interpreter will look for the type definition of “InitializeItemPart” (since it is triggered, its type is not a primitive operation and it is not a description).
The interpreter finds that “InitializeItemPart” is a method inherited from the class “Item” (since it occurs in the context of the method “InitializeClothe” in [Clothe@Clothe1.InitializeClothe], the interpreter checks first if the type “InitializeItemPart” represents a method defined in the class of the method “InitializeClothe” or in a super-class (direct or indirect) of the class).
The definition of “InitializeItemPart” is now located and the instantiation is done to produce a description for the referent of [InitializeItemPart] (Fig. 6). Its concept type becomes “Item.InitializeItemPart”, its state changes to “in-activation” and its description is now executed (interpreted) : the five primitive operations “:=” that constitute its body are triggered and then executed in parallel (Fig. 6).

To illustrate property inheritance and daemon invocation, let us consider now the execution of one of the five affectations contained in [Item.InitializeItemPart] :
[Int: in4| .] —in→ [:=]—out→[Int: Quantity.]

Co-reference resolution using inheritance. The interpreter has to determine first the defining concepts for the two arguments of the affecation : [Int: in4| .] and [Int: Quantity.].

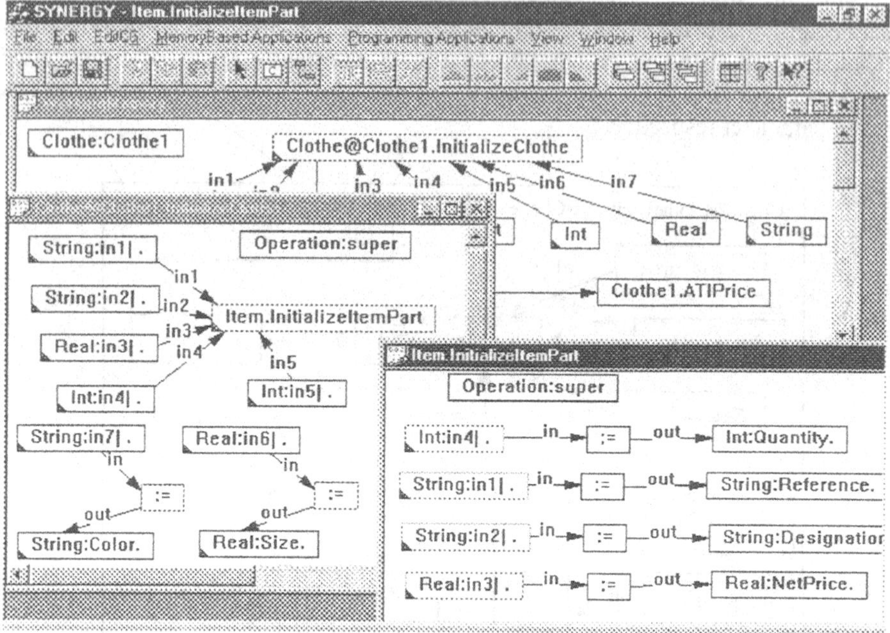

Fig. 6. Execution of the message [Clothe1.InitializeClothe]

To locate the defining concept of [Int: Quantity.] the interpreter initiates a search composed of two phases illustrated by Fig. 7 : an upward search in the embedded contexts starting from [Int: Quantity.] (Fig. 7.a) and then a downward search based on property inheritance (Fig. 7.b). Note that Fig. 7 shows only the parts that are in focus; the complete description of each context is not shown.

- *The upward search* (Fig. 7.a). The interpreter searches first in the current context [Item.InitializeItemPart] for a concept with referent "Quantity". Since no such a concept is specified in that context, the interpreter searches in the upward context : [Clothe@Clothe1.InitializeClothe]. Here too no such concept is found. Now and since the current context is a message (known from the form of its concept type : "Clothe@Clothe1.InitializeClothe"), the interpreter continues its search in the context of the instance ([Clothe: Clothe1]) instead of following the search in the upward context ([WorkingMemory]) of the previous current context ([Clothe@Clothe1.InitializeClothe]).
- *The downward search* (Fig. 7.b). First, if the current context [Clothe: Clothe1] has not a description yet, the interpreter will instantiate it. The result of the instantiation is similar to the CG in Fig. 3. Since no concept with referent "Quantity" is specified in the current context ([Clothe: Clothe1]) and since this latter has a super-concept ([Item: super]) with a defined type ("Item"), the search will continue in [Item: super] which becomes the current context. Property inheritance is in action now ! The search will continue in the context [Item: super]

once its description is created by instantiation. The current context [Item: super] contains a concept with referent "Quantity" : [Int: Quantity]. Thus the defining concept of [Int: Quantity.] has been located in [Item : super].

Fig. 7.b shows the co-reference link that connects now the two concepts [Int: Quantity.] and [Int: Quantity].

(a) The upward search step

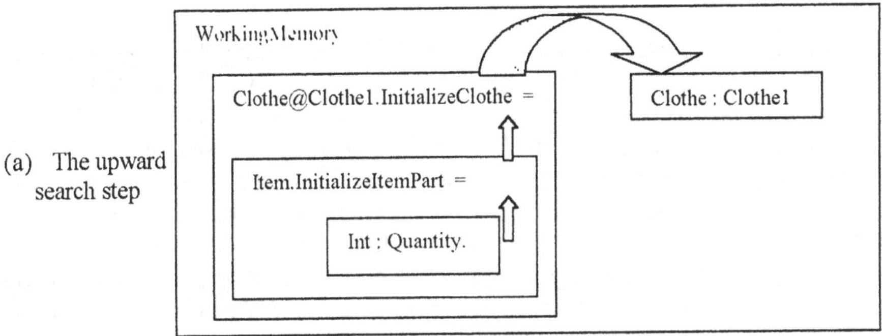

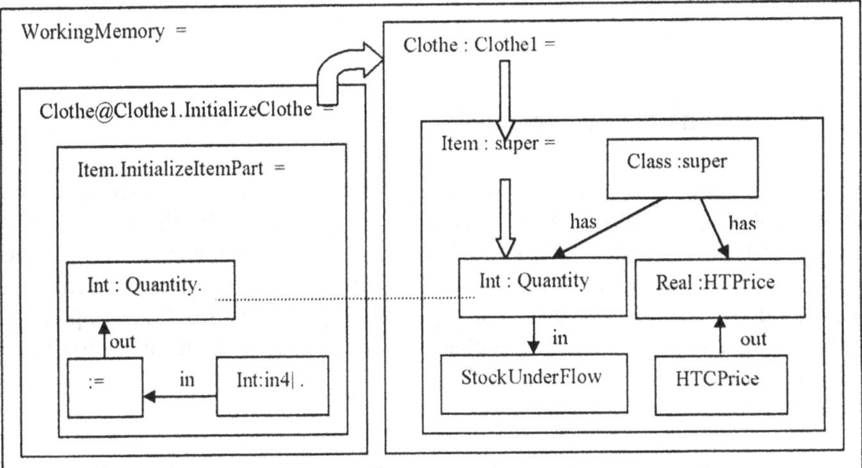

(b) : The backward search step

Fig. 7. Co-reference resolution using inheritance

Daemon invocation. Let us resume the execution of the affectation contained in [Item.InitializeItemPart] (Fig. 7): [Int: in4| .]—in→[:=]—out→[Int: Quantity.]. From the previous step (co-reference resolution), assume that the defining concept of [Int: Quantity.] is located and the defining concept of [Int: in4| .] is [Int = 100]; the fourth argument of [Clothe@Clothe1.InitializeClothe] (Fig. 4). After the execution of the affectation [:=] (Fig. 7), two events result : a) the value 100 is assigned to the defining concept [Int: Quantity] contained in [Item:super] (Fig. 7, the right part), b) the concept [Int: Quantity.] becomes in state "trigger" ("#?").

The state "trigger" of [Int: Quantity. #?] is then propagated, through the co-reference link, to the defining concept [Int: Quantity] and also to its upward contexts ([Item: super] and [Clothe: Clothe1]) (Fig. 7). The interpreter is now committed to interpret/activate those contexts and hence to interpret the concept [Int: Quantity #?] too (because it is contained in the active context [Item: super]). Finally, the state "trigger" is then propagated, through the relation "in", from [Int: Quantity #?] to [StockUnderFlow], leading to a forward invocation of the daemon [StockUnderFlow].

To summarize : the execution of the message [Clothe@Clothe1.InitializeClothe] (Fig. 6) involves the execution of the method "InitializeClothe" which leads to the parallel execution of the method [Item.InitializeItemPart] and two primitive operations ":=" (Fig. 6). Execution of [Item.InitializeItemPart] involves parallel execution of five primitive operations ":=", among them : [Int: in4| .] —in→ [:=]—out→[Int: Quantity.], which leads to the invocation of the daemon [StockUnderFlow] (Fig. 7).
When all the active concepts of [Item.InitializeItemPart] terminate their execution, the concept [Item.InitializeItemPart] terminates its execution too and because it is defined as an operation (not as a process), its description is erased. Again, when all the active concepts of [Clothe@Clothe1.InitializeClothe] terminate their execution, the message terminates its execution too.

5 Conclusion and future works

This paper presented the Synergy language as an hybrid object-oriented conceptual graph (HOO-CG) language. A detailed description is given for the formulation in Synergy of the basic concepts of the hybrid object-oriented paradigm: encapsulation, definition of a class with methods and daemons, method and daemon definitions, class hierarchy, instance and instantiation mechanism, inheritance (both property and method inheritance), method call, method execution and daemon invocation due to accessing data.
Those concepts have been illustrated by a detailed example.

Several projects within the SYNERGY group are in progress, including the development of complex object-oriented applications, the extension of Synergy to enable the construction of interfaces (composed of boxes, buttons, sliders, etc.) and the extension of Synergy to enable a connection with data bases (for efficient manipulation of a great number of instances).

References

1. Ellis G., Object-Oriented Conceptual Graphs, in Proc. of the Third Intern. Conf. on Conceptual Structures, ICCS'95, Santa Cruz, CA, USA (1995)
2. Hines T. R., J. C. Oh, and M. A. Hines, Object-Oriented Conceptual Graphs, in Proc. of the Fifth Annual Workshop on Conceptual Structures (1990)

3. Kabbaj A., C. Frasson, M. Kaltenbach and J-Y Djamen, A conceptual and contextual object-oriented logic programming : PROLOG++ language, in Proc. Second International Conference on Conceptual Structures, ICCS'94, Tepfenhart and al. (Eds), Springer-Verlag (1994)
4. Kabbaj A., Self-Organizing Knowledge Bases: The Integration Based Approach, in Proc. Of the Intern. KRUSE Symposium : Knowledge Retrieval, Use, and Storage for Efficiency, Santa Cruz, CA, USA (1995)
5. Kabbaj A. and C. Frasson, Dynamic CG: Toward a General Model of Computation, Proc. Third International Conference on Conceptual Structures, ICCS'95, Santa Cruz, CA (1995)
6. Kabbaj A., Un système multi-paradigme pour la manipulation des connaissances utilisant la théorie des graphes conceptuels, Ph.D Thesis, DIRO, Université de Montréal, June (1996)
7. Kabbaj A. and C. Frasson, An overview of SYNERGY : a conceptual graph activation language, accepted but not presented at the Fourth International Conference on Conceptual Structures (ICCS'96)
8. Kabbaj A., Rouane K. and Frasson C., The use of a semantic network activation language in an ITS project, Third International Conference on Intelligent Tutoring Systems, ITS'96, Springer-Verlag (1996)
9. Kabbaj A., Contexts, Canons and Co-references as a basis of a multi-paradigm language, submitted to Fifth International Conference on Conceptual Structures (ICCS'97)
10. Kabbaj A., A conceptual graph activation-based language : Synergy and its environment, submitted to Seventh International Conference on Conceptual Structures (ICCS'99), this volume
11. Kabbaj A., H. Er-remli and K. Mousaid, An integration-based approach to dynamic formation of a knowledge-base: The method, its implementation and its realizations, submitted to Seventh International Conference on Conceptual Structures (ICCS'99)
12. Kunz J. C., T. P. Kehler and M. D. Williams, Applications Development Using Hybrid AI Development System, The AI Magazine, 5:4, p. 41-54, (1984)
13. Lukose D., Executable conceptual structures, Proc. First International Conference on Conceptual Structures, ICCS'93, Quebec City, Canada (1993)
14. Lukose D., T. Cross, C. Munday and F. Sobora, Operational KADS Conceptual Model using conceptual graphs and executable conceptual structures, in Proc. of the Third Intern. Conf. on Conceptual Structures, ICCS'95, Santa Cruz, CA, USA (1995)
15. Masini G., A. Napoli, D. Colnet, D. Léonard and K. Tombre, Les langages à objets, Inter-Editions (1990)
16. Sowa J. F., Conceptual Structures : Information Processing in Mind and Machine, Addison-Wesley (1984)
17. Sowa J. F., Relating Diagrams to Logic, Proc. First International Conference on Conceptual Structures, ICCS'93, Quebec City, Canada (1993a)
18. Sowa J. F., Logical foundations for representing object-oriented systems, J. of Experimental and Theoretical AI, 5 (1993b)
19. Sowa J. F., Conceptual Graph Standard and Extensions, in Proc. Of the 6th International Conference on Conceptual Structures, ICCS'98, Springer (1998)
20. Wuwongse V., B. G. Ghosh, Towards Deductive Object-Oriented Databases Based on Conceptual Graphs, in Proc. of the 7th Annual Workshop on Conceptual Structures (1992)

Notio - A Java API for Developing CG Tools

Finnegan Southey[1] and James G. Linders[2]

[1]University of Waterloo, Dept. of Systems Design Engineering, Waterloo, Ontario, Canada, N2L 3G1
finnegan@pami.uwaterloo.ca
[2]University of Guelph, Department of Computing and Information Science, Guelph, Ontario, Canada, N1H 2G1
jgl@snowhite.cis.uoguelph.ca

Abstract. Notio [1] is a Java API for constructing conceptual graph tools and systems. Rather than attempting to provide a comprehensive toolset, Notio attempts to address the widely varying needs of the CG community by providing a platform for the development of tools and applications. It is, first and foremost, an API specification for which different underlying implementations may be constructed. A pure Java reference implementation is currently available for development and evaluation of the API, and to guide future implementations. An overview of the motivation, design, and features of Notio is provided.

Introduction

Since their introduction by John Sowa in 1984 [2] researchers have sought to create tools and applications for working with and harnessing conceptual graphs (CG). Many systems for representing, storing, retrieving, acquiring, and editing CG=s have been created over the years, some highly focussed on particular knowledge domains or applications, and others of more general utility. We present Notio, a new application programming interface (API) which presents a set of primitives for CG construction and manipulation, providing a platform for the development of CG tools and applications.

Motivation

Most of the general-purpose software developed for CG=s has been in the form of Atools@, software which provides a user interface or specialized language for manipulating graphs. Some of these interfaces have been graphical while others have employed special languages, command shells, or simple natural language interfaces to access the underlying graph implementation (e.g. Deakin Toolset [3] and CGKEE

[4]). The goal has often been to provide a general-purpose conceptual graph environment for knowledge engineers and researchers. Such systems are frequently described as *workbenches* (e.g. Peirce [5], CoGITo [6]).

The chief problem encountered with workbenches is a lack of consensus on the form they should take. No unifying, formal model for CG management currently exists. Thus, many different systems have been developed, each providing a set of tools for various tasks. These systems have been developed in many different languages and for many different platforms. The result is a considerable duplication of effort since most systems do not, or can not, use components from existing systems. In some cases this is inevitable, since researchers may wish to try new approaches and interfaces, but it is rare that the entire system is novel. At the same time, there are many differences in internal representations and algorithms for manipulating graphs. It is, therefore, no wonder that no one workbench satisfies everyone.

We decided to address the issue of generality, alternative implementations, and avoiding duplication of effort by adopting the approach used for the development of many major computing standards. We have elected to define an API that describes a number of operations for the construction and manipulation of CG=s without dictating the nature of the underlying implementation, or the applications and interfaces used to invoke the operations. While other CG programming libraries exist (e.g. CoGITo), we believe Notio is the only CG API offered first and foremost as a specification (although some documents on Peirce discuss a similar use of Abstract Data Types).

Design Goals for Notio

The Notio API was constructed with the following design goals in mind:

X Emphasis on ease-of-use for application authors
X Intuitive representation of commonly accepted CG abstractions
X API independent from underlying implementation=s internal representation
X Portability
X Extensibility by providing formal extension facilities
X Generality by minimizing specification of the structure of an application or knowledge base
X Flexibility by allowing varying levels of compliance and making various constraints optional
X Robust by providing an extensive set of exceptions for error handling

Notio: A Java Package

Notio[1] is a class library written in Java[2] using version 1.1 of the Java platform. Such libraries are known as *packages* in Java parlance. Java was selected as the development language for a variety of reasons:

- X Platform independence
- X Object-oriented language
- X Robust environment (no pointers, automatic garbage-collection)
- X Several free and commercial development environments available
- X Formal source-based documentation facilities (javadocs)

An object-oriented programming language was selected after a couple of prototypes using declarative languages (Prolog and CLIPS). While declarative languages have many attractive features and have been used to implement CG systems in the past (eg. CGPro [7]), it was decided that the object-oriented approach would find a broad acceptance both inside and outside of the knowledge representation community.

The Architecture of a Notio-based System

Notio is first and foremost an API specification, a document that specified a collection of Java class definitions and the behaviour expected of the methods in those classes. From this document a developer should be able to create an entire set of classes that implement the specified behaviour. Such an implementation is called an *implementation layer*. The internals of an implementation layer are entirely at the discretion of the developer. They can be a pure Java implementation of the API=s functionality, links through the Java Native Interface (JNI) to an existing CG-system written in another language, a client for a knowledge base system, or anything else the developer desires.

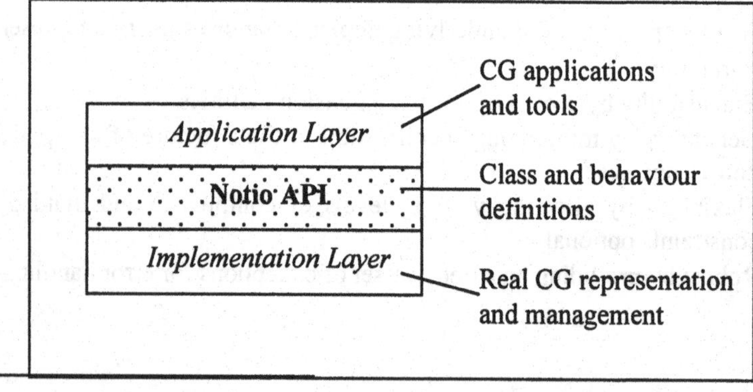

Figure 1: The structure of a Notio-based system.

[1] ANotio@ i

[2] JavaJ anc

Applications written to use the Notio API form what is known as an *application layer*. An application layer instantiates API classes and calls methods in them to accomplish its goals, the implementation layer responds by changing whatever internal structures it holds and presenting them when requested in the form of API classes. Thus, the API classes acts as representatives for graph structures in the implementation layer and need not comprise the actual representation. Only requested structures need be instantiated by the implementation layer. A diagrammatic view of the architecture of a Notio-based system is shown in **Figure 1**.

Our Approach to the Notio API

Having stated that an obstacle to the wide acceptance of any particular workbench or tool is the difficulty of satisfying enough of the CG community, it is fair to ask why we believe that a general-purpose API can fulfill a similar role. We do not claim that this API will satisfy all needs, nor that it will suit the development of any one application, but we do believe that this is correct level to generalize and standardize.

CG Applications and tools are often very specific to a knowledge domain or real-world application and thus are virtually impossible to generalize. Similarly, efficient CG representations and operations still form an area of active research in which a wide variety of approaches are taken and different aspects of performance emphasized. However, an interface between these two layers, based on the abstractions which the CG paradigm suggests, allows for either to be replaced or reused as necessary. A substantial level of general agreement on this set of abstractions does exist in the CG community.

We base this opinion on the recent efforts towards the establishment of the ANSI standard (dpANS) for CG=s [8], in which the CG community has worked together to reach a consensus. The standard seems to embody a set of properties and operations that are widely considered as integral to CG=s. The Notio API=s development began as the first publically available versions of the draft standard were circulating and it was decided to design the API under the dictates of the emerging standard. Notio is, therefore, an attempt to realize the abstractions presented in the dpANS. However, it is understood that this standard could never embrace all of the various extensions and subtleties related to CG=s, and so the API was also designed to allow extensions and varying levels of compliance.

While several CG tools have provided essentially the same set of abstractions, Notio provides them as an API specification. We believe an API to be an excellent generic method for presenting the abstractions and operations since it relies on a widely established language and programming paradigm, rather than custom languages or interfaces. This same principle guides the design of many development environments such as API=s for networking, graphical interfaces, and databases.

Design of the Notio API

The Basic Structures

Since the classes in the API are intended to provide an intuitive representation for working with CG=s we elected to create a fairly direct correspondence between the classes and the abstractions defined in the dpANS. Thus, we have a Graph class which is composed of instances of the Concept and Relation classes. The Concept class is in turn composed of instances of the ConceptType and Referent classes. Referents are composed of Quantifiers and Designators. Relation instances have a RelationType and a list of arguments which serve as the arcs in the graph.

Notio directly supports compound graphs. Concept=s may have a Referent with a DescriptorDesignator, which points to a nested graph. The CoreferenceSet class works with the Concept class to provide full support for coreferences, including the ability to enforce the scoping rules specified in the dpANS. The API allows for the traversal of compound graphs in either direction (top-down or bottom-up).

Types are handled via a ConceptTypeHierarchy instance or RelationTypeHierarchy instance as appropriate. They allow for labelled and unlabelled types, with or without signatures and type definitions. RelationTypes may optionally supply a valence (number of arguments) in the absence of a type definition. Changes to types are immediately reflected throughout existing graphs.

Knowledge Bases

Many tools require the use of a Aknowledge base@ construct that stores graphs and types. In developing the Notio API we wished to avoid dictating the overall structure of applications as much as possible and so the KnowledgeBase class, which corresponds to the abstraction specified in the dpANS, is entirely optional. While one must create type hierarchies in order to use types, there is no need to store them in a KnowledgeBase instance, and graphs may be created freely and tracked in any manner the developer sees fit. It is possible to create any number of entirely distinct type hierarchies, graphs, and knowledge bases, allowing for multiple, distinct Notio-based applications running concurrently and using one or more knowledge bases, either shared or unshared.

Operations

Several operations are defined in the Notio API. Operations are not described in terms of algorithm, but rather, in terms of the effect they produce. Thus, the focus is on defining a variety of operations rather than providing specific implementations. Currently defined operations include construction and deconstruction operations that combine components together to form graphs and take them apart respectively. The basic canonical operations are present: copy, restrict, simplify, and join. Other operations such as type expansion, path-finding, and subgraph extraction have been included or are under consideration. Many operations can be invoked in a few different ways. For example, simplification can be performed on a specific type of relation between specific concepts, on all instances of a given relation type within a graph, and on all relations in a graph, regardless of type. Joins can be single-point, multi-point, or maximal (join criteria can be controlled using the matching schemes described below). Restrictions can be performed on type, or referent, or both.

Copying deserves special notice since this apparently simple operation can actually become quite complex when applied to compound graphs, graphs with coreference sets, or to some subset of the components in a graph. In Notio, copy operations can be performed not only on graphs, but on concepts, relations, referents, and other components. The extent and behaviour of the copying operation is dictated by a *copying scheme*, an instance of the CopyingScheme class which contains flags that direct the copying process. For maximum control and flexibility, copying schemes can be nested within each other. Nested schemes are applied to correspondingly nested graphs, allowing for different copying techniques depending on context.

The Notio API also defines methods for graph matching. This is arguably one of the most complex operations commonly performed on graphs and it takes many different forms. Notio takes an approach to matching similar to that used for copying. Graphs and other components can be matched according to a *matching scheme*, a set of flags that determine how matching should be performed. The classification of matching operations was based on the approach presented by Nagel et al. [9]. That work has been extended to better describe the matching of compound graphs and coreference sets. The flags allow control over the comparison of elements such as types, referents, designators, markers, and graph structure. Refinement of the matching scheme structure is still ongoing.

Errors and Exceptions

Notio reports errors via Java=s exception mechanism. There are two basic exception types, OperationErrors and OperationExceptions. OperationErrors need not be explicitly handled in an application and are used to report problems which a Notio application may either confidently ignore or cannot handle. OperationExceptions must be explicitly handled, and are used chiefly to report problems in complex operations that are likely to throw exceptions in spite of the best planning by the developer. Using this mechanism means that Notio applications can handle errors at any level they choose but still keep their source code relatively simple. Implementation layers can also embed their own error information within Notio=s exception classes, thus providing a means for passing implementation-specific error messages through the API cleanly.

Translators

In addition to the basic conceptual graph structures and operations, Notio offers a standard interface for translators which parse or generate various CG formats such as the Conceptual Graph Interchange Format (CGIF) and the Linera Format (LF). This common interface allows pluggable translation modules so that new translators can be added without altering the API itself. Applications can use Java=s dynamic class loading capabilities to load new translators without recompilation. Translators can act on strings or streams, allowing use of files or network connections for translation.

Extensions

Naturally, researchers are going to want features not offered by the Notio API, either because they are too specific or are the result of new research. Instead of creating a separate library or modifying the API in incompatible ways, developers can add features using the formal extension mechanism. This allows new features to added to the API in such a way that they can be discovered, activated, and configured by applications in a programmatic fashion. It is hoped that commonly desired extensions will be standardized, and that multiple implementations of the same extension will be usable by applications in much the same manner as the core Notio API.

Omissions and Variable Compliance

While the Notio API tries in some ways to act as a lowest common denominator for CG operations, it was decided that requiring complete compliance from implementation layers would result in overly-demanding implementation projects or an excessively restricted API. It is not reasonable to expect all implementation layers to offer all of the operations defined in the Notio API. The most notable of these is graph matching which is extremely flexible and general in the API specification. A particular implementation layer may only be able to perform a subset of the possible matching operations. Thus, implementation layers may omit some operations and raise an exception indicating that the desired operation is not available. Applications should also be able to programmatically determine whether a given feature is available or not. These mechanisms have not been completely defined as we wish to relate them to the *conformance pairs* discussed but not defined in the currently available dpANS.

The Reference Implementation Layer

Along with the Notio API itself, we have developed an implementation layer that serves as the prototype and reference implementation for the Notio API. The *reference implementation layer* is written completely in Java and is correspondingly both portable and capable of running in web browsers. In developing the reference implementation, the emphasis has been on simplicity and clarity rather than performance. Efficiency concerns will be addressed once the API itself has achieved a high level of stability but it is hoped that alternate implementations will more effectively fill this need since it is desirable to keep the reference implementation fairly simple.

Another role for the reference implementation will be as an example and starting point for new implementation layers. One way to rapidly develop a new implementation layer would be to gradually replace functionality in the reference implementation. As far as possible, the reference implementation was designed with this possibility in mind, and uses standard API calls internally where possible rather than implementation specific calls.

At the time of writing, the reference implementation=s source code is not publically available but the compiled package is. This is because we wish to focus public attention on the API specification rather than the underlying implementation. The present plan is to release the reference implementation under GNU=s Library General Public License [3], which allows the use of the library in commercial applications. However, the best means for making the source available are still under consideration. Our intention is that the API and reference implementation should be as freely available as possible without compromising the goal of achieving a level of standardization. We anticipate releasing the source during 1999.
The API specification is generated directly from source of the reference implementation using the Java Development Kit=s (JDK) javadoc tool. This helps ensure that the API specification and reference implementation are synchronized and up to date. Bugs, ideas, and other details of the development process are handled in a similar manner.

The reference implementation is bundled with other packages which include translator implementations for CGIF and LF, a testing suite, a set of compiled examples, and a demo suite. The documentation consists of the API specification in HTML format, automatically generated lists of bugs and ideas, and source code for the examples.

Both the API documentation and the reference implementation are available from:
http://dorian.cis.uoguelph.ca/CG/projects/notio/index.html

[3] Information on GNU=s LGPL is available at http://www.gnu.org.

Using the API

To give an example of the kind of work that may be done with the API, consider a simple tool we have created as a testbed for Notio. This tool provides a GUI interface that manages a set of graphs. All graph display is accomplished using CGIF. Operations such as copy, join, and simplify are provided through lists of nodes and pull-down menus. Graphs can be entered in a window, or read from files and URL=s, all in CGIF format. They can also be saved in CGIF or LF files. The greatest effort in developing this tool was in building the user interface. The actual code involved in implementing operations and managing the objects was very small and quickly written.

Future Work

The specification of the Notio API is not complete although the essential structures are unlikely to change and we are currently engaged in a process of enhancement and refinement. The reference implementation is at the Aalpha@ stage and implements most of the specified features. It is already in use for three different projects at the University of Guelph including a database for storing and searching CG=s, a knowledge browser and summariser, and our own Ossa project, a conceptual modelling tool. Other institutions are also considering or using the Notio API in their own projects.

The present API is focussed on operations involving one or two graphs. We hope to add interfaces for working with larger sets of graphs. We are very interested to hear from the CG community about their needs and views on such an API and solicit any comments or criticism. We hope to start creating alternate implementation layers soon and encourage other groups to approach us about building implementations or an interface to existing systems. We are also starting work on a set of reusable JavaBean components based on the Notio API to handle common needs like graph editing and layout. These components will further our aim of helping to prevent unnecessary repetition of work in the CG community.

In conclusion, we encourage members of the CG community to examine, evaluate, criticize, contribute to, and above all, use the Notio API.

References

1. Finnegan Southey, AOssa: A Modelling System for Virtual Realities Based on Conceptual Graphs and Production Systems@, MSc. thesis, supervisor Dr. James G. Linders, University of Guelph, Canada, 1998.

2. John F. Sowa, AConceptual Structures: Information Processing in Mind and Machine@, Addison-Wesley, Reading, MA, 1984.
3. Brian Garner, Eric Tsui, and Dickson Lukose, ADeakin Toolset:Conceptual Graph Based Knowledge Acquisition, Management, and Processing Tools@, in *Proceedings of the Fifth International Conference on Conceptual Structures*, ICCS=97, pages 589-593, Springer, 1997.
4. Dickson Lukose, ACGKEE: Conceptual Graph Knowledge Enginnering Environment@, in *Proceedings of the Fifth International Conference on Conceptual Structures*, ICCS=97, pages 589-593, Springer, 1997.
5. G. Ellis and R. Levinson, AThe Birth of Peirce@, in *Proceedings of the Seventh Annual Workshop on Conceptual Structures*, pages 219-228, Springer-Verlag, 1993.
6. Michael Chein, AThe CORALI Project: From Conceptual Graphs to Conceptual Graphs via Labelled Graphs@, in *Proceedings of the Fifth International Conference on Conceptual Structures*, ICCS=97, pages 65-77, Springer, 1997.
7. H. Petermann, L.Euler, K.Bontcheva, ACGPro - a PROLOG implementation of Conceptual Graphs@, Technical Report FBI-HH-M-251/95, University of Hamburg, October 1995, 35 pages.
8. John Sowa et. al., ADraft Proposed American National Standard (dpANS) for Conceptual Graphs@, unpublished, February, 1999.
9. Timothy E. Nagle, John W. Esch, & Guy Mineau, AA notation for conceptual structure graph matchers@, in AConceptual Structures: Current Research and Practice@, Timothy E. Nagle, Janice A. Nagle, Laurie L. Gerholz and Peter W. Eklund (eds.), Ellis Horwood, 1992.

Multiperspective Analysis of the Sisyphus-I Room Allocation Task Modelled in a CG Meta-Representation Language

Thanwadee Thanitsukkarn[1], Anthony Finkelstein[2]

[1] Mahidol University Computing Center, Faculty of Science, Mahidol University
Rama 6 Road, Bangkok 10400, Thailand
cctth@mahidol.ac.th
[2] Department of Computer Science, University College London
Gower Street, London WC1E 6BT, England
A.Finkelstein@cs.ucl.ac.uk

Abstract. Different participants in system development often hold partial specifications of the knowledge relating to the system being developed. To reflect this, the system may be developed through 'perspectives' to make such information more manageable. The *ViewPoints* framework [1] offers a way of dealing with the partitioning and the organization of perspectives entailed in system development. ViewPoints represent "agents" having "roles-in" and "views-of" a problem domain. We are using *conceptual graphs* (CGs) as a *meta-representation language* to describe ViewPoints in order to facilitate reasoning about the partitioned knowledge that each ViewPoint represents. The established notations and operations of CGs provide a strong foundation for this purpose. This paper presents the applicability of ViewPoints together with a CG meta-representation language for the knowledge acquisition activities of the Sisyphus-I room allocation task. The paper provides an illustrative case study of our ongoing work on CG application to the area of software engineering [2]. Although the Sisyphus-I room allocation task is not a natural fit with the ViewPoints framework which aims at different tasks, we have made an attempt to apply it and believe that there are some interesting lessons.

1 Introduction

The rapid evolution of information and the increasing requirements in software development tends to make a software application large and complex. Different participants in system development often hold partial specification of the knowledge relating to the system being developed. Furthermore, a single system development techique may not be adequate to construct an application. As an alternative, it would be beneficial to decompose such an application into small manageable parts to which a number of appropriate methods and tools can be applied. This, in turn, enables an application to be described from a number of perspectives, each of which represents different aspects of the overall application. Such development of an application through multiple perspectives, or multiperspectives for short, facilitates separation of concerns, thereby offering a way to handle the complexity in the development. It also enables us to achieve the benefits that accrued from decentralization and distribution when constructing the application.

1.1 ViewPoint Analysis for the Problem of a Room Allocation Task

This paper explores the idea that the notion of ViewPoints together with the reasoning power of *conceptual graphs* (CGs) [3] can be applied to enhance knowledge acquisition activities. *ViewPoint-Oriented Software Engineering* (VOSE) is one method of achieving a suitable decomposition of a complex design into multiperspectives. The framework is presented in detail in [1] and initially realized as an implementation prototype in [4]. ViewPoints represent "agents" having "roles-in" and "views-of" a problem domain. A *ViewPoint* is defined as a loosely coupled, locally managed, self-contained object. It encapsulates *specification knowledge*, *representation knowledge* and *development knowledge* about a particular problem domain. Specification knowledge is about the aspects of the domain with which a ViewPoint is concerned. Representation knowledge is about the notations or representation styles deployed to express the specification of the domain. Development knowledge is about a sequence of steps or actions for constructing the specification.

In VOSE, a software specification is constructed by composing a set of ViewPoints. A software specification may be considered as being made of a number of domains. Each domain refers to a specific aspect of the problem. A domain is modelled as a collection of ViewPoints. Each ViewPoint represents different perspectives on the domain.

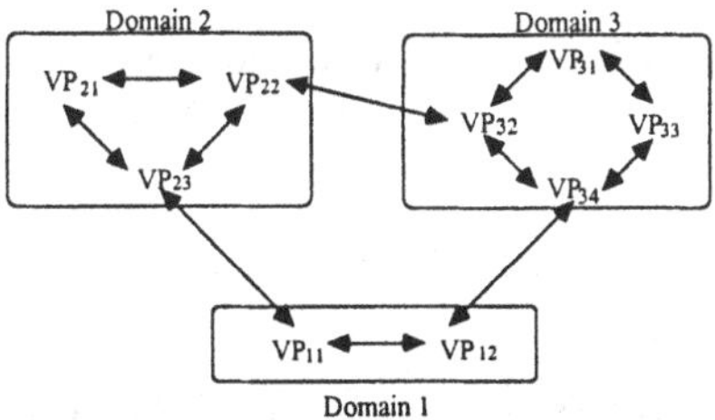

Fig. 1. A ViewPoint-Oriented Specification

Figure 1 exemplifies a ViewPoint-oriented specification which is composed of three domains. Bi-directional arrows in the figure depict the relationships between the ViewPoints within and across the domains. Such relationships enable explicit analysis of interaction and co-ordination among the ViewPoints so that the ViewPoints can be constructed independently, but managed jointly or interrelated through the relationships.

To apply VOSE to knowledge acquisition activities, different participants may be assigned to acquire different sets of knowledge separately in terms of ViewPoints.

For example, Figure 2 shows the separation of the acquisition activities for a situated classification method [5] into ViewPoints. As shown in the figure, different ViewPoints represent different perspectives which are the results of various modelling activities in the method. For instance, ViewPoints on domain ontology describe knowledge structure of the problem while ViewPoints on domain assertions describe the facts of a particular situation of the problem. As mentioned eariler, such a separation of concerns in modelling enables us to apply appropriate methods and tools to different parts of the problem. As ViewPoints can be constructed independently but managed jointly, these activities for knowledge acquisitions can be decentralized and distributed. By doing so, we achieve a way to manage complexity in modelling a large amount of knowledge.

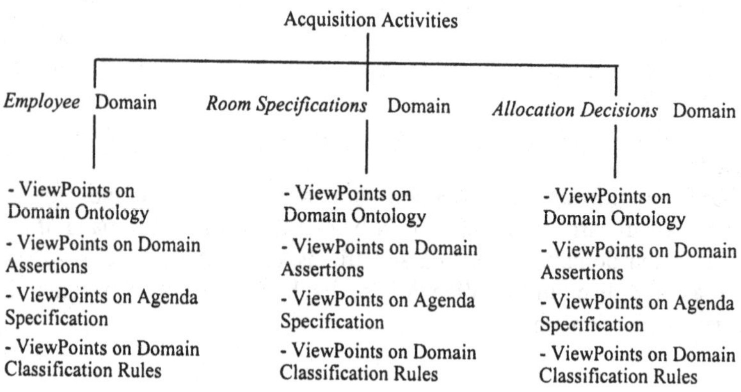

Fig. 2. Separation of Knowledge Acquisition Activities into ViewPoints

Figure 3 shows a sample set of ViewPoints for the assertions of the Sisyphus-I problem. In the figure, an ellipse denotes a ViewPoint. Assume that knowledge engineers *A*, *B* and *C* are responsible for constructing assertion ViewPoints for the domains of *employee, room specifications* and *allocation decisions* respectively. The knowledge engineer *A* is responsible for acquiring the details of each employee and monitoring that the allocation decisions made for an employee satisfy the preference of that employee. The knowledge engineer *B* is responsible for describing the specifications of the rooms that can be allocated in the problem. The knowledge engineer *C* is responsible for allocating the rooms by using information from the knowledge engineers *A* and *B*.

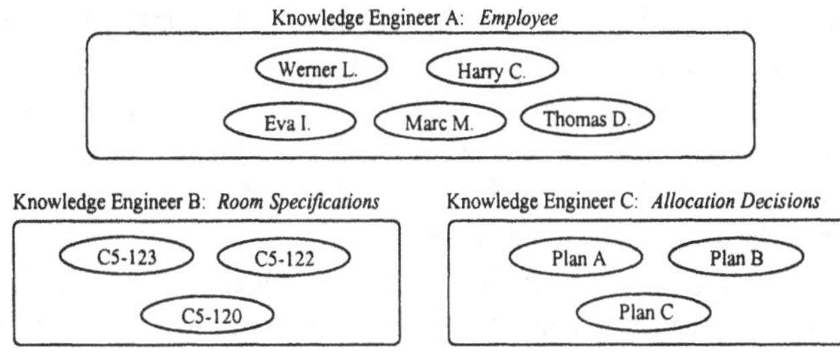

Fig. 3. ViewPoint Analysis for the Sisyphus-I Problem: Sample Solution I

Consider a more complex scenario of the problem. If there exists a large number of employees that must be taken into account, we could allocate the knowledge acquisition activities for different types of employee to a number of knowledge engineers. Figure 4 shows an alternative example of ViewPoint analysis for the problem in this scenario. Knowledge engineers *A*, *D* and *E* are assigned to take care of different types of employee, i.e. the knowledge about *heads of group*, *secretaries* and *researchers* respectively. Accordingly, such a complex problem becomes more manageable.

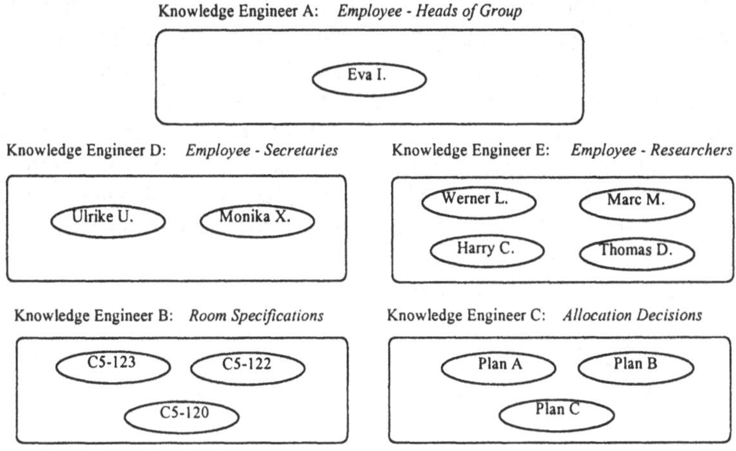

Fig. 4. ViewPoint Analysis for the Sisyphus-I Problem: Sample Solution II

Although the ViewPoints framework is an appealing way of managing the complexity inherent in knowledge modelling, the framework requires a systematic approach to ensure consistency in interaction and coordination of multiperspectives, i.e. ViewPoints. ViewPoints are not independent. They may constrain one another if they overlap, or interact, to represent part of a specification. Therefore, the relationships between ViewPoints need to be rigorously maintained to ensure that ViewPoints can work consistently together as an 'integrated' whole. Consistency of specifications or knowledge across ViewPoints thus becomes a central issue in applying ViewPoints. We have done a considerable amount of work in this area and this is where the notion of conceptual graphs has stepped in.

To enhance consistency checking in the ViewPoints framework, we adopt CGs as a *meta-representation language* to describe ViewPoints. Such enhancement to the framework is presented in detail in [6]. In summary, the features of the CG meta-representation language that we develop are:

- The language acts as an interpretation media which enables different ViewPoints to understand each other regardless of the diversity in their representation styles. To do so, the language allows a specialist, called the method engineer, to observe the definitions of representation styles on the basis of a limited set of constructs, i.e. concepts and conceptual relations. Such constructs relate syntactic terms of a representation style to the corresponding constructs in the language. Consistency rules for ViewPoints are also established by using the identified constructs of the representation styles.
- The automatic generation of a representation style translator and a consistency checking procedure are carried out via the representational types of the meta-representation language.

1.2 Outline of the Paper

This paper demonstrates the applicability of ViewPoints together with a CG meta-representation language on the knowledge acquisition activities, in particular for a solution of the Sisyphus-I room allocation task. For simplicity's sake, we will use only the examples of ViewPoints on domain assertions to illustrate our application of CGs in this paper. Section 2 describes the enhancement of CGs to the ViewPoints framework. The section shows how knowledge constructs and constraints of ViewPoints are modelled in our CG meta-representation language. Section 3 illustrates an example of ViewPoint consistency checking using CGs. Section 4 summarizes and discusses our results. The critique of applying the Sisyphus-I as our testbed is also given in the last section.

2 ViewPoint Modelling Constructs Using CGs

Figure 5 shows the roadmap of how to embed our CG meta-representation language in the ViewPoints framework. The *Viewer+CG* is the tool support that we have built to demonstrate the feasibility and applicability of the language. The tool is the modified implementation prototype of the *Viewer* [4]. The *Viewer+CG* is implemented in the Objectworks/SmallTalk Release 4.1 environment. The implementation modules for representing the CG structure and the graphical notations in the tool are adapted from the CG-editor [7].

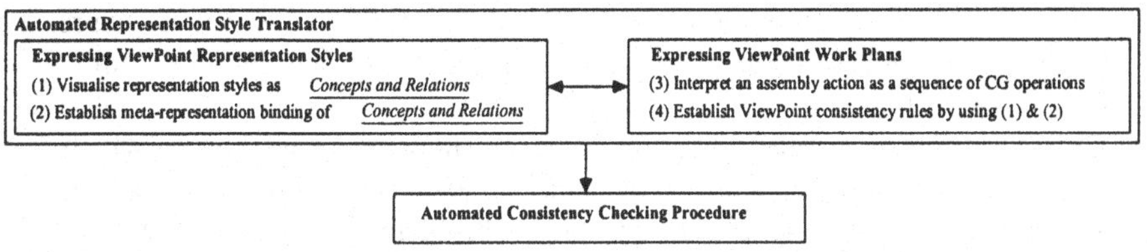

Fig. 5. CGs Enhancement to the ViewPoints Framework

The structure of a ViewPoint in the *Viewer+CG* consists of five slots, which are *style*, *work plan*, *domain*, *specification* and *work record* slots. As it is not our intention to discuss the notion of ViewPoints at length in this paper, the details of the ViewPoint structure is given in [1][4]. We have augmented the *Viewer+CG* with CGs for use in two 'dimensions'. Firstly, we employ CGs to describe representation styles and work plans of ViewPoints. In doing so, a ViewPoint representation style is described in terms of a set of CG concepts and conceptual relations. Such a CG definition constitutes a metamodel, namely a *meta-representation binding*, of the style. A meta-representation binding indicates a permissible structure of CG concepts and conceptual relations of knowledge that can be expressed within a ViewPoint. A ViewPoint work plan, which contains the definitions of assembly actions and consistency rules, is then expressed by using the resulting concepts and conceptual relations in the meta-representation binding. The expression of ViewPoint representation styles and work plans in this manner permit us to automate the translation of ViewPoint specifications into CGs. Secondly, we implement a procedure which utilizes such ViewPoint descriptions in terms of CGs to automate consistency checking in the ViewPoints framework.

In this paper, we show the ViewPoint-based models of the Sisyphus-I problem without elaborating the actions for constructing those models. Examples of such actions can be seen from our previous work [2][6].

2.1 Knowledge Constructs

A ViewPoint is created by instantiating a *ViewPoint template*. A single ViewPoint template is a description of development techniques that can be used to produce a ViewPoint. In other words, a ViewPoint template is a ViewPoint in which only the style and work plan slots have been elaborated. Accordingly, a ViewPoint template describes (i) the constructs of knowledge that can be expressed by a ViewPoint which is instantiated from that template and (ii) the activities (work plan) for such instantiation.

The following methodology specifies the steps to apply ViewPoints to knowledge acquisition activities for a problem context:

1. Classify scopes of knowledge to be acquired for the problem context into a set of ViewPoint templates;

2. For each ViewPoint template, identify (i) tangible entity types that are expressed by that template as CG concept types and (ii) the means that link those entities as CG conceptual relations;
3. Define consistency rules that are used to integrate ViewPoints which are produced for the problem context by using the identified ViewPoint templates.

To capture the knowledge for the Sisyphus-I problem, we construct the following types of ViewPoint template, namely *employee*, *room specifications* and *allocation decisions*. CGs are applied to express a permissible structure, namely meta-representation binding, of knowledge captured by each template.

- ViewPoint Template 1: *Employee*
- Representation Style

Style Component	Attribute	Graphical Notation
employee	name role is_smoker is_hacker	*name* *role* *is_smoker* *is_hacker*
project	name	*name*
work_with	-	
participate	-	

The *employee* template represents the details of employees. The meta-representation binding of this template describes the name of an employee, his or her role, the person with whom that employee works, the project in which he or she participates and other characteristics, such as whether he or she smokes and whether he or she implements a system. These descriptions are the criteria which are used to determine the room allocation for each employee. Such a meta-representation binding is illustrated as the CG below:

- Meta-Representation Binding

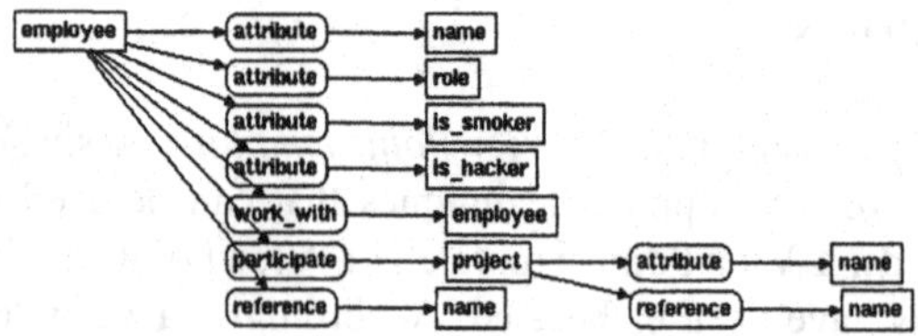

The following *partner* relation type refines the condition that an employee *X* works with an employee *Y*. An employee *X* is said to be a partner of an employee *Y* if there exists a relation *work_with* from the employee *X* to the employee *Y* or vice versa.

Relation Type *partner*(X,Y)

is

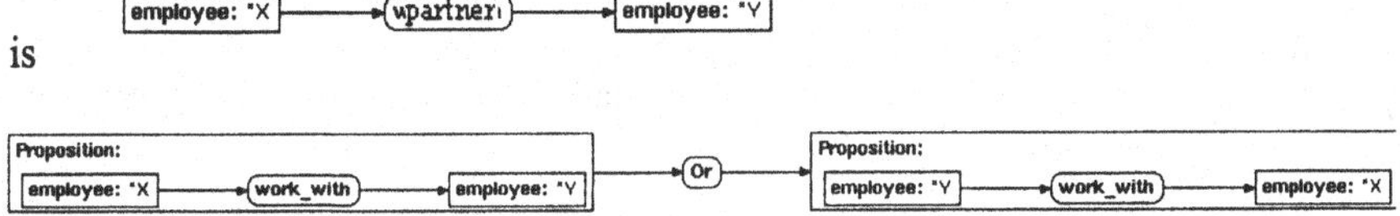

– ViewPoint Template 2: *Room Specifications*

• Representation Style

Style Component	Attribute	Graphical Notation
room	room number size is_central	*room number* *size* *is_central*
next to	-	———

The *room specification* template contains the descriptions of the rooms that are to be allocated. Accordingly, the meta-representation binding of this template describes room number, size and location. Such a meta-representation binding is illustrated as the CG below:

• Meta-Representation Binding

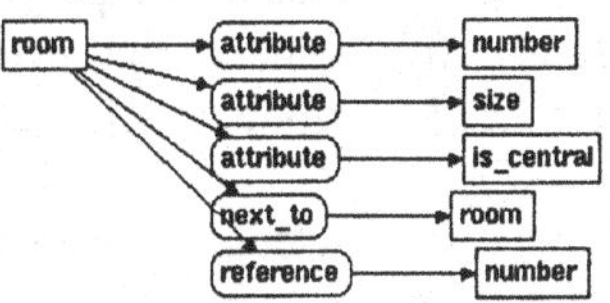

The following *neighbor* relation type refines the condition that a room *X* is next to a room *Y*. A room *X* is said to be a neighbor of a room *Y* if there exists a relation *next_to* from the room *X* to the room *Y* or vice versa.

Relation Type *neighbor(X,Y)*

room: *X → neighbor → room: *Y

is

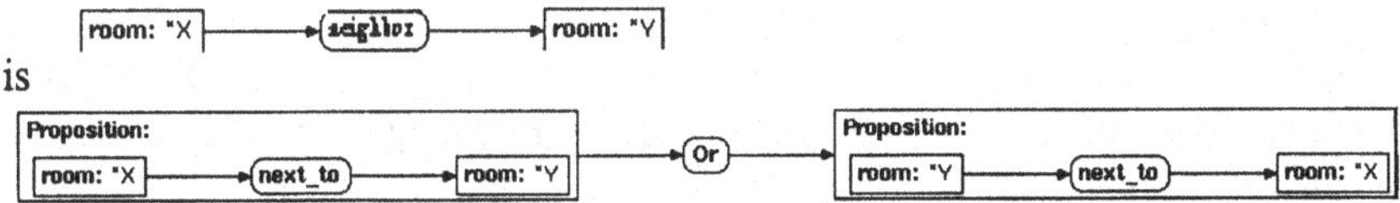

– ViewPoint Template 3: *Allocation Decisions*

• Representation Style

Style Component	Attribute	Graphical Notation
room	room number	*room number*
employee	name	*name*
allocate	-	——▶

The *allocation decisions* template specifies which room is allocated to which employee. The knowledge expressed in this template is constrained by the allocation rules in the problem and the knowledge in both the *employee* template and the *room specifications* template. The meta-representation binding of this template is illustrated as the following CG:

- Meta-Representation Binding

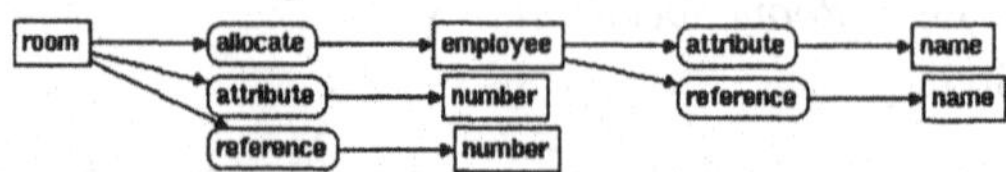

2.2 Knowledge Constraints

To identify a set of consistency rules for the ViewPoint templates specified in the previous section, we classify the knowledge constraints in the context of the Sisyphus-I problem into two types. Firstly, *consistency rules* specify what the system must achieve in order to maintain a consistent stage of its behaviour. In this problem context, such rules indicate the permissible conditions in which a room can be allocated to employees. Secondly, *optimization rules* suggest how things should be regardless of the system's behaviour. In this problem context, such rules represent the constraints which determine the best-suited solution for the allocation. For example, a consistency rule indicates that an employee who is the head of group must be allocated a large central office while an optimization rule indicates that the office which is allocated to the head of group should be next to all members of the group as possible.

Figure 6 shows the mapping of knowledge constraints for our solution of the Sisyphus-I problem. The direction of an arrow in the figure indicates from which source ViewPoint that a rule must be consistently hold when the rule is checked against to which destination ViewPoint. We define the mapping in Figure 6 such that (i) a consistency rule verifies whether the knowledge represented in *Allocation Decisions* ViewPoints is consistent with the knowledge represented in both *Employee* ViewPoints and *Room Specifications* ViewPoints and (ii) an optimization rule suggests that, based on the knowledge in *Employee* ViewPoints and *Room Specification* ViewPoints, what is the knowledge that should be hold in *Allocation Decisions* ViewPoints. In addition, a consistency rule may verify consistency of the knowledge within a ViewPoint itself.

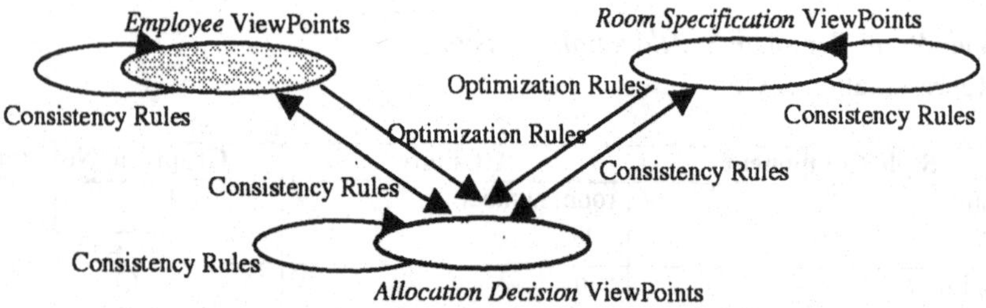

Fig. 6. Knowledge Constraints Between ViewPoint Templates for the Sisyphus-I Problem

Figure 7 shows the table of consistency rules and optimization rules that we address for different types of ViewPoint in the Sisyphus-I problem. The ViewPoints framework supports the consistency checking of such rules by means of in-ViewPoint and inter-ViewPoint checking. An *in-ViewPoint* rule is for consistency checking of data dependencies within a ViewPoint. An *inter-ViewPoint* rule is for consistency checking of data dependencies across ViewPoints.

We argue that it is straightforward to express a consistency rule in the ViewPoints framework as it fits to the ViewPoint consistency definition as we described earlier. In contrast, it is rather difficult to define an optimization rule in the same manner. This is simply because an execution of an optimization rule does not imply either consistency or inconsistency of the knowledge in ViewPoint specifications. Instead an optimization rule suggests what kind of decisions we should make as the solution of the problem. As the ViewPoints framework deals with the verification of knowledge but it is not set out to be a decision support system, we rather focus on only the expression of consistency rules for the solution to the Sisyphus-I problem.

The logical foundation for consistency rules using CGs that we apply to the ViewPoints framework is based on the definition of the ϕ operator, which translates CGs into first-order formulae, and on the definition of rules of inference in [3]. We define a general form of consistency rules as the following conditional clause, where *Conditions* and *Assertions* are conjunction and/or disjunction of CGs:

$$Conditions \rightarrow Assertions$$

The above form of consistency rules is equivalent to the following CG, where *If* and *Then* are ad-hoc concept types that we predefine for logical operations of the CGs in *Conditions* and *Assertions*:

[**If**: *Conditions* [**Then**: *Assertions*]].

We introduce a pre-defined concept type *Proposition* to emphasize the context of an individual CG in a consistency rule. Such a type is specified for implementation purposes in our tool. It imposes the *context* of CGs for assertions and conditions of ViewPoint consistency rules. As an example, given *x* as a CG, [**Proposition**: *x*] will be true if there exists a CG assertion matching with *x*. This form of *context* representation is equivalent to [*x*] in the CG notations defined in [3]. A negative context of CGs is represented as ¬[**Proposition**: *x*]. Such a form of a negative context is equivalent to ¬ [*x*] in the original CG notations.

ViewPoints	Consistency Rules	Optimization Rules
Employee	1. A researcher must participate in at least one project. 2. If two employees work together, they must participate in the same project.	1. The head of group should be allocated an office which is next to all members of the group as possible. 2. The secretariat's office should be located close to the office of the head of group. 3. A manager should have an office close to the head of group and secretariat. 4. The heads of large projects should have an office which is close to the head of group and the secretariat. 5. A large office already occupied by a secretary should be allocated to another secretary.
Room Specifications	1. A central room is a room which is next to more than one room. 2. A room cannot be specified next to itself. 3. Only only one person can occupy a small room.	1. A large office near the head of group is suitable for the secretary. 2. A large office near the head of group is suitable for the manager. 3. A single office near the head of group is suitable for the project leader. 4. Two people can occupy a large room.
Allocation Decisions	1. An employee cannot be allocated into more than one room. 2. The head of group should be allocated a large central office. 3. A manager should have a single central office. 4. If offices are shared, smoking preference should be the same. 5. Members of the same project should not share the same office. 6. A head of project should have a single office. 7. If two people occupy a room, that room must be large.	

Fig. 7. Classification of Knowledge Constraints in the Sisyphus-I Problem

To enhance readability of CGs in our tool, we introduce two pseudo relations, *And* and *Or*, to express the representations of CG conjunction and disjunction respectively. The following derivations represent the semantic translations of the conjunction and disjunction in logical formulae to those in our form of consistency rules.

$x \wedge y$ $\equiv$ [x] [y].
$\equiv$ [**Proposition**: x] → (**And**) → [**Proposition**: y].
$x \vee y$ $\equiv$ ¬ [¬ [x] ¬ [y]].
$\equiv$ [**Proposition**: x] → (**Or**) → [**Proposition**: y].

Further descriptions of built-in concept types and relation types which can be used to specify consistency rules for ViewPoints are defined at length in [6]. The following examples illustrate sample CGs which express consistency rules for ViewPoints.

– In-ViewPoint Rules

Rule *A*: If two employees work together, they must participate in the same project. This rule is illustrated as the CG below:

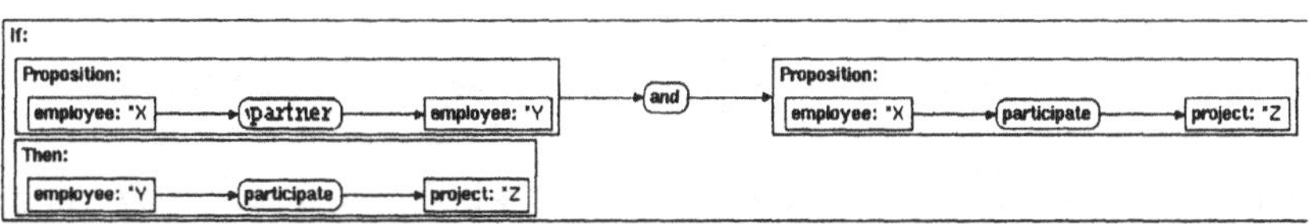

The CG states that if an employee *X* satisfies the relation type *partner* with an employee *Y* and the employee *X* participates in a project *Z*, then there must exist the fact that the employee *Y* also participates in the same project *Z*.

– Inter-ViewPoint Rules

The procedure for composing an inter-ViewPoint rule is similar to the one for composing an in-ViewPoint rule. A slight difference between these procedures is that, for an inter-ViewPoint rule, a method designer must specify the templates in which the assertion and the condition clauses of the rule must hold. Such template specification limits the set of ViewPoints which will engage in the checking of the rule accordingly.

Rule *B*: If offices are shared, smoking preference should be the same. In other words, smokers and non-smokers should not be allocated into the same room. This rule is illustrated as the following CG:

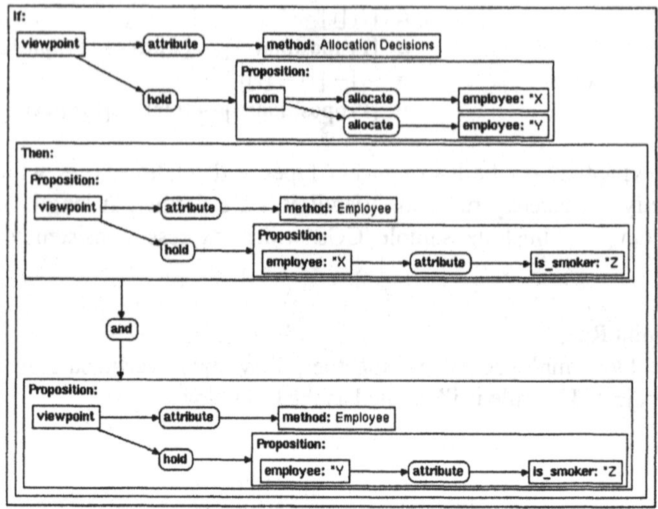

The CG states that if there exists a ViewPoint which is represented by using the *Allocation Decisions* template and that ViewPoint holds the fact that a room is allocated to both employees *X* and *Y*, then there must exist (i) a ViewPoint which is represented by using the *Employee* template and that ViewPoint holds the fact that the smoking preference of the employee *X* is *Z* and (ii) a ViewPoint which is represented by using the *Employee* template and that ViewPoint holds the fact that the smoking preference of the employee *Y* is also *Z*.

3. ViewPoint Consistency Checking Using CGs: A Case of Sisyphus-I

To demonstrate how our models in Section 2 work, we apply a sample data set to construct a set of ViewPoints showing our example solution of the Sisyphus-I problem. Appendix C shows the graphical representation of the example ViewPoints. Figure 8 illustrates an overview structure of the ViewPoints for the problem. The ViewPoints are modelled using the representation styles which are defined in Section 2.3.1. We deliberately construct these ViewPoints in such a way that they are inconsistent with the knowledge constraints that we define in Section 2.3.2.

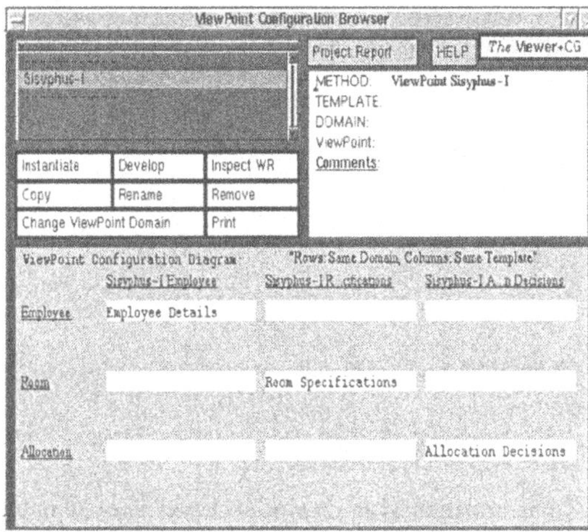

Fig. 8. ViewPoint Browser for the Solution to the Sisyphus-I Problem

The knowledge within a ViewPoint is transformed into CGs by applying CG canonical formation rules to the meta-representation binding of the ViewPoint template from which the ViewPoint is instantiated. Figure 10, Figure 12 and Figure 14 show the CG transformation of the *Employee Details* ViewPoint in Figure 9, the *Room Specifications* ViewPoint in Figure 11 and the *Allocation Decisions* ViewPoint in Figure 13 respectively. How such transformation is performed has been previously discussed in [2].

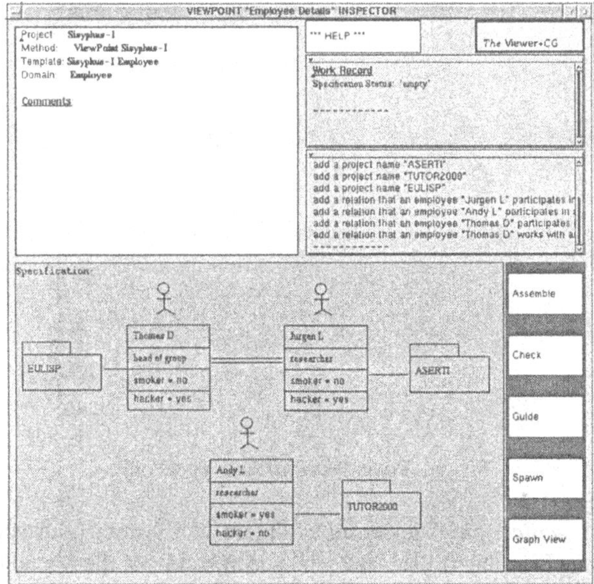

Fig. 9. *Employee Details* ViewPoint

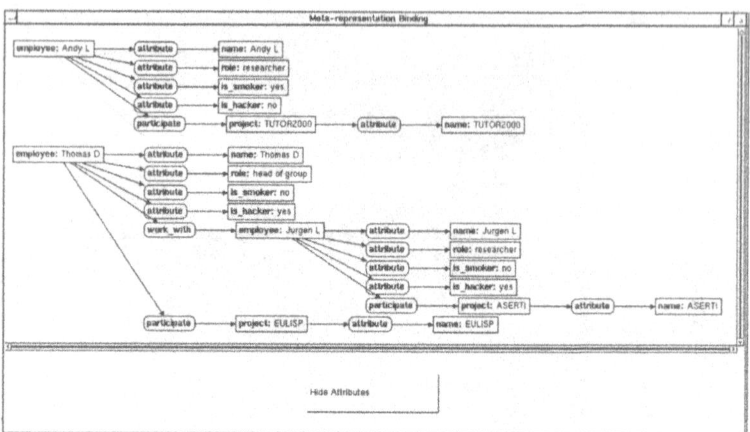

Fig. 10. Transformation of an *Employee Details* ViewPoint into CGs

Figure 9 shows the *Employee* ViewPoint which contains the records of three employees, which are *Thomas D*, *Jurgen L* and *Andy L*. The ViewPoint describes that (i) the employee named *Thomas D* participates in the project named *EULISP* (ii) the employee named *Jurgen L* participates in the project named *ASERTI* and (iii) the employee named *Andy L* participates in the project named *TUTOR2000*. The actions to specify all these descriptions invoke canonical formation rules onto the meta-representation binding of the *Employee* template in Section 2.3.1. The result of the actions is the CGs in Figure 10.

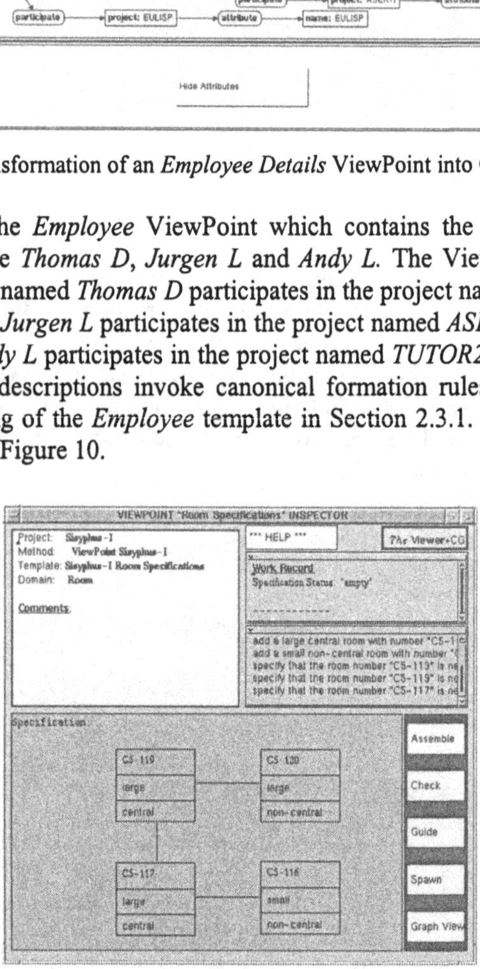

Fig. 11. *Room Specifications* ViewPoint

The *Room Specifications* ViewPoint in Figure 11 illustrates our specification of the rooms for the Sisyphus-I problem. The specification of the ViewPoint indicates that (i) the large central room numbered *C5-119* is suited next to the large non-central room numbered *C5-120* and the large central room numbered *C5-117* and (ii) the room numbered *C5-117* is suited next to the small non-central room numbered *C5-116*. Similar to the *Employee Details* ViewPoint, the actions to specify all these descriptions invoke canonical formation rules onto the meta-representation binding of the *Room Specifications* template in Section 1.3.1. The result of the actions is the CGs in Figure 12.

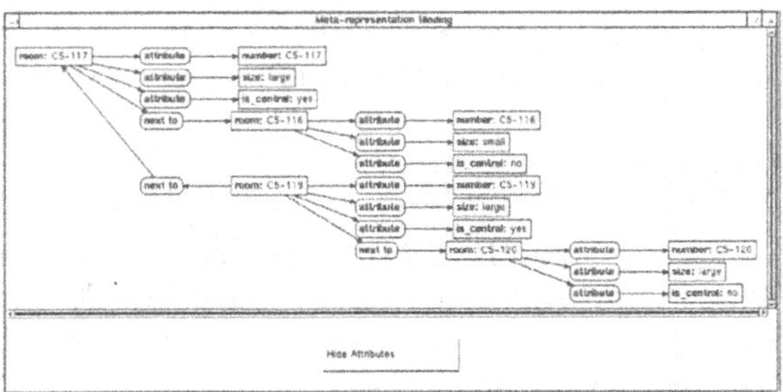

Fig. 12. Transformation of a *Room Specifications* ViewPoint into CGs

The *Allocation Decisions* ViewPoint in Figure 13 specifies that the room numbered *C5-119* is allocated to the employee named *Thomas D*, while the room numbered *C5-117* is allocated to two employees named *Jurgen L* and *Andy L*. The CGs in Figure 14 are the result of the CG transformation of this specification by performing canonical formation rules onto the meta-representation binding of the *Allocation Decisions* template in Section 2.3.1.

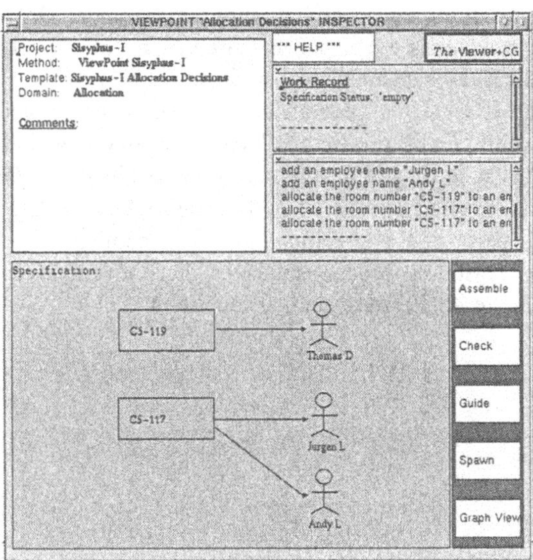

Fig. 13. *Allocation Decisions* ViewPoint

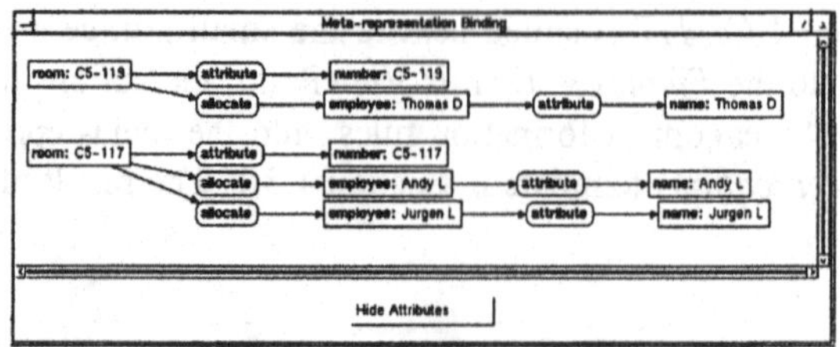

Fig. 14. Transformation of an *Allocation Decisions* ViewPoint into CGs

Consistency Checking

To facilitate reasoning process between partitioned knowledge in different ViewPoints, we perform such a process by using the CG form of ViewPoint specifications and the CG form of consistency rules. The following examples illustrate how we employ CGs for the process.

Example 1 In-ViewPoint Checking

This example shows the application of the CG meta-representation language for consistency checking within a ViewPoint.

Assume that we perform an in-ViewPoint rule, namely *Collaboration Check*, which is defined as the rule *A* in Section 2.3.2 on the *Employee Details* ViewPoint (c.f. Figure 9). Figure 15 shows the CGs that are involved in the reasoning process of this rule:

- Firstly, the CGs *R1* representing the *Collaboration Check* rule is applied to the CGs *VP1* representing the specification of the *Employee Details* ViewPoint. The relation type *partner* is expended to the checking of the relation *work_with* as defined for the *Employee* template (c.f. Section 2.3.1).
- Secondly, the CGs *G1* is generated as a goal which is applied to check the consistency of the *Employee Details* ViewPoint. In doing so, the variables of *R1* are instantiated according to the knowledge or fact held by *VP1* to compose *G1*.
- Finally, *G1* is verified against *VP1*. The highlighted concepts and relations in *VP1* indicate parts of the specification which conflict with the goal *G1*. The result of this reasoning process indicates that the employee named *Jurgen L* participates in the *ASERTI* project is inconsistent.

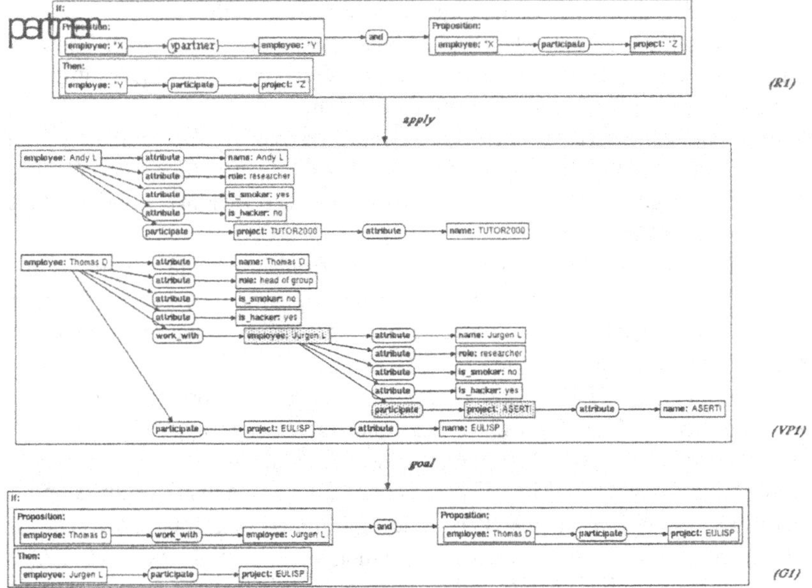

Fig. 15. Applying CGs to In-ViewPoint Checking

Figure 16 shows an output example of the reasoning process in Figure 15. The middle left window shows the list of the check that can be performed on the ViewPoint. The bottom left window shows the list of errors that are resulted from performing selected checks. The top right window shows possible inconsistency handling actions that, when performed, will lead the ViewPoint to a consistent stage. Such a handling action is generated from the assertion clause of consistency rules. In this example, the actions suggest two changes: firstly, the change to specify that the employee named *Jurgen L* participates in the *EULISP* project; or secondly, the change to specify that the employee named *Thomas D* participates in the *ASERTI* project. Either of these changes will make the two employees, *Thomas D* and *Jurgen L*, participate in the same project. Therefore, the *collaboration check* rule will not be violated.

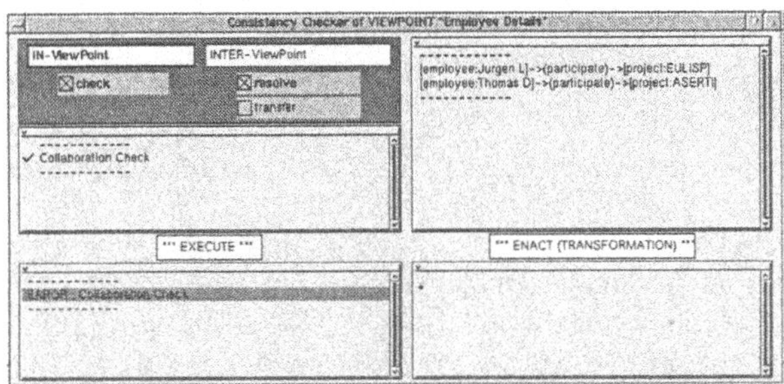

Fig. 16. An Output Example of In-ViewPoint Checking

Example 2 Inter-ViewPoint Checking

To ensure that different ViewPoints constructed for this testbed can work consistently together as an integrated whole, we must apply inter-ViewPoint checking to verify the relations between those ViewPoints. In this example, we illustrate such a reasoning process between ViewPoints.

Assume that we perform an inter-ViewPoint rule, namely *Smoking Preference*, which is defined as the rule *B* in Section 1.3.2 on the *Allocation Decisions* ViewPoint (c.f. Figure 13). Figure 17 shows the CGs that are involved in the reasoning process of this rule:

- Firstly, the CGs *R2* representing the *Smoking Preference* rule is applied to the CGs *VP2* representing the specification of the *Allocation Decisions* ViewPoint. As a result, the variables of *R2* are instantiated according to the knowledge or the fact held by *VP2* to generate the CGs *G1*.
- Secondly, the CGs *G1* is applied as a goal to check the condition of *R2* in the *Employee* type of ViewPoints. In this case, the goal is applied to verify *R2* against the CGs *VP1* which represents the CG form of the *Employee Details* ViewPoint (c.f. Figure 9). The CGs *G2* is generated as the result of this step.
- Finally, *G2* is verified against *VP1*. The highlighted concepts and relations in *VP1* indicate parts of the specification which conflict with the goal *G2*. The result of this reasoning process indicates that the fact that the employee named *Jurgen L* is not a smoker is inconsistent.

Figure 18 shows an output example of the reasoning process in Figure 17. The windows in the example represent the same information as those of the in-ViewPoint reasoning process in Figure 16. The possible inconsistency handling action in the example suggests that the change to specify that *Andy L* is not a smoker will make the two employees, *Andy L* and *Jurgen L*, have the same smoking preference. Therefore, the *Smoking Preference* rule will not be violated when we specify that *Andy L* and *Jurgen L* share the same office.

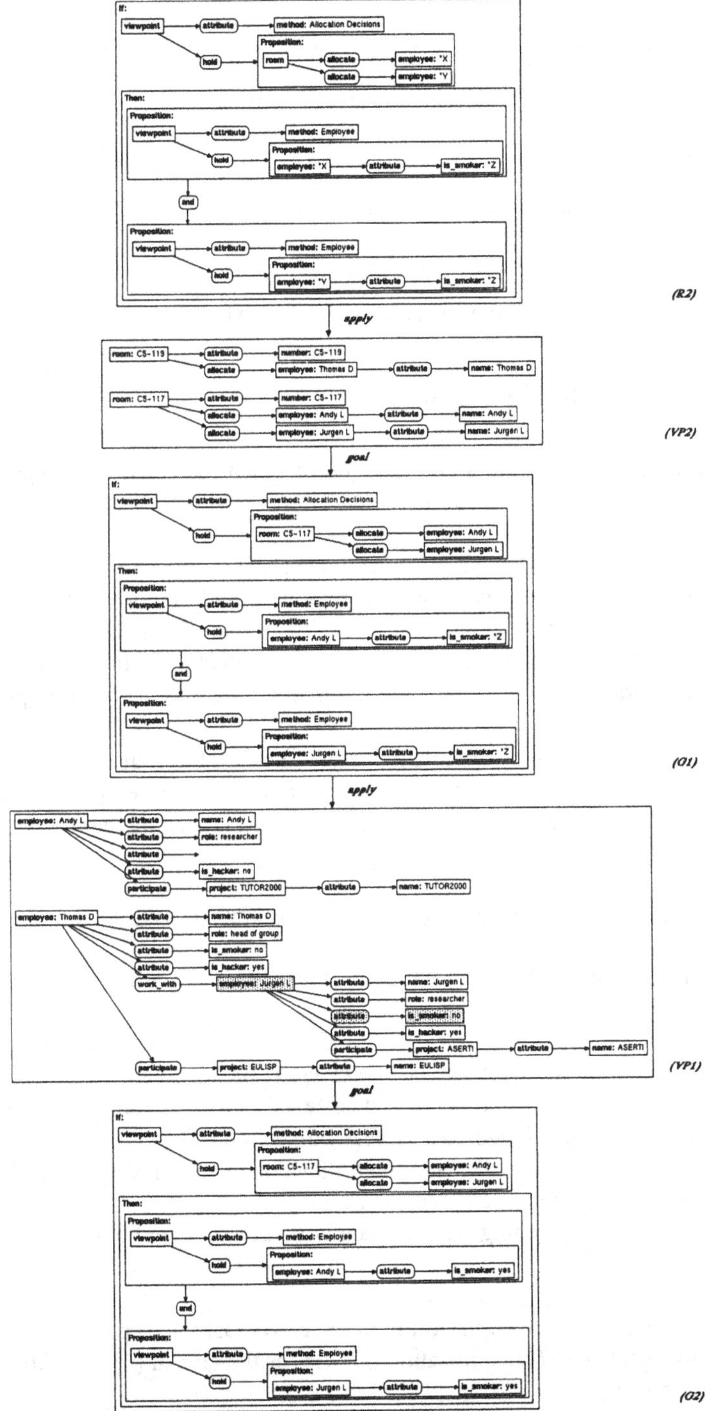

Fig. 17. Applying CGs to Inter-ViewPoint Checking

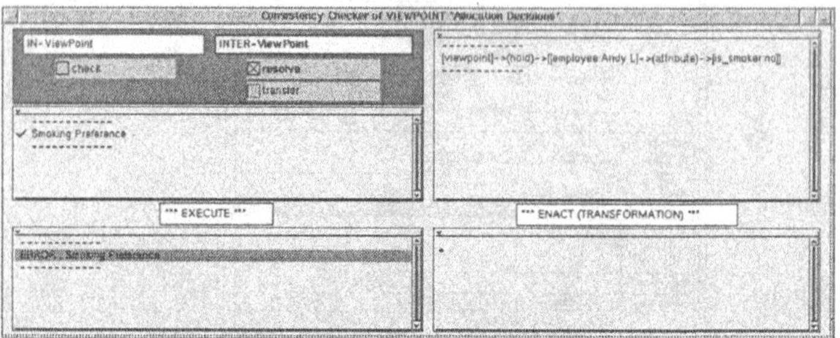

Fig. 18. An Output Example of Inter-ViewPoint Checking

4 Summary and Discussion

We select CGs to apply as a solution to our problem of consistency checking in multiperspective frameworks, in particular the ViewPoints framework, for the following reasons:

- The graphs serve as powerful *visual* consistency checking notations. The combination of the underlying logical reasoning and graph-based reasoning of CGs provides strong foundation for consistency checking mechanism for our framework. It is also more readable to map descriptive definitions of consistency rules in natural language to CGs than to map the definitions to logic predicates directly.
- The expression of a representation style in CG form also helps software developers to understand, and enable the developers to express clearly, the syntax and semantics of ViewPoint descriptions, including their relationships (consistency rules).

In summary, we apply CGs to perform the following tasks in the ViewPoints framework:

1. to visualize concepts and relations that are expressed by a method or a method fragment which is used to produce ViewPoints;
2. to establish the meta-representation binding or permissible structure of the concepts and relations which can be expressed by the method;
3. to interpret assembly actions as a sequence of CG operations to automate the transformation of knowledge hold in ViewPoints into CGs;
4. to establish in-ViewPoint rules and inter-ViewPoint rules of ViewPoints by using the resulting concepts and relations in (1); and
5. to perform logical reasoning on CGs transformation of ViewPoints specifications and the rules.

This paper illustrates the applicability of the above tasks, except the one in (3), to the Sisyphus-I problem. The application of CGs for the task in (3) is exemplified in our previous paper [2].

We evaluate the strengths and weaknesses regarding the applications of CGs in our models in terms of the following criteria:

- Soundness and Completeness

Although it has been argued that the definition of the ϕ operator and the definition of rules of inference in [3] that we adopt in our models are incomplete [8], the applications of such definitions lay essential groundwork of how CGs can facilitate reasoning procedure for knowledge modelling in the ViewPoints framework. Moreover, the combination of the underlying logical reasoning and graph-based reasoning of CGs provides a powerful consistency checking mechanism which is essential for ViewPoints. The modification to the theoretical work on CG reasoning is beyond the scope of our work.

- Simplicity

CGs offer intuitive, but expressive, notations to describe ViewPoint representation styles. Although the graphs provide the insight of formal reasoning techniques, their notations are not difficult to learn for those without a formal method background.

- Readability

Although logic-based consistency rules for ViewPoints have been proposed [9][10][11], the mapping of descriptive definitions of the rules in natural languages (i.e. English) to CGs is more readable than the mapping of those to logic predicates directly. Accordingly, CGs are used as a bridge between the logic-based rules and the descriptive rules in our models. The act of elaborating consistency rules precisely in terms of CG constructs also provides us with intangible gain of a deeper understanding of the rules. It is through the CG constructs that method designers distinguish clearly the tangible entities which can be expressed by a method or a method fragment and the means that link those entities. This leads to uncovering design flaws and ambiguities in the definition of the method.

- Implementation Issues

Our approach to use CGs as a meta-representation language is fundamentally different from the previous efforts in consistency checking and integration of multiperspectives. Such efforts are the implementation of pair-wise translation rules [12][13] and the application of canonical representations, including one using CGs [14][15][16]. We do not intend to apply CGs to interpret the entire semantics of data models in different ViewPoint representation styles. We contend that using a canonical form as such is not efficient. Firstly, it is difficult to identify a single canonical form which can express *all* possible representation styles. Secondly, it is difficult to implement a significant number of translation rules for the mapping between the canonical form and the representation styles. In particular, we argue that the deployment of a canonical representation in the ViewPoints framework restricts the extendibility of the framework, as all the ViewPoint representation styles would be tightly bounded to a particular canonical representation. If such a canonical representation is not expressive enough, it would be difficult to add a new representation style to the framework. The use of the CG meta-representation language helps us to avoid the difficulties associated with these previous approaches.

- Shortcomings

- There are some opportunities for further research with regard to our applications of CGs in the ViewPoints framework:
- It would be interesting and worthwhile to apply the notion of *actors* such as those defined in [17]. With the notion of actors, it is possible to represent an executable software specification. This might enable us to further investigate the applicability of other useful software modelling techniques, such as behavioural analysis [18], in the ViewPoints framework. For example, if we equip a *process* notation of a *dataflow* model [19] with the actor notion, then it may be possible to allow software developers to apply some test cases to their ViewPoints and execute the specifications of those ViewPoint to check whether some desired properties hold in their ViewPoints.
- As mentioned earlier, ViewPoints may interact with each other in order to perform consistency checking. The enhancement of the *conceptual graph interchange form* (CGIF) would be useful to the communication process between distributed ViewPoints.

Critique of the Sisyphus-I Testbed

Our application of CGs as a meta-representation language in the ViewPoints framework provides us with evidence of practical usefulness of the graphs in knowledge representation and reasoning, in particular for the modelling of software specifications. Towards that end, we have found that the Sisyphus-I domain is suitable for the modelling of problem-solving processes in knowledge-based systems rather than the modelling of software specifications. The domain contains some knowledge constraints, which we call optimization rules, that cannot be straightforwardly expressed in terms of consistency rules in the context of software specifications. As we discussed earlier, this is simply because the result of an execution of an optimization rule suggests what kind of decisions we should make, based on the given knowledge, but it does not imply either consistency or inconsistency of the knowledge. The current stage of the ViewPoints framework does not tackle such a decision support mechanism, but it would be both interesting and worthwhile to consider whether and how such an optimization rule can be handled within the framework.

The constraints in the modelling for the Sisyphus-I problem as given at the URL http://ksi.cpsc.ucalgary.ca/KAW/Sisyphus/Sis1/ are not clearly distinguished in the context of the problem domain itself. It would be useful to provide a more elaborated list of constraints that should be imposed on the problem so that we could compare how those constraints can be expressed in terms of CGs using different modelling techniques or definitions.

References

1. Finkelstein A., Kramer J., Nuseibeh B., Finkelstein L., Goedicke M.: Viewpoints: A Framework for Integrating Multiple Perspectives in System Development, Int. J. Software Engineering and Knowledge Engineering, Vol. 2(1). World Scientific Publishing Co. (1992) 31-58.
2. Thanitsukkarn T., Finkelstein A.: A Conceptual Graph Approach to Support Multiperspective Development Environments, In: Gaines B. R., Musen M. (eds.), Proceedings of the 11th Knowledge Acquisition For Knowledge-based Systems Workshop, Vol. 1, Banff, Canada (1998).
3. Sowa J. F.: Conceptual Structures: Information Processing in Mind and Machine, Addison-Wesley, Reading, MA (1984).
4. Nuseibeh B.: A Multi-Perspective Framework for Method Integration, Department of Computing, Imperial College, University of London, London, PhD. Thesis (1994).
5. Gaines B. R.: A Situated Classification Solution of a Resource Allocation Task Represented in a Visual Language, Int. J. Human-Computer Studies, Vol. 40(2), http://ksi.cpsc.ucalgary.ca/KAW/Sisyphus/Sis1/ (1994).
6. Thanitsukkarn T.: Multiperspective Development Environment for Configurable Distributed Applications, Department of Computing, Imperial College, University of London, PhD Thesis, February (1998).
7. Petermann H., Möller J. -U., Wiese D.: CG-Editor User's Guide, University of Hamburg (1995).
8. Wermelinger M.: Conceptual Graphs and First-Order Logic, In: Conceptual Structures: Applications, Implementation and Theory, Lecture Notes in Artificial Intelligence, Vol. 954, Springer-Verlag (1995) 323-337.
9. Nuseibeh, B., Kramer, J., Finkelstein, A: Expressing the Relationships Between Multiple Views in Requirements Specification, Proceedings of the 15th International Conference on Software Engineering, IEEE CS Press, Baltimore, USA, May (1993).
10. Finkelstein, A., Gabbay, D., Hunter, A., Kramer, J., Nuseibeh, B.: Inconsistency Handling in Multi-Perspective Specifications, IEEE Transactions on Software Engineering, Vol. 20(8), December (1994) 569-578.
11. Hunter, A., and Nuseibeh, B.: Managing Inconsistent Specifications: Reasoning, Analysis and Action, Technical Report Number 95/15, Department of Computing, Imperial College, London, UK, October (1995).
12. Bowman H., Derrick J., Steen M.: Some Results on Cross Viewpoint Consistency Checking, In: IFIP TC6 International Conference on Open Distributed Processing, Brisbane, Australia, Chapman and Hall (1995) 399-412.
13. Boiten E., Bowman H., Derrick J., Steen M.: Viewpoint Consistency in Z and LOTOS: A Case Study, In: Proceedings of the 4th Int. Symposium of Formal Methods Europe, FME'97: Industrial Applications and Strengthened Foundations of Formal Methods, Lecture Notes in Computer Science 1313, Graz, Austria, September, Springer-Verlag (1997) 644-664.
14. Delugach H. S.: Analysing Multiple Views of Software Requirements, In Nagle T. E., Nagle J. A., Gerholz L. L., and Eklund P. W. (eds.), Conceptual Structures: Current Research and Practice, Ellis Horwood (1992).
15. Delugach H. S.: Specifying Multiple-Viewed Software Requirements with Conceptual Graphs, J. System Software, Vol. 19 (1992) 207-224.
16. Delugach H. S.: An Approach to Conceptual Feedback in Multiple Viewed Software Engineering Models, Int. Workshop on Multiple Perspectives in Software Development, SIGSOFT'96 Workshops, San Francisco, USA, Vidal L., Finkelstein A., Spanoudakis G., and Wolf A. L. (Eds.), ACM Press (1996) 242-246.

17. Lukose D., Mineau, G. W.: A Comparative Study of Dynamic Conceptual Graphs, Proceedings of the 11th Knowledge Acquisition For Knowledge-based Systems Workshop, Gaines B. R., Musen M. (eds.), Vol. 1, Banff, Canada (1998).
18. Clarke E. M., Wing J. M.: Formal Methods: State of the Art and Future Directions, ACM Computing Surveys, Vol. 28(4) (1996) 626-643.
19. Yourdon E., Constantine L. L.: Structured Design: Fundamentals of a Discipline of Computer Program and Systems Design, Prentice-Hall, Englewood Cliffs (1979).

Using Conceptual Graphs to Solve a Resource Allocation Task

Svetlana Damianova[1] and Kristina Toutanova[2]

[1] Department of Computer Science, University College,
Belfield, Dublin 4, Ireland
Svetlana.Damianova@ucd.ie

[2] Linguistic Modeling Laboratory, Bulgarian Academy of Sciences
25A Acad. G. Bontchev Str., 1113 Sofia, Bulgaria
kris@lml.bas.bg

Abstract. This paper presents a solution of the Sisyphus-I room allocation task encoding the problem specific data in conceptual graphs (CG). We use extensively type definitions and relation definitions to represent knowledge about the domain and demons to represent allocation rules and dynamically changes in the knowledge base of conceptual graphs. It is assumed that an external system, which performs inference from the knowledge base, applying the standard conceptual graphs operations, is run before each allocation. We have to remark that at this stage we have only theoretical model. Using conceptual graphs we show a simple and readable solution of the Sisyphus-I problem.

1 The model

In modeling the Sisyphus-I domain, we tried to use as much as possible the basic constructs of the CG theory only. We also used extensively our experience in knowledge acquisition using conceptual graphs and the way we represented knowledge in other systems (DBR-MAT [1],[2] and CGLex [3]). Our design solutions are now explained in detail.

The CG knowledge base (KB) that we use in the solution of the Sisyphus-I task is separated into general knowledge and problem-specific data. We will jointly describe them, using the concrete formulation of the problem, given at
http://ksi.cpsc.ucalgary.ca/KAW/Sisyphus/Sisyphus1/.

1.1 Problem solving strategy

Analysing the given protocol lead us to the following problem-solving strategy:

1. Enter the knowledge structures.
2. Enter the facts of a specific situation.
3. Apply knowledge-based operations in attempt to trigger a demon execution, which makes an allocation of one or two persons to an office.
4. Loop back to 3.

We have adopted the following assignment rules, very similar to those from the example solution [4]. The rules are given in order of declining priority. The exact rules that we apply slightly differ from these below, because they capture constraints as to the compatibility of the researchers with respect to the projects they work on.

- *Allocation rule 1:* A large central office is suitable for the head of group (Protocol 1).
- *Allocation rule 2:* A large office near the head of group is suitable for the secretaries (Protocol 2).
- *Allocation rule 3:* A single office near the head of group is suitable for the Manager (Protocol 3).
- *Allocation rule 4:* A single office near the head of group is suitable for the Project Leaders (Protocol 4)
- *Allocation rule 5:* A large office should be allocated to smoking researchers (Protocol 5).
- *Allocation rule 6:* A large office should be allocated to non-smoking researchers (Protocol 6).

If the problem does not have a solution that satisfies the problem constraints, then the external system will not be able to trigger a demon execution after some stage in the problem solving process. In this case the external system raises an exception. This is a permission for breaking some of the constraints. The exceptions trigger demons, which are also prioritized. Below we list them in order of declining priority as well.

- *Exception rule 1*: A person without an office can be assigned to an office without occupants.
- *Exception rule 2*: A person without an office can be assigned to a large office with one occupant, if they have the same smoking habits.
- *Exception rule 3*: A person without an office can be assigned to a large office with one occupant, regardless of smoking habits.

The requirement that two researchers on the same project, if possible, should not share a room is treated in a similar fashion, by additional preconditions to the execution of the demons representing these rules.

1.2 Knowledge structures in our model

We apply the following knowledge structures:

- *Type hierarchy*, which defines the partition of domain concepts into subtypes;
- *Type definitions* and *relation definitions*. They are used for inference by type and relation expansion. Their syntax is defined in [7];
- *Demons* to maintain changes in the KB, which are used to represent rules as well as to assert and retract conceptual graphs;
- *Regular conceptual graphs*, which encode facts about particular individuals.

1.3 Type hierarchy

Fig. 1 shows an extract of our type hierarchy, which is relevant to the Sisyphus-I room allocation problem. We use the type hierarchy in the inference process, which includes mainly generalization and type expansion.

It turned out that to solve the Sisyphus-I task, we do not need to represent whether a partition is exhaustive or not, and whether it is into natural or role types.

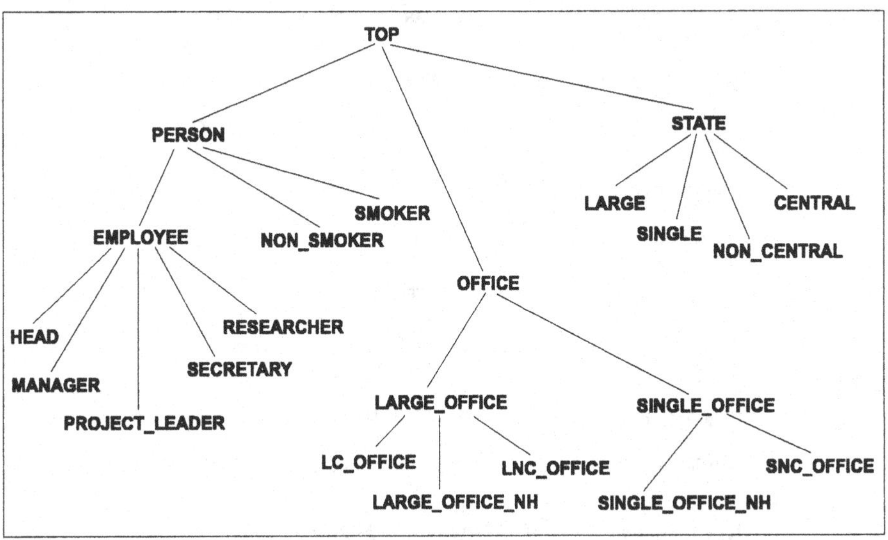

Fig. 1. The necessary branches of the type hierarchy.

1.4 Type definitions

We use type definitions extensively in the inference process (see Fig. 2 and Fig. 3). Fig. 2 shows the most often used type definitions. We need definitions of different kinds of offices, with respect to their size, location, and nearness to the office of the head of group. We represent the size and location characteristics of offices through relations with names that clearly show their semantics. LOCCHAR is a characteristic of location, and SIZECHAR is a characteristic of size. Unlike the model, represented in [4], we do not need to assert that every office has a characteristic of location and every office has a characteristic of size.

The recursive relation definition in Fig. 3 is extensively used. It is used to notify the nearest unoccupied office to the head of group. Later, when we introduce the rules, we will show with a particular example how this is achieved.

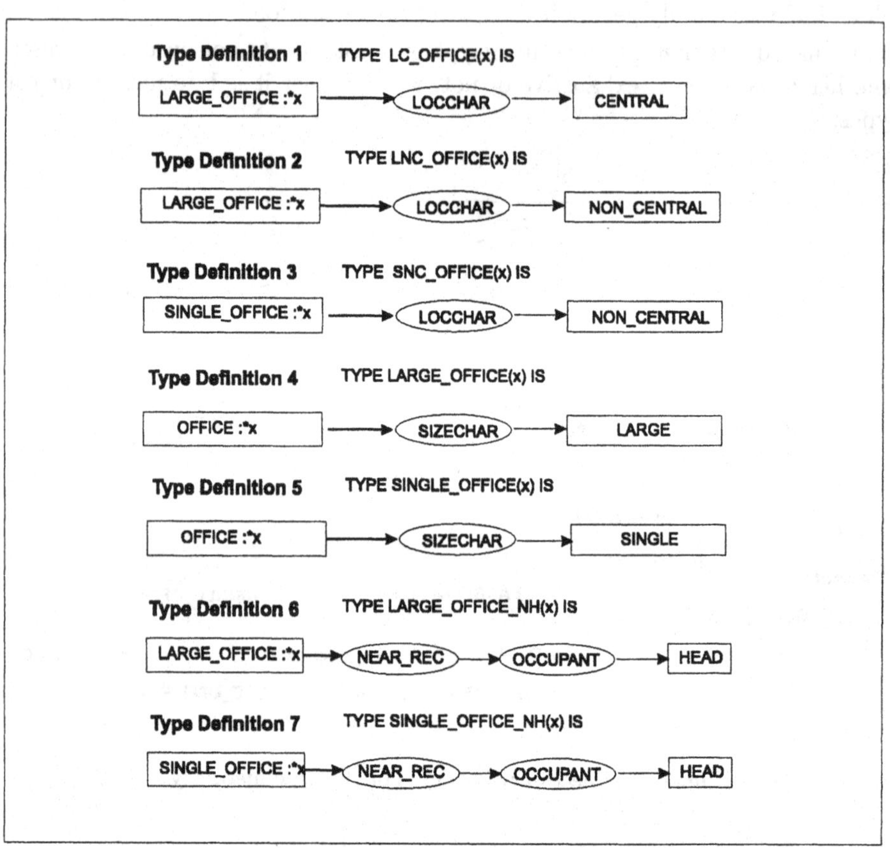

Fig. 2. Type definitions of the different types of offices.

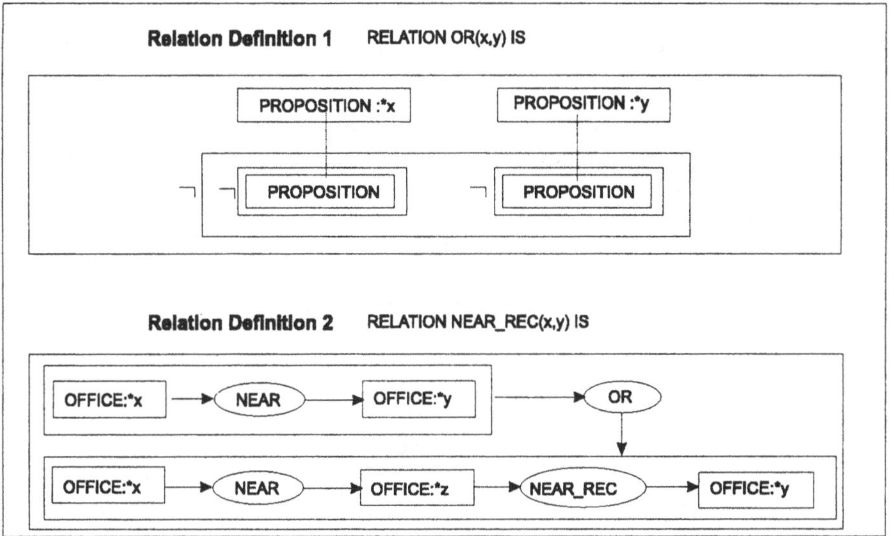

Fig. 3. Recursive relation definition of nearness to an office.

1.5 Representation of the rules

Each rule makes an allocation of one or two persons to a room. The allocation is done by retracting the condition that the person is without an office and asserting that this person has one. Every office has a NUMBER_OCCUPANTS feature, which is represented by a conceptual relation. The NUMBER_OCCUPANTS of every office is initially zero, and is incremented by retracts and asserts when an assignment is made.

Every rule is represented by a demon, which has one or more incoming arcs and one outgoing arc. The execution of the demon is triggered by an external system. If the conceptual graphs in the incoming arcs of the demon can be inferred from the knowledge base through the inference rules, and there is no demon with higher priority that can be executed, the rule is applied. The graphs in the contexts connected to the incoming arcs of the demon are retracted from the knowledge base, and the graph(s) in the context connected to the outgoing arc are asserted.

As in this process we remove from the knowledge base only initial facts, we keep the static facts for characteristics of the offices and people, that we inferred through type and relation expansion. The parameter passing is implemented through coreference links.

Bellow we discuss the rules in order of declining priority.

Fig. 4 displays the *Allocation rule 1*— the head of group should be allocated a large central office. It is the highest priority rule and is applied first. This assignment is defined first, because it strongly restricts the possibilities for subsequent assignments. It is interesting to note that after the assignment of the head of Group to an office, this office's number of occupants feature receives the

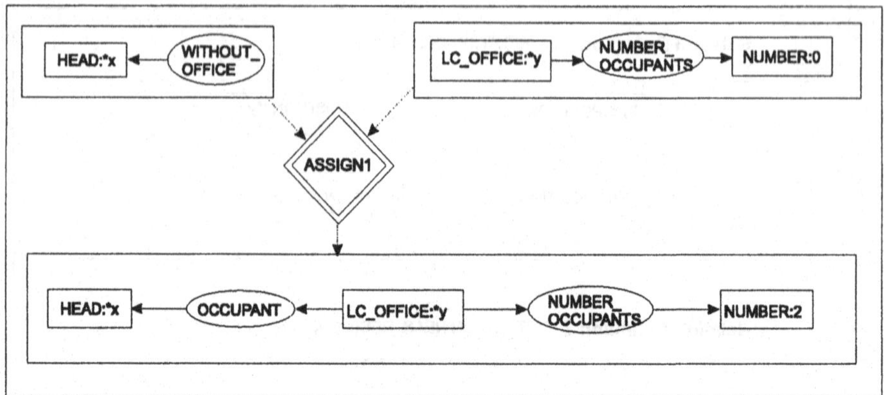

Fig. 4. Protocol 1.

value [NUMBER:2]. This is done to forbid subsequent assignments of people to this office. (Although this is not semantically correct solution, we prefer it, as otherwise we have to add additional feature AVAILABLE for every office and the model will be more complicated.)

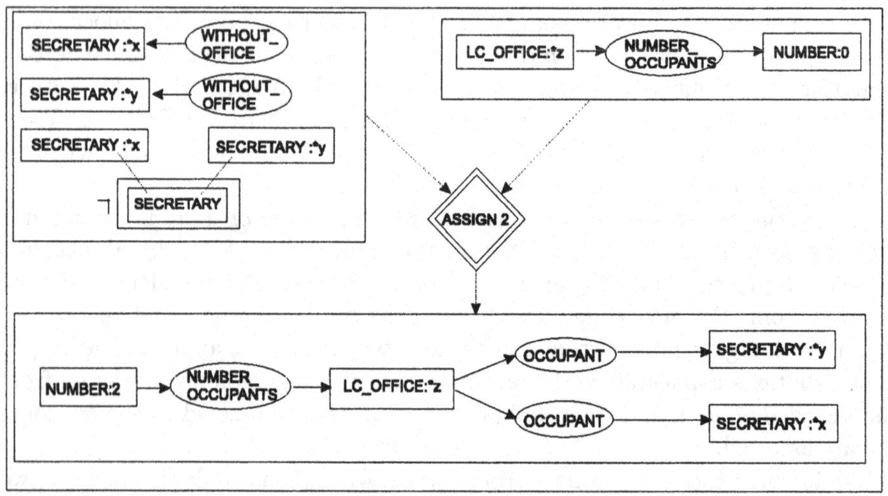

Fig. 5. Protocol 2.

Fig. 5 represents *Allocation rule 2* — two secretaries without offices should be assigned to a large office near the head of group. To express the constraint that the two secretaries should be different individuals, we use coreference links and negative contexts. For this assignment, an office near the office of the head is required. The nearness of offices is not a very clearly defined relation. In fact, in this case, the nearest large unoccupied office to the head of group is needed. We will show how the recursive definition of the relation NEAR_REC from Fig. 3 achieves this. When the inference engine tries to find an individual of type LARGE_OFFICE_NH, a type expansion is performed and the body of type definition 6 has to be satisfied. At the next step, a relation expansion of NEAR_REC is performed. It can be seen that the large offices which are asserted to be near the office of the head satisfy the resulting graph—the body of type definition 6. Therefore, they are individuals of type LARGE_OFFICE_NH. If some of them is unoccupied, i.e. it has a value [NUMBER:0] for its NUMBER_OCCUPANTS feature, then this office will be suitable for the secretaries. If there is no large unoccupied office near the office of the head, another relation expansion is performed in the body of the relation definition from Fig. 3, and the offices that are near offices near the head of group are considered. Thus, the inference engine will get the nearest to the head unoccupied large office.

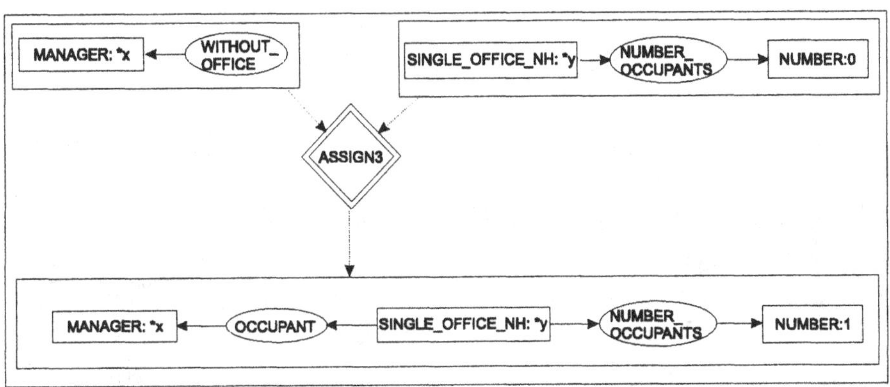

Fig. 6. Protocol 3.

Fig. 6 displays the ASSIGN3 demon, encoding the *Allocation rule 3*. It assigns the manager to a single office near the head of group.

The *Allocation rule 4*, represented by the demon ASSIGN4 assigns project leaders to single offices near the office of the head. It is very similar to the demon ASSIGN3 and is given here in linear form (Fig. 7).

Allocation rule 5 and *Allocation rule 6* are represented by demons ASSIGN5 and ASSIGN6 and allocate large offices to pairs of smoking and nonsmoking researchers respectively (see Fig. 8 and Fig. 9). They are very similar and thus the second one is given in linear form. It also takes in consideration the constraint that two researchers should not be allocated an office together, if they work on the same project.

Demons ASSIGN7,ASSIGN8 and ASSIGN9 handle the exception rules. Fig. 10 displays the demon ASSIGN7. ASSIGN8 and ASSIGN9 are given in linear form (Fig. 11, Fig. 12).

```
<<ASSIGN4>>-
  <-[[PROJECT_LEADER: *x]<-
      -(WITHOUT_OFFICE)]
  <-[[SINGLEOFFICE_NH: *y]->
      -(NUMBER_OCCUPANTS)->[NUMBER:0]]
  ->[[SINGLEOFFICE_NH: *y]->
      -(NUMBER_OCCUPANTS)->[NUMBER:1]
      -(OCCUPANT)->[PROJECT_LEADER: *x]]
```

Fig. 7. Protocol 4.

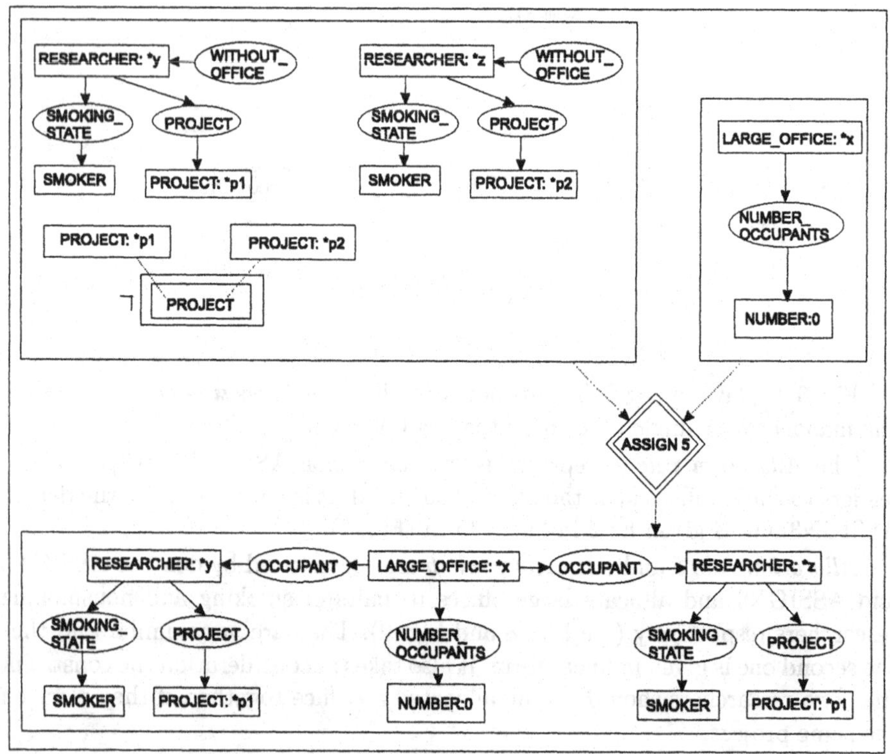

Fig. 8. Protocol 5.

```
<<ASSIGN6>>-
  <-[[RESEARCHER: *x]-
      -(WITHOUT_OFFICE)
      -(SMOKING_STATE)->[NON_SMOKER]
      -(PROJECT)->[PROJECT:*p1]
  <-[RESEARCHER: *y]-
      -(WITHOUT_OFFICE)
      -(SMOKING_STATE)->[NON_SMOKER]
      -(PROJECT)->[PROJECT:*p2]
  <-[[PROJECT:*p1] [PROJECT:*p2]  ~[PROJECT:*p1=*p2]]
  <-[[LARGE_OFFICE: *z]->(NUMBER_OCCUPANTS)->[NUMBER:0]]
  ->[[LARGE_OFFICE: *z]->
     -(NUMBER_OCCUPANTS)->[NUMBER:0]
     -(OCCUPANT)->[RESEARCHER: *x]-
          -(WITHOUT_OFFICE)
          -(SMOKING_STATE)->[NON_SMOKER]
          -(PROJECT)->[PROJECT:*p1]
     -(OCCUPANT)->[RESEARCHER: *y]-
          -(WITHOUT_OFFICE)
          -(SMOKING_STATE)->[NON_SMOKER]
          -(PROJECT)->[PROJECT:*p2]]
```

Fig. 9. Protocol 6.

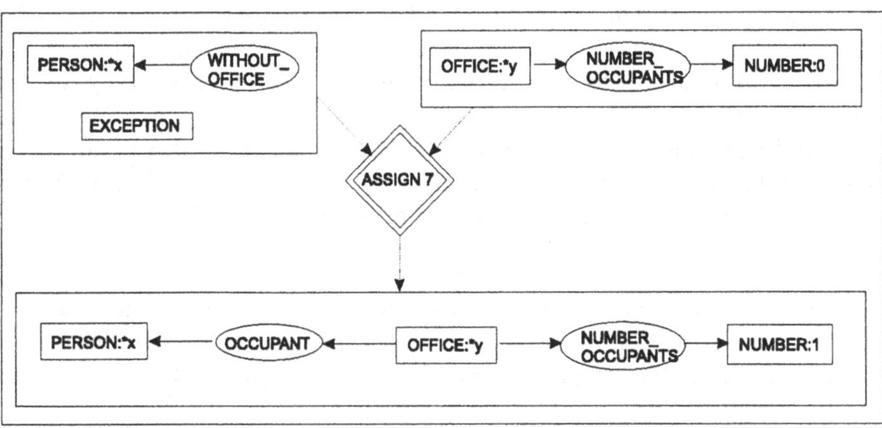

Fig. 10. Exception rule 1

```
<<ASSIGN8>>-
  <-[[PERSON:*x]-
      -(WITHOUT_OFFICE)
      -(SMOKING_STATE)->[SMOKER]
     [EXCEPTION]]
  <-[[LARGE_OFFICE:*z]->
      -(NUMBER_OCCUPANTS)->[NUMBER:1]
      -(OCCUPANT)->[PERSON:*y]->(SMOKING_STATE)->[SMOKER]]
  ->[[LARGE_OFFICE:*z]->
      -(NUMBER_OCCUPANTS)->[NUMBER:2]
      -(OCCUPANT)->[PERSON:*y]->(SMOKING_STATE)->[SMOKER]]
```

Fig. 11. Exception rule 2

```
<<ASSIGN9>>-
  <-[[PERSON:*x]-
      -(WITHOUT_OFFICE)
      [EXCEPTION]]
  <-[[LARGE_OFFICE:*z]->
      -(NUMBER_OCCUPANTS)->[NUMBER:1]
      -(OCCUPANT)->[PERSON:*y]]
  ->[[LARGE_OFFICE:*z]->
      -(NUMBER_OCCUPANTS)->[NUMBER:2]
      -(OCCUPANT)->[PERSON:*x]
      -(OCCUPANT)->[PERSON:*y]]
```

Fig. 12. Exception rule 3

1.6 Assertions about problem specific data

Assertions about the members of the YQT Research Center. We represent the data on the different members of the YQT Research Center through separate conceptual graphs, one for each member. Here we show only four of these conceptual graphs (see Fig. 13).

The relation names are quite straightforward. The data in these graphs is a direct translation of the problem data about the members of the YQT Research Center. Every characteristic is represented by a conceptual relation.

Assertions about the size and location of the rooms. The size and location characteristics of every office are specified, as well as the offices it is near to. The conceptual graphs for few offices are displayed on Fig. 14, the rest are very similar to them.

We also need to represent the initial conditions that no person has a room and no office has occupants. Fig. 15(a) shows part of the CG representing the initial state of the people, working in the YQT Research Center and Fig. 15(b) shows part of the conceptual graphs asserting the initial state of the offices.

These are all of the knowledge structures we have used. The constructs are in accordance with the CG theory and the CG standard. We have adopted the definition of demons from papers of Harry Delugach [5] and Guy Mineau [6]. We suppose the rest of the constructs we have used are well-known in the CG community.

2 The evaluation process

We have tested our model only empirically, on the two sample statements of the problem located at:
http://ksi.cpsc.ucalgary.ca/KAW/Sisyphus/Sisyphus1/.

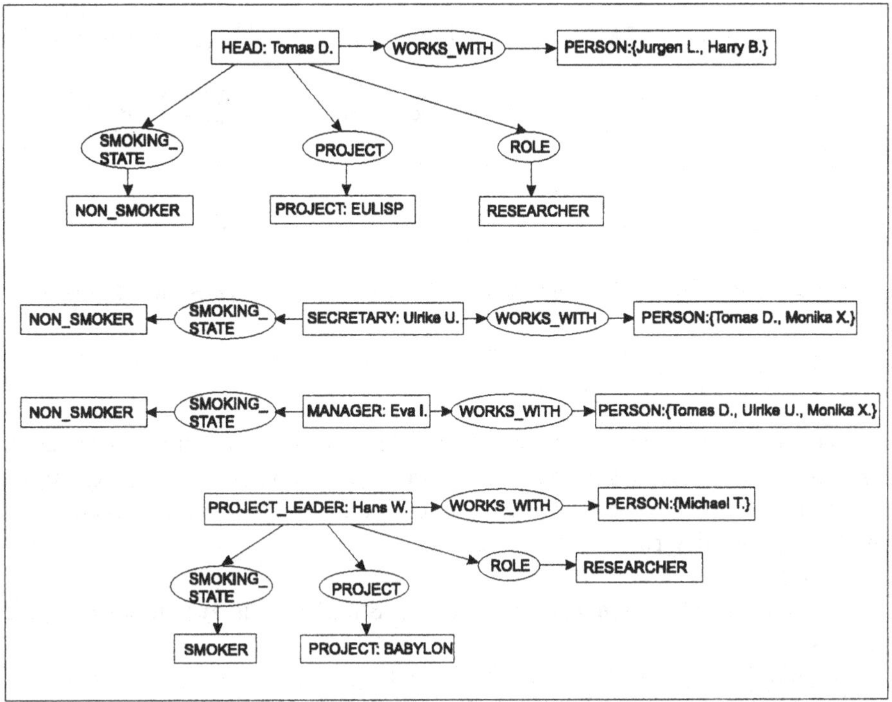

Fig. 13. Assertions about the members of the YQT Research Center

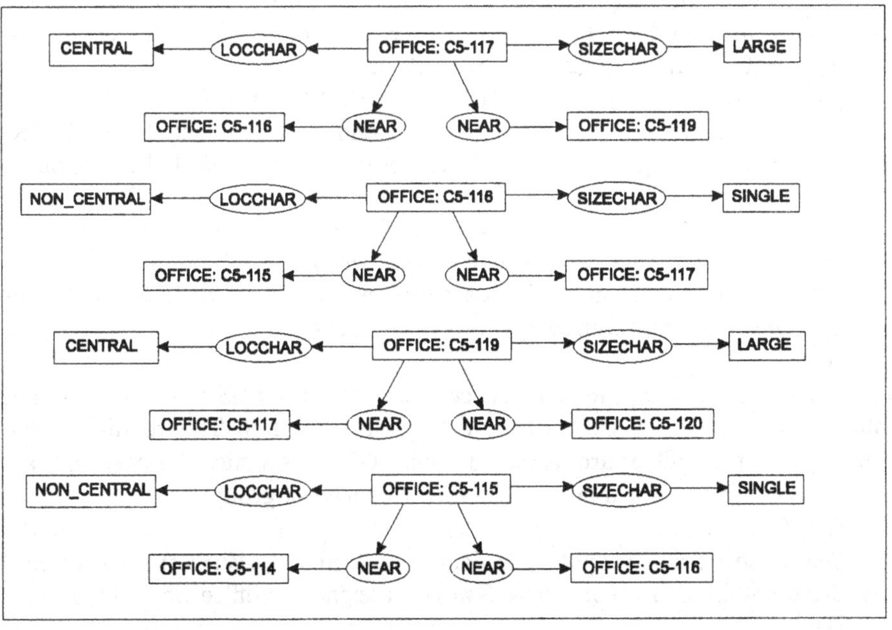

Fig. 14. Assertions about the size and location of the rooms.

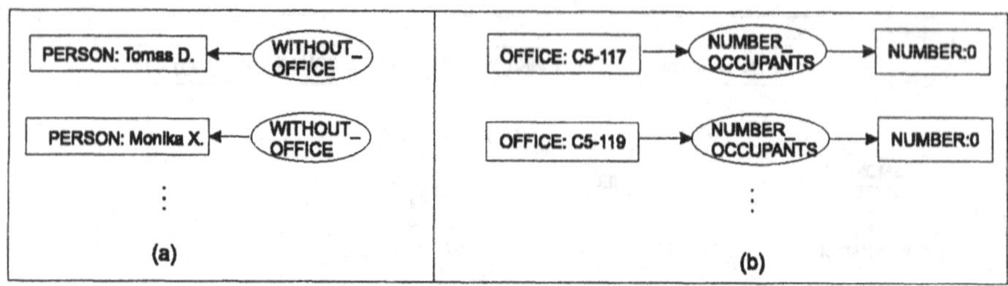

Fig. 15. Initial state of the employees and offices in the YQT Research Center.

2.1 First problem statement

Let us start by tracing the solution to the first problem statement. Having entered the facts of the particular situation, we wait for a proposal from the external system for an application of a rule. After each allocation we restart tracing from the highest priority rules.

Step 1

The demon with the highest priority is ASSIGN1. The system tries to infer the graphs
g1: [HEAD:*x]→(WITHOUT_OFFICE) and
g2: [LC_OFFICE: *y]→(NUMBER_OCCUPANTS)→[NUMBER:0].

An instantiated version of g1: [HEAD: Thomas D.]→(WITHOUT_OFFICE) is obtained by copy of a part of the graph asserted about Thomas D., representing the initial condition that Thomas D. is without an office— Fig. 15. Office C5-117 is classified as LC_OFFICE through type expansion and copy rules again. Thus an instantiated version of g2 looks like follows :

[LC_OFFICE : C5-117]→(NUMBER_OCCUPANTS)→[NUMBER:0].

Therefore ASSIGN1 can be triggered, g1 and g2 are retracted from the KB and the corresponding instantiated version of the graph connected to the outgoing arc of the demon is asserted. Thus we have assigned Thomas D. to C5-117. The external system could also try to find all solutions at this stage and let the user choose between the two possible choices which are C5-117 and C5-119 for Thomas D. The model can be elaborated to retrieve all solutions if a backtracking mechanism in the external machine is provided.

Step 2

The highest priority demon that can be trigered is ASSIGN2. At this stage Monika X. and Ulrike U. are assigned to office C5-119. The conditions that they are without offices are retracted. Office C5-119 satisfies the necessary and sufficient conditions for a large office near the head of group.

Step 3

The highest priority applicable demon is ASSIGN3. With the execution of the demon ASSIGN3 the manager Eva I. is assigned to office C5-116.

Step 4

The demon that can be triggered is ASSIGN4 and the project leader is assigned to office C5-115. There are three choices at this point for the project leader—Hans W., Joachim I., and Katharina N. The system can make a choice alone, or it may propose to the user the possible allocations and let her choose. Of instance, let us put Katharina into C5-115.

Step 5, Step 6

These two steps are also applications of demon ASSIGN4. They allocate project leaders to single offices near the head of group. Offices C5-114 and C5-113 are assigned to Joachim I. and Hans W. in an arbitrary order.

Step 7

The demon with the highest priority that can be triggered is ASSIGN5, which assigns two smoking researchers to a large office. The only possible researchers that are not yet assigned to offices and that are smokers, are Andy L. and Uwe T. They do not work on the same project. The possible large offices are C5-120, C5-121, C5-122, and C5-123. The allocation at this stage might be: Andy L. and Uwe T. to office C5-120.

Step 8, Step 9, Step 10

At these three steps, pairs of non-smoking researchers are allocated to large offices, so that researchers that work on the same project do not share an office. If the choice is made according to the order the conceptual graphs about the members of the research center are entered into the KB, the assignments will be: Werner L. and Jurgen L. into C5-121, Marc M. and Angi W. into C5-122, and Harry C. and Michael T. into C5-123. These assignments are made by the demon ASSIGN6.

2.2 Second problem statement

The KB for this problem statement will differ from the first one only in the conceptual graph for Katharina N., which will be now replaced with a CG for Christian I.

The graph for Christian I. is represented on Fig. 16.

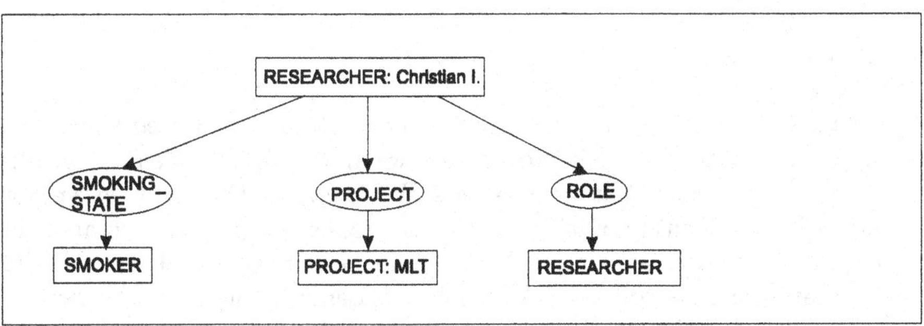

Fig. 16. Characteristics of Christian I.

The solution for this problem will differ from the solution to the first one after the assignment of the manager Eva I. to office C5-116.

Step 4

The demon that can be triggered is ASSIGN4 and a project leader should be assigned to office C5-115. There are two choices at this point for the project leader - Hans W. and Joachim I. Of instance, let us put Hans W. into C5-115.

Step 5

This step is also an application of ASSIGN4. It allocates a project leader to a single office near the head of group. Office C5-114 is assigned to Joachim I.

Step 6

The demon with the highest priority that can be triggered is ASSIGN5, which assigns two smoking researchers to a large office. Possible researchers that are not yet assigned to offices and that are smokers, are Andy L. and Uwe T. They do not work on the same project. The possible large offices are C5-120, C5-121, C5-122, and C5-123. The allocation at this stage might be: Andy L. and Uwe T. to office C5-120. Christian-I could also be put here with one of the others.

Step 7, Step 8, Step 9

Similarly to the first problem statement, the following assignments are made: Werner L. and Jurgen L. into C5-121, Marc M. and Angi W. into C5-122, and Harry C. and Michael T. into C5-123. These assignments are made by the demon ASSIGN6.

Step 10

After these assignments, the external system raises an exception. It raises the exception, when after all possible tracking to the choice points, there is no demon that can be triggered. Among the exception demons, the highest priority one is ASSIGN7. ASSIGN7 makes an allocation of a person to a room if the person is without a room and the office is without occupants. It can be triggered only when an exception is raised. Thus, the system suggests Christian I. to be assigned to office C5-113.

2.3 Evaluation

From the experiments we made, we came to the following conclusions about the strengths and weaknesses of our approach.

- *Soundness.* Our model for this specific room allocation task uses sound procedure for inference. The inference rules we apply are equivalence, generalization and type and relation expansion. We do not need negation, except for comparing the identity of individuals. The graphs and the type hierarchy in our knowledge base are simple. We have not tested the model theoretically and that is why we can not make absolute statements, but since the demons make only allowed transitions, contradictory information cannot be asserted. The definition of the demons guarantees that the obsolete information is first retracted and only then update is performed.

- *Completeness.* The problem-solving strategy that we assume in the external system ensure that if there is a solution that satisfies the problem constraints, at least on solution will be found. When the problem does not have any solution that satisfies the constraints, our model does not guarantee that the optimal solution will be found. The exception rules do not take in consideration, for example, that it is more important to put a secretary near the head of hroup than to break the constraint that researchers sharing an office should have the same smoking habits. This can be achieved if the exception rules are changed so that they can capture details like that. As we do not use backtracking mechanism, we can not derive all solutions.

- *Simplicity.* As to the simplicity of our proposal, this is one of its greatest advantages. It is pretty uncomplicated and models the problem domain in a quite straightforward manner. It is also extremely easy to maintain and change. When a new problem is solved, only the conceptual graphs about the particular people and rooms are asserted into the KB. The conceptual graphs we have used are readable, because they encode knowledge directly taken from the problem statement and the expert's protocol.

The shortcomings in the current model are the following:

- the constraint that two researchers being in an office together should not work on the same project is a desirable, and not a necessary condition; the current model does not take this in consideration and treats these constraints as the constraint for a smoker and a non-smoker being together (only the exception rules differentiate the two constraints). This could be easily corrected though, by introducing new demons with lower priorities that make allocations even if these constraints are not satisfied.

- the efficiency in terms of time and memory has not been a primary concern in the modeling of the domain; certainly, there could be service structures that make the algorithm more efficient.

- as already mentioned, we do not find the optimal solution to the problem in the case that there is no solution that satisfies all the constraints. Even more—we do not infer all solutions, altough as we noted before, this can be easily achieved using backtracking algorithm.

- we have not put maximal effort in generalizing the problem solving method so that it can handle situations which differ very much from the sample problem formulation; this in part relates to the previous point.

3 The Engineering Process

We have been working in the field of CG for nearly three years and we have used conceptual graphs for several tasks in two systems. The main purpose we have used them for is knowledge representation in a system for Machine Aided Translation, developed in the projects DB-MAT and DBR-MAT. This system uses operations on CG to extract relevant knowledge for generation of explanations of domain terms. We have significant experience in knowledge modeling with conceptual graphs. We also took part in the development of a tool for definition and browsing of conceptual graphs in the CGLex project. The tool provides a user-friendly interface to knowledge bases of conceptual graphs.

The model of the SCG-1 domain we have represented in this paper have been developed by us for about ten days. The process of development can be divided into several stages:

- *reading the documentation and relevant papers*— This took us about four days per person.
- *producing an initial vocabulary*— The vocabulary was clear nearly from the very beginning, though we needed some time to determine how we will represent the changes in the knowledge base made by allocations. This was the most difficult problem, because we did not have experience in representing nonmonotonic KBs.
- *identifying and representing semantic constraints*— We had several choices about representing semantic constraints. We could put most of the information in type definitions, but we ended up with the more simple solution of representing them only explicitly in the CG for the particular members of the YQT Research Center. This makes the inference procedure easier and faster. Each of us spent two days on this task.
- *devising special modeling constructs*— The more special constructs we have used are the demons, but they are very similar to those introduced in [5] and [6], so we have not devised completely new modeling constructs. The process of deciding on this type of representation took us about a day per person.
- *producing type definitions*— This was one of the easiest parts of the model and we needed a day for it.
- *testing the model*— Testing took us two days.
- *revising the model*— We revised the model simultaneously with testing it. This process took us about two days for two people.

4 Critique of the Sisyphus-1 Testbed

The Sisyphus-I testbed succeeds to fulfill most of the goals of the SCG-I initiative.

It is very useful for comparing the different approaches to CG that have evolved from the original theory of John Sowa. The testbed gives a great liberty on the choice of knowledge structures to be used, and allows for devising special modeling constructs.

It is also successful in providing a measure for usefulness of the modeling constructs and for the necessity of the ramifications connected with knowledge modeling using conceptual graphs. We, for example, used extensively type definitions, contexts, and demons in our solution.

A drawback in the formulation of the problem in the SCG-1 initiative, that hinders the achievement of the actual goals of the initiative is that it does not show clearly the level of detail, to which the solution should be developed; for example, in our approach, we talk about an external system, that makes inference from the KB, but we have not modeled this system. This freedom in the level of detailness of the model doesn't allow for assessing the actual engineering effort that must be put in the development of a conceptual graph based system.

Another drawback is that a solution to the Sisyphus-I problem doesn't have a measure of correctness. The domain allows for different solutions of the problem that cannot be compared. This is an obstacle to the comparing of different implementations and different tools.

Despite these shortcomings, the Sisyphus-I domain fulfills the two purposes mentioned in the above two paragraphs to a satisfying extent. It is also very good for showing original and relevant use of the conceptual graph theory.

The documentation is readable and easy to understand although some of the definitions of the constraints and processes are not absolutely clear. It is not always understandable which constraints are more important and which can be broken first.

According to us, the SCG-1 initiative will pour life into the CG community. It is an occasion where everyone can show his or her approach and compare it with the approach of the other members of the CG community. The state of the art of CG can be estimated and CG can be compared with other knowledge representation formalisms according to pertinent criteria.

References

1. Angelova, G., K. Bontcheva. *DB-MAT: Knowledge Acquisition, Processing and NL Generation using conceptual graphs.* Proc. ICCS-96, LNAI 1115, pp. 115–129. See also http://www.lml.bas.bg/projects/dbr-mat/
2. Galia Angelova, Kristina Toutanova, Svetlana Damianova *Knowledge Base of Conceptual Graphs for Generation of Natural Language Explanations in Knowledge Based Machine Aided Translation* Project DBR-MAT "Intelligent Translation System", Technical Report BG-3-1998, September 1998
3. Angelova, G., Damianova, Sv., Toutanova, K., Bontcheva, K. *Menu-Based Interface to Conceptual Graphs: The CGLex Approach* In Proc. ICCS-97, LNAI 1257, pp. 603–606. CGLex - funded by The Bulgarian National Science Fund under contract I- 420/94. See also http://www.lml.bas.bg/projects/cglex/
4. Gaines, Brian R.: *A Situated Classification Solution of a Resource Allocation Task Represented in a Visual Language.* See: http://ksi.cpsc.ucalgary.ca/KAW/Sisyphus/Sis1/
5. Delugach, Harry S.: *Dynamic Assertion and Retraction of Conceptual Graphs.* In: Proc. of the 6th Ann. Workshop on Conceptual Structures,1991, E. Way (Ed.), Binghamton, N.Y.: SUNY at Binghamton, pp. 15-24.

6. Mineau, Guy W.: *From Actors to Processes: The Representation of Dynamic Knowledge Using Conceptual Graphs.* In: Proc. of the 6th International Conference on Conceptual Structures, ICCS-98, Montpellier, France, August 1998, LNAI 1480, pp.65-79.
7. J. Sowa: *Conceptual Structures: Information Processing in Mind and Machine.* Addison-Wesley, Reading, MA, 1984.

WebKB and the Sisyphus-I Problem

Philippe Martin and Peter Eklund

Griffith University, School of Information Technology,
PMB 50 Gold Coast MC, QLD 9726 Australia
{p.eklund,philippe.martin}@gu.edu.au

Abstract. This article explains how we have used WebKB[1] - our Web-accessible and Conceptual Graph (CG) based knowledge acquisition and annotation tool - to provide *a solution and an interface*[2] to solve the Sisyphus-I office allocation problem[3], a problem intended to enable the comparison of knowledge acquisition tools.
We have exploited the WebKB CG assertion/query commands and script language to provide two procedural solutions to the problem. One is recursive and handles backtracking. We peresent the other which is simple and direct. The *script*, the *data set* and the *ontology* (i.e. structured set of types) may be stored in different Web-accessible documents, the URLs of which may be sent via a simple *interface* to the WebKB processor (a *CGI server*[4]) in order to solve the problem. Users may also provide their own files to test their solutions or use their own data sets.

1 The Model

To allow easy comparison with other knowledge acquisition tools we have followed the requirements and protocols given in the Sisyphus-I problem description[5] rather than invented our own solution (this was a proposed alternative). However, as in Brian Gaines' solution[6], we have added an additional constraint: in the data set the user may represent the fact that some offices must be allocated to some given employees. Thus, the office allocation *script* may still be best exploited in under-determined situations. Similarly, the user may modify the *data set* (or the script itself) to cope with situations that the script found over-determined (e.g. some offices may be added or some constraints for sharing offices relaxed).

The *data set* gives the characteristics of the offices, and employees to place within them, with commands for asserting individuals and CGs, and using the *ontology* the script relies on. Indeed, if the user wants to modify the ontology, s/he must also modify the script appropriately. The next subsection explains

[1] http://meganesia.int.gu.edu.au/~phmartin/WebKB/
[2] http://meganesia.int.gu.edu.au/~phmartin/WebKB/kb/sisyphus1.html
[3] http://ksi.cpsc.ucalgary.ca/KAW/Sisyphus/Sisyphus1/
[4] http://www.w3.org/CGI/
[5] http://ksi.cpsc.ucalgary.ca/KAW/Sisyphus/Sisyphus1/
[6] http://ksi.cpsc.ucalgary.ca/KAW/Sisyphus/Sis1/

why a textual interface is used for the ontology, data set and script, while the next three subsections respectively present the content of an ontology, a data set and a script.

1.1 A Textual Interface

Though WebKB proposes a CG graphic editor, we mainly use a *linear notation* to assert or query types, instances or CGs, and insert these commands in scripts or any other document (e.g. this article). There are three main reasons for this.

First, we find it easier and quicker to use a good text editor and an interpreter for building and exploiting a CG base than use the WebKB graphic editor (or any other CG graphic editor we have tried until now). Besides, the linear notation and script language in WebKB provide more features and combination possibilities.

Second, we think a linear notation is as good for readability as a graphic notation at least for tree-like structures (which is the case for all the structures we had to deal with to solve the Sisyphus-I problem). Third, since the linear notations take less place on the screen than graphic notations, a user may visually compare more CGs or more information including CGs (thus, for example, it is easier for a user to handle and debug programs including CGs).

For these reasons, and to remove ambiguities, in the following sections we also prefer to present the ontology, data set and script entirely in the form they are accepted by the WebKB processor rather than using any graphical form. It should however be noted that WebKB can handle HTML documents where type declarations or CG assertions are hidden behing images (more exactly, stored in the "alt" field of the "img" tags) and can present these images to the user when the types or CGs are results of queries.

As an example of WebKB flexibility, the URL of the HTML version of this document could also be sent to the WebKB CGI server to solve the problem since this file includes (and comments on) a script, a data set and the ontology they rely on. Such a feature is called "active/virtual document support" in the hypertext litterature[7].

Figure 1 shows an HTML&Javascript interface which sends an ontology, a data set and a script (plus possibly other commands) to the WebKB processor. Figure 2 shows the window which pops up to display the results of the execution of the ontology, data set and script by the WebKB processor.

1.2 An Ontology

The ontology below is composed of concept types, e.g. *Something* and *Location*, and relation types, e.g. *Agent* and *Project*. We had three goals in mind when designing it. By increasing order of importance: minimising the ontology length (only types used by the script are declared), minimising the data set length (to keep it simple and ease the updates), and maximising the efficiency of the script.

[7] ftp://ftp.inrialpes.fr/pub/opera/publications/ECHT92.ps.gz

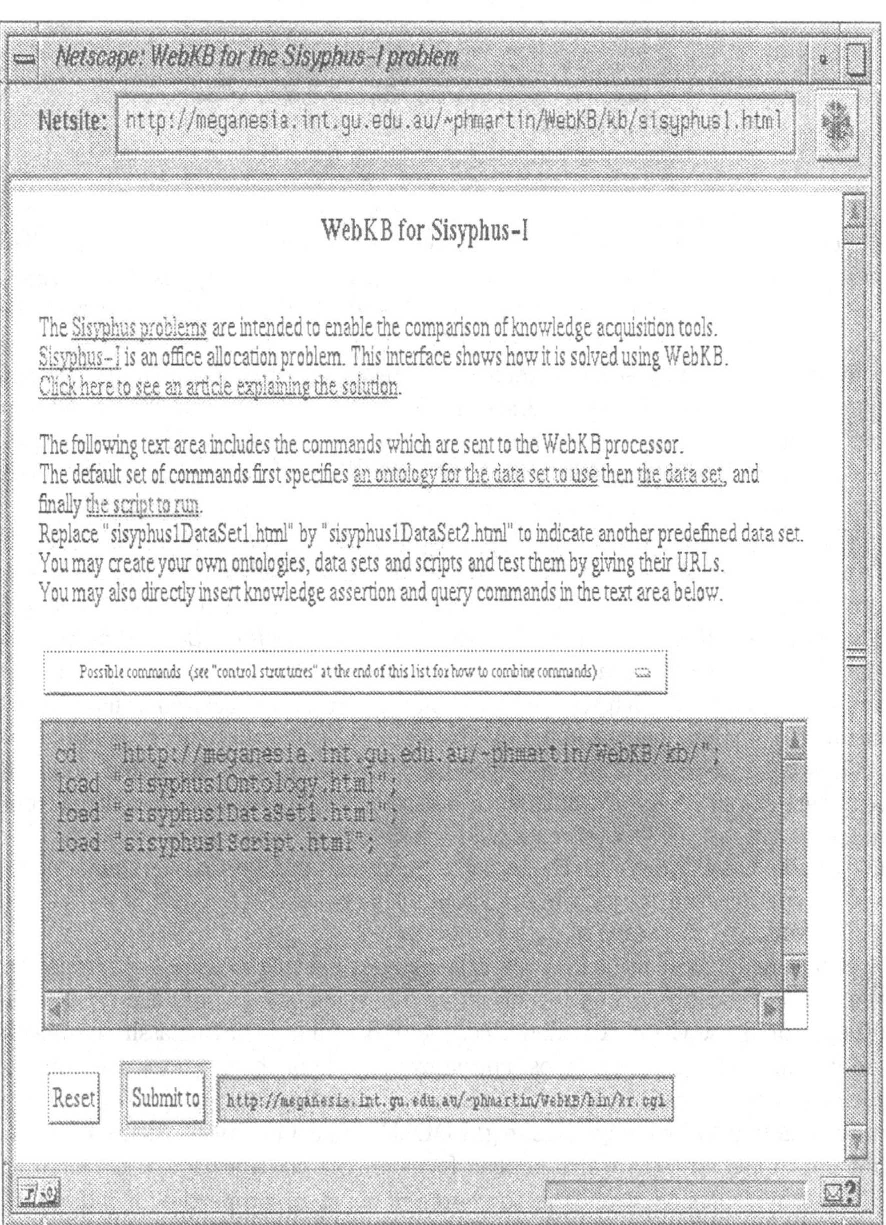

Fig. 1. An interface for solving the Sisyphus-I problem with WebKB

```
Something > (Organization  Person  Location  Process);

 Organization > Project  Research_group;

 Person > Employee (Smoker Non_smoker);
   Employee > (Manager  Secretary  Researcher);
     Researcher > (Group_head  Large_project_head  Simple_researcher);

 Location > Office;
   Office > Central_office (Small_office Large_office);
     Central_office > (Small_central_office Large_central_office);
     Small_office > (Small_central_office Small_and_not_central_office);
     Large_office > (Large_central_office Large_and_not_central_office);

 Process > (Hack Smoke);

 Agent (Process, Person);  Project (Person, Project);
 In (Person, Office);      Near (Location, Location);
```

The first line declares the concept type *Something* and four subtypes. The syntax is close to Sowa's usual linear notation for ordering types (Sowa, 1984). An extra feature is the pair of parenthesis around the subtypes which means that they belong to a same partition of exclusive types, i.e. they must not have common subtypes. WebKB enforces that property by refusing to create new types which would violate it. Since in WebKB - as in Sowa's CG model - an individual (i.e. an object conforming to a type) is instance of (is conforming to) one and only one type, exclusive types cannot have common instances. (Two non-exclusive types may have common instances if one specializes the other).

The second line declares *Project* and *Research_group* as non-exclusive subtypes of *Organisation*, and the third line declares *Smoker* and *Non_smoker* as two exclusive subtypes of *Person* but not exclusive with their sibling *Employee*. The types *Small_central_office*, *Small_and_not_central_office*, *Large_central_office* and *Large_and_not_central_office* have been introduced to simplify the data set and the script: thus, such size and location properties are declared or handled via instance assertion/query commands instead of CG assertion/query commands.

A similar representation could be used for storing the facts that some researchers smoke or hack, but in such a case there is no gain in simplicity compared to the use of CGs since a lot of "unnatural" types have to be declared and handled, e.g. types such as *Hacker_non_smoker* and *Non_hacker_non_smoker*, and a type for each project. Instead, the script generates the sets of individuals which would be instances of such types.

Conversely, some users may regret not to be able to *also* use CGs to represent the size and location properties of offices (i.e. some users would use types while others would use CGs). To allow this, WebKB would need to classify each individual under the adequate (most specialized) types of the ontology according to the way they are used in CGs. Such instance classification features are common in terminological logics systems, e.g. LOOM[8] and some object-oriented systems but not in CG systems which instead focus on CG classification, e.g. Peirce[9].

Another solution would be to represent most of properties of individuals using CGs, i.e. using conceptual relations types such as *Secretary* or *Manager* to relate a project to its employees, instead of using "instance-of" relations. This solution would need to introduce many relation types in the ontology and

---

[8] http://www.isi.edu/isd/LOOM/LOOM-HOME.html

[9] http://www.cs.rmit.edu.au/~ged/publications.html

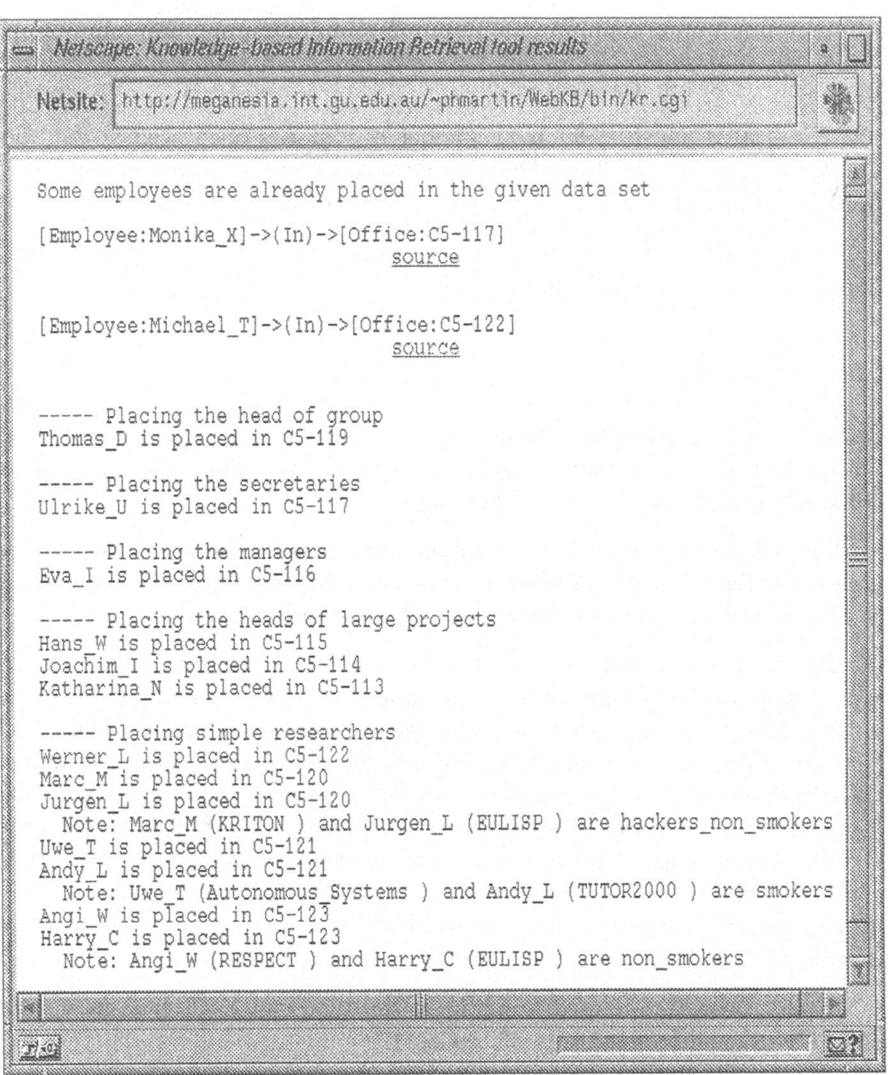

Fig. 2. Results from the WebKB processor in answer to the queries in Figure 1

therefore many queries on individuals to the script but could be useful if there were more than one research group to handle. In the given protocol, only the research group HQT is considered and this information is not used for allocating offices. Therefore, to simplify the data set and script, and show the exploitation of both CGs and instance relations, we have not chosen that solution.

To represent the fact that some researchers smoke or hack, the process types *Smoke* and *Hack* are introduced.

As indicated above, a minimal number of relations have been introduced: *Agent*, *Project*, *In* and *Near*. Their signatures are given between parenthesis. WebKB checks signatures are verified each time a CG is created.

The documentation of WebKB languages[10] describes additional ways to build an ontology, e.g. via type definitions.

1.3 A Data Set

Below is the data set referred to in Figure 1. The first line declares *C5-117* and *C5-119* as instances of the concept type *Large_central_office*. The other offices are similarly declared in the next two lines.

The neighbourhood of the offices is represented using conceptual relations of type *Near*. Since WebKB only handles oriented relations, a relation of type Near is used in both direction between each pair of neighbor offices.

Each project and employee is declared as an instance of *Project* or the relevant type of employee. Then, for each simple researcher, a CG is used to represent the membership of the employee to a project and if s/he uses to smoke or hack. Such facts are represented only for simple researchers since, according to the protocol, these facts are only used when offices have to be shared, and it seems that only simple researchers share offices. The requirements are ambiguous on this point. For the sake of simplicity, in the script presented below, only simple researchers can share offices. However, the representation of the smoking and hacking habits of all researchers wouldn't lead the script to false results but rather slow its execution.

Finally, two CGs are used to represent the facts that Michael T. must be in office C5-122 and Monika X. in office C5-117.

[10] http://meganesia.int.gu.edu.au/~phmartin/WebKB/doc/languages.html

```
Large_central_office: C5-117 C5-119;
Large_and_not_central_office: C5-120 C5-121 C5-122 C5-123;
Small_and_not_central_office: C5-113 C5-114 C5-115 C5-116;

[Office:C5-113]->(Near)->[Office:C5-114]->(Near)->[Office:C5-115]->(Near)->
   [Office:C5-116]->(Near)->[Office:C5-117]->(Near)->[Office:C5-119]->(Near)->
   [Office:C5-120]->(Near)->[Office:C5-121]->(Near)->[Office:C5-122]->(Near)->
   [Office:C5-123];
[Office:C5-113]<-(Near)<-[Office:C5-114]<-(Near)<-[Office:C5-115]<-(Near)<-
   [Office:C5-116]<-(Near)<-[Office:C5-117]<-(Near)<-[Office:C5-119]<-(Near)<-
   [Office:C5-120]<-(Near)<-[Office:C5-121]<-(Near)<-[Office:C5-122]<-(Near)<-
   [Office:C5-123];

Project: AutonomousSystems ASERTI BABYLON EULISP KRITON MLT RESPECT TUTOR2000;
Group_head: Thomas_D;   Secretary: Monika_X Ulrike_U;   Manager: Eva_I;
Large_project_head: Hans_W  Joachim_I  Katharina_N;
Simple_researcher:  Andy_L  Angi_W  Harry_C  Jurgen_L Marc_M Michael_T
                    Uwe_T  Werner_L;

[Simple_researcher:Andy_L]-{(Project)->[Project:TUTOR2000];(Agent)<-[Smoke];};
[Simple_researcher:Angi_W]- {(Project)->[Project:RESPECT];};
[Simple_researcher:Harry_C]- {(Project)->[Project:EULISP];  (Agent)<-[Hack];};
[Simple_researcher:Jurgen_L]- {(Project)->[Project:EULISP]; (Agent)<-[Hack];};
[Simple_researcher:Marc_M]-   {(Project)->[Project:KRITON]; (Agent)<-[Hack];};
[Simple_researcher:Michael_T]-{(Project)->[Project:BABYLON];(Agent)<-[Hack];};
[Simple_researcher:Uwe_T]-    {(Project)->[Project:AutonomousSystems];
                                         (Agent)<-[Smoke];  (Agent)<-[Hack];};
[Simple_researcher:Werner_L]- {(Project)->[Project:RESPECT];(Agent)<-[Hack];};

[Employee:Michael_T]->(In)->[Office:C5-122];
[Employee:Monika_X]->(In)->[Office:C5-117];
```

1.4 Problem-Solving Functions

To solve the Sisyphus-I problem, we first came up with a recursive solution[11] that handles backtracking by deleting CGs representing office allocations. However, we then realised that since the employees must be placed in a precise order (head of group, secretaries, managers, heads of large projects and finally simple researchers), backtracking is useless if (i) only simple researchers share offices, (ii) the head of group must have a large office whenever possible (even if because of this allocation, there are insufficient large offices for simple researchers), and (iii) offices may be left empty while simple researchers share large offices. This seems to be the case according to the protocols and requirements. Though additional constraints may voluntary be taken into account (such as the fact that some offices may be pre-allocated), we have preferred to take into account the above three hypothesis in order to present a script much simpler and more efficient than the recursive one. Here after, we present each function of this simple script.

Presentation Below is the main function. The WebKB scripting language is a language of commands simpler but similar to the C-shell on Unix. Control structures and pipes are used for combining commands. Variables need not to be declared and are global. However, function parameters are local to the function, and as a convention, the first letter of variables which are used in different functions is uppercase. The value of variables or function parameters can only be a string or a real and is accessed by prefixing the variable with the character '$'. The function "main" must be explicitly called such as any other function.

This function main() shows that there is a different function for placing each kind of employee. Each of these functions are called in the order recommended

[11] http://meganesia.int.gu.edu.au/~phmartin/WebKB/kb/sisyphus1RecursiveScript.html

by the expert and when necessary. To simplify and avoid backtracking, the script tries to place simple researchers together and then place the remaining employees alone. These remaining employees may be simple researchers or employees of a kind not included in the protocol.

```
void function main ()
{ check_if_already_placed_employees();
  if (possible_to_place_the_group_head())
  { if (possible_to_place_the_secretaries())
    { if (possible_to_place_the_managers())
      { if (possible_to_place_the_heads_of_large_projects())
        { print; print "----- Placing simple researchers";
          try_to_place_simple_researchers_by_pairs();
          try_to_place_remaining_employees_alone();
        } } } }
}

void function check_if_already_placed_employees()
{ spec [Employee:?]->(In)->[Office] | set Placed_employees;
  if ($Placed_employees)
  { print "Some employees are already placed in the given data set";
    spec [Employee]->(In)->[Office];  //list the relevant CGs
  }
  spec Employee : ?  |  set The_employees;
  subtract $The_employees $Placed_employees | set Employees_to_place;
}
```

Before trying to place employees, the function check_if_already_placed_employees() is called to print and take into account the employees which, in the data set, are already placed in an office.

The command "spec" searches the specializations of types or CGs. The results of a command (e.g. "spec") are printed unless it is followed by a pipe ('|'). In that case, the (possibly empty) list of results is automatically added to the parameters of the following command (if the results are CGs, the names of the CGs are added to the parameters).
If "spec" has a CG for parameter, it searches specializing CGs. However, if a question mark ('?') in the CG specifies that only some kinds of instances are to be presented (as in "`spec [Employee:?]->(In)->[Office]`"), the result is the list of instances belonging to the retrieved specializing CGs. A result is not added twice to the result list. This is important for loops iterating on a result list. If "spec" has only a type for parameter, it searches subtypes. If "spec" has for parameters a type, a semicolon and a question mark (as in "`spec Employee:?`"), it searches the instances of the type (including the instances of the type subtypes).

The command "subtract" takes two lists of white-separated strings as parameters, and returns as result the elements of the first list except those which appear in the second list. Thus, in the function check_if_already_placed_employees(), the global variable Employees_to_place is initialized to the list of employees not already placed in the data set.

Placing the Head of Group Below are the functions related to the placement of the head of group. The first two are simple models for the functions placing the other kinds of employees, while place_employee_in() is reused by all of them.

```
boolean function possible_to_place_the_group_head()
{ print; print "----- Placing the head of group";
  spec Group_head : ?  | set The_group_head;
  spec "[Employee:$The_group_head]->(In)->[Office:?]"
       | set Office_of_group_head;
  if ($Office_of_group_head == "")  //if the group head has no office
  {if (!possible_to_place($The_group_head,Large_central_office))
   {if (!possible_to_place($The_group_head,Small_central_office))
    {if (!possible_to_place($The_group_head,Large_and_not_central_office))
     {if(!possible_to_place($The_group_head,Small_and_not_central_office))
      {print "$The_group_head cannot be placed: no office left";
       return false; } } }
   }
   spec "[Employee:$The_group_head]->(In)->[Office:?]"
        | set Office_of_group_head;
  }
  return true;
}

boolean function possible_to_place (employee, kind_of_office)
{ spec $kind_of_office : ?  | set offices_of_this_kind;
  for o in $offices_of_this_kind
  { spec "[Employee]->(In)->[Office:$o]" | set occupied_office;
    if ($occupied_office == "") //$o is empty, the employee can be alone
    { place_employee_in($employee,$o); return true; }
  }
  return false;
}

void function place_employee_in (employee, office)
{name "#$employee$office" "[Employee: $employee]->(In)->[Office:$office]";
 print "$employee is placed in $office";
 subtractFrom Employees_to_place $employee;
}
```

The function possible_to_place_the_group_head() successively tries to place the head of group in each of the four exclusive kinds of offices, by decreasing order of preference: large and central office, small and central office, large and not central office, large and not central office. This order is not given by the protocol which only recommends the placement of the head of group in a large central office, but seems natural and suppresses the need for backtracking (provided that only simple researchers may share offices and the head of group must have a large office whenever possible).

The function possible_to_place() tries to place a given employee in the first unoccupied office which is an instance of the given kind of office. The script stops if it cannot find a way to place an employee.

The function place_employee_in() creates a CG for storing the allocation of an employee in an office. The name of the CG is built by concatening the name of the employee and the name of the office. Then, the employee is removed from the list of employees to place via the command "subtractFrom" which is similar to "subtract" except it takes a variable referring to a list as first argument, uses its value and then affects it with the result list.

Placing Employees Near the Head of Group Secretaries, managers and heads of large projects need to be placed in certain kinds of office and nearest possible to the office of the head of group. This is what the next function does. First, the offices adjacent to the office of the head of group are tested. Then, the distance is progressively augmented via a loop which adds a new relation of type *Near* to the query CG at each of its step.

```
boolean function possible_to_place_nearest_group_head (employee,kind_of_office)
{ set near_chain "->(Near)->";  //first, try very close offices
  while (true)                  //the loop progressively increases $near_chain
  { spec "[Office:?]$near_chain[Office:$Office_of_group_head]"
        | set offices_near_group_head;
    if ($offices_near_group_head == "") { break; } //no office at such distance
    for o in $offices_near_group_head
    { ? $kind_of_office : $o  | set office_is_suitable;
      if ($office_is_suitable)
      { spec "[Employee]->(In)->[Office:$o]" | set occupied_office;
        if ("$occupied_office" == "") //$o is empty
        { place_employee_in($employee,$o); return true; }
      }
    }
    set near_chain "$near_chain[Office]->(Near)->";
  }
  return false;
}
```

Placing Secretaries Secretaries must be placed nearest to the head of group and if possible in the same office. Thus, the following functions first search the offices containing a secretary and test if the office is not fully occupied. To simplify, we have considered that a large office may contain only two employees. This is in accordance with the protocol.

```
boolean function possible_to_place_the_secretaries()
{ print; print "----- Placing the secretaries";
  spec Secretary : ?  | set The_secretaries;
  for s in $The_secretaries
  { spec "[Employee:$s]->(In)->[Office]" | set office_of_this_employee;
    if ($office_of_this_employee == "")
    { if (! possible_to_place_with_another_secretary($s))
      { if (! possible_to_place_nearest_group_head($s,Large_office))
        { if (! possible_to_place_nearest_group_head($s,Small_office))
          { print "The secretary $s cannot be placed close to
                  $Office_of_group_head, the office of head of group
                  $The_group_head";
            return false;
          } } } }
  }return true;
}

boolean function possible_to_place_with_another_secretary (employee)
{for s2 in $The_secretaries
 { spec "[Employee:$s2]->(In)->[Office:?]"|set offices_with_a_secretary;
   for o in $offices_with_a_secretary
   { if (office_not_fully_occupied($o))
     { place_employee_in($employee,$o); return true; }
   }
 }return false;
}

boolean function office_not_fully_occupied (office)
{spec "[Employee:?]->(In)->[Office:$office]"|count|set number_of_occupants;
 if ($number_of_occupants == 0) { return true; }
 ? Large_office : $office | set office_is_large;
 if ($office_is_large) { if ($number_of_occupants < 2) { return true; } }
 return false;
}
```

Placing Managers and Heads of Large Projects The function possible_to_place_the_heads_of_large_projects() is similar to the function possible_to_place_the_secretaries except that no call is done to the function possible_to_place_with_another_secretary(). The function possible_to_place_the_managers() is also similar except that four types of offices may be tested instead of two. By decreasing priority, these four types are: Small_central_office, Small_and_not_central_office, Large_central_office and Large_and_not_central_office.

Placing Simple Researchers By Pairs Below are functions which allocate large offices to researchers. Three constraints govern the sharing of offices: (i) smokers and non-smokers should not be put together into an office, (ii) coworkers that are both hackers or both non-hackers can share an office, (iii) researchers in same projects are (if possible) not sharing an office.

```
void function try_to_place_simple_researchers_by_pairs()
{ if ($Employees_to_place == "") { return; }

  spec Large_office : ?  | set large_offices;
  spec [Employee]->(In)->[Office:?] | set occupied_offices;
  subtract $large_offices $occupied_offices | set Unoccupied_large_offices;

  spec [Simple_researcher:?]- { (Agent)<-[Smoke]; }  | set smokers;
  spec [Simple_researcher:?]- { (Agent)<-[Hack];  }  | set hackers;
  spec [Simple_researcher:?]- { (Agent)<-[Hack];
                                (Agent)<-[Smoke]; }  | set hackers_smokers;

  subtract "$smokers" "$hackers_smokers" | set smokers_non_hackers;
  subtract "$hackers" "$hackers_smokers" | set hackers_non_smokers;

  spec Simple_researcher : ?  | set the_simple_researchers;
  subtract "$the_simple_researchers" "$smokers" |set non_smokers;
  subtract "$non_smokers" "$hackers_non_smokers"|set non_hackers_non_smokers;

  if ($hackers_smokers)     {try_to_place_pairs_of(hackers_smokers);}
  if ($Employees_to_place) {try_to_place_pairs_of(smokers_non_hackers);}
  if ($Employees_to_place) {try_to_place_pairs_of(hackers_non_smokers);}
  if ($Employees_to_place) {try_to_place_pairs_of(non_hackers_non_smokers);}

  if ($Employees_to_place) {try_to_place_pairs_of(smokers);}
  if ($Employees_to_place) {try_to_place_pairs_of(non_smokers);}
}
```

Since we want to have a simple and non-recursive script, the script cannot place simple researchers alone in offices and then, when all offices are occupied, put the remaining researchers with already placed simple researchers: to best satisfy the sharing constraints, some already placed employees would have to be moved. Furthermore, it would then also be necessary to distinguish the employees that have been placed by the user from the employees that have been placed by the script. Hence, our solution is to search and place *pairs* of simple researchers that satisfy the sharing constraints, and then, if necessary, progressively relax the constraints while there are still simple researchers to place and places in large offices. The function try_to_place_pairs_of() perform the search and copes with the fact that one researcher of a suitable pair may have already been placed by the user (and indeed there is no problem when the user has placed two researchers in a same office). When necessary, this function also relaxes the third constraint: "researchers in same projects should rather not share an office".

The function try_to_place_simple_researchers_by_pairs() searches the instances of the four exclusive types of researchers suitable to share an office: smoker hackers, smoker non hackers, hacker non smokers, non hacker non smokers. When necessary, this function relaxes the constraint on the hacking habits.

The command "subtractFirstFrom" is similar to "subtractFrom" except that it has no second argument: the first element of the referred list is removed from it and returned as a result.

```
void function try_to_place_pairs_of (kind_of_employee)
{ set suitable_co-occupants $$kind_of_employee; //placed or not
  countIn $suitable_co-occupants | set number_of_suitable_co-occupants;
  if ($number_of_suitable_co-occupants < 2) { return; }

  for e1 in $suitable_co-occupants
  { spec "[Employee:$e1]->(In)->[Office: ?]"  | set o1;
    if ($o1 == "") //$e1 not placed
    { spec "[Employee:$e1]->(Project)->[Project:?]" | set p1;//project of $e1
      for e2 in $suitable_co-occupants
      { if ($e1 != $e2)
        { spec "[Employee:$e2]->(Project)->[Project:?]"  | set p2;
          if ($p1 != $p2)
          { spec "[Employee:$e2]->(In)->[Office: ?]"  | set o2;
            if ($o2) //e2 is already placed
            { if (office_not_fully_occupied($o2))
              { place_employee_in($e1,$o2); break; }
            }
            else
            { subtractFirstFrom Unoccupied_large_offices | set o2;
              if ($o2)
              { place_employee_in($e1,$o2); place_employee_in($e2,$o2);
                print " Note: $e1 ($p1) and $e2 ($p2) are $kind_of_employee";
                break;//the inner loop
              } } } } } } }

  if ($Employees_to_place == "") {return;}
  //Idem without the constraint of a different project
  for e1 in $suitable_co-occupants
  { spec "[Employee:$e1]->(In)->[Office: ?]"  | set o1;
    if ($o1 == "") //$e1 not placed
    { for e2 in $suitable_co-occupants
      { if ($e1 != $e2)
        { spec "[Employee:$e2]->(In)->[Office: ?]"  | set o2;
          if ($o2) //e2 is already placed
          { if (office_not_fully_occupied($o2))
            { place_employee_in($e1,$o2); break; }
          }
          else
          { subtractFirstFrom Unoccupied_large_offices | set o2;
            if ($o2)
            { place_employee_in($e1,$o2); place_employee_in($e2,$o2);
              print " Note: $e1 ($p1) and $e2 ($p2) are $kind_of_employee";
              break;//the inner loop
            } } } } } }
}
```

Placing Remaining Employees Alone The next function tries to place employees that are not yet placed, alone in an office . These are simple researchers or employees of a kind not foreseen by the protocol.

```
void function try_to_place_remaining_employees_alone()
{ spec Office : ?  | set the_offices;

  for e in $Employees_to_place
  { spec [Employee]->(In)->[Office:?] | set occupied_offices;
    subtract $the_offices $occupied_offices | set unoccupied_offices;
    if ($unoccupied_offices)
    { for o in $unoccupied_offices { place_employee_in($e,$o); } }
    else { print "$e cannot be placed: no office left"; return; }
  }
}
```

2 The Evaluation Process

2.1 Test of the Model

We have not tested our script theoretically but empirically. The reader is welcome to test it by visiting our Web-accessible Sisyphus-I interface[12] and adding queries or modifying the used data set or even the script. To give another test example, let us consider the second problem Katharina N. has left and Christian

[12] http://meganesia.int.gu.edu.au/~phmartin/WebKB/kb/sisyphus1.html

I. joined the group. Christian I. is a simple researcher who smokes, hacks and works in MLT. Below are the differences between the first data set and a data set for the second problem. The new results provided by WebKB are in Figure 3.

```
Large_project_head: Hans_W  Joachim_I; //Katharina_N left
Simple_researcher:  Andy_L  Angi_W  Harry_C  Jurgen_L   Christian_I
                    Marc_M  Michael_T  Uwe_T  Werner_L;

[Simple_researcher: Christian_I]- { (Project)->[Project:MLT];
                                    (Agent)<-[Smoke];  (Agent)<-[Hack];
                                  };

//no pre-placed employee: [Employee:Michael_T]->(In)->[Office:C5-122];
//no pre-placed employee: [Employee:Monika_X]->(In)->[Office:C5-117];
```

2.2 Strengths and Weaknesses of the Model

Our Web-accessible and script language based approach has several advantages. First, a user can easily test our programs or query our files and create new ones. As opposed to graphic interfaces, the linear notation for CGs and the script language allow the user to quickly create knowledge assertions or queries *and also* combine them to build complex queries or programs. We think the script discussed in this article is relatively readable and intuitive, at least for persons used to CGs and either the C, C-shell or shell language. Though they are not directly relevant to the Sisyphus-I problem, let us note that WebKB also proposes simpler knowledge representation notations[13] (a frame-oriented CG linear notation and a form of formalised English) and constructs (e.g. based on HTML constructs) in order to ease and improve the readability of knowledge in documents, the indexation of documents with knowledge, and finally the retrieval or generation of knowledge or parts of documents.

It seems interesting to compare our solution with Brian Gaines' solution[14] since the article that describes it is Web-accessible along with the Sisyphus-I problem description and shows how KSSn - a knowledge representation server with KL-ONE-like term subsumption representation capabilities and a graphic language - is exploited. (However, the solution itself is not Web-accessible for testing).

Our representation of the ontology and data set is relatively similar to the one with KSSn, though a bit less precise since WebKB does not have constructs to represent and check (i) the number or maximum number of concepts related to a concept of a certain type via a relation of a certain type, and (ii) the fact that all the concepts related via a relation of a certain type to a certain concept have been represented. (However, these constructs would not have been useful to us for the Sisyphus-I problem). Besides, our linear notation may also appear less intuitive than the visual language used for KSSn.

The real difference is in the problem-solving description. Brian Gaines' solution uses 18 rules which rely on a rule-based system in KSSn. Each rule adds

[13] http://meganesia.int.gu.edu.au/~phmartin/WebKB/doc/languages.html#FrameCG
[14] http://ksi.cpsc.ucalgary.ca/KAW/Sisyphus/Sis1/

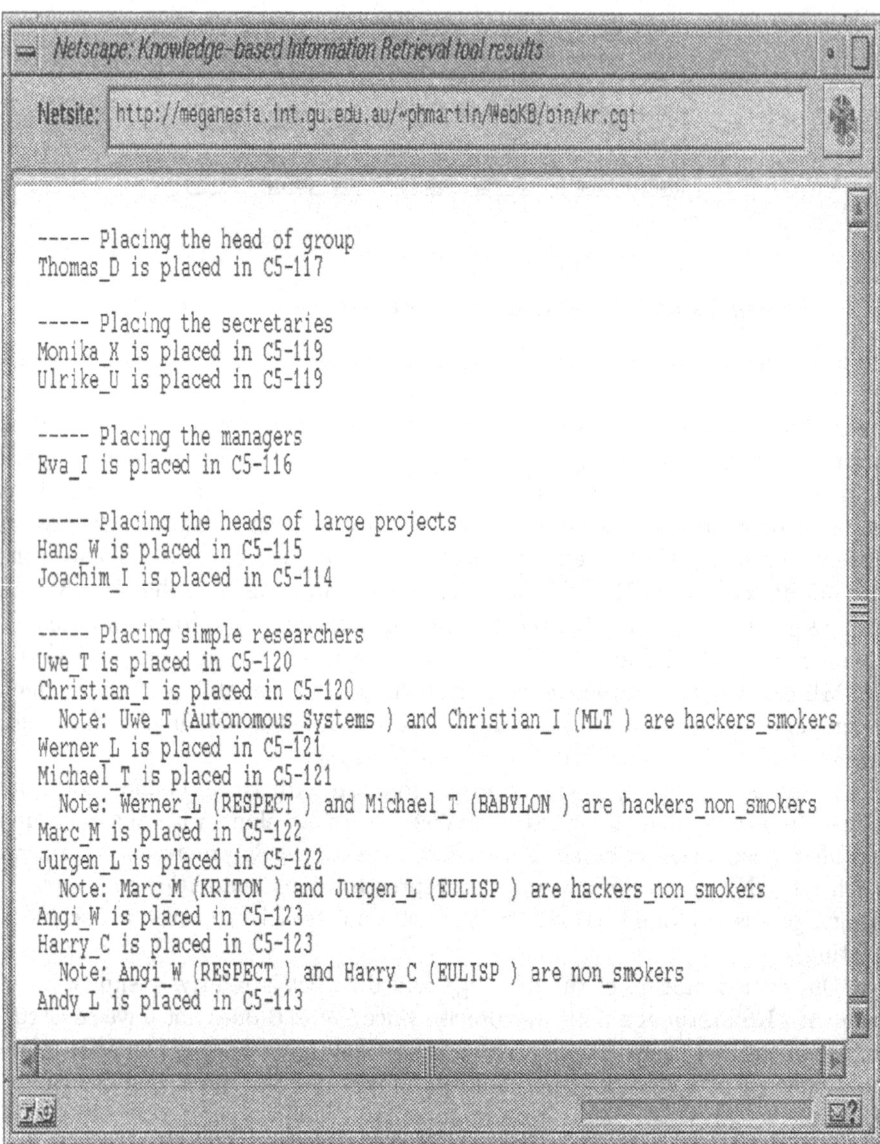

Fig. 3. Results from the WebKB processor for the second data set

relations to some objects and a truth maintenance system handles the propagation of the consequences of those new assertions. The solution, which is based on the simutaneous search for ready-to-be-allocated offices (8 rules) and ready-to-be-triggered office allocation tasks (7 rules), is certainly ingenious and short but we found it quite difficult to understand and, despite the trace examples, still do no know what is exactly going on and how all the rules combine. As a small stand-alone example, the reader can look at the last 3 other rules[15] which search for an unoccupied office nearest to the office of the group of head. Besides, the solution does not seem to be as complete as ours because the 8 rules which decide to whom an office should be allocated, only follow the expert protocol and do not generalize it. Here are these 8 rules in natural language and declining priority:
- if an office is large and central and the head of group is without an office, suggest the allocation;
- a large office already occupied by a secretary should be allocated to another secretary;
- a large office near the head of group is suitable for the secretaries;
- a single office near the head of group is suitable for the manager;
- a single office near the head of group is suitable for the heads of large projects;
- a large office with a smoking researcher occupant should be allocated to another smoking researcher;
- a large office should be allocated to smoking researchers;
- a large office with a non-smoking researcher occupant should be allocated to another non-smoking researcher;
- a large office should be allocated to non-smoking researchers.

Therefore, for example, how are the head of group and secretaries placed if there is no large offices? Similarly, how are the manager(s) and heads of large projects placed if there is no available single offices?

In our procedural approach, the underlying rules of office allocation are quite clear, the order of execution is explicit and the constraints are progressively relaxed in each allocation case.

The question of efficiency is complex and is not really important in this problem when the data are as small as those in the given examples. Though our approach includes a lot of searches for specializations of types or CGs and though WebKB is not optimised for these searches, it loads itself (6Mb), gets the files (ontology, data set and script) and solves the problem in only a second or two when transmission latency on the Web is minimum (and about two or three seconds when the URL of this article is sent to WebKB instead of URLs for separate small files). Generally speaking, rule-based approaches are often said to be slower to run but easier to implement than procedural approaches. It is true in this case that we took a lot of time to implement our procedural solutions (recursive and direct) but this is mainly for the following reasons: (i) the Sisyphus-I problem is under-specified so we had many alternatives to explore, (ii) it was the first time we developed a genuine (long) program with

[15] http://ksi.cpsc.ucalgary.ca/KAW/Sisyphus/Sis1/#Section5.6

CGs as data structures (again, there were many alternatives), (iii) we had to add some code in WebKB for handling some special constructs and commands (e.g. for supporting functions and the commands subtract, subtractFrom and subtractFirstFrom). More details are given in the next section.

Regarding the second reason, our initial problem was to find an elegant way to deal with the absence of what are called objects in object-oriented languages or structures in classic procedural languages. For example in those languages, employees could simply have been represented by objects of type Employee with the attributes (or slots) *name, role, project, smoker, hacker* and *works-with*, i.e. exactly as they are described in the Sisyphus-I problem description. We could have used conceptual relations for representing such attributes but:
(i) CGs such as `[Simple_researcher:Uwe_T]- { (Smoker)->[Boolean:yes]; (Hacker)->[Boolean:yes]; }` sound artificial (why using CGs instead of an object-oriented language for such representations?),
(ii) testing if a researcher smokes or not still involves a search for specializations which is not a quick or elegant way to test an object attribute, e.g.:
`spec "[Simple_researcher:$o]->(isSmoker)->[Boolean:?]" | set is_a_smoker.`
Consequently, our initial problem was to find a rationale for deciding what to represent in the type hierarchy and what to represent with CGs. We discussed the results in section 1.2.

Though our combination of searches for specializations and "for" loops seems quite natural, it remains that the implementation of a solution in a classic object-oriented language would have been quite easy for the Sisyphus-I problem and, especially if the language is compiled, more efficient than our solution. For large-scale problems, this is an issue.

The WebKB processor has been developed in C on top of the CG workbench CoGITo[16]. The CG workbench Peirce was also used at one stage but this configuration was slower since Peirce has to be exploited as a separate process. The W3C library[17] (5 Mb over the 6Mb of WebKB) is reused to access Web files. The interfaces for WebKB have been written in HTML and Javascript[18].

3 The Engineering Process

The way we solved the Sisyphus-I problem did not draw much upon our experience in CGs or knowledge acquisition but rather in programming! Indeed, even though we have worked on knowledge modeling, visualisation and ordering using CGs for more than 5 years (because CGs are a rather general and intuitive knowledge representation language which exploits type ordering), we think solving the Sisyphus-I problem with WebKB did not required such an experience with CGs:
(i) the Sisyphus-I knowledge structures are already overt (as Brian Gaines notes)

[16] http://www.lirmm.fr/~salvat/cogito.html
[17] http://www.w3.org/pub/WWW/Library/
[18] http://developer.netscape.com/docs/manuals/communicator/jsref/index.htm

and very simple (only type ordering and simple existential CGs are necessary);
(ii) the protocol and the requirements strongly suggest a certain order in the allocation tasks (thus, a procedural solution comes naturally),
(iii) WebKB does not have a rule-based system but it has a scripting language and commands for declaring, defining, ordering, asserting, searching and joining types, individuals and CGs.
Thus, for us the Sisyphus-I problem was not a knowledge acquisition problem (we did not follow a methodology or reused an ontology) but a programming problem: how to use the WebKB commands to find an elegant and (if possible) efficient solution? The advantage we had as designers of WebKB is that we already knew well the WebKB commands, and that we could add other ones when necessary.

As required, we now give an approximate schedule for the tasks that led to the implementation of our solution. We use a "day" or "half a day" as units of time, half a day being 3 to 4 hours. Since we had other tasks to do, these units were often split over several actual days.

First, we read the Sisyphus-I problem description and, to better understand what should or could be done, the associated Web-accessible article of Brian Gaines. This took at least a day since the article is long and, as we pointed out, the meaning and combination of the rules complex to understand.

Then, during half a day, we tested a translation of some of the representations used in Gaines' article into CGs. The result was not really satisfying since sets and value constraints were used, and though they can be represented in CGs, these constructs cannot not be enforced by WebKB. Another half day was spent to decide if a classification and agenda mecanism similar to the one described in the article was worth simulating with the WebKB script language. We found it much simpler to implement the rules via functions and use recursivity to handle backtracking.

We first implemented the interface in about half a day.

We designed and entered the bulk of the recursive script (and the relevant ontology and data set for it) in a day or a day and a half. Two and a half days were then devoted to progressively debug and refine the script, intertwined with two days for adding a few things to the WebKB code, mainly functions (with parameters, instruction return, etc.) and the command "subtract".

By then we understood much better the protocols and the requirements. We had realised that a solution without backtracking would be simpler, more efficient and nearly as complete. It took us a day to design and enter the bulk of the direct script, half a day to add the commands "subtractFrom" and "SubtractFirstFrom", and two days to progressively debug and refine the script.

4 Critique of the Sisyphus-I Testbed

The documentation is ambiguous and incomplete on the rules to follow. Here are some examples we noted earlier. Can the head of group, manager(s) and heads of large projects share their office if necessary? (The requirements just

note they are "eligible for large offices"). May offices be left empty while simple researchers share large offices? Are all large offices twin offices? Does the expression "coworkers that work on related subjects" mean "employees that have the same hacking habits"?
Here are some other points noted by Brian Gaines. What should be done to place the head of group if there is no large central office? What should be done to place the secretaries if there is no office close to the head of group? What should be done if there are several secretaries, not all of whom can be close to the head of group? What should be done if one secretary is a smoker and another not? What should be done to place the manager if there is no single office close? What should be done if the head of group and secretariat are split? What should be done if there is more than one manager?

We made a decision on each of the above points. Since other people may make other decisions, the documentation ambiguity and incompleteness may make it harder to compare the possibilities of different knowledge acquisition tools. As the previous section shows, the ambiguity and incompleteness of the problem does not give more freedom but creates more work to the knowledge engineer.

We do not think the application domain of Sisyphus-I is very appropriate for the comparison of knowledge acquisition tools (CG-based or not) since (i) the problem has a natural procedural solution and the ontology is small and fixed (i.e. it is not updated interactively by the users), therefore classic object-oriented language are adequate to solve it, (ii) as opposed to a lot of real-world problems, efficiency is not an issue (it is unlikely that a lot of employees have to be placed and that the problem solver has to be run frequently).

Conversely, this problem allowed us to show that CGs may be used and mixed in a script language to elegantly solve a problem which has a procedural solution.

5 Conclusion

Though WebKB was not developped to solve problems but rather to ease the modeling, organisation, retrieval and generation of knowledge or parts of documents indexed by knowledge, this application showed the generality and interest of its features:
- the integration of CG search/handling commands and a script language;
- the insertion of commands (e.g. knowledge assertions) inside documents;
- its Web-accessibility and the exploitation of Web-accessible documents.

To our knowledge, procedural languages have not been associated with CGs or description logics. The use of scripts and their insertion in documents has a broad range of applications and is common in the hypertext litterature but WebKB seems to be the first Web-based knowledge-oriented hypertext system to exploit this facility. Web-based knowledge-oriented systems are currently either ontology servers (e.g. Loom Ontosaurus[19]) or information-retrieval/brokering tools (e.g. Ontobroker[20]).

For a long-term project, we are also studying how the Sisyphus-I problem-solving knowledge could be directly represented by a set of CGs[21] which would be used by a constraint satisfaction problem solver. Thus, these CGs would replace the script. The ontology and data set presented here would not have to be changed.

Acknowledgments

This work is supported by a research grant from the Australian Defense, Science and Technology Organisation.

---

[19] http://www.isi.edu/isd/ontosaurus.html

[20] http://www.aifb.uni-karlsruhe.de/WBS/broker/

[21] http://meganesia.int.gu.edu.au/~phmartin/WebKB/kb/sisyphus1Rules.html

Constraints and Goals under the Conceptual Graph Formalism: One Way to Solve the SCG-1 Problem

Guy W. Mineau
Dept. of Computer Science
Université Laval
Quebec City, Quebec
Canada, G1K 7P4
mineau@ift.ulaval.ca

Abstract. The Sisyphus-I problem is a *room allocation problem* used as a common testbed to compare different knowledge acquisition and problem solving methodologies. This paper shows how it can be represented and solved using conceptual graphs (CGs). Since CGs offer a graphical representation of knowledge for, among other things, facts, constraints and goals, this paper shows how the subsumption relation defined on CGs help reformulate the problem in terms of a classification problem. It also shows how a graphical representation of this classification structure can be helpful in the generation of explanations pertaining to the behavior of the system, or as support to a knowledge engineer who must assist the system in its task. Thus we claim that such a classification structure, which can be visualized, may contribute: 1) to solving the problem, 2) to keeping track of the behavior of the system, and 3) to interpreting possible solutions.

1 The Model

1.1 The Representation of Semantic Constraints using CGs

In [1], we introduced a simple way to describe semantic constraints under the CG formalism. In their proposal, false graphs, i.e., graphs whose truth-value is known to be false, are never stored in a CG system. In fact, a semantic constraint should prevent some graphs to be expressed, those that violate the constraint. This allows constraints to be represented as sets of false graphs. That means that any graph explicitly acquired by the system should never be a specialization of a false graph (otherwise the constraint would be violated). Since all graphs in a CG system are ordered according to a partial order of subsumption, producing what is called the *generalization hierarchy* of the system [2], a set of false graphs represents subspaces of this hierarchy in which no graph should ever appear, i.e., in which no graph should be classified. Our previous work [1,3] shows that all semantic constraints found in database literature are covered by this mechanism, even transition constraints which restrict the evolution of a dynamic system [4].

For example, Fig. 1 below shows the following constraint: *no two researchers working on the same project should share a room* (so that researchers sharing a room work on different projects, allowing synergy among projects). In order to prevent any graph from making the graph of Fig. 1 true, no graph lower than the graph of Fig. 1 (in terms of the subsumption operation) in the generalization hierarchy of the system should be acquired. Let us define the graph of Fig. 2 as always false (which in fact it is since no individual complies with the absurd concept type, by definition), and let us define it so that it is the most specific graph in the system (the lowest in the generalization hierarchy). Then the above constraint could be expressed as the set of all graphs between u_1 to v_0, which we call a *non-validity interval*. As proposed in [4], the definition of this constraint would simply be: **false** $\mathbf{u_1}$, where u_1 is the graph of Fig. 1.[1]

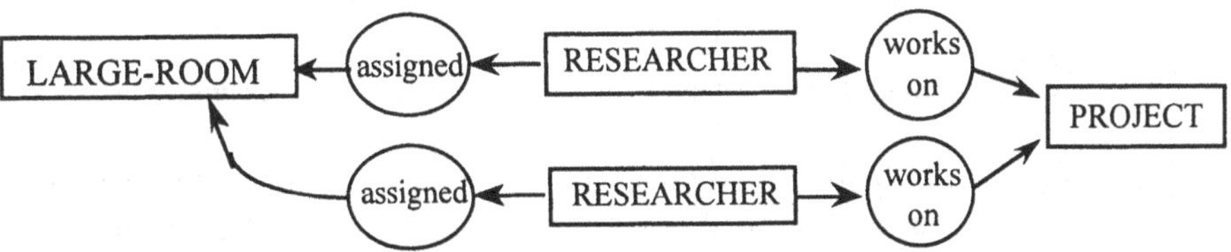

Fig. 1. A false graph u_1: a constraint on the assignment of researchers to rooms.

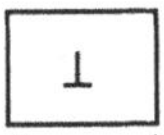

⊥

Fig. 2. The universally false graph v_0, where $\perp$ is the absurd concept type.

If we define the graph of Fig. 3 as being universally true (which it is in any non-empty CG system since any individual complies with the universal concept type), and if we define it as the most general graph in the system (which it is by definition of the canonical formation operations), then we can represent the generalization hierarchy of this example graphically as done in Fig. 4 using a compact representation called a *validity space* (VS) [1].

T

Fig. 3. The universally true graph u_0, where T is the universal concept type.

[1] An optional **excluded** keyword may appear after **false**. This would indicate that graph u_1 is not part of the non-validity interval that it identifies. By default, it is.

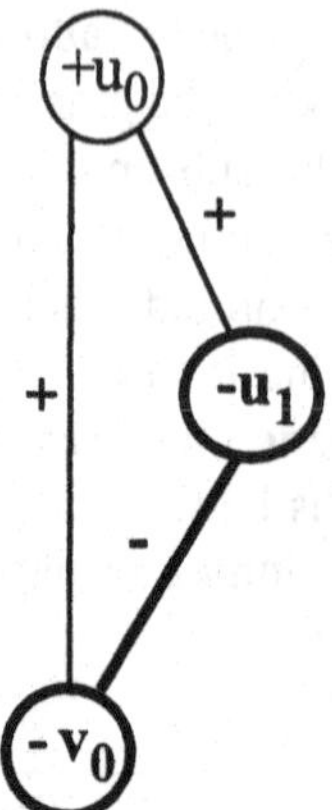

Fig. 4. A graphical representation of the validity space for our example.[2]

The representation of Figure 4 shows that any acquired graph classified in subspaces marked as + are plausible (assumed to be true), while those classified in subspaces marked as – are known to be false (and are therefore not kept in the system).[3] So the validation of any new graph is based on its classification in the VS of the system. When one - mark is associated with it, we know that the graph violates at least one constraint and should therefore be rejected.

Furthermore, we may introduce exceptions to a constraint. For instance, we could have the following constraint: *every researcher must be assigned to some room*, which can also be expressed as: *there is no researcher who is not assigned to a room*. This could appear as the following constraint: **false u_2 unless v_2**, where u_2 and v_2 are the graphs of Fig. 5 and Fig. 6 respectively.[4] Fig. 7 shows the VS of Fig. 4 where graphs u_2 and v_2 were added. Here variables are used to symbolize coreferencing.

RESEARCHER: *x1

Fig. 5. A false graph u_2 representing any researcher.

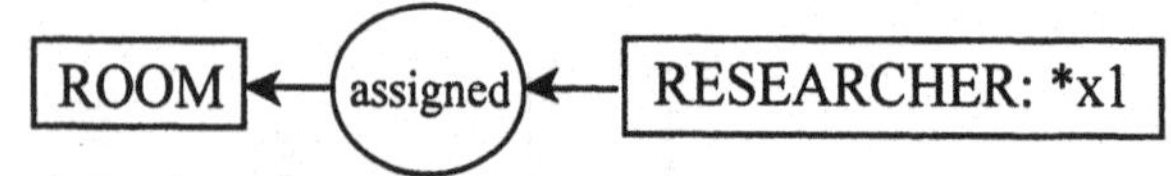

Fig. 6. Graph v_2 representing an exception to the constraint of Fig. 5.

---

2 Of course, if space would allow it, we would have a validity space with the actual conceptual graphs in the nodes instead of using references such as ui and vi.

3 We assume that all conceptual graphs acquired and stored in the system are kept in normal form [5], i.e., all graphs which can be joined because they share joinable concepts in a truth-preserving manner are effectively joined.

4 We could have an optional included keyword associated with the unless clause, which would signify that graph v2 is part of the non-validity interval. By default, it is not.

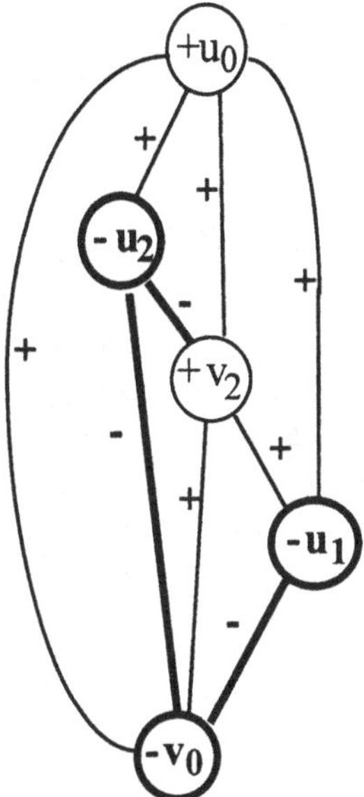

Fig. 7. The updated version of the VS of Fig. 4.

According to Fig. 7, the definition of a researcher must include a relation *assigned* to some room. However, this room must not be shared with another researcher working on the same project. All other graphs are not submitted to any constraint. From this simple example, one can see that the validation of a newly acquired graph is based on a classification procedure[5]. Of course, the user may query the VS to see if or what constraint is reported as being violated, and how to modify the faulty graph to miss the non-validity interval (subspace) where it was previously classified. So, with the appropriate graphical interface and query tools, the VS of the system can provide explanations and support to either a computer program or knowledge engineer responsible for constraint satisfaction when describing the application domain.

However, the Sisyphus-I problem is not only a constraint satisfaction problem but also implies a goal-driven heuristic search among different possibilities towards acceptable solutions.[6] These goals are in no way compulsory to obtaining an acceptable solution, but rather guide the search engine towards *preferred* solutions. For instance, *geographical proximity between teammates should be sought when rooms are allocated* is one such goal. *The head of a research project should be assigned to a single room, if possible,* is another example. However, the previous constraint stating that: *researchers assigned to the same room should not work on the same project* should overwrite these goals (when teammates are in fact researchers). Consequently, goals should be expressible in the CG formalism in such a way as to

5 We assume that a graph is classified in the most specific subspaces where it belongs.

6 In this paper, *constraints* refer to integrity constraints of database literature, and *goals* refer to what is called constraints in traditional planning literature. One should see that the former represents strict constraints which must never be violated; while the latter, often called "weak constraints", are goals to be satisfied, if possible, but without any obligation to do so.

be integrated with constraints and as to be prioritized. The next section introduces our proposal with regard to the representation of goals using the CG formalism.

1.2 The Representation of Goals

Goals are assertions that the system should try to make, but without having the obligation to. Contrarily to constraints, the system should seek to produce graphs which are specializations of goals. So, a goal can be represented by the set of graphs more specific than (or equal to) itself (in terms of the subsumption relation between all graphs). Like constraints, a goal can be represented as a subspace of the generalization hierarchy. In order to ensure uniformity of representation, we will integrate them to the VS of the system. For instance, the following goal: *employees working on the same project (teammates) should be assigned to rooms close to one another* could be represented by the following three goal definitions: **goal u_3 priority 1** (see Fig. 8), **goal u_4 priority 1.1** (see Fig. 9), and **goal u_5 priority 1.1.1** (see Fig. 10).

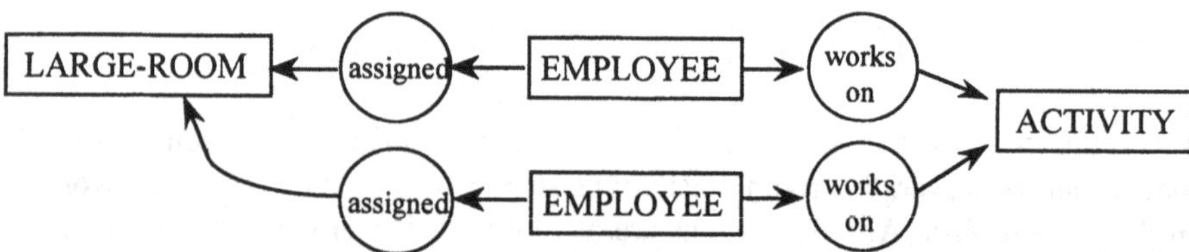

Fig. 8. Goal u_3: *if possible, assign teammates to the same room.*

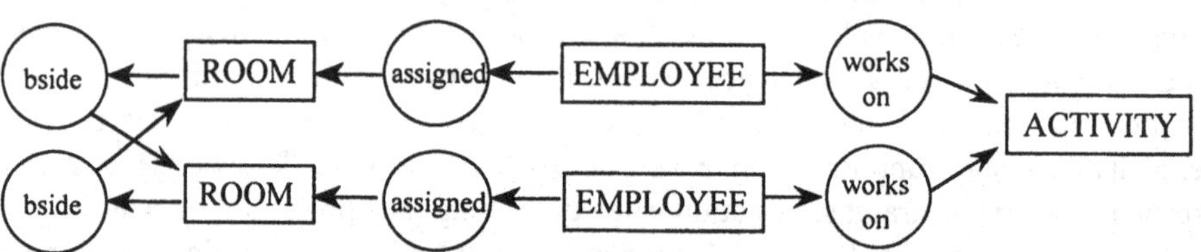

Fig. 9. Goal u_4: *if u_3 is not reachable, then try to assign teammates to adjacent rooms.*

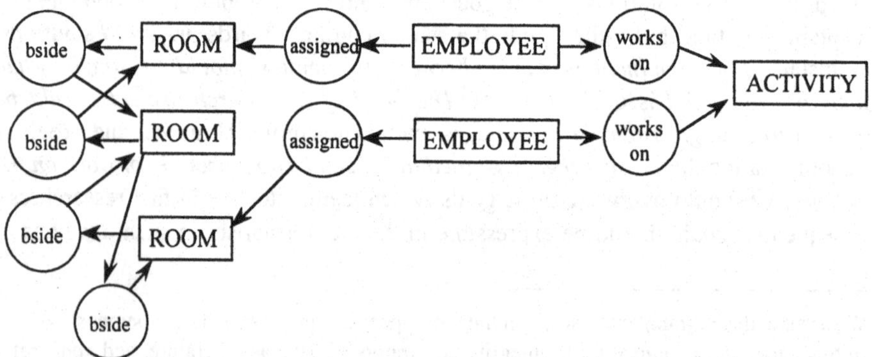

Fig. 10. Goal u_5: *if u_4 is not reachable, try to assign teammates to closely located rooms*

If the system fails to meet u_5, then the teammates could be assigned anywhere, as no other goal is related to goal u_5. The hierarchical numbering of the goals renders a priority ranking among them. Of course, a more sophisticated ranking system could be devised but this falls outside the scope of this paper and will not be discussed here.

Fig. 11 shows the VS of Fig. 7 where the three previous goals were added. Here the priorities of the goals label the nodes where they appear. This prioritizes the entire subspace identified by the node. In other words, the different + marked subspaces will be scrutinized by the classification algorithm which will try to satisfy as many as possible. By looking at Fig. 11, one can see that the subspaces identified by u_3 and u_1 intersect ($u_3 > u_1$). This means that the goal u_3 will not be reachable if the employees are researchers and the activity is a research project.[7] If we had $u_1 > u_3$, that would mean that goal u_3 would never be reachable, and could thus be considered useless and be deleted from the system. Such interactions among constraints and goals can be reported back to the knowledge engineer as soon as they are detected, helping in the validation of the VS itself. By doing so, the resolution strategy (for the classification algorithm) as expressed by the set of sought-for goals can be revised while being developed. The VS acts as a support tool in the determination of a sound resolution strategy.

Of course, some goals could have the same priority. For instance, we could have goals u_4 and u_5 of equal priority (then labeled 1.1 and 1.2 since on the same level of priority), letting a random process (or the knowledge engineer) intervene in the subsequent classification of graphs under either one of them (when goal u_3 fails to be met). Otherwise, the classification algorithm can automatically maximize priority satisfaction, avoiding conflicts introduced by constraints. For example, let us define the following goal: **goal u_6 priority 2**, where graph u_6 is shown in Fig. **12**. This goal states that the head of the research group should be physically close to his/her administrative employees (in adjacent rooms). It has the same priority as goal u_3 since it is on the same level. So no priority ranking exists between u_6 and u_3. If this goal were to have subgoals like u_3, then they would be labeled as 2.x. Again we present a simple ordering scheme in this paper, sufficient to solve the Sisyphus-I problem. More elaborate schemes could be devised, but fall outside the scope of this paper. The updated version of the VS of Fig. 11 is shown in Fig. 13.

In summary, when a new graph needs to be represented in the CG system, the VS is searched in order to identify all the appropriate subspaces (marked as +) where the new graph could be classified. The subspace(s) with the highest priority is (are) identified. When more than one subspace is identified and when these goals can not be satisfied simultaneously, some decision making mechanism must determine which one(s) is (are) to be preferred. Again, the knowledge engineer could contribute to the reaching of a solution if asked to make that decision. To do so, 1) the possible alternatives can be presented as CGs (which is a graphical format usually more readable than linear notations), 2) previous classification of graphs can

---

7 Here we suppose that in the type hierarchy of the CG system, we have: RESEARCHER < EMPLOYEE, and PROJECT < ACTIVITY. See Appendix A.

be explained (either by CGs or natural language sentences derived from them), and 3) relevant subspaces (parts) of the VS can be visualized along with their associated priority[8]. With the appropriate graphical interface and query tools, this could provide an interactive environment to assist the knowledge engineer in the exploration of the possible solutions to this classification problem that knowledge acquisition now represents under this framework.

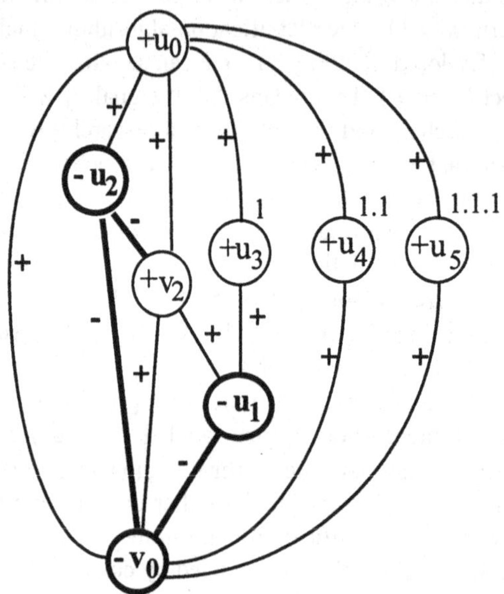

Fig. 11. The updated version of the VS of Fig. 7.

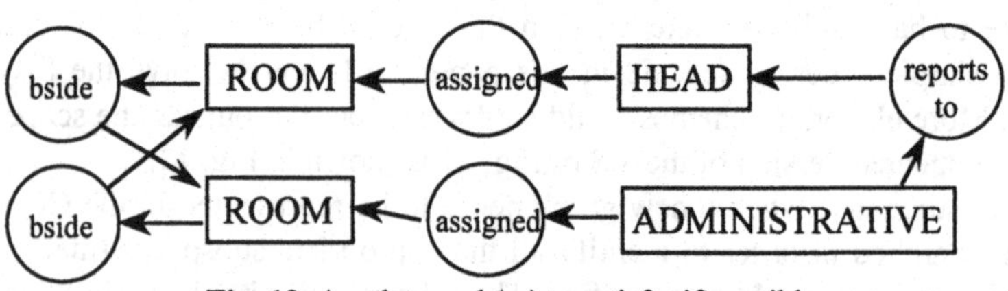

Fig. 12. Another goal (u_6) to satisfy, if possible.

8 One can easily imagine that a coloring scheme could be used to represent different levels of priority for instance.

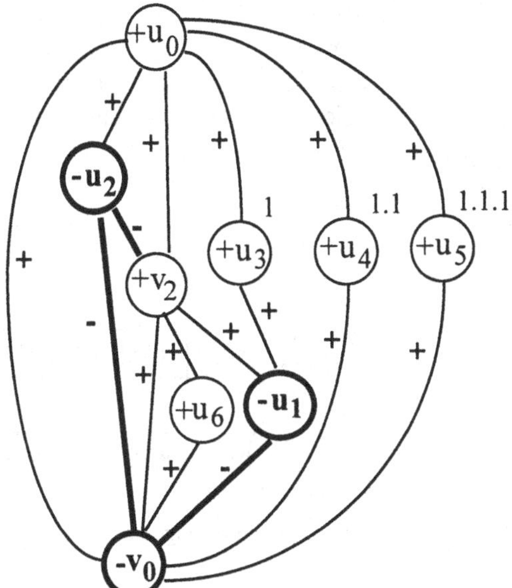

Fig. 13. The VS of Fig. 11 where the graph of Fig. 12 was added.

1.3 Solving the Sisyphus-I Problem

As mentioned before, the Sisyphus-I problem can be reformulated as a classification problem in a VS where constraints and sought-for goals are represented. The objects to be classified will be instantiated graphs like the one in Fig. 14, either automatically produced by the algorithm which seeks to solve the problem, or by the knowledge engineer (see Appendix C for a brief discussion on the different resolution methods that could be adopted). These graphs are those which will eventually solve the problem: none should violate any constraint, and together they should maximize priority satisfaction.

Fig. 14. A new asserted graph: the assignment of a researcher to some room.

Once asserted, we keep all graphs in normal form, that is, all joinable concepts that can be joined in a truth-preserving manner are joined. The graph of Fig. 15, which is part of the description of the domain, would be joined with the graph of Fig. 14, producing the graph of Fig. 16. Whenever the graph of Fig. 16 is joined with a graph assigning either Jürgen or Thomas to room #120, the resulting graph would be a specialization of graph u_1, falling in a non-validity interval. Consequently, this prevents the system from assigning the same room to researchers working on the same project. So the system will avoid assignments which would violate the constraints of the domain, while maximizing priority satisfaction. With our example, the exploration of the VS of Fig. 13 could be used to propose viable alternatives. For example, since goal u_3 does not apply for researchers, then we would go one priority level lower to goal u_4 (priority 1.1). Satisfying goal u_4 implies assigning Jürgen or Thomas to room 121 or 119, as indicated by the background knowledge that we have on room proximity (as partially shown in Fig. 17).

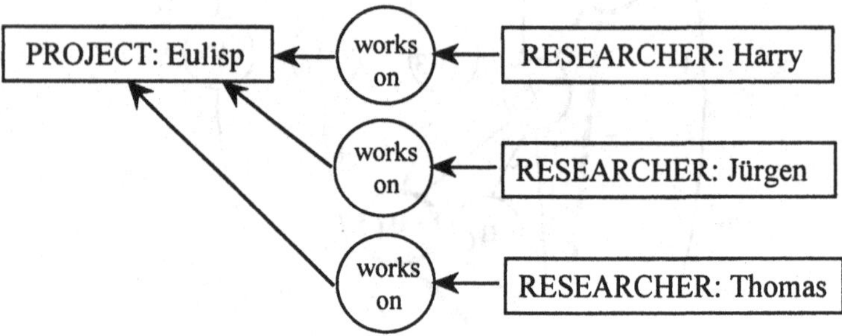

Fig. 15. Background knowledge on Harry.

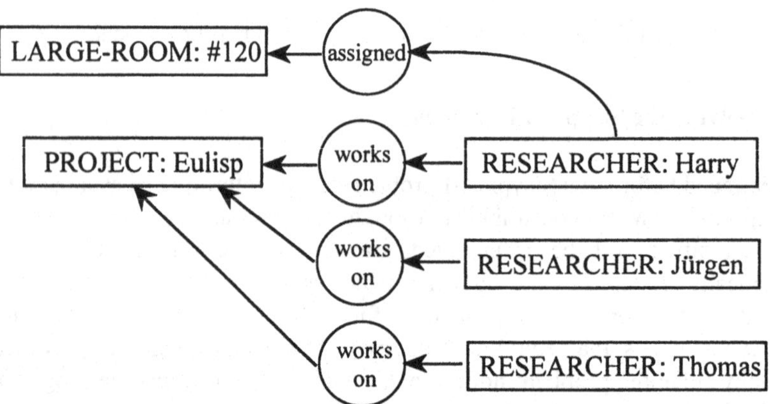

Fig. 16. Joining the graphs of Fig. 14 and Fig. 15.

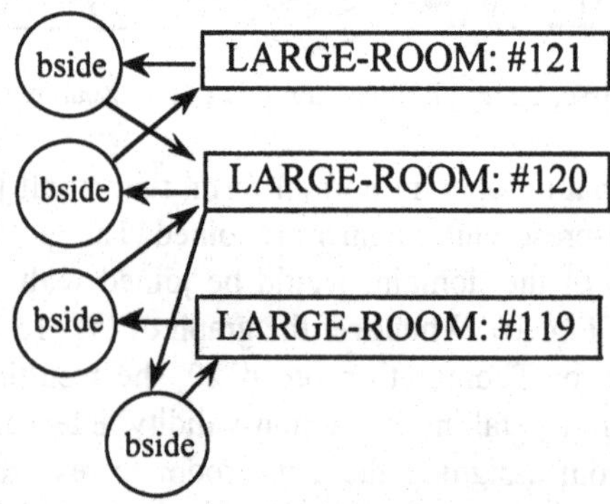

Fig. 17. Room proximity (partially shown).

Assuming that room #119 is assigned to Thomas (arbitrary choice at this point), Jürgen could then be assigned to either room #121 or room #117 in order to satisfy goal u_4 and avoid constraint u_1. However, since the system is trying to maximize priority satisfaction, goal u_6 could also be satisfied if some administrative employee working under Thomas would be assigned to an adjacent room. Knowing that administrative employees work on the same activity, the system could satisfy goal u_3 as well by allowing two administrative employees to be assigned to a room adjacent to room #119. Consequently, maximizing priority satisfaction would result in the assignment of two administrative employees in room #117 and in Jürgen being assigned to room #121.

In order to make this choice, the system must test hypotheses about priority satisfaction, which requires a backtracking mechanism to be implemented so that the system can assess and revise its strategy. Of course, at any given time the knowledge engineer could query the VS in order to assess the potential of the partial solution being developed, or to resolve conflicts. Once again, one can imagine a graphical interface where assignments could be changed interactively and where updates on the VS could be shown in real-time.

In summary, factual knowledge about the domain was acquired, i.e., knowledge describing room proximity, research projects, researchers, administrative employees, the hierarchical organization of the research group, their activities, and the constraints pertaining to the domain (like constraints u_1 and u_2). Below we list the five constraints that we represented and used to solve the Sisyphus-I problem.

1. Large rooms contain at most 2 employees.
2. Small rooms contain at most 1 employee.
3. Researchers assigned to a large room do not work on the same research project.
4. Smokers and non-smokers are never assigned to the same room.
5. The head of the research group does not share his/her office with anyone.

Then five sought-for goals were defined; some had subgoals (as indicated by the hierarchical ordering scheme). Together, this represented a set of 10 CGs.

1. Employees working on the same activity should share a room (priority: 1).
2. Employees working on the same activity should be assigned to adjacent room (priority: 1.1).
3. Employees working on the same activity should be assigned to nearby rooms (priority: 1.1.1).
4. A manager and his/her administrative employees should be assigned to the same room (priority: 2).
5. A manager and his/her administrative employees should be assigned to adjacent rooms (priority: 2.1).
6. A manager and his/her administrative employees should be assigned to nearby rooms (priority: 2.1.1).
7. The head of the research group and administrative employees should be assigned to adjacent rooms (priority: 3).

8. The head of the research group and administrative employees should be assigned to nearby rooms (priority: 3.1).
9. A researcher should share his office, if assigned to a large room, with another researcher (priority: 4).
10. Heads of large research projects should be assigned single rooms, if possible (priority: 5).

Then we gave a starting point to the system: *Thomas, being the head of the research group, must have a central office*. We did not let the system derive what room could be assigned to Thomas. We decided that it would be in room #119 (which parts the set of rooms in two sections, one of seven people, the other of eight). This starting point was decided in order to simplify the resolution process of the problem, and in order to follow the sample protocol given with the description of the Sisyphus-I problem. So Fig. 18 shows this initial starting point.

Finally, we gave the system assignment graphs like the one in Fig. 19, one for each employee (except for Thomas).[9] In [1], the #? referent symbolizes an unknown individual, which must be known before asserting the graph where it appears. This enforces referential integrity. In terms of subsumption, we have the following order: x > #? > i, for x, an existentially quantified variable, and i, an element of the set of known individuals I. In order to assert these graphs, the system must instantiate the ? variable for each of these graphs. To do that, the knowledge engineer can provide values for each ?, or the system may try to resolve these references on its own. Doing so will solve the Sisyphus-I problem.

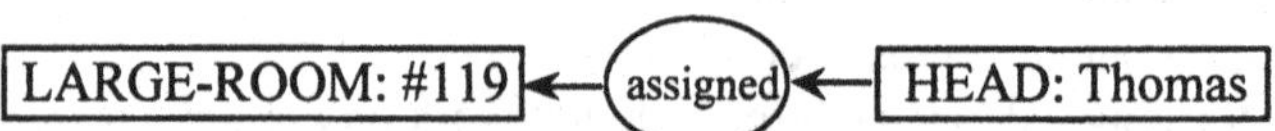

Fig. 18. The first assignment graph given to the system as starting point.

Fig. 19. Assignment graphs which will prompt the resolution of the problem.

The type hierarchy used to compute subsumption among graphs is given in Appendix A. Below we present the best two solutions that we found. Other solutions are presented in Appendix B.

1.4 Our Solutions

Since we have no software currently implementing the necessary functionalities needed to create and update a validity space as described in this paper, all simulations were done by hand, following a depth-first search strategy (from the

---

9 In all 14 assignment graphs, the i is actually replaced by an individual name, since all names are assumed to be different and can thus be used as referents.

given starting point of Fig. 18). From these, some proved to be more interesting than others. In all cases, a system that would solve the problem according to our methodology could justify its decisions every step of the way. No solution violates any constraint; but no solution could satisfy all goals simultaneously. This section presents the two solutions that maximized priority satisfaction according to the list of goals given in Section 1.3 above.

Fig. 20 below shows the assignments that the method proposes in order to maximize priority satisfaction. From this figure, one can see that Joachim, though a head of a large project, is not assigned a single room (see goal #10 above). This is the only goal not satisfied. In the resolution of this problem, Joachim could be assigned to either room #122 or room #115. If assigned to room #115, it would imply that Michael be moved to room #122. Because of our proximity goal between teammates, this would move Hans to room #123. Hans would then be sharing a room even though he is also the head of a large project (see Fig. 21). Since the description of the problem seems to imply that Joachim and Katharina are of lesser importance in this situation, we would prefer the solution of Fig. 20 over the one in Fig. 21 though both are technically of equal quality in terms of priority satisfaction. This ranking was decided by the knowledge engineer once the simulations were run.

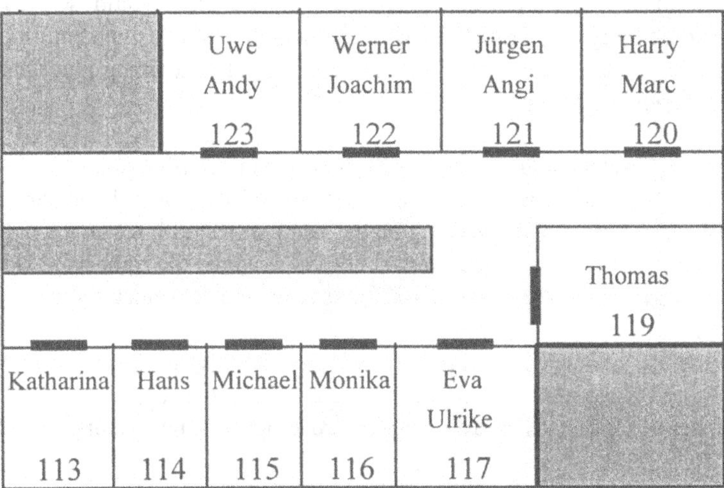

Fig. 20. Our best solution.

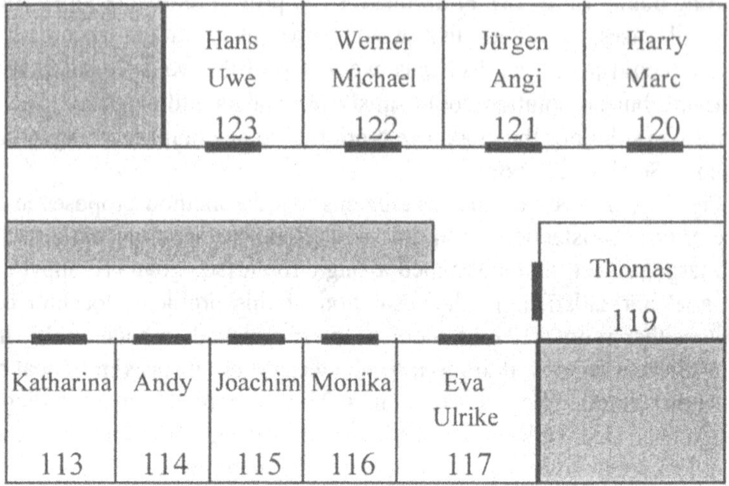

Fig. 21. Our second best solution.

In fact, from the solutions that we produced, a learning algorithm could extract the following generic features characterizing the solutions that are classified as being good solutions by the knowledge engineer [6,7,8]. This would summarize the behavior of the system in terms of the goals which guide the assignment process. For instance, with the two solutions of Fig. 20 and Fig. 21, a learning algorithm would extract the following features.

1. Thomas is always assigned room #119 (which is our starting point).
2. Thomas has his administrative staff on one side of his office (the lower side), and his colleague researchers (working on the same project as him) on the upper side.
3. The other team of three researchers is always assigned the right-hand side rooms of the upper side (#122, #121, #120), leaving the left-most room to smokers (#123).
4. Smokers are assigned one large and two small rooms, the left-most upper room (#123), and two out of three of the left-most lower rooms (#113, #114, #115).
5. Three out of four heads of large projects are assigned single rooms.

Fig. 22 below shows a plan that could be inferred from the two previous plans. Based on this plan, the behavior of the system can be illustrated. This could help the knowledge engineer understand the different interactions between the constraints, the goals and the priority structure, interactions which are normally too complex to predict. By running a learning algorithm on many solution plans, the recurrent components will identify the heavily weighed elements of the strategy once the interactions act up in the resolution algorithm. Once again, this could help the knowledge engineer to validate the strategy that he/she encoded (and thought would be used) in order to solve the problem.

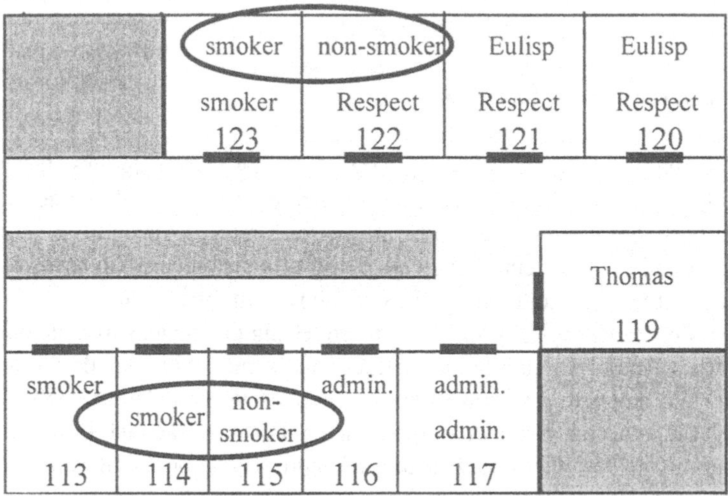

Fig. 22. A generalization of the plans of Fig. 20 and Fig. 21.[10]

In summary, the reader should notice that the strategy described by a set of prioritized goals represented by CGs is less precise than any strategy described in an algorithmic-like language. The main advantages of this are that: 1) the system can explore, on its own, unsuspected possibilities, 2) the knowledge engineer is not required to come up with a complete resolution strategy, 3) the description of a strategy is done through the sole identification of relevant goals (simplicity of expression: *say what you wish for, not how to get it!*), 4) these relevant goals are somewhat independent from the resolution method (which can then be considered generic since it is not tied to the application domain), 5) since these goals are integrated with the constraints and background knowledge of the system, explanations can be derived at any time, either in conceptual graph or natural language format, to justify the actions of the system, 6) through a visual interface, the description of the system available as a set of CGs is probably more meaningful to a lot of people who would otherwise have to use a linear notation, and consequently, 7) the knowledge engineer may intervene more easily when appropriate to do so.

However, the main disadvantages are that: 1) computing resources may be needed in such a way as to quickly bound the set of solutions, 2) this set of solutions, found in a finite amount of time, heavily depends on the search strategy (depth-first, breath-first, etc.), 3) the solutions explored by the system may represent local optima, and 4) with a set of CGs, the description of an application domain is focused on bits and pieces of knowledge, the whole picture may be lost because it was

10 In this plan, the name of the project is used to represent a researcher working on the project. Also, the ovals represent adjacent rooms where the two researchers working on the BABYLON Project could be assigned.

broken down into small CGs (resulting in a potential cognitive limitation to further model the domain correctly and efficiently).

2 The Evaluation Process

As mentioned above, there is no software producing these solutions; the proposed methodology was simulated by hand on the dataset provided with the description of the Sisyphus-I project, using a depth-first search strategy. Appendix B shows other solutions found by this methodology, all being of worse quality (in terms of priority satisfaction) than the solutions presented above. In brief, only the solutions of Section 1.4 maximize priority satisfaction. So testing the model was done only in this application domain, for the small dataset that comes with the definition of the problem. We stopped the simulations as soon as we found local optima (which would be the expected behavior of the system). For now, the lack of a tool refrains our capability to test the model in more length. Theoretically at least, the model seems to hold. Naturally, it suffers from the same complexity normally associated with subsumption based problem solving. However, efficient algorithms used to compute subsumption relations between conceptual graphs were devised over the years [9,10] and can help speed up the process by precompiling order relations among graphs. These algorithms are incremental, which allows the knowledge engineer to use the system to fine-tune the formulation of the problem. Also, factual graphs describing the domain are mainly grounded graphs, reducing the complexity related to the computation of the subsumption relation.

Like with other logic programming environments, provided that the model is noise-free and renders an adequate, sufficient and complete image of the application domain, one can assume that the methodology proposed in this paper is sound and complete. Furthermore, provided that recursive assignments of people to rooms are detected and blocked, i.e., when the assignment of someone to some room depends on the assignment of the same person to some other (or possibly the same) room, then the methodology is decidable. Asserting assignment graphs like the one in Fig. 18 as soon as a hypothesis is made, provides this desirable decidability property. This simplifies the resolution procedure, giving better performance results. In any case, empirical testing of the methodology would be required to fully assess: 1) its viability (soundness, completeness and efficiency) under different datasets and workloads, and 2) the usefulness of its graphical components (CGs and VS) in the description of the model, its validation and the resolution of the problem.

3 The Engineering Process

One very important problem when developing a knowledge-based system using conceptual graphs lies in the fact that concepts and relations can be expanded. Consequently, a graph u may have multiple representations which are all logically equivalent. This complexifies the comparison of graphs when determining subsumption between them. In order to avoid this efficiency pitfall, the knowledge engineer must come up with a representation of the world that does not rely on expansion (or contraction) to be meaningful, so that graphs can be compared efficiently. This was particularly true with our model since the problem is reformulated as a classification problem where graph comparison is at the heart of the resolution methodology. Consequently, a few days work (about 3 days/person) were put in the reading of the documentation and the elaboration of a framework to describe the basic elements of our CG system: concept types and type hierarchies, so that similar facts would be represented in syntactically similar graphs. In order to test and validate a few definitions, the semantic constraints of the domain were represented as CGs as soon as possible (included in the 3 days/person).

Then, goals were identified both from the description of the problem and from the sample protocol given with the documentation of the problem. Of course, not all goals were kept, only those that we listed in Section 1.3 above. The goals had to be represented in the same framework used for constraints and factual knowledge. We used reverse engineering: once we had the goals we adjusted the representation of constraints and factual graphs. This is always necessary, but much particularly true in this application domain where most goals refer to physical proximity. In short, we had to define (sub)graphs representing physical proximity in a precise manner. We realize that the model we came up with has a very limited definition of proximity. Since physical proximity is normally defined in a continuum that needed to be translated to precise CGs, we had to make choices about the kind of proximity that the system should prefer. A more complete description of proximity would have resulted in the addition of other goals into the model. CGs aim at representing discrete objects; continuous objects like *distance* introduces a representational bias that the knowledge engineer must take into account in the representation of all other graphs. Two more days/person were necessary to propose goals and revise the representation framework accordingly.

A final representation problem we had to solve was the classification of every employee as either smoker or non-smoker. We chose to identify these two complementary classes with individual concepts: [SMOKING: Yes] and [SMOKING: No] attached to the EMPLOYEE concept using a *chrc* relation. This would normally imply an additional constraint stating that *nobody can be both a smoker and a non-smoker*. We did not incorporate this constraint to the system for simplicity reasons and made sure by hand that this was the case for all 15 employees. However, Werner and Jürgen are employees for which we do not have this information. For both of them, we allowed the concept of type SMOKING to be generic. For these two employees, the system could not verify the constraint that states that *a non-smoker and a smoker should not share a room*. In the event that this information is acquired later on and that the constraint is then violated, revisions would be necessary. Basically, a new solution would need to be computed from scratch since the actual solution may now become of much lower quality than if the system had known this information from the start.

Finally, we proposed an extension to the constraint representation framework presented in [1]. The development and refinement of these ideas took about three days work. This work is not part of the engineering effort to solve the

Sisyphus-I problem and we do not include it in the global days/person figures that we mention in this section.

Once all the information was represented, the model was tested with the dataset that came with the problem. By examining sample solutions we found that administrative employees could share a room with a researcher. Again to comply with the goal of optimizing synergy among projects, we then introduced goal #9 (see Section 1.3 above) which favors the allocation of rooms to employees of the same type (in this case: researchers). This last goal was found to be useful thereafter. The problem was that proximity between researchers was biased by the fact that there was a strong restriction on researchers assigned to a large room: they had to work on different projects. With our model, administrative employees are considered to work on different activities than researchers, making them candidate to sharing large rooms with them, which is in contradiction with the *synergy among different projects* goal. That is why we added goal #9. This elicitation process was a direct result of applying the methodology on our dataset and validating the results. Testing our methodology and revising it took 1 day/person. We produced about 10 different solutions which helped validate the model. In summary, we would need about one week of work to represent and solve the Sisyphus-I problem if we had a CG workbench available to us, workbench which would also implement all the necessary functionalities to manage the validity space of the system so described.

4 Critique of the Sisyphus-I Testbed

The Sisyphus-I problem poses challenging issues to the Conceptual Graph formalism. First, physical proximity, which is normally expressed in a continuum, must be modeled with predicates. This implies the addition of many graphs to the system, one for each precise way of representing proximity. With problems of more considerable size than this one, this may have overloaded the system, and consequently slowed it down since more alternatives would have been explored in order to produce a solution.

Second, the complete classification of all individuals in terms of *smokers* and *non-smokers* needed to be represented by constants. Using a negation operator would have had an impact on graph subsumption: the projection operator would have needed to be redefined, this would have opened the door to additional computational complexity as well.

Finally, incomplete knowledge is a problem. The fact that we don't know whether Werner and Jürgen are smokers or not casts a doubt on our solution. Also, it is not clear in the data if Marc works on the same project as Angi an Werner. The data states that they work together, but the project on which Marc works is KRITON and not RESPECT (which is the project that Angi and Werner work on). We assumed that KRITON was a subproject of RESPECT, but that's only an assumption. This would need to be clarified.

In brief, this problem helped us identify ways of representing goal-oriented heuristic knowledge and constraints in a unified framework under the CG formalism.

As such this exploration makes it worthwhile since it improves the range of problems that the CG formalism not only can represent but can also help to solve.

References

1. Mineau, G.W. & Missaoui, R.: The Representation of Semantic Constraints in CG Systems. In: Lukose, D., Delugach, H., Keeler, M., Searle, L., Sowa, J.F. (eds.): Lecture Notes in Artificial Intelligence, vol. 1257. Springer-Verlag, (1997) 138-152
2. Sowa, J.F.: Conceptual Structures: Information Processing in Mind and Machine. Addison-Wesley, (1984)
3. Mineau, G.W., Missaoui, R.: Semantic Constraints in Conceptual Graph Systems. DMR Consulting Group. R&D Division. Internal Report #960611A. (1996)
4. Mineau, G.W.: Constraints on Processes: Essential Elements for the Validation and Execution of Processes. In this issue, (1999)
5. Mugnier, M.-L. & Chein, M.: Représenter des connaissances et raisonner avec des graphes, Revue d'Intelligence Artificielle, vol. 10-1. (1996) 7-56
6. Mineau, G.W.: Induction on Conceptual Graphs: Finding Common Generalizations and Compatible Projections. In: Nagle, T., Nagle, J., Gerholz, L., Eklund, P. (eds.): Conceptual Structures: current research and practice. Ellis Horwood, (1992) 295-310
7. Miranda, C.J.: Interface graphique pour l'acquisition de descriptions conceptuelles à partir de plans architecturaux.. M.Sc. dissertation. Dept. of Computer Science, Université Laval, (1996)
8. Levinson, R.: USD: A Universal Data Structure. In: Lecture Notes in Artificial Intelligence, vol. 835. Springer-Verlag, (1994) 230-250
9. Ellis, G.: Efficient Retrieval from Hierarchies of Objects Using Lattice Operations. In: Lecture Notes in Artificial Intelligence, vol. 699. Springer-Verlag, (1993) 274-293
10. Ellis, G. & Lehmann, F.: Exploiting the Induced Order on Type-Labeled Graphs for Fast Knowledge Retrieval. In: Lecture Notes in Artificial Intelligence, vol. 835. Springer-Verlag, (1994) 293-310

Appendix A

Fig. 23 presents four parts of the type hierarchy that needs to be defined before the system can solve the problem; they define subsumption relations between concepts which need to be compared in order to compute subsumption among graphs in which they appear.

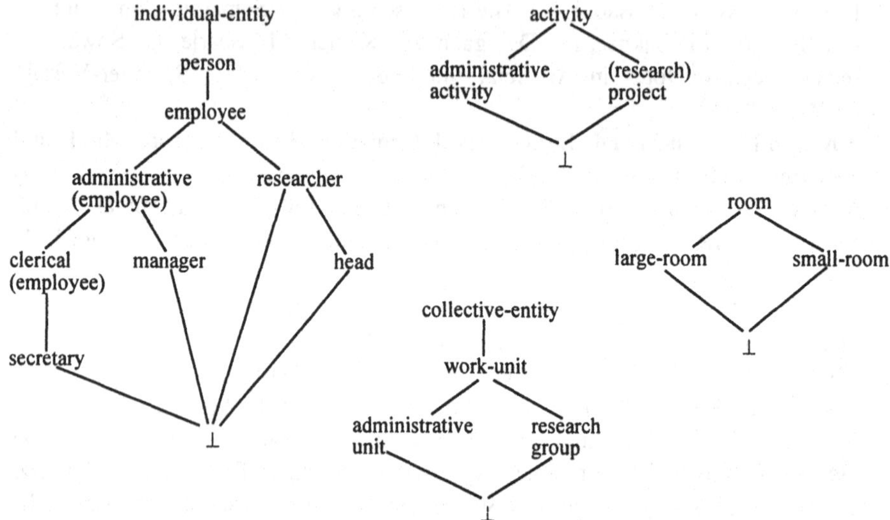

Fig. 23. Parts of the type hierarchy used to compute subsumption among graphs.

Appendix B

This section presents three other solutions that were found using our methodology. They are all of lower quality than those of Section 1.4 (in terms of priority maximization). The first one uses the same starting point of Fig. 18; while the other two starts with Thomas in room #117. The first one (Fig. 24) presents four goals which could not be satisfied: three heads of large projects could not be assigned small (individual) rooms (Katharina, Hans and Joachim), and there is a researcher (Michael) sharing his office with a secretary.

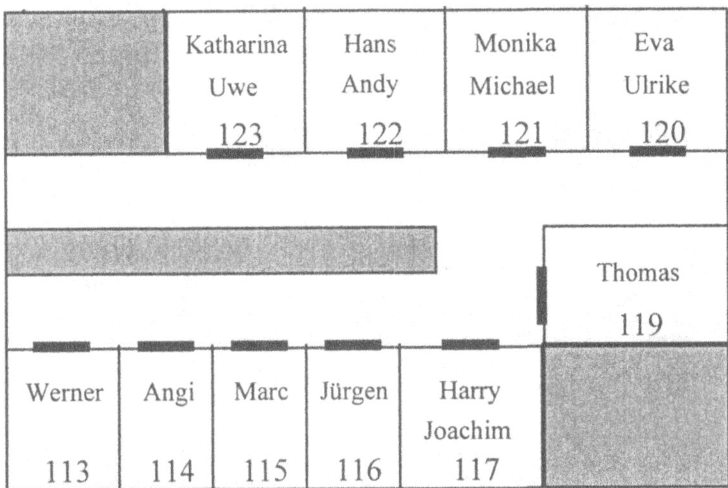

Fig. 24. A typical solution where the administrative staff is in the upper part of the plan.

The other solution (shown in Fig. 25) starts with Thomas in room #117 and assigns the administrative staff in the upper part of the plan. This also creates some problems: Marc, a researcher, shares a room with a secretary; Hans and Joachim, head of large projects, are not assigned individual rooms. But overall, this solution is better than the one in Fig. 24.

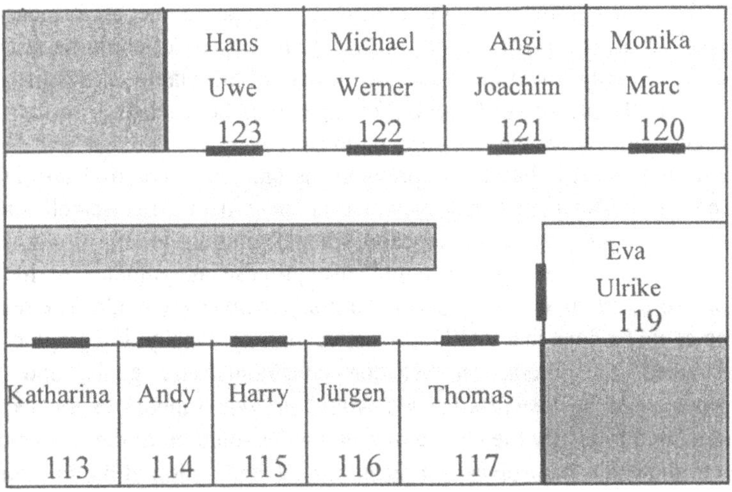

Hans
Uwe
123
Michael
Werner
122
Angi
Joachim
121
Monika
Marc
120
Eva
Ulrike
119
Katharina
113
Andy
114
Harry
115
Jürgen
116
Thomas
117

Fig. 25. Starting with Thomas in room #117 with his administrative staff to his right.

Finally, our last example, Fig. 26 shows a solution where Thomas is assigned to room #117, and where his administrative staff is assigned to his left. This leaves two goals to be desired: Hans and Katharina are not assigned individual rooms. Despite this, this solution is better, in terms of non-satisfied goals, than the previous one.

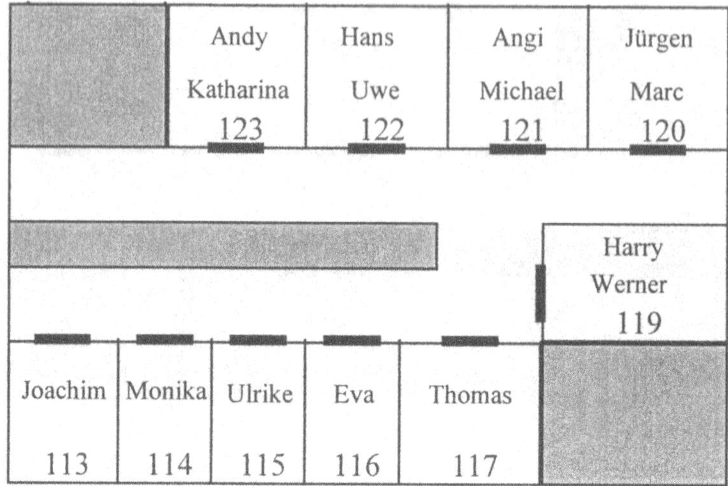

Fig. 26. Starting with Thomas in room #117 with his administrative staff to his left.

Appendix C

The framework proposed in this paper is independent of the resolution algorithm used to obtain a solution. For instance, assignments of researchers to rooms could be done in batch, then constraint satisfaction and priority maximization could be assessed for each combination. Furthermore, to improve the performance of the system, only those combinations not violating any constraint could be selected and assessed for priority maximization. By scanning all viable alternatives, global optimization of the priority satisfaction function could be reached.

Alternatively, since constraints must never be violated and goals are prioritized, the system, based on previous assignments, could determine which assignments to make next. That way, based on a depth-first search strategy as proposed in this paper, we obtain a locally optimal solution.

The knowledge engineer could also propose assignments of his own or could query the system as to why particular assignments were made. The reaching of a solution could be done in a totally interactive manner. Finally, if a graphic interface were provided, the interaction between constraints and goals, and previous assignments could be visualized. This would provide support to the explanation module required to justify the choice of a particular solution, to the validation of the constraints and goals themselves, and to additional exploration of the solution space.

A Pure Graph-Based Solution to the SCG-1 Initiative

J.F. Baget, D. Genest, and M.L. Mugnier

LIRMM (CNRS and Université Montpellier II)
161, rue Ada, 34392 Montpellier - France
{baget, genest, mugnier}@lirmm.fr

Abstract. This paper answers the SCG-1 initiative. The room allocation problem provided has been solved in a *generic* and *automatic* way. The solution is based on a totally *declarative* formal model. Basic constructs are simple graphs and the fundamental operation for doing reasonings is the graph morphism known as projection. The other formal constructs are rules and constraints defined in terms of simple graphs. The modeling framework built upon the formal model allows one to describe a problem with asserted facts, rules representing implicit knowledge about the domain, validity constraints and rules transforming the world. A prototype implementing this framework has been built upon the tool CoGITaNT. It has been used to test our modelization of the room allocation problem.

1 Introduction

Since 1991, our team has been working on CGs along a specific approach [4]. We study CGs as a *graphical* knowledge representation model, where "graphical" is used in the sense of [14], i.e. a model that "uses graph-theoretic notions in an essential and nontrivial way". Indeed, not only CGs are displayed as graphs but also reasonings are based on graph operations. The formal bases of our work are the following. Basic objects are simple CGs, i.e. labeled graphs. The fundamental operation for doing reasonings is the projection, which is a morphism between simple CGs. Reasonings are logically founded, since projection is sound and complete w.r.t. deduction in first order logic. We built extensions of this kernel — namely CG rules [13] and nested CGs [5] — keeping its basic properties. We also developed the software CoGITo [11, 3] (and its extension CoGITaNT [10]), a workbench for building knowledge-based applications, where every piece of knowledge is described by CGs. The theoretical framework and CoGITo have been used in several applications (see [2] and [9] for the applications we are currently involved in).

The Sisyphus-I problem, is a resource allocation task, therefore basically a constraint satisfaction problem. In our previous applications, reasonings were mainly based on deduction (given a knowledge base and a request, find whether the request can be deduced from the knowledge base). Therefore, at least superficially, i.e. in its formulation, the Sisyphus problem is not of same nature as the problems we are used to solve in our framework.

When studying the SCG-1 problem, two immediate questions arose :

- How much could the problem be solved within our framework ? More specifically, how could we represent constraints in our graph-based approach ?
- What additional programming upon CoGITaNT was required ?

The initial requirement for us was to build a solution which operates as much as possible with the existing theoretical framework. The first decision was to keep a completely *declarative* model.

Another decision was to build a *generic* solution, general enough to include at least the family of the "resource allocation task" problems.

A third architectural decision was to design a system able to solve the problem in an *entirely automatic* way. The reason for this choice was not that we did not attach importance to human involvement in the problem solving process. On the contrary, we find it is unrealistic in many constraint satisfaction problems to hope to find a solution satisfying the user without offering him the possibility to add, remove and reformulate constraints during the solving process. But our main goal was to prove that our formal tools were able to solve the problem. Cooperation with the user should come later.

Finally, it seemed important to empirically test our theoretical construction. For that purpose we built a prototype upon CoGITaNT. It is not – yet – a usable tool.

The first section is devoted to the theoretical framework. We then propose a modelization of the Sisyphus problem within this framework. In the last section, we present the prototype and the obtained experimental results.

2 The Formal Model

In order that the paper is self-contained, we will recall all definitions and results needed to understand the modelization, except for the very basic notions (namely basic CGs and their logical translation by the semantics Φ). For more details about the simple graph model see [5]. For rules, see [13, 12]. The notion of a constraint is similar to that of [8] and the notion of a constrained derivation is new. Let us specify that all objects and sets considered here are finite.

2.1 Simple Graphs and Projection

Support The basic ontology is encoded in a structure we traditionally call a *support*. We consider here a simplified version of a support, containing:

- T_C, a partially ordered set of concept types whose greatest element is $\top$;
- T_R, a partially ordered set of relation types. For this problem, we only use binary relations. Each relation type has a signature, which gives the greatest possible concept type for each argument. The partial ordering on relation types may decrease the signature (the more specific a relation type, the more restrictive its signature);

- I, a countably infinite set of individual markers. The following partial order is defined on the set $I \cup \{*\}$, where $*$ is the generic marker: $*$ is the greatest element and elements of I are pairwise non-comparable.

Simple Graphs Basic CGs, without negation or nesting, are the basic constructs. They are used as such to represent asserted facts. They are also the basic structure for more complex constructs such as rules and constraints.

Simple graphs may have co-reference links and difference (or non-co-reference) links. A co-reference link between two nodes says these two nodes represent the same entity. On the contrary, a difference link between two nodes says these nodes represent distinct entities.

Individual concept nodes with the same marker implicitly represent the same entity. Formally we say that two nodes are "co-identical" if either they are related with a co-reference link or they have the same individual marker. Co-identity is an equivalence relation on the set of concept nodes of a graph (see [5]). We say that two nodes are "non-co-identical" if either they are related with a difference link or they have different individual markers.

A simple graph is said to be *valid* if the intersection between the co-identity relation and the non-co-identity relation is empty. In other words, they are no nodes being both co-identical and non-co-identical, nor one node non-co-identical with itself.

CoGITaNT does not implement co-reference and difference links in simple graphs, so we actually simulated these links with two relations types, "equal" and "diff". "equal" is provided with rules saying that it is a reflexive, symmetric and transitive relation [1]. "diff" is provided with one rule saying that it is a symmetric relation. The validity of a graph is defined by means of a constraint, which expresses that two entities cannot be "equal" and "diff" at the same time.

Projection Projection is the basic operation on simple graphs. Its definition is recalled for completeness reasons. A *projection* Π from G to H is a mapping (not necessarily one-to-one) from the nodes of G to the nodes of H which:

1. preserves the graph bipartition (it maps relation nodes to relation nodes and concept nodes to concept nodes);
2. preserves adjacency and order on edges (if a concept node c is the i-th neighbor of a relation node r then $\Pi(c)$ is the i-th neighbor of $\Pi(r)$;
3. may decrease node labels (the order on relation types is that of T_R; the order on concept types is the product of the order on T_C and the order on $I \cup \{*\}$).

Note that the empty graph can be projected into any graph. Since we simulate co-reference and difference links by first-class objects (relations, rules and constraints) previous definition does not take these links into account. Otherwise, some conditions should have been added saying that projection has to preserve co-identity and non-co-identity.

Let us recall that the following results hold when considering the classical logical semantics Φ. Projection is sound with respect to deduction in FOL. It is also complete when the graph H is in normal form, i.e. each node is co-identical only to itself. A graph can be put in normal form by merging co-identical nodes (nodes with the same individual markers or related by co-reference links) provided that they have the same type [5].

2.2 Rules and Derivation

Simple Graphs Rules A rule "If G_1 then G_2" is a couple of simple graphs, G_1 and G_2, respectively called hypothesis and conclusion of the rule, which share some concept nodes. See [13] for precise definitions of a rule, its logical translation, forward and backward chaining operations, and logical soundness and completeness results about these operations. We will use here the notations of [1], which provides a graphical visualization of the rules nicer to see.

In this framework, a *rule* is a simple graph provided with a coloring of its nodes with two colors, say 0 and 1. In drawings, 0-colored nodes are painted in white, and the others in grey. To differentiate rules from the constraints defined below, we mark their graphical representation with the symbol ⇥. The subgraph induced by the color 0 nodes must be a syntactically correct simple graph. Nodes with color 0 are the hypothesis nodes and make up the *hypothesis* of the rule. Concept nodes of color 0 with at least one neighbor outside the hypothesis part are the *frontier* nodes. Nodes with color 1 are the conclusion nodes (when the rule is applied to a graph in forward chaining, these nodes are added to the graph) and, together with the frontier nodes, they make up the *conclusion* of the rule.

A rule R is *applicable* to a simple graph G if there is a projection, say Π, from the hypothesis of R to G. In this case, the result of the application of R to G following Π is the simple graph G' obtained from G and the conclusion of R by restricting the label of each frontier node c in the conclusion to the label of its image $\Pi(c)$ in G, then joining c to $\Pi(c)$. In this case, we say that G' is *immediately derived* from (G, R). Let $\mathcal{R}$ be a set of rules. G' is said to be *immediately derived* from $(G, \mathcal{R})$ if there exists a rule R in $\mathcal{R}$ such that G' is immediately derived from (G, R).

Derivation and Deduction A graph G' is said to be *derived* from $(G, \mathcal{R})$ if there exists $G = G_0, G_1, \ldots, G_k = G'$ such that each G_{i+1} is immediately derived from $(G_i, \mathcal{R})$. A simple graph H is said *deducible* from $(G, \mathcal{R})$ if there exists a derivation from $(G, \mathcal{R})$ to a graph G' such that H can be projected into G'.

Note that when a rule is applicable to a graph following a projection, it can be indefinitely applied to the resulting graph following the same projection, but all graphs obtained are equivalent to the graph built by the first application. In what follows a rule is applied *only once* to a graph following a given projection. A graph G is said to be *closed* w.r.t. a set of rules $\mathcal{R}$ if all information that can be added by a rule is already present in G (i.e. more formally, for each rule $R \in \mathcal{R}$, each projection from the hypothesis of R into G can be extended to a projection from R as a whole into G). The problem of deciding whether a given graph is deducible from a given knowledge base is a semi-decidable problem. This problem becomes decidable for some specific sets of rules. In particular it is the case when the set of rules is such that every graph can be closed by a finite number of rule applications. In such a case the set of rules is said to be a *finite expansion set*. A *closure* of a simple graph G by a finite expansion set of rules $\mathcal{R}$ is a simple graph derived from $(G, \mathcal{R})$, closed w.r.t. $\mathcal{R}$. All closures are equivalent with respect to projection, and the minimum element of this equivalence class is called *the* closure of G w.r.t. $\mathcal{R}$ and is denoted by $G^*_{\mathcal{R}}$. Fig. 1 presents a graph G

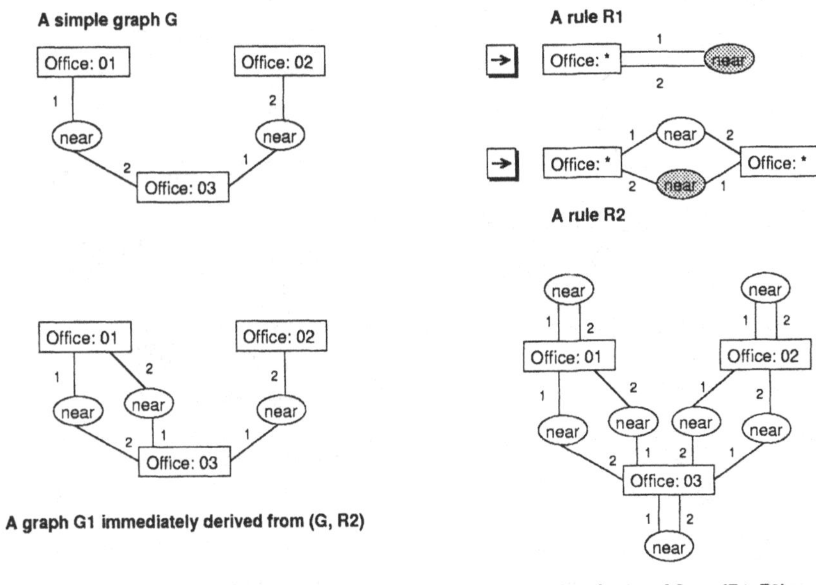

Fig. 1. Applying a rule and closing a graph

expressing that the office #1 is near the office #3 which is itself near the office #2, a rule R_1 expressing that the relation *near* is reflexive, a rule R_2 expressing that the relation *near* is symmetrical, a graph G_1 immediately derived from (G, R_2), and the closure $G^*_{\{R_1 \cup R_2\}}$ of G w.r.t. the rules R_1 and R_2.

For the Sisyphus problem, we use rules in which all concept nodes belong to the hypothesis (in other words the application of a rule only add relation nodes, including the "equal" and "diff" relations). Any set of such rules is obviously a finite expansion set.

2.3 Adding Constraints to the Model

Positive and Negative Constraints We define two kinds of *constraints*: *positive* constraints (very similar to the ones presented in [8]) and *negative* constraints. In the same way as for rules, we will mark positive constraints by the symbol ⊞ and negative constraints by the symbol ⊟. The intuitive semantics that can be attached to these constraints are respectively: "whenever we find the information A in the graph, we must also find the information B", and "whenever we find the information A in the graph, we must not find the information B". Though they have different semantics, their formal definition is identical. A constraint is defined in the same way as a rule, as a simple graph provided with a coloration of its nodes with the two colors 0 and 1. The subgraph generated by nodes whose color is 0 is called the *condition* of the constraint. The condition must be a valid simple graph. In particular, the condition can be an *empty graph*.

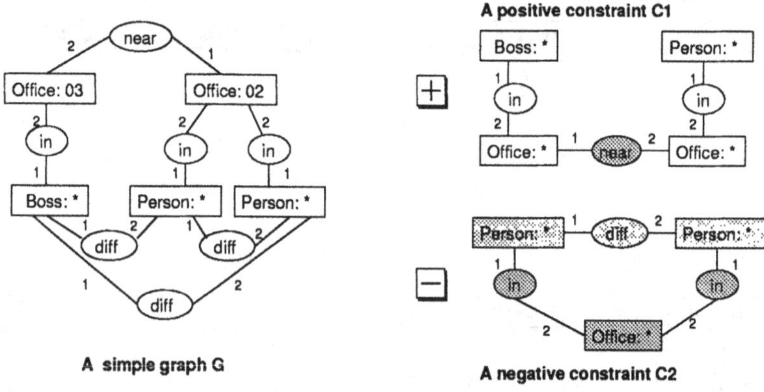

Fig. 2. Example of positive and negative constraints

Let G be a simple graph, and C be a positive constraint. We say that G *satisfies* C if every projection Π of the condition of C into G can be extended to a projection of C as a whole.

Let G be a simple graph, and C be a negative constraint. We say that G *satisfies* C if no projection Π of the condition of C into G can be extended to a projection of C as a whole.

Validity of a Graph Let G be a simple graph, $\mathcal{C}$ be a set of positive and negative constraints. The graph G is said to be *valid* with respect to $\mathcal{C}$ if G satisfies every constraint of $\mathcal{C}$.

The positive constraint C_1 in fig. 2 can be read as "if a boss is in an office and a person is in an office, then these offices must be *near*". The negative constraint C_2 in the same figure can be read as "No office can host two different persons". The graph G in the same figure violates the two constraints :

– For each of the two possible projections of the condition of C_1 into G, we cannot extend the projection to "a good orientation" of the relation *near*.

– The condition of the negative constraint C_2 is the empty graph, which can be projected into G. And this constraint is violated since it exists a projection of C_2 as a whole into G.

Validity of a Graph Given Implicit Knowledge Rules can be used to factorize information, in this case we say that they represent *implicit knowledge*. For the SCG-1 problem, we will restrict these rules to finite expansion sets.

Let G be a simple graph, $\mathcal{R}$ be a finite expansion set of rules representing implicit knowledge, and $\mathcal{C}$ be a set of positive and negative constraints. We say that the knowledge base $(G, \mathcal{R})$ is valid with respect to $\mathcal{C}$ if and only if the closure $G^*_{\mathcal{R}}$ of G is valid with respect to $\mathcal{C}$.

As an example, let G be the graph defined in fig. 2, $\mathcal{R}$ be the set of rules defined in fig. 1, and $\mathcal{C}$ be the set of constraints defined in fig. 2. $(G, \mathcal{R})$ is not valid with respect to $\mathcal{C}$ since $G^*_{\mathcal{R}}$ violates the negative constraint C_2. But also note that $G^*_{\mathcal{R}}$ satisfies now the positive constraint C_1.

Constrained Derivation Consider that *asserted facts* (simple graphs) and *implicit knowledge* rules describe a *world*. Constraints determine whether this world is valid or not. Let us now introduce another set of rules, called *transformation rules*. These rules are used to generate new worlds, which may be valid or not. Given a valid *initial world* $(G, \mathcal{R})$, a set of constraints $\mathcal{C}$ and a set of transformation rules $\mathcal{T}$, answering a request H consists in building a sequence of successive valid worlds issued from the initial one, such that the last one is an answer to H.

Let us give a formal definition of such a derivation.

Let G be a simple graph, $\mathcal{R}$ be a set of rules representing implicit knowledge, $\mathcal{C}$ be a set of positive and negative constraints, and $\mathcal{T}$ be a set of *transformation rules*. We say that a graph G' is a *valid immediate transformation* of G with respect to $(\mathcal{R}, \mathcal{C}, \mathcal{T})$ if and only if:

- $(G, \mathcal{R})$ is valid with respect to $\mathcal{C}$ (i.e. $G^*_{\mathcal{R}}$ is valid in the case of $\mathcal{R}$ being a finite expansion set).
- G' is immediately derived from $(G^*_{\mathcal{R}}, \mathcal{T})$.
- $(G', \mathcal{R})$ is valid with respect to $\mathcal{C}$.

We say that a graph G' is a *valid transformation* of G with respect to $(\mathcal{R}, \mathcal{C}, \mathcal{T})$ if and only if there exists a sequence $G = G_0, G_1, \ldots, G_k = G'$ of graphs such that each G_{i+1} is a valid immediate transformation of G_i with respect to $(\mathcal{R}, \mathcal{C}, \mathcal{T})$. We also say that this sequence is a *constrained derivation* from G to G'.

We say that a simple graph H is *deducible* from $(G, \mathcal{R}, \mathcal{C}, \mathcal{T})$ iff there exists a constrained derivation from G to a graph G' such that H can be projected into G'.

By example, let G be the graph defined in fig. 3, $\mathcal{R}$ be the set of rules defined in fig. 1, $\mathcal{C}$ be the constraints defined in fig. 2, $\mathcal{T}$ be the set of transformation rules containing only the rule R in fig. 3, expressing that we can assign an office

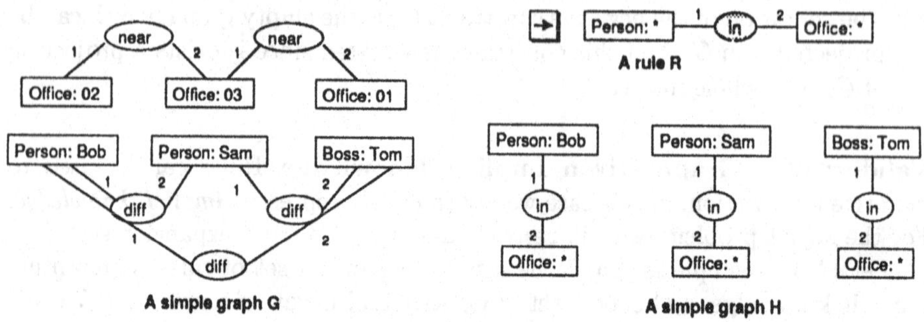

Fig. 3. A graph G, a transformation rule R, and a request H

to each person, and the request H be the graph defined in fig. 3. This request expresses that we want every person to be placed into an office. In order to solve the problem, we first compute the closure of G w.r.t. $\mathcal{R}$, which is a valid graph, then apply the transformation rule once, say by placing *Bob* into the third office. Now, we verify that this graph is valid, and try another assignment, until the request is satisfied. Fig. 4 traces the applications of the transformation rules leading to a correct answer.

Tom → 1
- Sam → 1 Violation: two in the same office
- Sam → 2 Violation: far from boss
- Sam → 3
  - Bob → 1 Violation: two in the same office
  - Bob → 2 Violation: far from boss
  - Bob → 3 Violation: two in the same office

Tom → 2
- Sam → 1 Violation: far from boss
- Sam → 2 Violation: two in the same office
- Sam → 3
  - Bob → 1 Violation: far from boss
  - Bob → 2 Violation: two in the same office
  - Bob → 3 Violation: two in the same office

Tom → 3
- Sam → 1
  - Bob → 1 Violation: two in the same office
  - Bob → 2 **Solution found**

Fig. 4. A trace of the backtrack leading to the solution

3 Representing the Problem

In this section, we present the support, fact graphs, implicit knowledge rules, constraints, transformations rules and the request we used to model and solve the Sisyphus-I problem.

3.1 The Support

The support defined in fig. 5 and 6 represents all types used in our modelization. Relation types "equal" and "diff" represent co-identity and its negation. Each relation type in fig. 6 is provided with its signature.

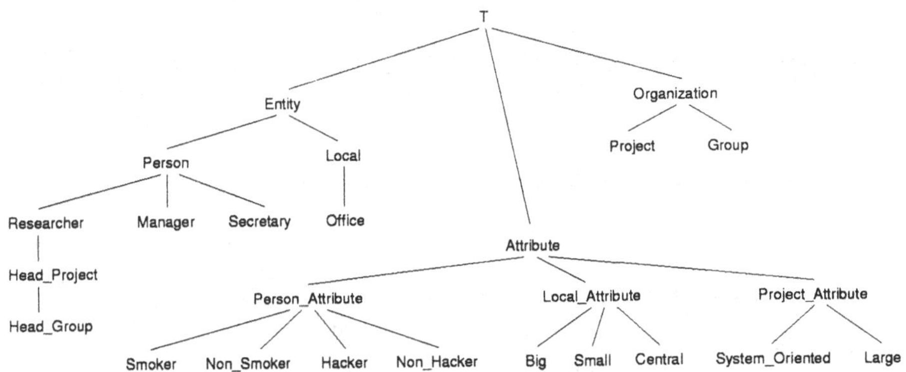

Fig. 5. Hierarchy of concept types

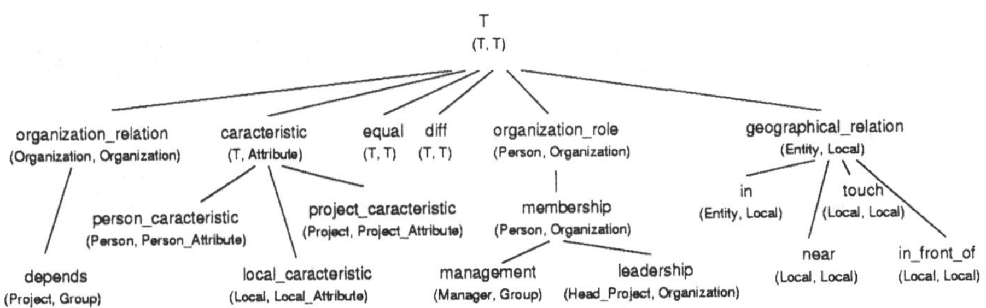

Fig. 6. Hierarchy of relation types

3.2 Fact Graphs

The initial graph is obtained by making the disjoint union of the graphs represented in figs. 7, 8, 9 and 10. These graphs have been separated for better readability. To ensure completeness of our computation, we had the choice between using the rules assigned to the *equal* relation, or to compute the *normal form* of the resulting graph, by merging concept nodes having the same individual marker. The latest solution was chosen for a performance purpose. To present more readable graphs, we did not represent *diff* relations in these graphs. In the graph of fig. 7, all pairs of concept nodes typed *Office* and having different individual markers are linked by a relation node typed *diff*. The same assumption is done for the concept nodes typed *Project* in graph of fig. 8, and for concept nodes whose type is more specific than *Person* in graph of fig. 10.

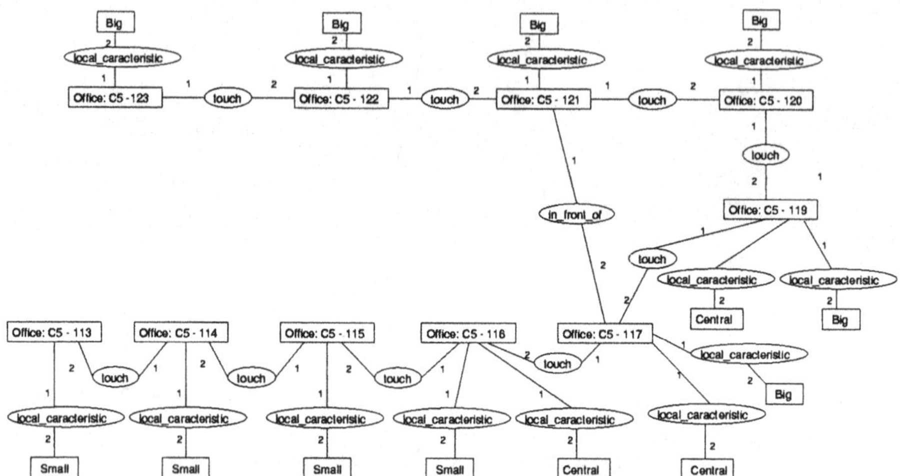

Fig. 7. Geographical information for the first floor of the château of HNE

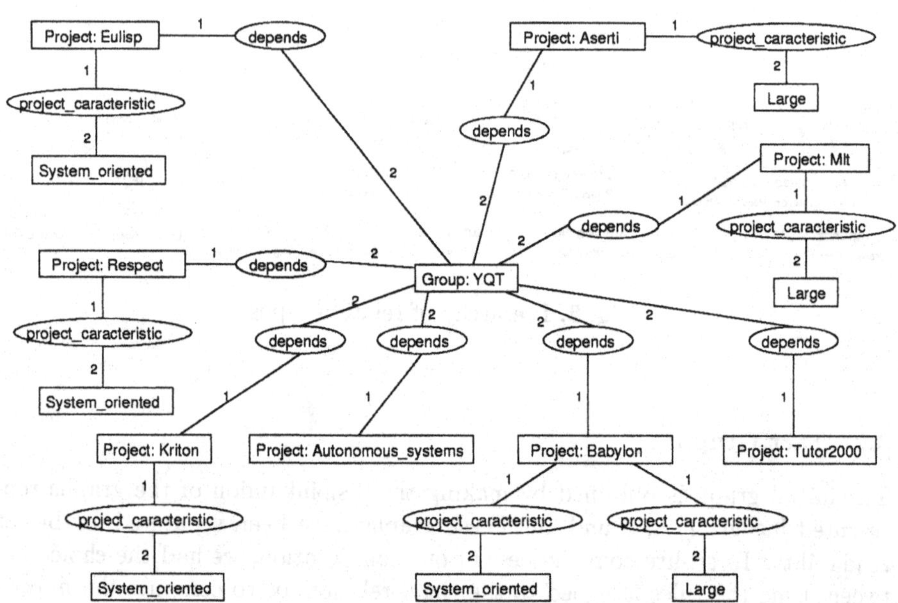

Fig. 8. Organizational structure of the YQT group

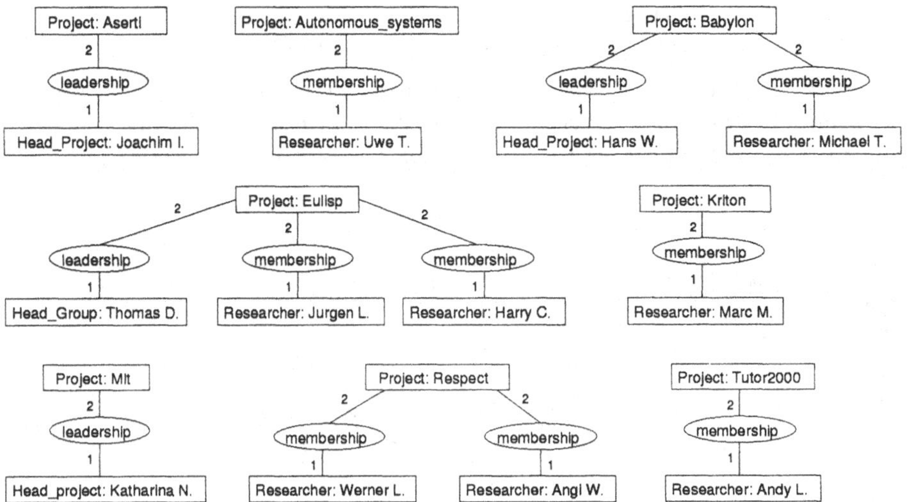

Fig. 9. Team members of the YQT group

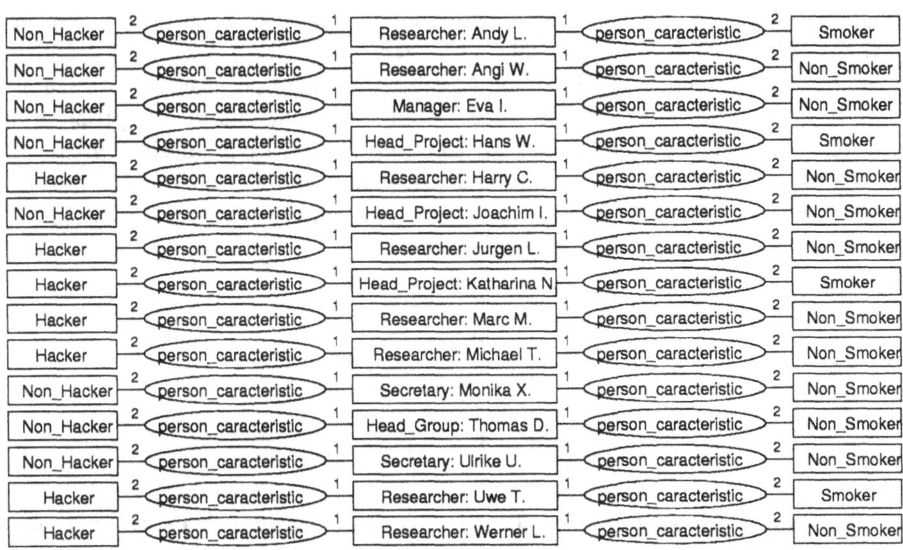

Fig. 10. Personal data about members of the YQT group

3.3 Implicit Knowledge – the Set of Rules $\mathcal{R}$

The five rules represented in fig. 11 express that:

A. The relation *near* is symmetrical.
B. Two locals that *touch* the same one are considered *near*.
C. The relation *touch* is symmetrical.

D. Two locals that *touch* each other are considered *near.*
E. The same for locals being *in_front_of* each other.

Note that the two last rules could be replaced by expressing that *in_front_of* and *touch* are two relation types more specific than *near.* We did not represent, but use the rules expressing that *in_front_of* and *diff* are symmetrical.

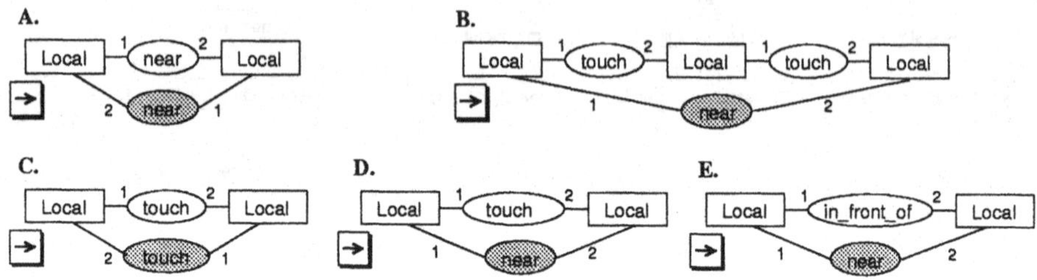

Fig. 11. Rules representing implicit knowledge

3.4 Constraints – the Set $\mathcal{C}$

The constraints related to possible assignments can be either positive or negative. We also sorted these constraints in three categories of decreasing priority: absolute, strong and weak. It is not possible to violate an *absolute* constraint. *Strong* constraints are to be satisfied by any solution, but, if the problem is over-constrained, the user may accept to reformulate it by modifying this set of constraints. *Weak* constraints represent preferences.

Every constraint used is described below, but whenever we have very similar constraints, only one of them is represented.

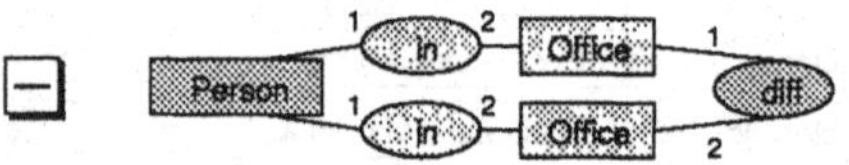

Fig. 12. Ubiquity ?

The graph represented in fig. 12 is a negative constraint. It expresses that a person cannot be into two different places at the same time. This is an *absolute* constraint.

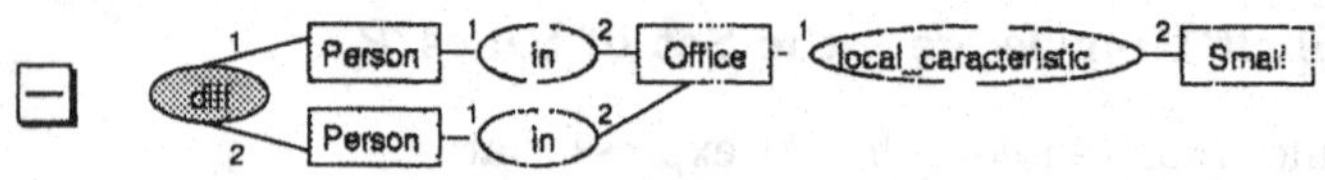

Fig. 13. Number of persons in small offices

The graph represented in fig. 13 is a negative constraint. It expresses that a small office cannot host more than one person. This is a *strong* constraint.

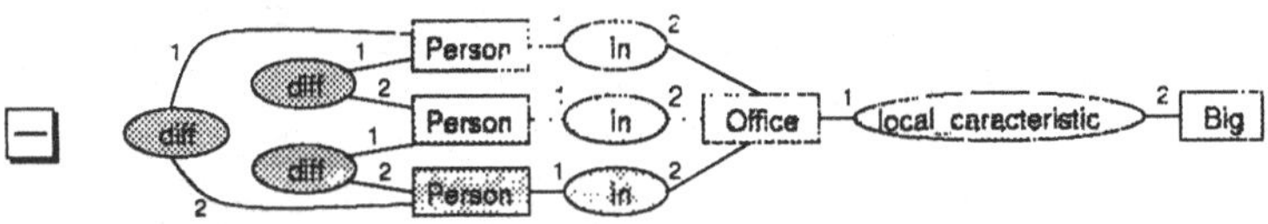

Fig. 14. Number of persons in big offices

The graph represented in fig. 14 is a negative constraint. It expresses that a big office cannot host more than two persons. This is a *strong* constraint.

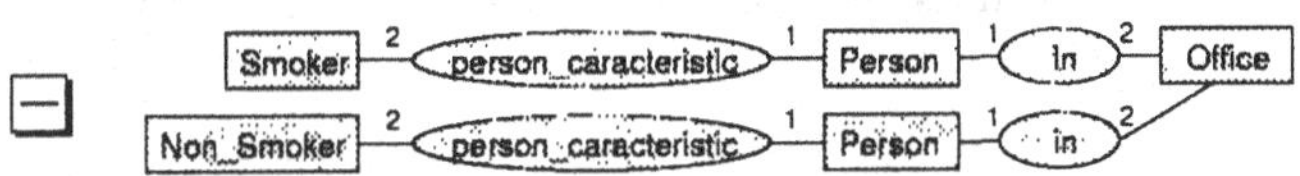

Fig. 15. Smoker – Non-Smoker antagonism

The graph represented in fig. 15 is a negative constraint. It expresses that no office can host a *Smoker* and a *Non_Smoker*. This is a *strong* constraint.

Fig. 16. Head of group accessibility

The graph represented in fig. 16 is a positive constraint. It expresses that the head of group needs a central office (if the head of group is in an office, then this office has to be a central one). This is a *strong* constraint. In the same way, the manager would be pleased to have a central office, but we consider this as only a *weak* constraint.

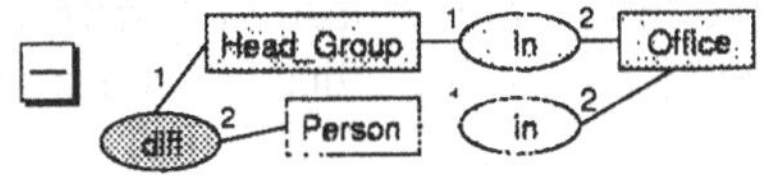

Fig. 17. Privileges of the head of group (1)

The graph represented in fig. 17 is a negative constraint. It expresses that the head of group needs to be alone in his office. This is a *strong* constraint. In the same way, the manager as well as heads of large projects would be pleased to be alone in their office, but we consider this as only *weak* constraints. We also added a positive constraint expressing that the head of a large project should be in a small office. This *weak* constraint is a consequence of the others, and is present for optimization purposes.

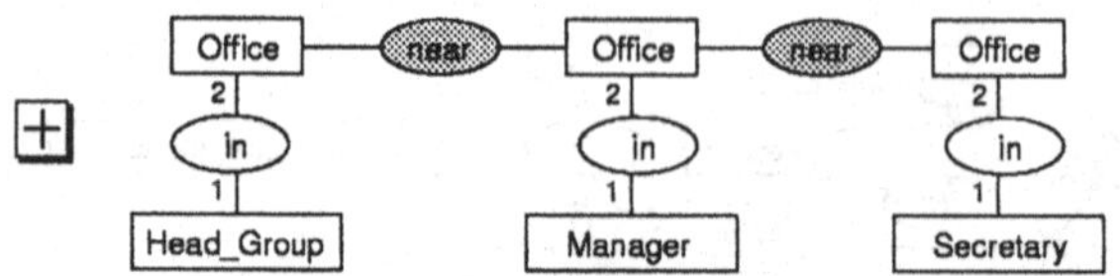

Fig. 18. Manager's neighborhood

The graph represented in fig. 18 is a positive constraint. It expresses that the manager needs to be near the head of group as well as the secretariat. This is a *strong* constraint. In the same way, heads of large projects should be close to the head of group as well as the secretariat, but we only consider this as only a *weak* constraint.

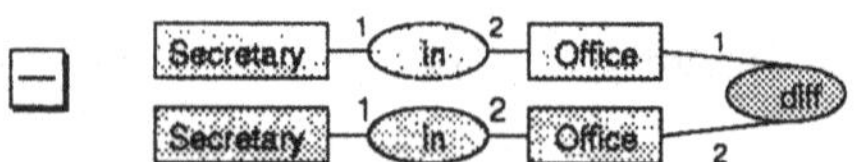

Fig. 19. Secretariat holds secretaries

The graph represented in fig. 19 is a negative constraint. It expresses that no two secretaries can be in different offices. Note that this constraint can be satisfied since there are only two secretaries. Otherwise, should we want "secretaries-only" offices, we could express it by a positive constraint: "if a person is in the same office as a secretary, this person must also be a secretary". This is a *weak* constraint.

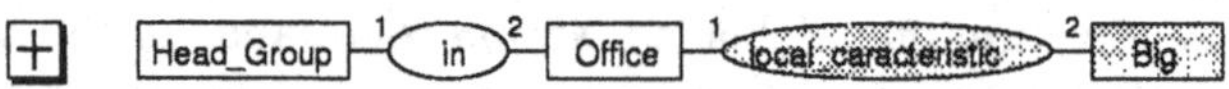

Fig. 20. Privileges of the head of group (2)

The graph represented in fig. 20 is a positive constraint. It expresses that the head of group would like to have a big office. This is a *weak* constraint.

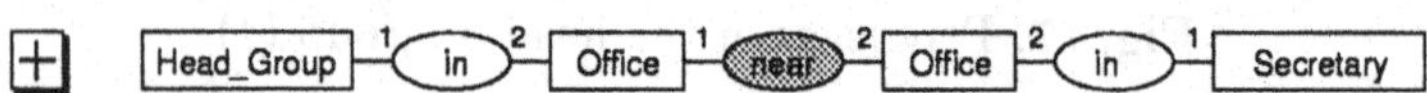

Fig. 21. Secretariat's accessibility

The graph represented in fig. 21 is a positive constraint. It expresses that the secretaries' office should be located close to the office of the head of group. This is a *weak* constraint.

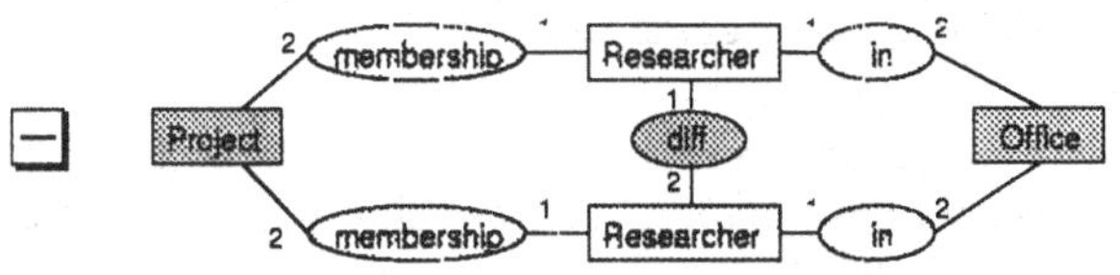

Fig. 22. Synergy

The graph represented in fig. 22 is a negative constraint. It expresses that members of a same project should not share offices. This is a *weak* constraint.

3.5 Transformations – the Set of Rules $\mathcal{T}$

There is only one rule adding information to the base graph, it is the rule represented in fig. 23 which tries to assign a given person into a given office.

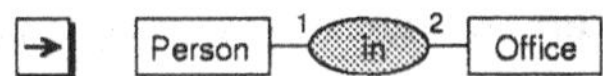

Fig. 23. Placing a person into an office

3.6 The Request

The graph we want to deduce is the solution to the SCG-1 problem: we want every person being placed into an office. This graph is represented in fig. 24.

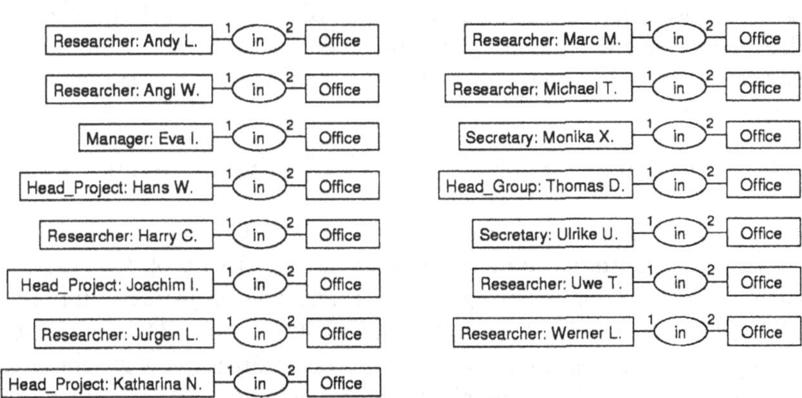

Fig. 24. The request

4 The Evaluation Process

This section first presents some considerations about the computational complexity of the solving process. A deeper study of this complexity is yet to be done. We then present the implementation of the *Constrained Derivation Engine* on top of CoGITaNT and results of the computation.

4.1 Combinatorial Considerations

The existence of a projection is a NP-complete problem [6]. The existence of a deduction from a simple graph G and a set of rules $\mathcal{R}$ is a semi-decidable problem [7] (by analogy with TGDs). In case of a finite expansion set of rules, the problem is obviously decidable. In particular, when the rules only have relation nodes in their conclusion, the problem is NP-complete.

The problem of knowing if a graph is valid with respect to a set of constraints is a co-NP-complete problem (as we need to exhibit two projections in order to prove that a constraint is violated).

If the set of rules $\mathcal{R}$ representing implicit knowledge is a finite expansion set, then the problem of deduction with constrained derivation is semi-decidable. If the set of rules $\mathcal{T}$ representing the possible transformations of the world is also a finite expansion set, then the problem is decidable. This is the case in our modeling of the Sisyphus-I problem.

We consider now the tree representing the possible applications of the transformation rule. For each step, we have to choose between *# of persons* × *# of offices = 150* possible assignments of a person in an office. There are *# of persons = 15* such steps, and this lead us to the study of the validity of 150^{15} different worlds. Even if we consider that violations of the constraints will prune many branches of this tree, this is not reasonable.

So we have chosen, for this problem only, to force a particular person to be assigned at each step, and this choice leads to *only* 10^{15} possible worlds. We also forced these assignments to follow a specific order, which is more or less the one suggested by the wizard Siggi D. These restrictions, however incomplete in a general case, are well-founded in the case of our modelization of the SCG-1 problem. We are currently studying the possibility of using heuristics to compute automatically the best order with respect to the rules and constraints involved.

4.2 Implementation of the Constrained Derivation engine

We developed constrained derivation on top of the CoGITaNT platform [10]. Our work is yet a prototype library and not a definitive tool. CoGITaNT can be seen as a tool for manipulating graphs, computing projections and applying rules, all graphs involved being read in the BCGCT format [11]. As can be seen in fig. 25, our work is based upon a client–server architecture. A graphical editor has been designed, which communicates with the *Editor Server* built on top of CoGITaNT. This tool allows the user to edit and modify simple graphs, rules and constraints which form the knowledge base located on a distant server.

We have also added on top of CoGITaNT a constrained derivation engine, which is not yet provided with the necessary optimizations. This engine is intended to communicate with a user-friendly *Problem Manager* client, which will enable the user to select rules and constraints that he/she needs for a particular problem solving and visualize a step by step evolution of the base graphs by interacting with the graph editor. This *Problem Manager* is not yet ready, and we had to hard-code the SCG-1 problem on top of the constrained derivation engine.

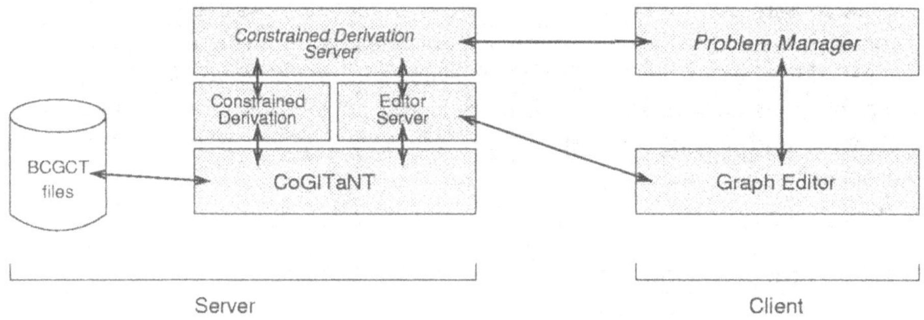

Fig. 25. System architecture

In order to develop a generic constrained derivation engine, we defined several new classes built on top of the the C++ classes of the CoGITaNT library. These new classes are a straightforward implementation of the model presented in section 2, and no particular work has yet been done to optimize the research process.

- The class `rule`, which provides methods such as the application of a rule on a graph given a projection, has been updated to compute the closure of a graph with respect to a single rule. The class `set_of_rules` mainly provides a calculus of the closure of a graph by a (finite expansion) set of rules.
- The class `constraint` represents positive and negative constraints and a method of this class verifies whether a given graph satisfies the constraint. The class `set_of_constraints` determines whether a given graph is valid with respect to this set of constraints.

These classes are used to define the core of the engine. The class `cd_problem` allows the user to add to the problem the BCGCT files containing the fact graph[1], rules representing implicit knowledge, constraints, transformation rules and the request. Each graph is added to the problem in the same way, we just have to indicate the nature of the graph contained in the file such as "negative weak constraint" or "transformation rule". Once these graphs have been loaded, the method `execute` can be called and computes *one* or *every* solution.

Of course, this engine can be easily used on the modeling described in section 3. The `scg1_problem` class, inheriting from `cd_problem` benefits from the optimization presented in section 4.1.

To take into account the different priorities assigned to constraints, we used the following algorithm. We sorted all constraints used along a total order such

[1] A single graph or several graphs can be loaded. In the latter case, the fact graph is automatically computed as the normal form of the disjoint sum of these graphs.

that we have ***weak*** constraints first, then ***strong*** constraints, then ***absolute*** constraints. If we find a solution satisfying every constraint, then we return this solution. Otherwise, we remove the weakest constraint from C and try to find a solution to this modified problem. If we can find a solution by removing only weak constraints, we answer "yes, there is a solution if we remove the following constraints ...". If the only solution is obtained by removing at least one strong constraint, then we answer: "no, there is no solution unless you accept to remove the following constraints ...". If there is still no solution after having removed all weak and strong constraints, then we answer "there is no solution to the problem".

4.3 Solutions Found

The research engine found 2880 solutions to the SCG-1 problem. These solutions satisfy all constraints excepted a weak one which expresses that heads of large projects should be next to the secretariat and the head of group. Obviously, this constraint cannot be satisfied. Given our modeling, the number of solutions can be explained as follows:

- The head of group can only be in the office 117 or 119 ***(2 solutions)***
- The manager can only be placed in a small central office, i.e. the office 116. ***(1 solution)***
- The positions of the head group and manager determine the position of the secretariat (the office which has not been assigned to the head of group) ***(1 solution)***
- There are only three small offices left, and they must be assigned to the three heads of large projects, which should be alone in their office ***(6 solutions)***
- The two smokers must be together and have the choice between the 4 big offices left ***(4 solutions)***
- There are 15 possible sets of couples for the 6 researchers left, of them 5 are impossible (they are in the same project). There are 6 possible assignments for these couples in the 3 last big offices ***(60 solutions)***

Finally, we have the $2 \times 1 \times 1 \times 6 \times 4 \times 60 = 2880$ different solutions exhibited by our research engine. Two of the computed solutions are presented in fig. 26. The first solution we find is the solution A in fig. 26. The beginning of the research tree used to find this solution is presented in fig. 27. We can see there one backtrack, as the first secretary is assigned a small office (no explicit constraint forbids it), and problems arise only when we try to assign the same small office to the second secretary. We explored 85 different worlds before generating the first valid solution (so we had 70 backtracks), and we must remind here that each of these worlds required to solve the problem of its validity, which is a co-NP-complete problem. However, the length of the computation is counted in seconds.

Office	Solution A
113	Hans W. (Head of Large Project)
114	Katharina N. (Head of Large Project)
115	Joachim I. (Head of Large Project)
116	Eva I. (Manager)
117	Thomas D. (Head of group)
119	Monika X. Ulrike U. (Secretaries)
120	Andy L. Uwe T. (Researchers, smokers)
121	Michael T. Mark M. (Researchers, non smokers)
122	Jurgen L. Werner L. (Researchers, non smokers)
123	Angi W. Harry C. (Researchers, non smokers)
Office	**Solution B**
113	Katharina N. (Head of Large Project)
114	Hans W. (Head of Large Project)
115	Joachim I. (Head of Large Project)
116	Eva I. (Manager)
117	Monika X. Ulrike U. (Secretaries)
119	Thomas D. (Head of group)
120	Angi W. Mark M. (Researchers, non smokers)
121	Harry C. Werner L. (Researchers, non smokers)
122	Jurgen L. Michael T. (Researchers, non smokers)
123	Andy L. Uwe T. (Researchers, smokers)

Fig. 26. Two solutions to the SCG-1 problem

```
Thomas D. → 113 Violation: Not Central
Thomas D. → 114 Violation: Not Central
Thomas D. → 115 Violation: Not Central
Thomas D. → 116 Violation: Small Office
Thomas D.→ 117
        Eva I. → 113 Violation: Not Central
        Eva I. → 114 Violation: Not Central
        Eva I. → 115 Violation: Not Central
        Eva I. → 116
                Monika X. → 113 Violation: Far from Manager
                Monika X. → 114 Violation: Far from Manager
                Monika X. → 115
                        Ulrike U. → 113 Violation: Far from Manager
                        Ulrike U. → 114 Violation: Not with other Secretary
                        Ulrike U. → 115 Violation: Two persons in a small office
                        Ulrike U. → 116 Violation: Two persons in a small office
                        Ulrike U. → 117 Violation: Head of group must be alone
                        Ulrike U. → 119 Violation: Not with other Secretary
                        [ ... ] (120, 121, 122, 123) Violation: Not with other Secretary
                Monika X. → 116 Violation: Two persons in a small office
                Monika X. → 117 Violation: Head of group must be alone
                Monika X. → 119
                        Ulrike U. → 113 Violation: Far from Manager
                        Ulrike U. → 114 Violation: Not with other Secretary
                        Ulrike U. → 115 Violation: Not with other Secretary
                        Ulrike U. → 116 Violation: Two persons in a small office
                        Ulrike U. → 117 Violation: Head of group must be alone
                        Ulrike U. → 119
                                Hans W. → ...
```

Fig. 27. A trace of the backtrack leading to the solution *A*

4.4 Coping with Changes in Data

In order to cope with the slightly modified data set, we only have to modify the base graph and the request for our problem.

- The graph represented in fig. 9 (team membership) is modified by removing `[Head_Project:Katharina N.]→(leadership)→ [Project:Mlt]`, replacing it by `[Researcher:Christian I.]→(membership)→[Project:Mlt]`
- The graph represented in fig. 10 (personal information) is modified by removing the connected component containing the concept node whose marker is `Katharina N.` and replacing it by: `[Hacker]←(person_caracteristic)← [Researcher:Christian I.]→(person_caracteristic)→[Smoker]`.
- The request presented in fig. 24 is modified in such a way that the concept node `[Head_Project:Katharina N.]` is replaced by the concept node `[Researcher:Christian I.]`.

As we have less constraints about Christian I. than about Katharina N., there are now 8640 different solutions to the problem (once again, according to our modelization). As in the first problem, the last weak constraint cannot be satisfied.

If the objective is to modify as few as possible the existing solution, one can proceed in the following way. Weak constraints can be added, expressing the fact that everybody (or some persons) would like to stay at the same place. This is done by transforming each or some of the previous assignments in a negative constraint, such as the one represented in fig. 28. These constraints express that a person should not be in an office different from the one he/she has already been assigned to.

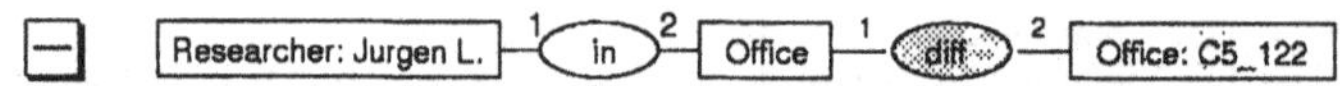

Fig. 28. Minimum displacement constraint (case of Jurgen L.)

By introducing all such constraints in our problem (nobody want to change), it is not surprising that the only one solution found replaces Katharina N. by Christian I.

Conclusion

The fundamental objective of the Corali project [3] is to develop conceptual graphs as a graphical knowledge representation model (in the sense given in the introduction of this paper). The research works are based on a four-stroke experimental methodology: build a theoretical formal model, build software tools implementing the formal model, use the two preceding points to build real-world applications, then evaluate the systems built, and loop through this four-step process.

This paper presents a *generic* way of solving the Sisyphus-I room allocation problem, in the sense that the modeling framework introduced is general enough to enable the representation of any resource allocation problem. This framework is based on a *graphical* formal model. Basic constructs are simple conceptual graphs and the basic operation for doing reasonings is projection. Two more complex constructs are defined in terms of simple graphs: rules and constraints. They are processed by operations based on projection. To summarize the different kinds of knowledge represented by these constructs, we can say that :

- asserted *facts* provided with *implicit knowledge* about the domain define *worlds*;
- *constraints* express conditions for a world to be considered as valid;
- *transformation rules* define possible changes in worlds;
- *the request* represents a question.

Answering a request consists in finding a *constrained derivation* from the initial world to one satisfying the request.

Using the modeling constructs to produce a modelization of the given problem is done in a rather "natural" way: simple graphs assert facts using an initial vocabulary limited to primitive types (concepts and relations) and individuals, rules represent implicit knowledge resulting from the asserted facts, constraints translate obligations and interdictions for possible solutions of the problem in a rather straightforward way, and transformation rules describe the way solutions are constructed.

Let us add that all CG applications we were involved in seem to confirm that the formal constructs are really understandable and usable by an end-user. Graphs in their graphical form are easy to read and operations are not difficult to understand because they are "matching" operations and can be graphically represented.

Dealing with the modified set of data was straightforward: we only had to slightly modify asserted facts and the request. Using constraints we were also able to specify which existing room assignments should be kept in the new solutions.

A *prototype* implementing this framework has been built upon CoGITaNT. It has been used to test our modelization of the Sisyphus-I problem. Constraints have been classified into three clusters of different priority, allowing more flexibility in the answers given by the system. This prototype is not yet a really usable tool. Further developments are needed in order to enable communication with the user in a friendly way, give the user the possibility to intervene at several stages of the solving process, and improve computational complexity of the solving process.

References

1. J.-F. Baget. A simulation of co-identity with rules in simple and nested graphs. In *Proceedings of the 7th ICCS*, 1999.
2. C. Bos, B. Botella, and P. Vanheeghe. Modeling and simulating human behaviors with conceptual graphs. In *Conceptual Structures: Fulfilling Peirce's Dream, ICCS'97 Proc., LNAI 1257*, pages 275–289. Springer Verlag, 1997.
3. B. Carbonneill, M. Chein, O. Cogis, O. Guinaldo, O. Haemmerlé, M.L. Mugnier, and E. Salvat. The COnceptual gRAphs at LIrmm Project. In *Proc. of the first CGTOOLS workshop*, pages 5–8, 1996.
4. M. Chein. The CORALI project: From conceptual graphs to conceptual graphs via labelled graphs. In *Conceptual Structures: Fulfilling Peirce's Dream, ICCS'97 Proc., LNAI 1257*, pages 65–79. Springer Verlag, 1997.
5. M. Chein, M.-L. Mugnier, and G. Simonet. Nested Graphs: a Graph-based Knowledge Representation Model with FOL semantics. In *Proc. KR'98*, pages 524–534. Morgan Kaufmann, 1998.
6. M. Chein and M.L. Mugnier. Conceptual Graphs: Fundamental Notions. *Revue d'Intelligence Artificielle*, 6(4):365–406, 1992.
7. S. Coulondre and E. Salvat. Piece Resolution: Towards Larger Perspectives. In *Conceptual Structures: Theory, Tools and Applications, ICCS'98 Proc., LNAI 1453*, pages 179–193. Springer Verlag, 1998.
8. J. Dibie, O. Haemmerlé, and S. Loiseau. A Semantic Validation of Conceptual Graphs. In *Conceptual Structures: Theory, Tools and Applications, ICCS'98 Proc., LNAI 1453*, pages 80–93. Springer Verlag, 1998.
9. D. Genest and M. Chein. An experiment in Document Retrieval Using Conceptual Graphs. In *Conceptual Structures: Fulfilling Peirce's Dream, ICCS'97 Proc., LNAI 1257*, pages 489–504. Springer Verlag, 1997.
10. D. Genest and E. Salvat. A Platform Allowing Typed Nested Graphs: How CoGITo Became CoGITaNT. In *Conceptual Structures: Theory, Tools and Applications, ICCS'98 Proc., LNAI 1453*, pages 154–161. Springer Verlag, 1998.
11. O. Haemmerlé. *CoGITo : une plate-forme de développement de logiciels sur les graphes conceptuels*. PhD thesis, Université Montpellier II, 1995.
12. E. Salvat. Theorem proving using graph operations in the conceptual graph formalism. In *Proceedings of the 13th European Conference on Artificial Intelligence (ECAI'98), Brighton, UK*, 1998.
13. E. Salvat and M.L. Mugnier. Sound and complete forward and backward chainings of graph rules. In *Conceptual Structures: Knowledge Representation as Interlingua, ICCS'96 Proc., LNAI 1115*, pages 248–262. Springer, 1996.
14. L.K. Schubert. Semantic Networks are in the Eye of the Beholder. In J. F. Sowa, editor, *Principles of Semantic Networks*, pages 95—108. Morgan Kaufmann, 1991.

Contextual Attribute Logic

Bernhard Ganter and Rudolf Wille

Technische Universität Dresden, Institut für Algebra
D-01062 Dresden,
ganter@math.tu-dresden.de
Technische Universität Darmstadt, Fachbereich Mathematik
D-64289 Darmstadt,
wille@mathematik.tu-darmstadt.de

Abstract. *Contextual Attribute Logic* is part of Contextual Concept Logic. It may be considered as a contextual version of the Boolean Logic of Signs and Classes. In this paper we survey basic notions and results of a Contextual Attribute Logic. Main themes are the clause logic and the implication logic of formal contexts. For algorithmically computing bases of those logics, a common theory of cumulated clauses is presented.

1 Contextual Concept Logic

"Contextual Logic" has been introduced with the aim to support knowledge representation and knowledge processing (cf. [17], [18], [14]). It is developed by mathematizing the doctrines of concepts, judgments, and conclusions as they have evolved in philosophical logic since the 16th century. According to I. Kant, Elementary Logic has to be "the theory of the three essential main functions of thinking - *concepts*, *judgments*, and *conclusions*" ([10], p.6), since human thinking is based on concepts as basic units of thought, on judgments as combinations of concepts, and on conclusions as entailments between judgments. In the actual development of Contextual Logic, the doctrine of concepts is mathematized by using *Formal Concept Analysis* [9], and the doctrines of judgments and conclusions by incorporating the *Theory of Conceptual Graphs* [So84], [16]. In this way, Contextual Logic shall be worked out in three parts, a *"Contextual Concept Logic"*, a *"Contextual Judgment Logic"*, and a *"Contextual Conclusion Logic"*.

As fundamentals of a Contextual Concept Logic, we survey in this paper basic notions and results of a *"Contextual Attribute Logic"* that may be considered as a contextual version of the Boolean Logic of Signs and Classes. In his book *"An Investigation of the Laws of Thought"* [3], G. Boole grounded his analysis of human reasoning on symbols and signs representing things and classes of things (objects) as subjects of human conceptions, and on signs of operations standing for those operations of language by which conceptions of things are combined or resolved so as to form new conceptions. Although Boole considered the limits of a discourse or process of reasoning, he introduced a class, called "the Universe" and represented by **1**, consisting of all the individuals that exist in any class. Much later, in particular, when computers were used for natural-language processing, it became clear that such a universal approach is not fully appropriate, because it does not support the fact that human conceptions depend on context (cf. [4]). We therefore suggest a contextual approach that also respects these limits of formal representations.

	even	odd	prime	square	sum2sq
1		×		×	
2	×		×		×
3		×	×		
4	×			×	
5		×	×		×
6	×				
8	×				×
15		×			
25		×		×	×
45		×			×
100	×			×	×

Fig. 1. Some properties of numbers. sum2sq stands for "is the sum of two squares of natural numbers".

Contextual Concept Logic is grounded on the mathematical notion of a *formal context*, which is defined as a set structure (G, M, I) consisting of two sets G and M and a binary relation I between G and M, i.e., $I \subseteq G \times M$. The elements of G are called the *(formal) objects*, those of M the *(formal) attributes* of the formal context (G, M, I), and gIm (i.e., $(g, m) \in I$) is read "the object g has the attribute m". For demonstrating notions and results presented in this paper, we choose the formal context $(\mathbb{N}, M_\mathbb{N}, I_\mathbb{N})$ where $\mathbb{N} := \{1, 2, 3, ...\}$ is the set of all natural numbers, $M_\mathbb{N} :=$ {even, odd, prime, square, sum2sq} is a set of specific properties of natural numbers ("sum2sq" stands for "is the sum of two squares of natural numbers"), and $I_\mathbb{N}$ indicates which natural number has which property out of $M_\mathbb{N}$. A subcontext of $(\mathbb{N}, M_\mathbb{N}, I_\mathbb{N})$ is represented in Figure 1 by a cross table. We will argue later that the formal context in Figure 1 has the same clause logic as the formal context $(\mathbb{N}, M_\mathbb{N}, I_\mathbb{N})$.

Contextual Attribute Logic focusses, in formal contexts (G, M, I), on the formal attributes and their extents, understood as the formalizations of the extensions of the attributes: The *extent* m' of an attribute $m \in M$ is the set of all objects of (G, M, I) that have this attribute:

$$m' := \{g \in G \mid gIm\} \tag{1}$$

For a set $A \subseteq M$ of attributes the *extent* A' of A is the set of all objects of (G, M, I) that have all the attributes in A:

$$A' := \{g \in G \mid \forall_{m \in A}\, gIm\} = \bigcap \{m' \mid m \in A\}. \tag{2}$$

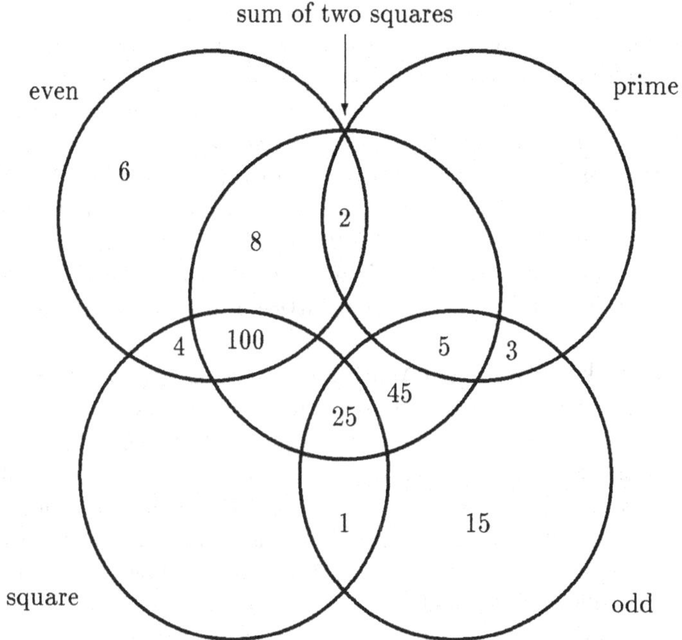

Fig. 2. The extensions of the attributes in Figure 1.

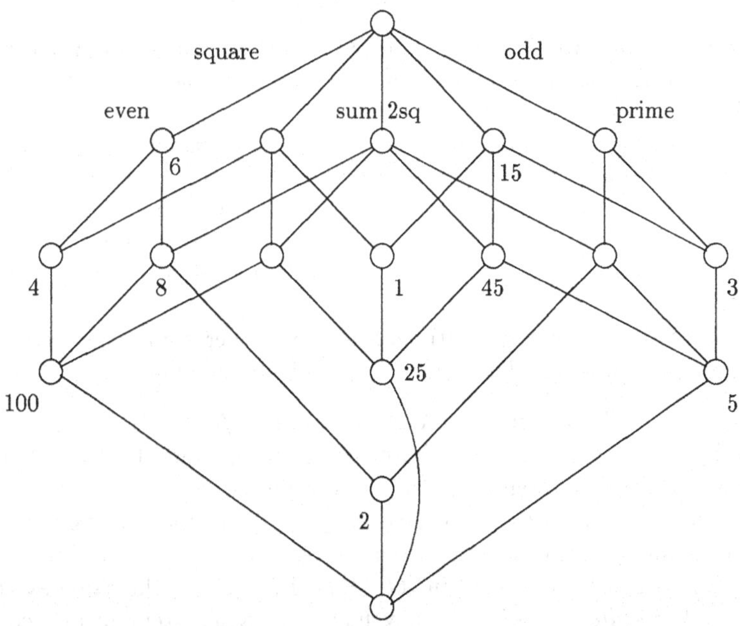

Fig. 3. The concept lattice of the formal context in Figure 1.

A consequence of the second equality is worth noting: The extents of a formal context are the intersections of the attribute extents (convention: $\bigcap\emptyset = G$). Thus, the extents of a formal context (G, M, I) form a closure system on G and therefore a complete lattice with set intersection as meet operation; this lattice is called the *extent lattice* of (G, M, I) and denoted by $\underline{\mathfrak{U}}(G, M, I)$. Figure 2 represents the extent lattice of the formal context of Figure 1 by a Venn diagram in which the attribute extents are depicted by circles. Contextual Concept Logic deals more generally with the *concept lattices* of formal contexts where a formal concept represents a concept as a unit of thought constituted by its extension and its intension (see [9],[2]); the concept lattice of the formal context of Figure 1 is shown in Figure 3.

The central task of Contextual Attribute Logic is the investigation of the "logical relationships" between the attributes of formal contexts and, more generally, between combinations of attributes, such as implications and incompatibilities. They constitute for each formal context a "local logic" (cf. [1]). Tools for the analysis of such local logics may be taken from Mathematical Logic (cf. [17]), and we shall gratefully make use of its definitions and results. But since the aim of Contextual Logic differs from that of Mathematical Logic, we will not fully adopt the terminology of that field.

The logical relationships between formal attributes will be expressed via their extents. For example, we will say that an attribute m *implies* an attribute n if $m' \subseteq n'$, and that m and n are *incompatible* if $m' \cap n' = \emptyset$. In order to have more expressivity in Contextual Attribute Logic, we introduce *compound attributes* of a formal context (G, M, I) by using the operational elements $\neg$, $\bigwedge$, and $\bigvee$:

- For each attribute $m \in M$ we define its *negation*, $\neg m$, to be a compound attribute, which has the extent $G \setminus m'$. Thus, g is in the extent of $\neg m$ if and only if g is not in the extent of m.
- For each set $A \subseteq M$ of attributes, we define the *conjunction*, $\bigwedge A$, and the *disjunction*, $\bigvee A$, to be compound attributes, where the extent of $\bigwedge A$ is $\bigcap\{m' \mid m \in A\}$ and the extent of $\bigvee A$ is $\bigcup\{m' \mid m \in A\}$. Thus, g is in the extent of $\bigwedge A$ if and only if g is in the extent of all attributes $m \in A$, and g is in the extent of $\bigvee A$ if and only if g is in the extent of some attribute $m \in A$.
- Iteration of the above compositions leads to further compound attributes, the extents of which are determined in the obvious manner.

Note that for $A = \emptyset$ the extent of $\bigvee A$ is $\emptyset$ and of $\bigwedge A$ is G. Since it seems natural to interprete a set A of attributes as the compound attribute $\bigwedge A$, the operational element $\bigwedge$ is often omitted (but not here).

Compound attributes of a formal context (G, M, I) may be extensionally represented in any context with the attribute set M. Two such attributes are said to be *extensionally equivalent* in (G, M, I) if they have the same extent in (G, M, I), and *globally equivalent* if they have the same extent in any context with the attribute set M. Note that for $m \in M$ the compound attributes $\bigwedge\{m\}$ and $\bigvee\{m\}$ are both globally equivalent to m, so that we may include the original attributes (those of M) when speaking about compound attributes. For testing global equivalence, we consider the *test context* $(\mathfrak{P}(M), M, \ni)$ (where $\mathfrak{P}(M)$ denotes the power set of M) because of the following fact:

Proposition 1. *Two compound attributes of an attribute set M are globally equivalent if and only if they are extensionally equivalent in $(\mathfrak{P}(M), M, \ni)$.*

The introduction of the basic notions of Contextual Attribute Logic might have made clear how Contextual Concept Logic differs from Propositional Logic. G. Boole distinguishes in [3] between "primary" propositions relating to things and "secondary" propositions relating to propositions, and therefore treats primary propositions first and secondary later. Similarly, in Contextual Attribute Logic we first try to clarify logical relationships between formal objects, attributes, and concepts of formal contexts and then use the results for studying formal judgments formed by formal concepts. Nevertheless, already on the level of Contextual Attribute Logic, a connection to Propositional Logic can be established by considering formal contexts $(S, P, \models)$ where S is a set of situations, P is a set of propositions, and $s \models p$ indicates when a proposition p is true in a situation s. Then the "attribute" p is all-extensional in $(S, P, \models)$ iff the proposition p is true in all situations out of S.

2 Clause Logics of Formal Contexts

Since we have permitted iterations of the composition process, compound attributes may be arbitrarily complex. Up to equivalence, however, they can be represented by rather simple ones:

A *sequent* (A, S) over M is a compound attribute, represented by two sets $A, S \subseteq M$, that is globally equivalent to the compound attribute

$$\bigvee(S \cup \{\neg m \mid m \in A\}). \tag{3}$$

(Such expressions are called *clauses* in Mathematical Logic.) An object g is in the extent of (A, S) if and only if $(g, m) \in I$ for at least one $m \in S$ or $(g, m) \notin I$ for at least one $m \in A$. A sequent (A, S) is *disjoint* if $A \cap S = \emptyset$. We will consider only disjoint sequents here. Sequents can be ordered by componentwise inclusion: We write

$$(A_1, S_1) \leq (A_2, S_2) :\Longleftrightarrow A_1 \subseteq A_2 \text{ and } S_1 \subseteq S_2. \tag{4}$$

A sequent (A, S) is *finite* if A and S are finite sets, and *full* if $A \cup S = M$. A full sequent has a simple extent:

Proposition 2. *The extent of a full sequent $(A, M \setminus A)$ in the test context is $\mathfrak{P}(M) \setminus \{A\}$.*

It is immediate from this proposition that the extent of a conjunction of full sequents $(A_t, M \setminus A_t)$, $t \in T$, is just $\mathfrak{P}(M) \setminus \{A_t \mid t \in T\}$. This gives the following well known result:

Theorem 1 ([CONJUNCTIVE NORMAL FORM]). *Every compound attribute is globally equivalent to a unique conjunction of full disjoint sequents.*

A compound attribute is *all-extensional* in (G, M, I) if its extent is the set G of all objects. The *Boolean attribute logic* of a formal context is the class of its all-extensional compound attributes.

As can be seen from this definition, we interprete compound attributes not only as "generalized attributes", but also as logical rules that may or may not hold in the given context. It is often natural to express this in the terminology. For example, a sequent (A, S) may be interpreted as an implication

$$\bigwedge A \to \bigvee S. \tag{5}$$

In fact, the compound attribute $\bigwedge A$ implies $\bigvee S$ in (G, M, I) if and only if the sequent (A, S) is all-extensional.

A *clause set* over M is a set of sequents over M. The *clause logic of a formal context* (G, M, I) is the set of all sequents that are all-extensional in (G, M, I). A clause set $\mathcal{C}$ is *regular* if it satisfies the following conditions:

1. If $(A, S) \in \mathcal{C}$ and $(A, S) \leq (B, T)$, then $(B, T) \in \mathcal{C}$.
2. If for each sequent (B, T) satisfying $(A, S) \leq (B, T)$ there is some sequent $(C, U) \in \mathcal{C}$ with $(B, T) \leq (C, U)$, then $(A, S) \in \mathcal{C}$.

It is easy to see that the two conditions are equivalent to the following:

3. $(A, S) \in \mathcal{C} \iff \forall_{X \subseteq M} ((A, S) \leq (X, M \setminus X) \Rightarrow (X, M \setminus X) \in \mathcal{C})$.

In words: (A, S) belongs to $\mathcal{C}$ iff each full sequent that contains (A, S) belongs to $\mathcal{C}$. This immediately yields the next proposition:

Proposition 3. *Two regular clause sets containing the same full sequents are equal.*

Theorem 2. *A clause set is regular iff it is the clause logic of some formal context.*

Proof. It is easy to verify that the clause logic of any formal context satisfies the two conditions of regularity. To construct a context for a given regular clause set $\mathcal{C}$, let $\{A_t \mid t \in T\}$ denote the set of first components of full sequents $(A_t, M \setminus A_t)$ in $\mathcal{C}$, and let

$$\mathcal{G} := \mathfrak{P}(M) \setminus \{A_t \mid t \in T\}. \tag{6}$$

$\mathcal{G}$ is the extent of the conjunction of these full sequents in the test context. Therefore these full sequents are exactly the ones which are all-extensional in $(\mathcal{G}, M, \ni)$. By the preceding proposition, the clause logic of this context is equal to $\mathcal{C}$.

The theorem is not new. Our presentation resembles that of [1], cf. Corollary 9.34. In the language of Propositional Logic, the elements of $\mathcal{G}$ would correspond to those truth assignments that make all clauses in $\mathcal{C}$ true.

It is sometimes helpful to replace the clumsy formulation "the extent of a compound attribute c (or of the conjunction of a clause set $\mathcal{C}$) in the test context" by something shorter. We shall use the term *free extent* of c (or of $\mathcal{C}$) instead. The elements of the free extent of c (or of $\mathcal{C}$) are the subsets *consistent* with the attribute c (or the clause set $\mathcal{C}$). From the above propositions we know that a set $A \subseteq M$ is consistent with a clause set $\mathcal{C}$ if and only if the sequent $(A, M \setminus A)$ is not contained in the regular closure of $\mathcal{C}$.

Compound attributes do not distinguish between objects that have the same attributes. We may introduce an equivalence relation $\sim$ on G by

$$g{\sim}h :\iff \forall_{m \in M}\ (gIm \iff hIm). \tag{7}$$

A compound attribute is all-extensional in (G, M, I) iff it is all-extensional in the *object-clarified* context $(G/{\sim}, M, I/{\sim})$, which is obtained from (G, M, I) by merging objects which are in relation $\sim$. Therefore, (G, M, I) has the same clause logic as its "object clarification" $(G/{\sim}, M, I/{\sim})$. Its clause logic determines a formal context "up to object clarification", as the following proposition shows:

Proposition 4. *Two contexts having the same clause logic have isomorphic object clarifications.*

Another easy consequence of the definition of regularity is that the intersection of regular clause sets is again regular. Note that the set of all sequents over M is a regular clause set (it is the clause logic of the formal context $(\emptyset, M, \emptyset)$ with no objects). Therefore, the regular clause sets form a closure system and thus each clause set is contained in a smallest regular one. We call this the regular clause set *generated* by the given one, or its *regular closure.* For reasons of simplicity it is often of interest to describe a regular clause set by few clauses (in particular when, as we will assume from now on, the set M of attributes is finite). We may ask for *minimal* or at least *irredundant* generating sets of regular clause sets. Unfortunalety these problems are notoriously intractable (in terms of computational complexity), see e.g. [11]. A natural generating set of a regular clause set is the set of its minimal sequents, called the *prime implicands.* In general, the set of prime implicands may be redundant. Nevertheless these are often useful. Note that the regular closure of a clause set is also generated by the prime implicands contained in the given sequents.

For the example in Figures 1 and 2 it is not difficult to determine the prime implicands:

$$\begin{array}{ll} (\emptyset, \{\text{even}, \text{odd}\}), & (\{\text{even}, \text{odd}\}, \emptyset), \\ (\{\text{prime}, \text{square}\}, \emptyset), & (\{\text{even}, \text{prime}\}, \{\text{sum2sq}\}). \end{array} \tag{8}$$

Written as clauses, these are

$$\begin{array}{ll} \emptyset \to \text{even} \vee \text{odd}, & \text{even} \wedge \text{odd} \to \emptyset \\ \text{prime} \wedge \text{square} \to \emptyset, & \text{even} \wedge \text{prime} \to \text{sum2sq}. \end{array} \tag{9}$$

It is straightforward to check that these clauses hold for *all* natural numbers, not only for those given in the example. The clause logic of the formal context in Figure 1 therefore is the same as the clause logic of the extended formal context

$$(\mathbb{N}, M, I)$$

with the same attribute set and the obvious extension of the incidence relation. If we change the meaning of the attribute sum2sq to "sum of two squares" (admitting zero as a summand) the example context changes by only two crosses: the objects 1 and 4 then have the attribute sum2sq. The new context is no longer object clarified, since 1 has the same attributes as 5, so the object 1 can be omitted. The new clause logic is that of the remaining subcontext and therefore contains the old one. The prime implicands given above remain prime implicands, but we find an additional one:

$$\text{square} \to \text{sum2sq}. \tag{10}$$

Again, this is a law that holds for all natural numbers, and for zero..

Another algorithmically difficult problem is that of *deduction*: to decide for a given clause set and a given sequent (A, S) if that seqent is in the generated regular clause set. A related problem is to decide if a given clause set is *inconsistent*, that is, if it generates the set of all sequents and therefore has no consistent sets. A powerful algorithm (which nevertheless cannot overcome the fact that the problem is $\mathcal{NP}$-hard) is based on a single derivation rule, called *resolution*: If $\mathcal{C}$ is a regular clause set and if $m \in M$ is an attribute and $(A_1, S_1), (A_2, S_2)$ are sequents in $\mathcal{C}$ with $A_1 \cap S_2 = \emptyset$ and $A_2 \cap S_1 = \{m\}$, then $((A_1 \cup A_2) \setminus \{m\}, (S_1 \cup S_2) \setminus \{m\}) \in \mathcal{C}$. The closure under resolution of a given clause set $\mathcal{C}$ is the smallest set that contains $\mathcal{C}$ and is closed under this derivation rule. It is contained in the regular clause set generated by $\mathcal{C}$. The following well known result is the basis of the resolution algorithm (see [11]):

Proposition 5. *A clause set on a finite set M of attributes is inconsistent if and only if its closure under resolution contains the empty sequent $(\emptyset, \emptyset)$.*

3 Implications and Inconsistencies

It is to be expected that the free extent of a clause set $\mathcal{C}$ has nice structural properties if $\mathcal{C}$ is of a simple form, and vice versa. We give some examples of this phenomenon. Let us call a collection $\mathcal{S}$ of subsets of M an *order ideal* if it is closed under the formation of subsets, that is, iff

$$A \in \mathcal{S}, B \subseteq A \Rightarrow B \in \mathcal{S}, \tag{11}$$

an *order filter* if it is closed under the formation of supersets, that is, iff

$$A \in \mathcal{S}, A \subseteq B \Rightarrow B \in \mathcal{S}, \tag{12}$$

a *closure system* if it is closed under intersections, that is, iff

$$\mathcal{F} \subseteq \mathcal{S} \Rightarrow \bigcap \mathcal{F} \in \mathcal{S}, \tag{13}$$

a *kernel system* if it is closed under unions, that is, iff

$$\mathcal{F} \subseteq \mathcal{S} \Rightarrow \bigcup \mathcal{F} \in \mathcal{S}. \tag{14}$$

Let $A, S \subseteq M$ be nonempty sets of attributes and let $m \in M$ be some attribute. We call a sequent of the form

$(A, \emptyset)$ an *inconsistency* (because the sets inconsistent with this clause are precisely the sets containing A),
$(\emptyset, A)$ a disjunction (this is in accordance with the definition above),
$(A, \{m\})$ an *implication* with singleton conclusion and
$(\{m\}, S)$ a *surmise relationship* (cf. [5]).

We say that $\mathcal{S}$ is the *free extent of inconsistencies* if $\mathcal{S}$ is the free extent of some clause set consisting of inconsistencies only. In the other cases, the formulation is similar.

Proposition 6. *A set $\mathcal{S}$ of subsets of M is*

- *an order ideal iff it is the free extent of inconsistencies,*
- *an order filter iff it is the free extent of disjunctions,*
- *an closure system iff it is the free extent of implications,*
- *an kernel system iff it is the free extent of surmise relationships.*

If $\mathcal{C}_1, \mathcal{C}_2$ are clause sets, then the free extent of $\mathcal{C}_1 \cup \mathcal{C}_2$ is just the intersection of the free extents of $\mathcal{C}_1$ and of $\mathcal{C}_2$. This allows to combine the results of the proposition to give further characterizations, e.g.

- *$\mathcal{S}$ is convex iff it is the free extent of inconsistencies and disjunctions,*
- *$\mathcal{S}$ is a convex subset of a closure system iff it is the free extent of inconsistencies, disjunctions and implications.*

With a little extra argument we obtain

Proposition 7. *A set $\mathcal{S}$ of subsets of M is closed under arbitrary unions and intersections if and only if $\mathcal{S}$ is the free extent of sequents $(\{m\}, \{n\})$, $m, n \in M$.*

Proof. Sequents of the form $(\{m\}, \{n\})$ are both implications and surmise relationships. By Proposition 6 their consistent sets are closed under unions and under intersections. Every prime implicand of a clause set generated by implications is an implication, and similarly for surmise relationships. Therefore the prime implicands of a regular clause set both generated by implications and by surmise relationships must be of the form $(\{m\}, \{n\})$.

If $\mathbb{K} := (G, M, I)$ is some formal context, its *complementary context* is defined as

$$\mathbb{K}^c := (G, M, G \times M \setminus I). \tag{15}$$

So $\mathbb{K}^c$ has the same objects and attributes as $\mathbb{K}$, but the complementray incidence relation: an object g has an attribute m in $\mathbb{K}^c$ iff g does not have m in $\mathbb{K}$. There is a simple connection between the clause logic of $\mathbb{K}$ and that of $\mathbb{K}^c$:

Proposition 8. *A sequent (A, S) is all-extensional in $\mathbb{K}$ if and only if the sequent (S, A) is all-extensional in $\mathbb{K}^c$.*

4 Cumulated Clauses

We have already mentioned that it is algorithmically difficult to find a "nice" generating set for a given clause logic. It may be rather tedious to read off, from a given formal context, the prime implicands of its clause logic, and these are not always well structured. We therefore discuss another generating set. It has similar disadvantages: it may be difficult to find, and there may be smaller generating sets. But it is always irredundant, and because of its recursive nature it can be used for knowledge acquisition algorithms. In the important case of implications it can be shown to be of minimal size.

This generating set consists of *cumulated clauses*, which are compound attributes of the form

$$\bigwedge A \to \bigvee_{t \in T} \bigwedge A_t, \tag{16}$$

where T is some index set and A and the A_t, $t \in T$, are subsets of M. On the first glance these expressions may seem complicated, but it is easy to describe their extents: An object $g \in G$ is in the extent iff it satisfies the following condition:

> If g has all the attributes from A, then g has all attributes from at least one A_t, $t \in T$.

A *choice set* for a given family A_t, $t \in T$, of subsets of M is a set $C \subseteq M$ which has exactly one element in common with each of the A_t. If is straightforward to verify that a cumulated clause $\bigwedge A \to \bigvee_{t \in T} \bigwedge A_t$ is globally equivalent to the conjunction of the sequents (A, C), where C runs over all choice sets. In the case that all the sets A_t are singletons, say $A_t = \{a_t\}$ for all $t \in T$, the cumulated clause becomes an implication

$$\bigwedge A \to \bigwedge \{a_t \mid t \in T\}. \tag{17}$$

To describe a generating set, we use a notion from Contextual Concept Logic: Given a formal context (G, M, I) and an object $g \in G$. Then the *object intent* g' of g is the set of all attributes in M that g has:

$$g' := \{m \in M \mid gIm\}. \tag{18}$$

The reader should not be irritated by the fact that the object intents of a formal context are precisely the elements of the free extent of its clause logic.

To each set $A \subseteq M$ of attributes we can associate a cumulated clause c_A that obviously is all-extensional in (G, M, I), namely

$$\bigwedge A \rightarrow \bigvee (\bigwedge \{g' \setminus A \mid A \subseteq g'\}). \tag{19}$$

Let (G, M, I) be a formal context with finite attribute set M. A *pseudo object intent* of (G, M, I) is a subset $P \subseteq M$ having the following properties:

- P is not an object intent of (G, M, I),
- for each pseudo object intent $Q \subseteq P$ with $Q \neq P$ there exists some object $g \in G$ with $Q \subseteq g' \subseteq P$.

An algorithm to find all pseudo object intents has been given in [12]. We have the following theorem (cf. [7]):

Theorem 3. *The set of cumulated clauses*

$$\{c_P \mid P \subseteq M \textit{ pseudo object intent}\} \tag{20}$$

is an irredundant generating set for the clause logic of (G, M, I).

This generating system was first studied in the case of an implicational logic (i.e., if the free extent is a closure system), cf. [6], where it in fact is minimal, see [9] for details.

References

1. J. Barwise, J. Seligman. *Information Flow: The Logic of Distributed Sytems.* Cambridge University Press, Cambridge 1997.
2. H. Berg. *Terminologische Begriffslogik.* Diplomarbeit. FB Mathematik, TU Darmstadt 1997.
3. G. Boole. *An Investigation of the Laws of Thought On Which Are Founded the Mathematical Theories of Logic and Probabilities.* Dover, New York 1958 (first 1854).
4. K. Devlin. *Goodbye Descartes: The End of Logic and the Search for a New Cosmology of the Mind.* Wiley, New York 1997.
5. J.-P. Doignon, J.-C. Falmagne. *Knowledge Spaces.* Springer, Berlin–Heidelberg–New York 1999.
6. J.-L. Guigues and Vincent Duquenne. *Familles minimales d'implications informatives resultant d'un tableau de données binaires.* Math. Sci. Humaines 95 (1986), 5–18.
7. B. Ganter. *Attribute exploration with background knowledge.* Preprint, TU Dresden 1996. To appear in Theoretical Computer Science, 1999.
8. B. Ganter. *Begriffe und Implikationen.* In: G. Stumme, R. Wille (eds.). *Begriffliche Wissensverarbeitung: Methoden und Anwendungen.* Springer, Berlin–Heidelberg (in preparation)

9. B. Ganter, R. Wille. *Formal Concept Analysis: Mathematical Foundations.* Springer, Berlin–Heidelberg–New York 1999.
10. I. Kant. *Logic.* Dover, New York 1988.
11. H. Kleine Büning, T. Lettmann. *Aussagenlogik: Deduktion und Algorithmen.* B.G. Teubner, Stuttgart 1994.
12. R. Krauße: *Kumulierte Klauseln als ausagenlogische Sprachmittel für die Formale Begriffsanalyse.* Diplomarbeit. Institut für Algebra, TU Dresden 1998.
13. S. Prediger. *Terminologische Merkmalslogik in der Formalen Begriffsanalyse.* In: G. Stumme, R. Wille (eds.). *Begriffliche Wissensverarbeitung: Methoden und Anwendungen.* Springer, Berlin–Heidelberg (in preparation)
14. S. Prediger. *Kontextuelle Urteilslogik mit Begriffsgraphen. Ein Beitrag zur Restrukturierung der mathematischen Logik.* Dissertation, TU Darmstadt. Shaker Verlag, Aachen 1998.
15. St. Read. *Thinking About Logic.* Oxford University Press 1995.
J. F. Sowa. *Conceptual structures: information processing in mind and machine.* Addison-Wesley, Reading 1984.
16. J. F. Sowa. *Knowledge representation: logical, philosophical, and computational foundations.* PWS Publishing Co., Boston (to appear)
17. R. Wille. *Restructuring mathematical logic: an approach based on Peirce's pragmatism.* In: A. Ursini, P. Agliano (eds.). *Logic and Algebra.* Marcel Dekker, New York 1996, 267–281.
18. R. Wille. *Conceptual Graphs and Formal Concept Analysis.* In: D. Lukose, H. Delugach, M. Keeler, L. Searle, J. Sowa (eds.). *Conceptual Structures: Fulfilling Peirce's Dream.* Springer, Berlin–Heidelberg–New York 1997, 290–303.

Algorithms for Creating Relational Power Context Families from Conceptual Graphs

Bernd Groh and Peter Eklund

School of Information Technology
Griffith University
PMB 50 Gold Coast Mail Centre
Queensland 9726
AUSTRALIA
{b.groh,p.eklund}@gu.edu.au

Abstract. This paper reports on the implementation of Wille's algorithm for creating power context families from conceptual graphs. It describes how to extend FCA tools, such as TOSCANA, to accommodate the expressibility of CG's. It further reports on a tool where Wille's Algorithm, along with an algorithm to create SQL queries from power context families, has been implemented and applied. These algorithms are given a detailed presentation in this paper and their complexity provided.

Introduction

Wille [7, 8] proposes an algorithm to transform Conceptual Graphs into a set of formal contexts, the Power Context Family (PCF). It takes a Conceptual Graph as input and creates formal contexts, where the instances are the objects and the types and the relations are the attributes. For a concept the object is the instance of that specific type, for a relation, the object is an ordered pair of two instances – object instances as arguments to the specific relation. Because of the differences between concepts and relations in CG theory, these two sets of objects are stored in different formal contexts. This set of formal contexts is the PCF. The first of these contexts, for concepts and instances, is called $\mathbb{K}_1$. The second, for the binary relations and the ordered pairs of instances is called $\mathbb{K}_2$. We can generalize for n-ary relations and the n-tuples of instances as arguments called $\mathbb{K}_n$. This set of formal contexts $\mathbb{K}_1$ to $\mathbb{K}_n$ is called the Power Context Family.

This paper presents and analyses an algorithm that connects Formal Concept Analysis (FCA) and Conceptual Graphs (CG's) using the PCF. The algorithm to transform a given simple CG into the PCF, as proposed by Wille [7], has been implemented and is described. Furthermore, the algorithm has been expanded to accommodate positive nested CG's as proposed in Wille [8]. The format in which the formal context are stored is ConScript [6]. The paper examines the complexity results of this algorithm.

The presented software is embedded into an ODBC environment and has a SQL interface for querying relational databases – this is the way most software tools for Formal Concept Analysis [5] have been developed. This tool is thought of as a possible extension of TOSCANA to deal with Conceptual Graphs. As in TOSCANA, a Conceptual File is opened which conforms to a Conceptual Graph database. From these simple or positive nested CGs, n tables (n contexts) in a relational database, representing the Power Context Family, are created. TOSCANA conceptual scales allow a selected view on the data which permits a particular view of the loaded CG. This subgraph is the input to a query. Our experimental environment creates a SQL statement to query the database for instances that fit into the query graph and communicates via ODBC with the database to retrieve results.

Storing Nested Conceptual Graphs in Formal Contexts

The method of storing nested CG's in formal contexts differs slightly from the approach in Wille [8], although it is an extension of the simple CG approach described in [7]. Differences result from the restriction of ConScript to diadic contexts. It is therefore necessary to treat nested Conceptual Graphs as simple Conceptual Graphs. We will show the solution to that problem on the sentence "Mary thinks that Peter loves her" shown in Figure 1. Using ConScript it is not possible to express such a sentence in the power context family because the instance of STATEMENT is a Conceptual Graph and not just an instance of a type. However, the two simple CG's given in Figure 2 are expressable in ConScript.

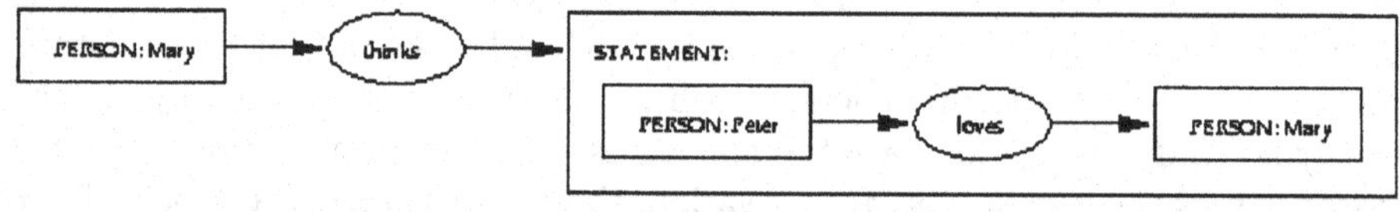

Fig. 1. Conceptual Graph for the sentence "Mary thinks that Peter loves her."

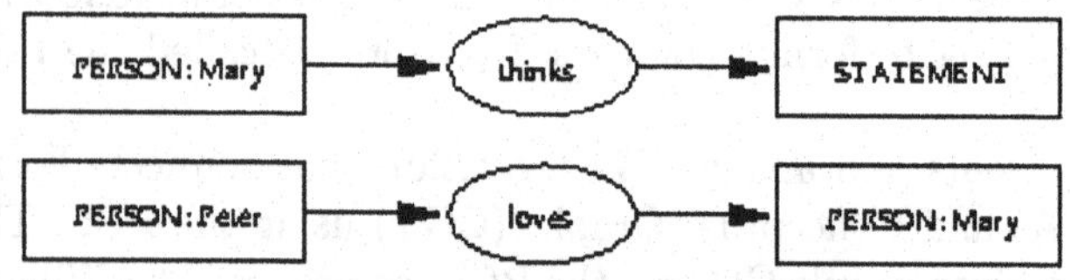

Fig. 2. Simple Conceptual Graphs for the example in Figure 1

Furthermore the CG in Figure 2 is not true as "Peter loves Mary", represented by the lower subgraph is only valid in the situation that "Mary thinks" and not necessarily in any universal sense. As shown by Chein and Mugnier [2], is it possible to represent nested Conceptual Graphs by trees of simple Conceptual Graphs. This is shown in Figure 3.

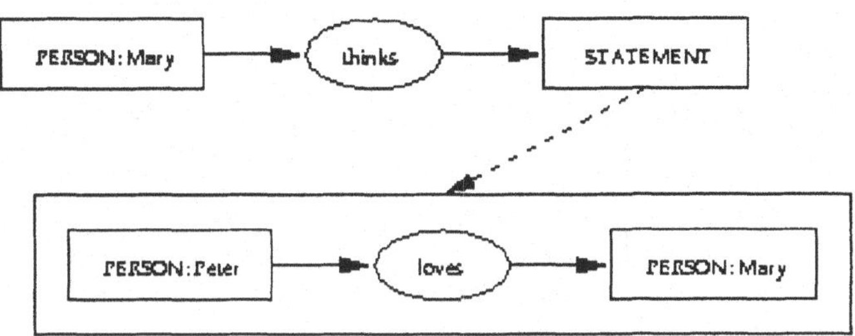

Fig. 3. Tree structure of the simple Conceptual Graphs in Figure 2 to represent the Conceptual Graph in Figure 1

The question is how to represent this tree structure within simple Conceptual Graphs. The answer is in the restriction of concept types and relation types to specific situations. A similar approach is adopted in the CORALI project [1]. A relational signature is no longer described by a relation name and number of arguments, but also an additional attribute including the situation in which the relation holds. The situation is nothing but a reference to the upper neighbour in the tree – also see Chein and Mugnier [2].
The resulting Conceptual Graph is shown in Figure 4. Also shown is that concept types interpreted as unary relations provide a similar treatment as relations. The pair (*Person,#1*) expresses that *Peter* is of type *Person*, but only in the situation *#1*. And situation *#1* is that situation which *"Mary thinks"*. This does not necessarily mean that *Peter* is of type *Person* in all situations and does not even have to mean that *Peter* is of type *Person* in any other situation. The relation *loves* only exists in situation *#1*, expressed by the pair (*loves,#1*). If in a relation no situation is given, the relation is interpreted as valid in *#0*, the T-Element in the tree.

Note that the scope of concept types and relations is within a specific situation. The same object can appear in more than one situation, independent of the relations they participate in. *Mary*, for example, exists in the given example in the situation *#0* and the situation which "Mary thinks". In both cases it is the same *Mary*, and this therefore means there is a **coreference link** between the two. In the present notation identical instance names are identical. Figure 5 shows an alternative representation of the Conceptual Graph in Figure 4.

The power context familiy resulting from the Conceptual Graph in Figure 4 or Figure 5 is shown in Table 1.

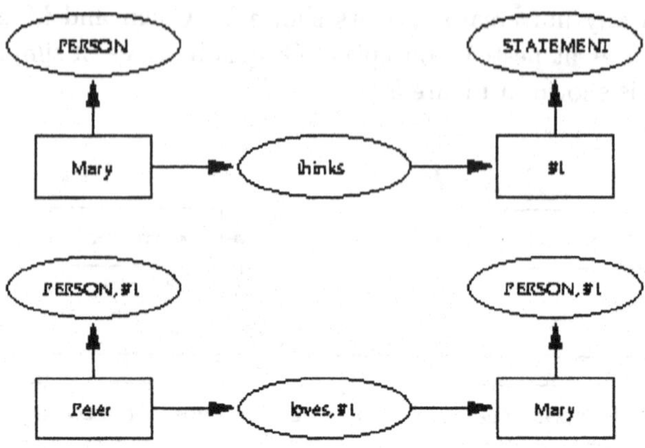

Fig. 4. Representation of the nested Conceptual Graph in Figure 1 as simple Conceptual Graphs

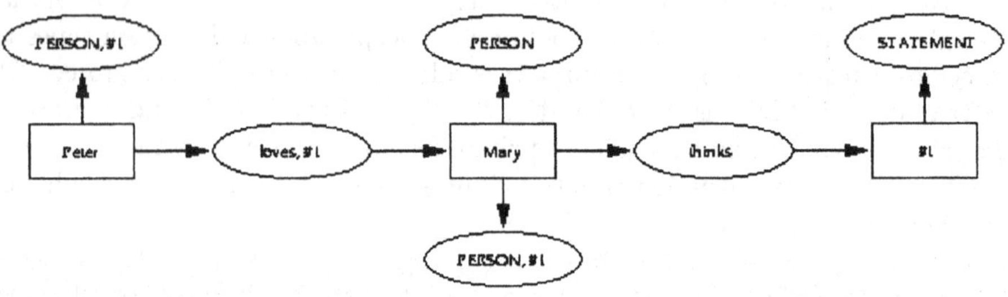

Fig. 5. Alternative representation of the Conceptual Graph in Figure 4

Table 1. Power context family of the CG in Figure 4

K1	PERSON	(#1,PERSON)	STATEMENT
Mary	X	X	
Peter		X	
#1			X

K2	THINKS	(#1,LOVES)
(Mary,#1)	X	
(Peter,Mary)		X

Treating Power Context Families as a Relational Database

Any extension to the formal concept analysis toolkit TOSCANA to make use of conceptual graphs requires our algorithm and system to create SQL statements that query the concrete objects for a given scale – a given subgraph from a relational database. This is shown on an example taken from the work of Mineau and Gerbe [3]. The Conceptual Graph is shown in Figure 6.

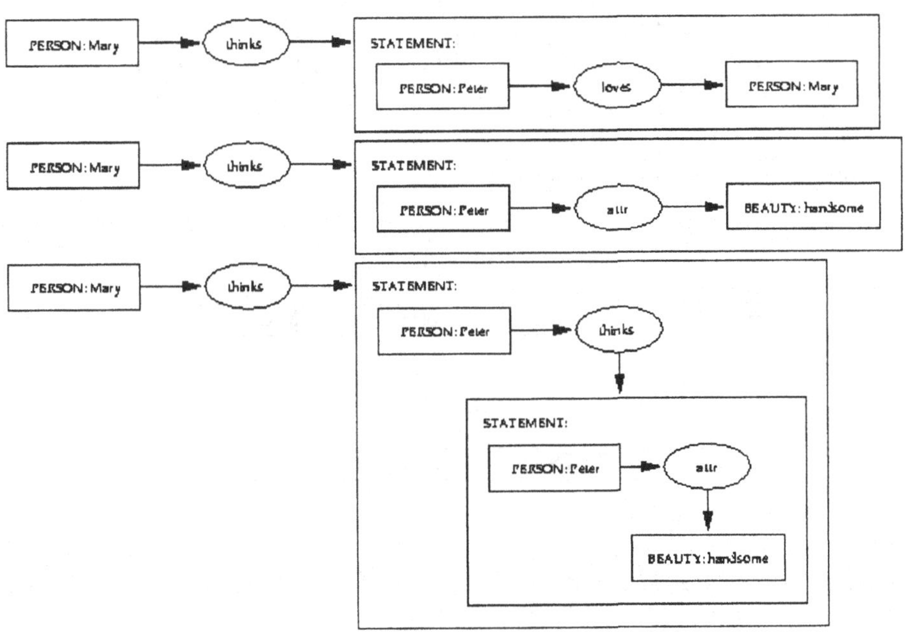

Fig. 6. Conceptual Graph about a part of a world that "Mary thinks"

Representation of the Power Context Family as database relations

Table 2 shows the tables, representing the power context family of the Conceptual Graph in Figure 6. The pair (*Relation*, *Situation*) is split into the columns **Situation** and **Relation** for faster access. This is the same reason for splitting pairs of instances into separate colums, as seen at the second table of Table 2. The column name of the instances indicates the position of the tuple in which the instance appears. The relation *loves(Peter,Mary)* is represented by the row in Table 3.

From the row in Table 4 it is readable that #1, the situation in which the relation *loves(Peter,Mary)* exists, something is that "Mary thinks".

Table 2. Power context family as tables for the Conceptual Graph in Figure 6

Situation	Relation	1
	PERSON	Mary
	STATEMENT	#1
#1	PERSON	Peter
#1	PERSON	Mary
	STATEMENT	#2
#2	PERSON	Peter
#2	BEAUTY	Handsome
	STATEMENT	#3
#3	PERSON	Peter
#3	STATEMENT	#4
#4	PERSON	Peter
#4	BEAUTY	Handsome

Situation	Relation	1	2
	THINKS	Mary	#1
#1	LOVES	Peter	Mary
	THINKS	Mary	#2
#2	ATTR	Peter	handsome
	THINKS	Mary	#3
#3	THINKS	Peter	#4
#4	ATTR	Peter	handsome

Table 3.

#1	LOVES	Peter	Mary

Table 4.

	THINKS	Mary	#1

From the given tables the concrete objects for any subgraph, any scale in TOSCANA, can be retrieved with the specific SQL statement.

Generating a SQL query from a given subgraph

Generating a SQL query from a simple Conceptual Graph

Fig. 7. Simple Conceptual Graph

An interest in what a person *v1* is thinking generates the query graph shown in Figure 7. The implementation would generate the following SQL statement from this query:

```
SELECT DISTINCT Kv1.[1] AS v1, Kv1v2.[2] AS v2
```

```
FROM K1 AS Kv1 INNER JOIN K2 AS Kv1v2 ON Kv1.[1]=Kv1v2.[1]
WHERE (Kv1.[Relation]="PERSON" AND Kv1v2.[Relation]="THINKS");
```

The query result table is shown in Table 5. *Mary* THINKS #1, #2, #3 and *Peter* THINKS #4. If querying with a simple Conceptual Graph, the situation in which the objects exist is not considered. If the situation were taken into account, the Conceptual Graph will need to be nested.

Table 5. Retrieved table from the query generated from the Conceptual Graph in Figure 7

v1	v2
Mary	#1
Mary	#2
Mary	#3
Peter	#4

Generating a SQL query from a nested Conceptual Graph

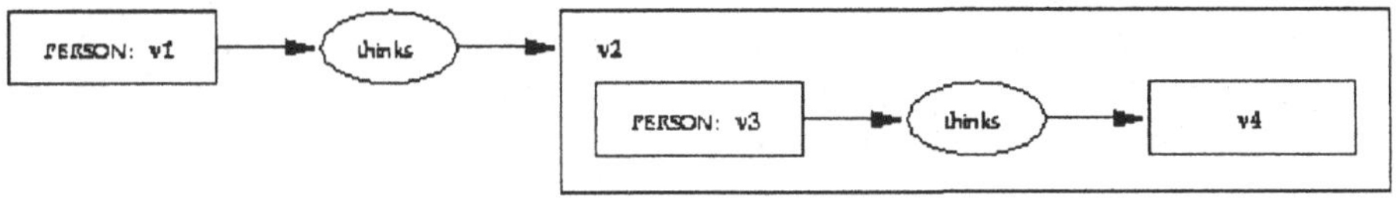

Fig. 8. Nested Conceptual Graph

A general interest in which persons think other persons are thinking something can be formulated by the query graph in Figure 8. The implementation would generate the following SQL statement from this query graph:

```
SELECT DISTINCT Kv1.[1] AS v1, Kv3.[1] AS v3, Kv1v2.[2] AS v2,
 Kv3v4.[2] AS v4
FROM K1 AS Kv1 INNER JOIN K2 AS Kv1v2 ON Kv1.[1]=Kv1v2.[1],
 K1 AS Kv3 INNER JOIN K2 AS Kv3v4 ON Kv3.[1]=Kv3v4.[1]
WHERE (Kv1.[Relation]="PERSON" AND Kv3.[Relation]="PERSON"
 AND Kv1v2.[Relation]="THINKS" AND Kv3v4.[Relation]="THINKS"
 AND Kv1v2.[2]=Kv3.[Situation] AND Kv1v2.[2]=Kv3v4.[Situation]);
```

The retrieved table is shown in Table 6. *Mary* is the only Person who thinks that a Person, *Peter*, is thinking something, #4. The situation in which she is thinking this is #3.

Table 6. Retrieved table from the query generated from the Conceptual Graph in Figure 8

v1	v2	v3	v4
Mary	#3	Peter	#4

If there is now further interest in what Peter is actually thinking, what #4 actually is, a further SQL query could be created and the whole Conceptual Graph could be explored from this starting point.

Wille's Mathematization

Before we can start to analyse the algorithms, we now need to have a look at Wille's mathematization, as our terminology is based on it:

"The mathematization starts by defining an *abstract concept graph* as a structure $B := (V, E, \nu, C, \kappa, \theta)$ for which

1. V and E are finite sets and ν is a mapping of E to $\bigcup_{k=1}^{n} V^k$ $(n \geq 2)$ so that (V, E, ν) can be considered as finite directed multi- hypergraph with vertices from V and edges from E (we define $\mid e \mid = k :\Leftrightarrow \nu(e) = (v_1, \ldots, v_k)$),
2. C is a finite set and κ is a mapping of $V \bigcup E$ to C such that κ $(e_1) = \kappa$ (e_2) always implies $\mid e_1 \mid = \mid e_2 \mid$ (the elements of C may be understood as abstract concepts),
3. θ is an equivalence relation on V.

Next, abstract concept graphs shall be related to formal contexts and their concept lattices. For this, we introduce a *power context family* $\vec{\mathbb{K}} := (\mathbb{K}_1, \ldots, \mathbb{K}_n)$ $(n \geq 2)$ with $\mathbb{K}_k := (G_k, M_k, I_k)$ $(k = 1, \ldots, n)$ such that $G_k \subseteq (G_1)^k$. Let $C_{\vec{\mathbb{K}}} := \bigcup_{k=1}^{n} \mathfrak{B}(\mathbb{K}_k)$. Now, we call an abstract concept graph $B := (V, E, \nu, C, \kappa, \theta)$ an *abstract concept graph over the power context family* $\vec{\mathbb{K}}$ if $C = C_{\vec{\mathbb{K}}}$, $\kappa(V) \subseteq \mathfrak{B}(\mathbb{K}_1)$, and $\kappa(e) \in \mathfrak{B}(\mathbb{K}_k)$ for all $e \in E$ with $\mid e \mid = k$. A *realization* of such abstract concept graph $B \in \vec{\mathbb{K}}$ is defined to be a mapping ρ of V to the power set of G_1 for which $\emptyset \neq \rho(v) \subseteq Ext(\kappa(v))$ for $v \in V$ and $\rho(v_1) \times \ldots \times \rho(v_k) \subseteq Ext(\kappa(e))$ for $e \in E$ with $\nu(e) = (v_1, \ldots, v_k)$, and $v_1 \theta v_2$ always imply $\rho(v_1) = \rho(v_2)$. Then, the pair $\underline{B} := (B, \rho)$ is called a *realized concept graph* of $\vec{\mathbb{K}}$ or, shortly, a *concept graph* of $\vec{\mathbb{K}}$."

The algorithm to create the power context families

The input of the algorithm is a set of facts. As a fact we consider a tuple of either concept type and individual reference or relation type and an ordered tuple of individual references. We call both concept types and relation types relations and we call individual references or tuples of individual references objects. We write for a concept type $\kappa(v)$ where $v \in V$ and V is a set of vertices. For a relation type we write $\kappa(e)$ where $e \in E$ and E is a set of edges. We call R a set of relations where $r \in R$ and $r = \mu r = \kappa(v)$ and $|r| = 1$ for concept types and $r = \mu r = \kappa(e)$ and $|r| = |e|$ for relation types. We write for an individual reference $\rho(v)$ and for an ordered tuple of individual references $\nu(e)$. We call O a set of objects where $o \in O$ and $o = \rho(v)$ for individual references and $o = \nu(e)$ for ordered tuples of individual references. In the Algorithm, see Figure 9, we have a loop over all facts. The number of iterations is therefore $\geq |R|$ as in the best case $|O| = |R| = |F|$ and the complexity therefore is $O(|R|)$ as each relation contains only one object and therefore appears only once as a fact. In general the complexity is $O(|F|)$. In each iteration we add the current fact $f = (o, r)$ to the appropriate context $\mathbb{K}_{|r|}$. After the loop we will have a set of contexts $\vec{\mathbb{K}} = (\mathbb{K}_1, \ldots, \mathbb{K}_n)$ $(n \geq 2)$ as output.

Inputs

Let $F = (O, R)$ be a set of facts
Let $O = \rho(V) \bigcup \nu(E)$ and $R = \kappa(V) \bigcup \kappa(E)$
Let $|r| = 1$ iff $o \in \rho(V)$ and $|r| = |e|$ iff $o \in \nu(E)$

Outputs

Let $\vec{\mathbb{K}}$ be a set of contexts

Algorithm

```
FOR f ∈ F LOOP
    K_|r|.Add(o,r)
END LOOP
```

Fig. 9. Algorithm to create the power context family

The function *Add* has been introduced to add a fact to the specific context and is in a time efficient implementation of constant time.

The extension to positive nested conceptual graphs

A fact no longer consists of a tuple of relation and object, it now consists of a triple of situation, relation and object or as we said before a relation no longer consists of a concept type or a relation type, it now consists of a tuple of situation and concept type or a tuple of situation and relation type. The input of the algorithm is therefore still a set of facts and the complexity is still $O(|F|)$.

The algorithm to create SQL queries

The algorithm is shown in Figure 10. For each vertex we have to add a field to our SELECT statement. The first two FOR statements is a loop over all objects in all contexts $o = \rho(v)$ in $\mathbb{K}_1$ and $o = \nu(e) = (v_1, ..., v_k)$ in $\mathbb{K}_k$. For each two vertices $(v_x, v_y) \in \theta$ in the two objects (o_k, o_i) we have to add an entry to our FROM statement as we have to express the join in the SQL query. The next two FOR statements is a loop over all objects that have been passed by the first two FOR statements to be able to determine if a vertex of the current object is together with a vertex in a passed object $\in \theta$. A much more time efficient way would be to store, with each vertex v_x in an object o_k in $\mathbb{K}_k$, a reference to each vertex v_y where $(v_x, v_y) \in \theta$. Even if this takes up much more memory, it overcomes the need for a second run over the objects within the loop over the objects. As ConScript should be extended for the use of triadic contexts anyway, such an extension is strongly recommended. Whenever there is an entry added to the SELECT or the FROM statement there is an entry added to the WHERE statement.

Inputs
    Let $\vec{\mathbb{K}}$ be an array of contexts
Variables
    Let O_k be an array of objects
    Let O_i be an array of objects
Outputs
    Let *sql* be a string containing a SQL query
Algorithm
    FOR $\mathbb{K}_k \in \vec{\mathbb{K}}$ **LOOP**
        $O_k = \mathbb{K}_k.Objects()$
        FOR $o_k \in O_k$ **LOOP**
            FOR $\mathbb{K}_i \in \vec{\mathbb{K}}(i \leq k)$ **LOOP**
                $O_i = \mathbb{K}_i.Objects()$
                FOR $o_i \in O_i(i < k$ or $(i == k$ && o_i before $o_k))$ **LOOP**
                    $sql := ExtendSQLQuery(sql, \mathbb{K}_k, o_k, \mathbb{K}_i, o_i)$
                END LOOP
            END LOOP
        END LOOP
    END LOOP

Fig. 10. Algorithm to create the SQL statement for the query

The function *ExtendSQLQuery* is a function to add entries to those parts of the SQL query where needed, regarding the two objects or the two vertices. With a time efficient implementation of the data structures in ConScript this should be done in constant time. The output of this algorithm is a string containing an SQL statement. If we consider $O^E = \bigcup O^n_{k=2}$ then the complexity of the algorithm, not considering the function ExtendSQLQuery, is $(\mid O^E \mid^2) \div 2 + n\times \mid O_1 \mid$. As $(\mid O^E \mid^2) \div 2 + n\times \mid O_1 \mid \leq \mid \bigcup O^n_{k=1} \mid^2$ we can therefore say the complexity is

$O(| \bigcup O_{k=1}^{n} |^2)$.
Figure 11 shows a modified version of the algorithm, if for each vertex v_x the reference to each v_y where $(v_x, v_y) \in \theta$ is stored.

Inputs
Let $\vec{\mathbb{K}}$ be a set of contexts
Variables
Let O be a set of objects
Let θ_v be a set of equivalent vertices
Outputs
Let sql be a string containing a SQL query
Algorithm
FOR $\mathbb{K}_k \in \vec{\mathbb{K}}$ **LOOP**
$O = \mathbb{K}_k.Objects()$
FOR $o \in O$ **LOOP**
FOR $v \in \nu(e)$ **LOOP**
$\theta_v = Equivalents(v)$
FOR $t \in \theta_v$ **LOOP**
$sql := ExtendSQLQuery(sql, \mathbb{K}_k, o, v, t)$
END LOOP
END LOOP
END LOOP
END LOOP

Fig. 11. Alternative algorithm to create the SQL statement for the query

The function *ExtendSQLQuery* has been adapted to the different needs. The complexity is now $| O_1 | + | O^E | \times k \times | \theta_v |$ in worst case. θ_v is depended on the structure of the conceptual graph. If we consider θ_v as constant, we can say the time complexity of the algorithm is $O(| \bigcup O_{k=1}^{n} |)$.

Conclusion

It has been shown how the power context family can be represented as database tables in a relational database to allow easy access to the data. Furthermore, it has been demonstrated how concrete objects can be retrieved from those tables by formulating a Conceptual Graph query. The algorithm to generate the SQL queries, as well as the other parts described above, have been implemented and their complexity analysed.

References

1. Chein, M. The CORALI Project: From Conceptual Graphs to Conceptual Graphs via Labelled Graphs. In D. Lukose, H. Delugach, M. Keeler, L. Searle & J.

SOWA (Eds.), Conceptual Structures: Fulfilling Peirce's Dream, pp. 65-79, Springer-Verlag, 1997.

2. CHEIN, M. and MUGNIER, M.-L. Positive Nested Conceptual Graphs. In D. LUKOSE, H. DELUGACH, M. KEELER, L. SEARLE & J. SOWA (Eds.), Conceptual Structures: Fulfilling Peirce's Dream, pp. 95-109, Springer-Verlag, 1997.
3. MINEAU, G. W. and GERBE O. Contexts: A Formal Definition of Worlds of Assertion. In D. LUKOSE, H. DELUGACH, M. KEELER, L. SEARLE & J. SOWA (Eds.), Conceptual Structures: Fulfilling Peirce's Dream, pp. 80-94, Springer-Verlag, 1997.
4. SOWA, J. F. Conceptual Graphs Summary. In T. E. NAGLE, J. A. NAGLE, L. L. GERHOLZ, P. W. EKLUND (Eds.), Conceptual Structures: Current Research and Practice, pp. 3-51, Ellis Horwood, 1992.
5. VOGT, F. and WILLE, R. TOSCANA - a graphical tool for analyzing and exploring data. In R. TAMASSIA, I. G. TOLLIS (Eds.), Graph Drawing 94, pp. 226-233, Springer-Verlag, 1995.
6. VOGT, F. Formale Begriffsanalyse mit C++. Springer-Verlag, 1996.
7. WILLE, R. Conceptual Graphs and Formal Concept Analysis. In D. LUKOSE, H. DELUGACH, M. KEELER, L. SEARLE & J. SOWA (Eds.), Conceptual Structures: Fulfilling Peirce's Dream, pp. 290-303, Springer-Verlag, 1997.
8. WILLE, R. Triadic Concept Graphs. In M.-L. MUGNIER & M. CHEIN (Eds.), Conceptual Structures: Theory, Tools and Applications, pp. 194-208, Springer-Verlag, 1998.

The Lattice of Concept Graphs of a Relationally Scaled Context

Susanne Prediger and Rudolf Wille

Technische Universität Darmstadt, Fachbereich Mathematik
Schloßgartenstr. 7, D–64289 Darmstadt,
{prediger,wille}@mathematik.tu-darmstadt.de

Abstract. The aim of this paper is to contribute to Data Analysis by clarifying how concept graphs may be derived from data tables. First it is shown how, by the method of relational scaling, a many-valued data context can be transformed into a power context family. Then it is proved that the concept graphs of a power context family form a lattice which can be described as a subdirect product of specific intervals of the concept lattices of the power context family (each extended by a new top-element). How this may become practical is demonstrated using a data table about the domestic flights in Austria. Finally, the lattice of syntactic concept graphs over an alphabet of object, concept, and relation names is determined and related to the lattices of concept graphs of the power context families which are semantic models of the given contextual syntax.

Contents

1 Introduction

Conceptual Graphs have been introduced by J. F. Sowa as a system of logic "to express meaning in a form that is logically precise, humanly readable, and computationally tractable" [So92]. In [Wi97] and [Pr98b] Sowa's Theory of Conceptual Graphs has been used in combination with *Formal Concept Analysis* [GW99] to design a mathematical logic of judgment as part of a contextual logic. This *"Contextual Logic"* is understood as a mathematization of the traditional philosophical logic which is based on "the three essential main functions of thinking - concepts, judgments, and conclusions" [Ka88]. Contextual Logic, which is philosophically supported by Peirce's pragmatic epistemology, is grounded on families of related formal contexts whose formal concepts allow a mathematical representation of the concepts and relations of conceptual graphs. Such representation of a conceptual graph is called a *"Concept Graph"* of the context family from which it is derived. To indicate the specific relationship between the considered contexts, such a family of contexts is named a *"Power Context Family"*.

The aim of this paper is to contribute to Data Analysis by clarifying how concept graphs may be derived from data tables. Since data tables mostly relate objects, attributes, and atttribute values, our approach starts with many-valued contexts which have been introduced to formalize such data tables (see [Wi82], [GW99]). In Section 2 we explain first how to turn a many-valued context into a power context family. This transformation, called *"Relational Scaling"*, is guided by the specific purpose of the data analysis to perform. We demonstrate the method of relational scaling by data about the domestic flights in Austria. In Section 3 we show how to determine all concept graphs of a power context family. For this task it is useful that the concept graphs of a power context family form a lattice which is isomorphic to a subdirect product of specific intervals of the concept lattices of the power context family (each extended by a new top-element). How this may become practical is demonstrated by the example of the domestic flights in Austria. In Section 4, more generally, the lattice of syntactic concept graphs, defined on a contextual alphabet, is determined. Its connection to the lattices of concept graphs of power context families is also clarified. Further research is discussed in the final section.

The following explanations presuppose some knowledge about Formal Concept Analysis for which we refer to the monograph [GW99]. For a better understanding of the connection between Conceptual Graphs and Formal Concept Analysis, the papers [Wi97] and [Pr98a] might be helpful.

2 From a Many-valued Context to a Power Context Family

Data tables representing relationships between objects, attributes, and attribute values are mathematized by many-valued contexts. A *many-valued context* is defined as a set structure (G, M, W, I) where G is a set of *(formal) objects*, M is a set of *(formal) attributes*, W is a set of *(formal) attribute values*, and I is a ternary relation between G, M, and W (i.e. $I \subseteq G \times M \times W$) for which $(g, m, v) \in I$ and $(g, m, w) \in I$ always imply that $v = w$; $(g, m, w) \in I$ is read: the object g has the attribute value w for the attribute m. The data table in Figure 1 (see [OAG98]) may be understood as a many-valued context (G, M, W, I) for which G is the set of all listed flights, M is the set consisting of the attributes "Airline", "Departure Airport", "Departure Time", "Arrival Airport", "Arrival Time", "Days", and "Aircraft", while W is a set containing all attribute values described by the entries in the columns of the table.

Conceptual Scaling [GW89] has been established as a useful method for the conceptual analysis of many-valued data contexts. For such data analysis it is desirable to make available, besides structures of formal concepts, also structures

		Departure		Arrival			
Flight	Airline	Airport	Time	Airport	Time	Days	Aircraft
070	VO	Vienna	07.50	Innsbruck	08.40	1-6	F70
071	VO	Innsbruck	06.25	Vienna	07.20	1-5	F70
072a	VO	Vienna	10.20	Innsbruck	11.35	6	DH8
072b	VO	Vienna	10.50	Innsbruck	12.05	1-5, 7	DH8
073a	VO	Innsbruck	08.35	Vienna	09.45	67	DH8
073b	VO	Innsbruck	09.05	Vienna	09.55	1-5	F70
074	VO	Vienna	13.55	Innsbruck	15.10	2-5	DH8
075	VO	Innsbruck	11.40	Vienna	12.50	1-5	DH8
076a	VO	Vienna	17.45	Innsbruck	18.40	1-6	F70
076b	VO	Vienna	18.40	Innsbruck	19.55	7	DH8
077	VO	Innsbruck	15.35	Vienna	16.45	2-5	DH8
078a	VO	Vienna	20.35	Innsbruck	21.25	1-4	F70
078b	VO	Vienna	21.30	Innsbruck	22.45	7	DH8
078c	VO	Vienna	21.40	Innsbruck	22.35	5	CRJ
330	VO	Linz	06.20	Salzburg	06.50	1-6	CRJ
331	VO	Salzburg	11.20	Linz	11.45	1-5	CRJ
332	VO	Linz	16.05	Salzburg	16.35	1-5	CRJ
333	VO	Salzburg	21.50	Linz	22.15	1-5, 7	CRJ
409	VO	Graz	12.10	Linz	12.45	1-5	CRJ
410	VO	Linz	16.10	Graz	16.50	1-5	CRJ
412	VO	Linz	10.35	Graz	11.10	1-5	CRJ
413	VO	Graz	06.15	Salzburg	06.50	1-5	CRJ
415	VO	Graz	17.30	Salzburg	18.10	1-5	CRJ
416	VO	Salzburg	21.50	Graz	22.25	1-5, 7	CRJ
417	VO	Graz	17.15	Linz	17.45	7	CRJ
501	VO	Klagenfurt	06.00	Salzburg	06.45	1-5	DH8
502	VO	Salzburg	21.55	Klagenfurt	22.40	1-5, 7	DH8
531*	VO-OS	Linz	06.00	Vienna	06.45	1-6	DH8
532*	VO-OS	Vienna	10.40	Linz	11.20	1-5, 7	DH8
533*	VO-OS	Linz	08.35	Vienna	09.25	1-7	DH8
534*	VO-OS	Vienna	22.15	Linz	23.00	1-5, 7	DH8
536a*	VO-OS	Vienna	17.10	Linz	17.55	5	DH8
536b*	VO-OS	Vienna	17.15	Linz	17.55	1-4, 7	DH8
537*	VO-OS	Linz	12.00	Vienna	12.50	1-5, 7	DH8
538*	VO-OS	Vienna	20.30	Linz	21.15	1-7	DH8
539*	VO-OS	Linz	18.15	Vienna	19.00	1-5, 7	DH8
540*	VO-OS	Vienna	10.45	Graz	11.30	1-7	DH8
541*	VO-OS	Graz	06.05	Vienna	06.45	1-6	DH8
542*	VO-OS	Vienna	13.50	Graz	14.35	1-5	DH8
543*	VO-OS	Graz	08.50	Vienna	09.35	1-7	DH8
544*	VO-OS	Vienna	17.20	Graz	18.00	1-7	DH8
545*	VO-OS	Graz	11.55	Vienna	12.35	1-5	DH8
546*	VO-OS	Vienna	19.40	Graz	20.20	1-7	DH8
547*	VO-OS	Graz	15.30	Vienna	16.15	1-5, 7	DH8
548*	VO-OS	Vienna	22.30	Graz	23.10	1-5, 7	DH8
549*	VO-OS	Graz	15.30	Vienna	16.15	1-5, 7	DH8
550*	VO-OS	Vienna	07.25	Klagenfurt	08.15	1-5	DH8
551*	VO-OS	Klagenfurt	06.00	Vienna	06.50	1-6	DH8
552*	VO-OS	Vienna	10.40	Klagenfurt	11.30	1-7	DH8
553*	VO-OS	Klagenfurt	08.40	Vienna	09.30	1-7	DH8
554*	VO-OS	Vienna	13.55	Klagenfurt	14.50	1-5	DH8
555*	VO-OS	Klagenfurt	11.55	Vienna	12.45	1-7	DH8
556*	VO-OS	Vienna	17.10	Klagenfurt	18.00	1-7	DH8
557*	VO-OS	Klagenfurt	15.15	Vienna	16.10	1-5	DH8
558*	VO-OS	Vienna	19.50	Klagenfurt	20.45	1-7	DH8
559*	VO-OS	Klagenfurt	18.20	Vienna	19.10	1-7	DH8
560*	VO-OS	Vienna	22.30	Klagenfurt	23.20	457	DH8
561*	VO-OS	Klagenfurt	21.00	Vienna	22.00	457	DH8
590*	VO-OS	Vienna	10.25	Salzburg	11.20	1-7	DH8
591*	VO-OS	Salzburg	17.15	Vienna	18.10	7	DH8
593*	VO-OS	Salzburg	08.15	Vienna	09.15	1-7	DH8
594*	VO-OS	Vienna	17.35	Salzburg	18.35	1-7	DH8
595*	VO-OS	Salzburg	11.45	Vienna	12.40	1-7	DH8
596a*	VO-OS	Vienna	20.25	Salzburg	21.20	6	DH8
596b*	VO-OS	Vienna	20.35	Salzburg	21.30	1-5, 7	DH8
597*	VO-OS	Salzburg	19.05	Vienna	20.00	1-7	DH8
1557	VO	Klagenfurt	16.00	Vienna	16.50	7	DH8
1583	VO	Innsbruck	15.10	Graz	15.55	7	CRJ
1596	VO	Vienna	14.05	Salzburg	15.05	5	DH8
2980	VO	Innsbruck	06.10	Salzburg	06.40	1-7	DH8
2981	VO	Salzburg	12.30	Innsbruck	13.00	1-7	DH8
2983	VO	Salzburg	16.40	Linz	17.05	1-5	CRJ
2984	VO	Innsbruck	14.35	Salzburg	15.10	1-7	DH8
2985	VO	Salzburg	21.40	Innsbruck	22.05	1-7	DH8
2986	VO	Innsbruck	10.20	Salzburg	10.55	7	DH8

Fig. 1. A Many-valued Context about *Domestic flights in Austria*

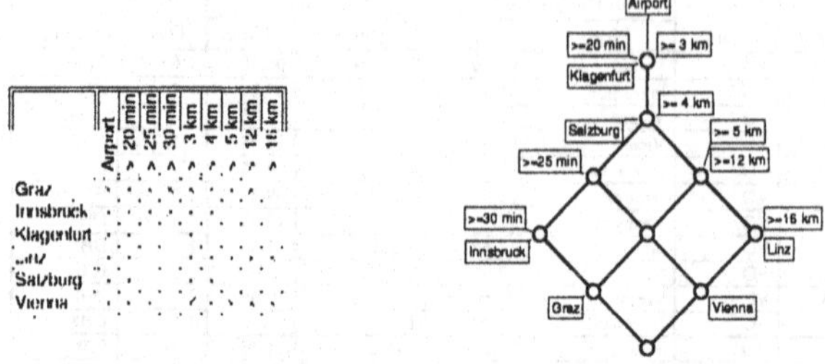

Fig. 2. Conceptual Scale *Airports* and its Concept Lattice

of formal judgments. Therefore we extend the method of conceptual scaling to that of *"Relational Scaling"* for deriving structures of concept graphs from many-valued contexts in addition to concept lattices.

Before describing this in general, we explain the idea of relational scaling by our example of the domestic flights in Austria. The aim is to turn the many-valued context represented in Figure 1 into a family of formal contexts whose formal concepts also yield binary relations. Such transformation should be guided by some purpose which we assume to be the support of flight information. In the formal context represented in Figure 2, for the six airports given in Figure 1, the minimum connecting time between domestic flights and the distance from the airport to the city are indicated (in this scale, additional information from [OAG98] is coded); the concept lattice of this formal context is shown next to it. The concept lattice in Figure 4 yields, for the 75 domestic flights given in Figure 1, the information about the airline, the aircraft, and the days of operating. The lattice of time intervals is represented in Figure 3. Connections between formal attributes in Figure 1 are coded by the binary relations "Flight -

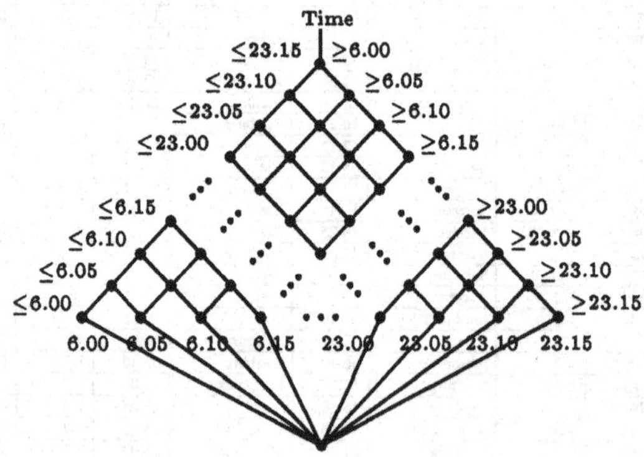

Fig. 3. Concept Lattice of the Conceptual Scale *Time*

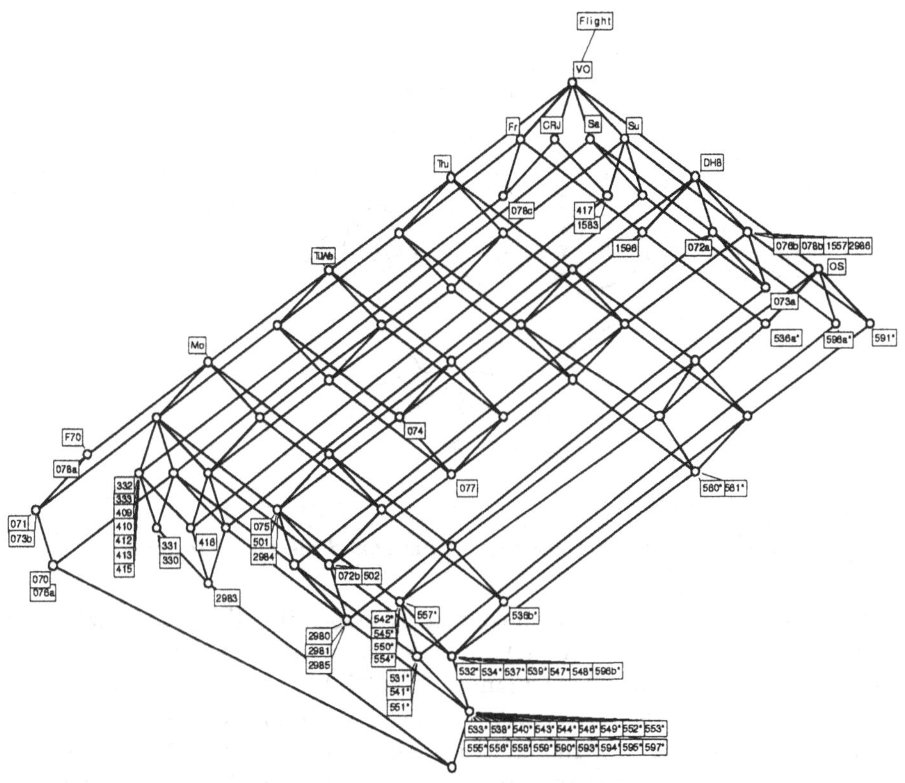

Fig. 4. Concept Lattice of the Conceptual Scale *Flights*

Departure Airport", "Flight - Departure Time", "Flight - Arrival Airport", "Flight - Arrival Time", and represented with respect to the attributes "From", "To", "a.m.", "p.m.", "Graz", "Innsbruck", "Klagenfurt", "Linz", "Salzburg", and "Vienna" by the nested line diagram in Figure 5.

If the underlying contexts of the concept lattices in Figure 2, 3, and 4 are associated as a union $\mathbb{K}_0$ and if $\mathbb{K}_2$ is the underlying context of the concept lattice in Figure 5, then the contexts $\mathbb{K}_0$ and $\mathbb{K}_2$ form a power context family. In general, a *power context family* is a sequence $\vec{\mathbb{K}} := (\mathbb{K}_k)_{k=0,\ldots,n}$ of formal contexts $\mathbb{K}_k := (G_k, M_k, I_k)$ $(k = 0, \ldots, n)$ with $G_k \subseteq (G_0)^k$ for $k \geq 1$. The formal concepts of $\mathbb{K}_k$ with $k = 1, \ldots, n$ represent by their extents k-ary relations on the object set G_0; they are therefore called *"relation concepts"*.

To derive a power context family from a many-valued context (G, M, W, I), several contexts are formed (guided by some purpose) in combining elements of G and W to object sets of formal contexts where the objects of each one of these contexts have always to be k-tuples for a fixed natural number k; these contexts are called *relational scales.* Object sets of formal contexts may also be formed by single elements of G and W; those contexts are called *conceptual scales.* Then the many-valued context together with the chosen conceptual and relational scales is said to be a *relationally scaled context.* Now, a power context family can be

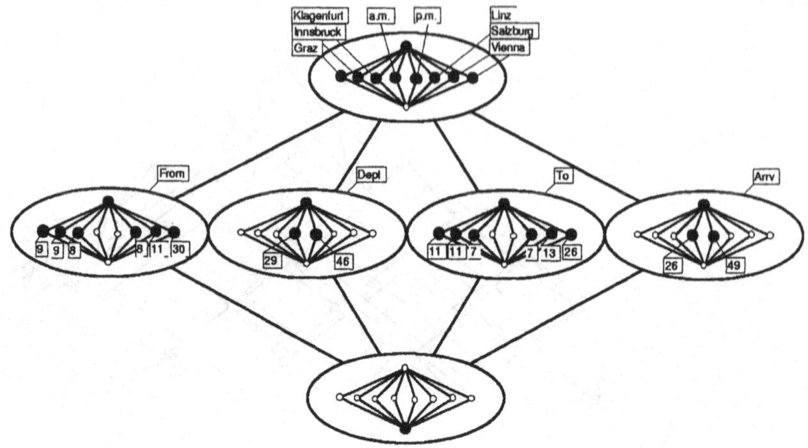

Fig. 5. Concept Lattice of the Relational Scale *Flight Schedule*

derived from a relationally scaled context by associating the conceptual scales to a formal context $\mathbb{K}_0$ and the relational scales to formal contexts $\mathbb{K}_1, \ldots, \mathbb{K}_n$ where the resulting power context family $\vec{\mathbb{K}} := (\mathbb{K}_k)_{k=0,\ldots,n}$ should represent the same information as the relationally scaled context from which it is obtained.

In order to analyze the structure of the data given in the resulting power context family, concept lattices and concept graphs are derived from it. In [VW95], it is shown how the software tool TOSCANA helps to determine and visualize concept lattices from conceptually scaled contexts. For deriving concept graphs from power context families, the first steps are done in [Wi97] and [Pr98b]. In the next section it is explained how *all* concept graphs and specially their natural order can be derived from a power context family.

3 The Lattice of Concept Graphs of a Power Context Family

Concept graphs of a power context family are finite directed multi-hypergraphs whose vertices and edges are specifically labelled by concepts and objects taken from the given power context family. A finite *directed multi-hypergraph* is defined as a set structure (V, E, ν) consisting of two finite sets V and E and a mapping $\nu : E \to \bigcup_{k=1}^{n} V^k$ $(n \geq 2)$; the elements of V and E are called *vertices* and *edges*, respectively, and, if $\nu(e) = (v_1, \ldots, v_k)$, we say that $v_1, \ldots, v_k$ are the *adjacent vertices* of the *k-ary edge* e. We write $|v| = 0$ for $v \in V$ and $|e| = k$ for $e \in E$ with $\nu(e) = (v_1, \ldots, v_k)$.

A *(simple) concept graph of a power context family* $\vec{\mathbb{K}} := (\mathbb{K}_k)_{k=0,\ldots,n}$ with $\mathbb{K}_k := (G_k, M_k, I_k)$ for $k = 0, \ldots, n$ is a set structure $\mathfrak{G} := (V, E, \nu, \kappa, \rho)$ where (V, E, ν) is a finite directed multi-hypergraph, κ assigns to each vertex v a formal concept $\kappa(v)$ of $\mathbb{K}_0$ and to each k-ary edge e a formal concept $\kappa(e)$ of $\mathbb{K}_k$, and ρ

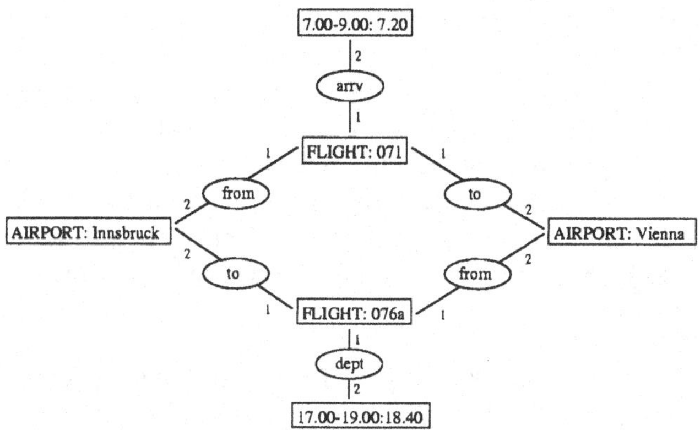

Fig. 6. Concept Graph *Commuter Flight Connection*

yields finite sets $\rho(v)$ of references from the extents of the $\kappa(v)$ so that the extents of the $\kappa(e)$ consist of the k-tuples formed by the references of the adjacent vertices of e, respectively; more precisely,

- $\kappa\colon V \cup E \to \bigcup_{k=0,\ldots,n} \underline{\mathfrak{B}}(\mathbb{K}_k)$ is a mapping such that $\kappa(u) \in \underline{\mathfrak{B}}(\mathbb{K}_k)$ for all $u \in V \cup E$ with $|u| = k$, and
- $\rho\colon V \cup E \to \bigcup_{k=0,\ldots,n} \mathfrak{P}_{fin}(G_k)\backslash\{\emptyset\}$ is a mapping such that $\rho(u) \subseteq Ext(\kappa(u))$ for all $u \in V \cup E$ and, if $|u| = k > 0$ and $\nu(u) = (v_1, \ldots, v_k)$, $\rho(u) = \rho(v_1) \times \cdots \times \rho(v_k)$.

An example for a concept graph of the derived power context family presented above, is given in Figure 6 where the flight connection of a commuter living in Innsbruck and working in Vienna is formalized. The concepts of this concept graph are taken from the conceptual scales described above and the relations are relation concepts of the relational scale in Figure 5 (the concepts having the same name as the attributes in the scales are the attribute concepts of the corresponding attribute). The commuter has to be in Vienna at about 8 o'clock (we took the concept representing the interval 7.00 - 9.00) and can go back as soon as possible after 17 o'clock. Therefore, he takes the flights no. 071 and 076a. For the organization of the flight, it might be interesting to know more specialized concept graphs, having all information given in the concept graph above and additional information (for example about the distances of the airports to the city). For finding a suitable concept graph, a characterization of the generalization is needed.

Therefore, we consider the natural quasi-order $\lesssim$ of *generalization* on the set $\Gamma(\vec{\mathbb{K}})$ of all concept graphs of a power context family $\vec{\mathbb{K}}$. For two concept graphs $\mathfrak{G}_1 := (V_1, E_1, \nu_1, \kappa_1, \rho_1)$ and $\mathfrak{G}_2 := (V_2, E_2, \nu_2, \kappa_2, \rho_2)$, we say $\mathfrak{G}_1$ is *more general* than $\mathfrak{G}_2$ (in symbols: $\mathfrak{G}_1 \lesssim \mathfrak{G}_2$) if for all $u \in V_2 \cup E_2$ there exist $u_1, \ldots, u_j \in V_1 \cup E_1$ with $|u| = |u_1| = \cdots = |u_j|$ and $\kappa(u_1), \ldots, \kappa(u_j) \leq \kappa(u)$ and $\rho(u) \subseteq \rho(u_1) \cup \cdots \cup \rho(u_j)$. The concept graphs $\mathfrak{G}_1$ and $\mathfrak{G}_2$ are said to be *equivalent* in $\vec{\mathbb{K}}$ (in symbols: $\mathfrak{G}_1 \sim \mathfrak{G}_2$) if $\mathfrak{G}_1 \lesssim \mathfrak{G}_2$ and $\mathfrak{G}_2 \lesssim \mathfrak{G}_1$. The class of all concept graphs of $\vec{\mathbb{K}}$ which are equivalent to a given concept graph $\mathfrak{G}$ is denoted by $\underset{\sim}{\mathfrak{G}}$.

The set of all equivalence classes of concept graphs in $\vec{\mathbb{K}}$ together with the order induced by the quasi-order $\lesssim$ is an ordered set denoted by $\underset{\sim}{\Gamma}(\vec{\mathbb{K}})$. For the purpose-oriented search of suitable concept graphs, humanly readable representations of $\underset{\sim}{\Gamma}(\vec{\mathbb{K}})$ are desirable. For this it is useful that $\underset{\sim}{\Gamma}(\vec{\mathbb{K}})$ is always a lattice which is isomorphic to a subdirect product of specific sublattices of the concept lattices $\underline{\mathfrak{B}}(\mathbb{K}_k)$ $(k = 0, \dots, n)$, each extended by a new top element, as Proposition 1 states.

Proposition 1. *Let* $\vec{\mathbb{K}} := (\mathbb{K}_k)_{k=0,\dots,n}$ *be a power context family with* $\mathbb{K}_k := (G_k, M_k, I_k)$ *for* $k = 0, \dots, n$*; furthermore, for each* $g \in G_k$*, let*

$$L_k^g := \{\mathfrak{c} \in \underline{\mathfrak{B}}(\mathbb{K}_k) \mid g \in Ext(\mathfrak{c})\} \cup \{\top_k^g\}$$

be the interval of all superconcepts of (g'', g') *in* $\underline{\mathfrak{B}}(\mathbb{K}_k)$*, together with a new top-element* $\top_k^g$*. Then* $\underset{\sim}{\Gamma}(\vec{\mathbb{K}})$ *is isomorphic to the subdirect product of the lattices* L_k^g *with* $k \in \{0, \dots, n\}$ *and* $g \in G_k$ *consisting of all elements* $\vec{\mathfrak{a}} := (\mathfrak{a}_k^g)_{k=0,\dots,n}^{g \in G_k}$ *of the directed product with only finitely many non-top components satisfying the following condition:*

$(\star)$ *If* $\mathfrak{a}_k^g \neq \top_k^g$ *and* $g = (g_1, \dots, g_k)$ *then* $\mathfrak{a}_0^{g_i} \neq \top_0^{g_i}$ *for* $i = 1, \dots, k$.

Proof. By definition, a concept graph $\mathfrak{G} := (V, E, \nu, \kappa, \rho)$ is equivalent to the disjoint union of all its elementary subgraphs consisting of at most one edge. If an elementary subgraph $\mathfrak{H}$ has an edge e and an adjacent vertex v with $\kappa(v) \neq \top_0 := (G_0, G_0')$, then $\mathfrak{H}$ is equivalent to the disjoint union of the subgraph that is only consisting of the single vertex v and of $\mathfrak{H}$, modified by setting $\kappa(v) := \top_0$. This argument shows that a concept graph $\mathfrak{G}$ is always equivalent to the disjoint union of all its elementary subgraphs consisting of only one vertex and of all concept graphs derived from the elementary subgraphs with exactly one edge by replacing the images of the adjacent vertices under κ by $\top_0$. Further, we use that a concept graph consisting of only one vertex v with object set $\rho(v)$ is equivalent to the disjoint union of $|\rho(v)|$-many of its copies having just one object out of $\rho(v)$ as reference; analogously, a concept graph consisting of only one edge (having adjacent $\top_0$-vertices) with object set $\rho(e)$ is equivalent to the disjoint union of $|\rho(e)|$-many of its copies having just one object out of $\rho(e)$ as reference. In this way we obtain that the concept graph $\mathfrak{G}$ is equivalent to the disjoint union of the derived *atomic concept graphs* which are either single vertices with only one reference or single edges whose adjacent vertices have assigned only one reference and concepts $\top_0$.

Now, for $g \in G_k$, let $\mathfrak{c}_k^g(\mathfrak{G})$ be the element $\top_k^g$ if there is no $u \in V \cup E$ with $g \in \rho(u)$, and let it otherwise be the infimum of all $\kappa(u)$ with $g \in \rho(u)$ (and $|u| = k$). In the second case we construct the concept graph consisting only of $\hat{u}$ with $\kappa(\hat{u}) = \mathfrak{c}_k^g(\mathfrak{G})$ and $\rho(\hat{u}) = \{g\}$ and, if $g = (g_1, \dots, g_k)$, also of $\hat{v}_1, \dots, \hat{v}_k$ with $\nu(\hat{u}) = (\hat{v}_1, \dots, \hat{v}_k)$, $\kappa(\hat{v}_i) = \top_0$ and $\rho(\hat{v}_i) = g_i$ for $i = 1, \dots, k$. The constructed concept graph is equivalent to the disjoint union of all the derived atomic concept graphs having g as reference.

To sum up, the concept graph $\mathfrak{G} := (V, E, \nu, \kappa, \rho)$ is equivalent to the disjoint union of the atomic concept graphs $\mathfrak{G}(\mathfrak{c}_0^g(\mathfrak{G})) := (\{v\}, \emptyset, \nu_0^g, \kappa_0^g, \rho_0^g)$ ($g \in G_0$ and $v \in V$) with $\kappa_0^g(v) = \mathfrak{c}_0^g(\mathfrak{G})$ and $\rho_0^g(v) = \{g\}$ and of the atomic concept graphs $\mathfrak{G}(\mathfrak{c}_k^g(\mathfrak{G})) := (\{v_1, \ldots, v_k\}, \{e\}, \nu_k^g, \kappa_k^g, \rho_k^g)$ (with $k \in \{1, \ldots, n\}$, $g \in G_k$, $e \in E$ and $\nu(e) = (v_1, \ldots, v_k)$) with $\kappa_k^g(e) = \mathfrak{c}_k^g(\mathfrak{G})$ and $\rho_k^g(e) = \{g\}$. It follows that

$$\begin{aligned} \eta^{\vec{\mathbb{K}}}: \ \underset{\sim}{\Gamma}(\vec{\mathbb{K}}) &\to \prod (L_k^g \mid k \in \{0, \ldots, n\} \text{ and } g \in G_k) \quad \text{with} \\ \eta^{\vec{\mathbb{K}}}(\underset{\sim}{\mathfrak{G}}) &:= (\mathfrak{c}_k^g(\mathfrak{G}) \mid k \in \{0, \ldots, n\} \text{ and } g \in G_k) \end{aligned}$$

is a mapping, the image of which consists of all elements of the direct product with only finitely many non-top components satisfying condition ($\star$). It can be easily seen that this image is a subdirect product. For concept graphs $\mathfrak{G}_1$ and $\mathfrak{G}_2$, we have the equivalences

$$\begin{aligned} \mathfrak{G}_1 \lesssim \mathfrak{G}_2 \quad &\Leftrightarrow \quad \forall k \in \{0, \ldots, n\}\ \forall g \in G_k\colon\ \mathfrak{c}_k^g(\mathfrak{G}_1) \leq \mathfrak{c}_k^g(\mathfrak{G}_2) \\ &\Leftrightarrow \quad \eta^{\vec{\mathbb{K}}}(\underset{\sim}{\mathfrak{G}}_1) \leq \eta^{\vec{\mathbb{K}}}(\underset{\sim}{\mathfrak{G}}_2). \end{aligned}$$

Therefore, $\eta^{\vec{\mathbb{K}}}$ is an injective homomorphism from the set $\underset{\sim}{\Gamma}(\vec{\mathbb{K}})$ into the product $\prod(L_k^g \mid k \in \{0, \ldots, n\}$ and $g \in G_k)$. Thus, the assertion of the proposition is proved. $\square$

Proposition 1 yields, for the concept graphs of a power context family, a system of representatives for their equivalence classes which makes the ordering of generalization transparent. These representatives are described by the elements of all the direct products

$$\prod_{(k,g)\in U} L_k^g \setminus \{\top_k^g\}$$

for which U is a finite subset of $\bigcup_{k=0,\ldots,n}\{k\} \times G_k$ satisfying the implication

$$(g, k) \in U \text{ and } g = (g_1, \ldots, g_k) \Rightarrow (0, g_1), \ldots, (0, g_k) \in U.$$

An element $\vec{\mathfrak{a}} := (\mathfrak{a}_k^g)_{(k,g)\in U}$ of the product represents the concept graph which is the disjoint union of the atomic concept graphs $\mathfrak{G}(\mathfrak{a}_k^g)$ with $(k, g) \in U$. In the full direct product $\prod(L_k^g \mid k \in \{0, \ldots, n\}$ and $g \in G_k)$ this concept graph is represented by the element which coincides with $\vec{\mathfrak{a}}$ on U and has outside U the corresponding top-elements as components (the top-element $\top_k^g$ as a component indicates that g is not a reference in the represented graph).

Proposition 1 clarifies how to deduce concept graphs from the concept lattices of a power context family. This shall be demonstrated by our flight example. In Figure 7 the lattice of concept graphs is shown so far that the representation of the concept graph in Figure 6 becomes visible, i.e. for each vertex and edge of the concept graph in Figure 6, we have considered its reference g and represented the lattice L_k^g (where $k = 0$ if g is reference of a vertex and $k = 2$ if g is a pair of objects, i. e. a reference of an edge). Theses lattices are obtained by adding a new top element to the interval of all superconcepts of the object concept (g'', g')

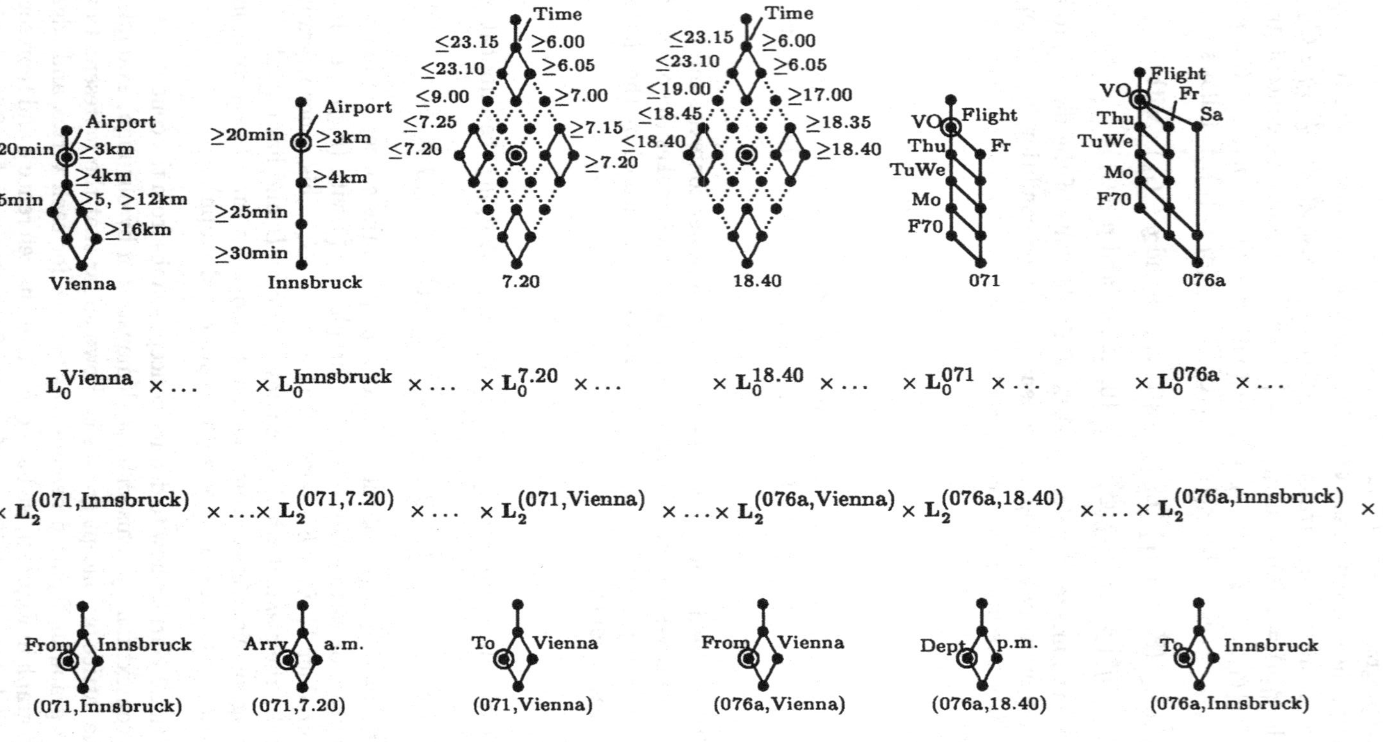

Fig. 7. Lattice of Concept Graphs of the Relationally Scaled Context *Flights in Austria*

of the reference g in $\underline{\mathfrak{B}}(\mathbb{K}_k)$. For example, for the vertice of the concept graph in Figure 6 being labbeled with Flight and referenced with 071, the lattice L_0^{071}, without the top element, is a sublattice of the lattice in Figure 4.

For deducing the concept graph from the product of all lattices L_k^g, we take, for each g, a component $\mathfrak{a}_k^g$ that is represented as an element of L_k^g. For example, the component Flight_0^{071} is represented by the attribute concept of Flight in the lattice L_0^{071}. For our concept graph in Figure 6, we choose the following components (and the top elements for all other g):

$$\mathsf{AIRPORT}_0^{\mathsf{Vienna}}, \mathsf{AIRPORT}_0^{\mathsf{Innsbruck}}, \mathsf{7.00\text{-}9.00}_0^{\mathsf{7.20}}, \mathsf{17.00\text{-}19.00}_0^{\mathsf{18.40}},$$

$$\mathsf{FLIGHT}_0^{\mathsf{071}}, \mathsf{FLIGHT}_0^{\mathsf{076a}}, \mathsf{from}_2^{(\mathsf{071,Innsbruck})}, \mathsf{arrv}_2^{(\mathsf{071,7.20})},$$

$$\mathsf{to}_2^{(\mathsf{071,Vienna})}, \mathsf{from}_2^{(\mathsf{076a,Vienna})}, \mathsf{dept}_2^{(\mathsf{076a,18.40})}, \mathsf{to}_2^{(\mathsf{076a,Innsbruck})}.$$

Their representations in the corresponding lattices are marked by a second circle. This may justify hopes that line diagrams could support the finding of meaningful concept graphs for interpreting data. If, for example, the commuter wants a concept graph that is more specified than the one in Figure 6, he can take each graph that is represented by elements being under the marked components. Since the lattices L_k^g can be easily represented as concept lattices of suitable contexts so that their product becomes the concept lattice of the sum of those contexts, it is possible to use the TOSCANA software [VW95] for navigating visually through the lattice of concept graphs of a given power context family (corresponding to a relationally scaled data context).

4 The Lattice of Concept Graphs of a Contextual Syntax

So far, we have studied the lattice of concept graphs of a power context family. The results change slightly when we start with a contextual syntax and examine all concept graphs over the given alphabet, independent of a concrete model. For this, we consider concept graphs as syntactical constructs with semantics in power context families. This has been presented in [Pr98a]; here, we only repeat briefly the main definitions.

Considering a contextual syntax, we start with a *conceptual alphabet* which is a triple $(\mathcal{G}, \mathcal{C}, \mathcal{R})$ where $\mathcal{G}$ is a finite set of *object names*, $(\mathcal{C}, \leq_{\mathcal{C}})$ is a finite ordered set of *concept names*, and $(\mathcal{R}, \leq_{\mathcal{R}})$ is a set, partitioned into finite ordered sets $(\mathcal{R}_k, \leq_{\mathcal{R}_k})$ of *relation names* (with $k = 1, \ldots, n$). These orders are determined by the taxonomies of the domains in view; they formalize background knowledge.

A *(syntactic) concept graph over the alphabet* $(\mathcal{G}, \mathcal{C}, \mathcal{R})$ is a structure $\mathfrak{G} := (V, E, \nu, \kappa, \rho)$ where κ and ρ map into $\mathcal{G}, \mathcal{C}$ and $\mathcal{R}$. Thus, it is a directed multi-hypergraph with vertices and edges labelled by object, concept and relation names. More precisely,

- $\kappa\colon V \cup E \to \mathcal{C} \cup \mathcal{R}$ is a mapping such that $\kappa(V) \subseteq \mathcal{C}$ and $\kappa(E) \subseteq \mathcal{R}$, and all $e \in E$ with $\nu(e) = (v_1, \ldots, v_k)$ satisfy $\kappa(e) \in \mathcal{R}_k$, and

- $\rho : V \cup E \to \bigcup_{k=0,\dots,n} \mathfrak{P}(\mathcal{G}^k)\backslash\{\emptyset\}$ is a mapping with $\rho(V) \subseteq \mathfrak{P}(\mathcal{G})$ and $\rho(e) = \rho(v_1) \times \cdots \times \rho(v_k)$ for all $e \in E$ with $\nu(e) = (v_1, \dots, v_k)$.

For this syntactical construct, a *semantics* is given in a power context family $\vec{\mathbb{K}} := (\mathbb{K}_k)_{k=0,\dots,n}$ with $\mathbb{K}_k := (G_k, M_k, I_k)$ for each k. The object names are interpreted by objects of G_0, the concept names by concepts of $\mathbb{K}_0$ and the relation names of $\mathcal{R}_k$ by relation concepts of $\mathbb{K}_k$. This interpretation is described by an order-preserving mapping $\iota : \mathcal{G} \cup \mathcal{C} \cup \mathcal{R} \to G_0 \cup \mathfrak{B}(\mathbb{K}_0) \cup \bigcup_{k=1,\dots,n} \mathfrak{B}(\mathbb{K}_k)$. The context-interpretation $(\vec{\mathbb{K}}, \iota)$ is called a model if ι is consistent with all information given by the concept graph, i.e. all $u \in V \cup E$ satisfy $\iota(\rho(u)) \subseteq \operatorname{Ext}(\iota(\kappa(u)))$.

An interesting model for our purpose is the so-called standard model of a given concept graph. We recall the definition: The *standard model* of the concept graph $\mathfrak{G} := (V, E, \nu, \kappa, \rho)$ over the alphabet $(\mathcal{G}, \mathcal{C}, \mathcal{R})$ is defined by $\vec{\mathbb{K}}^{\mathfrak{G}} := (G_k^{\mathfrak{G}}, M_k^{\mathfrak{G}}, I_k^{\mathfrak{G}})_{k=0,\dots,n}$ with $G_0^{\mathfrak{G}} := \mathcal{G}, M_0^{\mathfrak{G}} := \mathcal{C}$, $G_k^{\mathfrak{G}} := \mathcal{G}^k$, and $M_k^{\mathfrak{G}} := \mathcal{R}_k$ for all $k = 1, \dots, n$. The incidence relation $I_0^{\mathfrak{G}}$ is defined in such a way that all $g \in \mathcal{G}$ and $c \in \mathcal{C}$ satisfy $g I_0^{\mathfrak{G}} c$ if there exists a $v \in V$ with $\kappa(v) \leq_{\mathcal{C}} c$ and $g \in \rho(v)$. Analogously, all $(g_1, \dots, g_k) \in \mathcal{G}^k$ and all $R \in \mathcal{R}_k$ with $k = 1, \dots, n$ satisfy $(g_1, \dots, g_k) I_k^{\mathfrak{G}} R$ if there exists an $e \in E$ with $\kappa(e) \leq_{\mathcal{R}} R$ and $(g_1, \dots, g_k) \in \rho(e)$. In this power context family, the object names of $\mathfrak{G}$ are interpreted by themselves, the concept and relation names by the corresponding attribute concepts.

We can use the standard model as an interesting tool for the problem of entailment of concept graphs. By definition, $\mathfrak{G}_1$ entails $\mathfrak{G}_2$ if $\mathfrak{G}_2$ is valid in every model for $\mathfrak{G}_1$. The notion of entailment corresponds to a sound and complete set of derivation rules (see [Pr98a]) and the definition of generalization given above. Thus, entailment of two concept graphs can be characterized by subsumption of the incidence relations of their standard models. This is stated in the following proposition that is proved in [Pr98b].

Proposition 2. *Let $\mathfrak{G}_1$ and $\mathfrak{G}_2$ be two concept graphs over the same alphabet and let $(G_k^{\mathfrak{G}_i}, M_k^{\mathfrak{G}_i}, I_k^{\mathfrak{G}_i})_{k=0,\dots,n}$ for $i = 1, 2$ be their standard models. Then, we have*

$$\mathfrak{G}_1 \models \mathfrak{G}_2 \quad \iff \quad I_k^{\mathfrak{G}_1} \supseteq I_k^{\mathfrak{G}_2} \textit{ for all } k = 0, \dots, n.$$

The proposition implies that equivalent concept graphs have equal standard models. With it, we can characterize the lattice $(\underline{\Gamma}(\mathcal{A}), \models)$ of (equivalence classes of) concept graphs of a given alphabet $\mathcal{A}$ by means of the incidence relation in the corresponding standard models. The greatest lower bound of the two equivalence classes $\underset{\sim}{\mathfrak{G}}_1$ and $\underset{\sim}{\mathfrak{G}}_2$ is the equivalence class of the concept graph $\mathfrak{G}_1 \dot{\cup} \mathfrak{G}_2$ which is the disjoint union of both parts. The incidence relations of its standard model are $I_k^{\mathfrak{G}_1 \dot{\cup} \mathfrak{G}_2} = I_k^{\mathfrak{G}_1} \cup I_k^{\mathfrak{G}_2}$ for all $k = 0, \dots, n$. The lowest upper bound (i.e. the least common generalization) of two concept graphs (resp. their equivalence classes) is given by the intersection of the incidence relations. By this means, we can easily get that the least common generalization is obtained by the disjoint union of all common generalizations.

Furthermore, we can derive from Proposition 2 that the lattice of equivalence classes of concept graphs of a given alphabet is complete and distributive because intersection and union are. For finite distributive lattices, we know from Birkhoff's Theorem (cf. eg. [GW99]) that they are isomorphic to the lattice of the orderfilters of the ordered set of all $\wedge$-irreducible elements of the lattice. Since the $\wedge$-irreducible elements of $\underset{\sim}{\Gamma}(\mathcal{A})$ are exactly the equivalence classes of the atomic concept graphs (those that consist of only one isolated vertex with a single reference or one edge with its adjacent vertices labelled with $\top$ and only one reference), we obtain the following result:

Proposition 3. *The lattice $(\underset{\sim}{\Gamma}(\mathcal{A}), \models)$ of all equivalence classes of concept graphs of an alphabet $\mathcal{A}$ is the free $\wedge$-semilattice over the ordered set of the atomic concept graphs of the contextual syntax.*

This proposition helps us to clarify how the lattice $\underset{\sim}{\Gamma}(\mathcal{A})$ of concept graphs of a given alphabet $\mathcal{A}$ is related to the lattice $\underset{\sim}{\Gamma}(\vec{\mathbb{K}})$ of concept graphs of a power context family $\vec{\mathbb{K}}$ for a context-interpretation $(\vec{\mathbb{K}}, \iota)$ of $\mathcal{A}$. With Proposition 1 and $\underset{\sim}{\Gamma}(\mathcal{A})$ being the free $\wedge$-semilattice over the ordered set of atomic concept graphs, we obtain that there exists a ι-faithful $\wedge$-homomorphism from $\underset{\sim}{\Gamma}(\mathcal{A})$ to $\underset{\sim}{\Gamma}(\vec{\mathbb{K}})$. Thus, whereas the order of $\underset{\sim}{\Gamma}(\mathcal{A})$ is only determined by the orders $\leq_{\mathcal{C}}$ and $\leq_{\mathcal{R}}$ of the alphabet, the $\underset{\sim}{\Gamma}(\vec{\mathbb{K}})$ is restricted by additional dependencies given in $\vec{\mathbb{K}}$.

To sum up, the lattice of concept graphs of a power context family as well as the lattice of concept graphs of a contextual syntax can completely be described by structural considerations. This gives us additional methods to determine logical dependencies of concept graphs besides the inference tools presented in [Pr98a].

5 Further Research

During the work on this paper, "Relational Scaling" of many-valued contexts has been created as an extension of Conceptual Scaling. It seems worth to investigate the range of applications of this new method in Data Analysis. Relational Scaling especially allows to activate the Theory of Conceptual Graphs for analyzing and interpreting data. This stimulates many research questions. An important one asks for methods of finding meaningful concept graphs contributing to the fulfillment of specific purposes of data analysis. In particular, graphical methods are desirable for which the TOSCANA software should be further elaborated.

Another research problem is how to use conceptual graphs for purpose-oriented retrieval on relationally scaled databases. For this, the theory of concept graphs with quantifiers should be further developed.

From such a development, the general research program of establishing a comprehensive contextual logic would benifit too. Contextual Logic in its support of conceptual knowledge representation and processing will gain from further research about concept graphs of relationally scaled contexts. In particular, the development of conceptual knowledge and information systems should integrate such research on relational scaling and concept graphs.

References

[GW89] B. Ganter, R. Wille: Conceptual scaling. In: F. S. Roberts (ed.): Applications of Combinatorics and Graph Theory to the Biological and Social Sciences. Springer, Berlin-Heidelberg-New York 1989, 139–167.

[GW99] B. Ganter, R. Wille: Formal Concept Analysis: Mathematical Foundations. Springer, Berlin-Heidelberg-New York 1999.

[Ka88] I. Kant: Logic. Dover, New York 1988.

[OAG98] OAG Pocket Flight Guide - Europe/Africa/Middle East. Reed Elsevier, July 1998.

[Pr98a] S. Prediger: Simple concept graphs: a logic approach. In: M.-L. Mugnier. M. Chein (eds.): Conceptual Structures: Theory, Tools and Applications. Springer, Berlin-Heidelberg-New York 1998, 225–239.

[Pr98b] S. Prediger: Kontextuelle Urteilslogik mit Begriffsgraphen. Ein Beitrag zur Restrukturierung der mathematischen Logik. Dissertation, Shaker Verlag, Aachen 1998.

[So84] J. F. Sowa: Conceptual structures: information processing in mind and machine. Addison-Wesley, Reading 1984.

[So92] J. F. Sowa: Conceptual graphs summary. In: T. E. Nagle, J. A. Nagle, L. L. Gerholz, P. W. Eklund (eds.): Conceptual Structures: Current Research and Practice. Ellis Horwood 1992, 3–51.

[VW95] F. Vogt, R. Wille: TOSCANA - a graphical tool for analyzing and exploring data. In: R. Tamassia, I. G. Tollis (eds.): Graph Drawing. LNCS 894. Springer, Berlin-Heidelberg-New York 1995, 226–233.

[Wi82] R. Wille: Restructuring lattice theory: an approach based on hierarchies of concepts. In: I. Rival (ed.): Ordered sets. Reidel, Dordrecht, Boston 1982, 445–470.

[Wi97] R. Wille: Conceptual Graphs and Formal Concept Analysis. In: D. Lukose, H. Delugach, M. Keeler, L. Searle, J. Sowa (eds.): Conceptual Structures: Fulfilling Peirce's Dream. Springer, Berlin-Heidelberg-New York 1997, 290–303.

Contexts in Information Systems Development

Ryszard Raban, Brian Garner

University of Technology, Sydney
Deakin University
Richard@socs.uts.edu.au, brian@deakin.edu.au

Abstract. This paper looks at information systems development as a process of gathering and unifying user requirements that usually carry only partial and conditional descriptions of the system. It has been suggested to use context dependent requirements specifications to formalize the requirements and conditional conceptual graphs to create a global system model. A procedure for context dependent requirements unification has been described and an example given.

1 Introduction

Information systems development incorporates activities that bring together many parties in order to

- Gather requirements for information services,
- Transform the requirements into designs of different components of the system,
- Deliver a system that meets all the original requirements.

The main challenge of the information systems development is to transform the original user requirements into a working system that meets the complete set of requirements.

This is a complex process that produces a great number of information systems representations. J. Sowa together with J. Zachman [1] have created taxonomy of possible systems representations. It is built around two dimensions:

- the perspectives (the planner's scope description, the owner's business model, the designer's system model, the builder's technology model, the sub-contractor's component models), and
- the system focuses (data – *what* is to be processed, function - *how* is to be processed, network - *where* is to be processed, people - *who* is involved, time - *when* it will take place, motivation - *why* the processing is to be done),

as shown in Fig. 1. They also suggested the use of conceptual graphs to deal with the multiplicity of modeling techniques used to represent different facets of the system and to enable effective unification of these representations.

However, there is an additional level of complexity in the information systems development process. All the information systems requirements come from different parties, as there are usually many planners, owners and designers that deal with different parts of the system. They create partial descriptions of the system, and also they might use different terminologies to state their requirements. And yet, there is a common underlying goal for all the participants, which is to build a sharable system meeting all the individual requirements. In order to deal with this situation the information systems development has to be a process of constant goal setting, requirements solicitation, consistency checks, conflict resolutions and requirements unification.

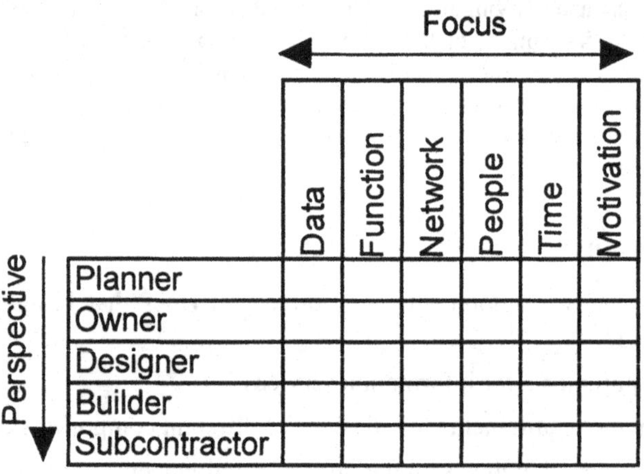

Focus
Data
Function
Network
People
Time
Motivation
Perspective
Planner
Owner
Designer
Builder
Subcontractor

Fig. 1. Zachman's taxonomy of IS models.

The paper suggests using contexts for specifying the situation in which each requirement is formulated. Then, it discusses how the contexts can be unified under possible incompleteness and conflict conditions in order to create the goal setting model of the whole system. Conceptual graphs are used to represent the contexts and requirements, and a procedure for the context dependent requirements unification is presented.

2 Information System Contexts

Let M be a set of propositions defining the structure of information system.

Let W be a set of assertions representing facts stored in information system such that $\forall p\ (p \in M) \wedge (W \models p)$.

Let $\mathcal{W}_M$ be a set of all sets of assertions such that
$\forall W\ \forall p\ (W \in \mathcal{W}_M) \wedge (p \in M) \wedge (W \models p)$.

Information system transactions change the set of assertions by adding, deleting and/or updating existing facts. In fact, every transaction creates a new set of assertions W'. It is the responsibility of the information system processing engine to ensure that every new set of assertions belongs to W_M. In this way, the information system is said to maintain its integrity.

In the information system development process, analysts gather requirements from the users, and unify them in order to create information system model M. However, as it has been explained earlier, different participants of the process are interested in different parts of the system. Software designers are concerned with processing while database designers with data. Also, many of those requirements are relevant only in specific situations determined by a task, location, time, type of data, etc. These specific situations are called requirement contexts.

This definition of context is different to what has been discussed in [2] or [3]. A context here is not to define a world of assertions, but rather to define a relevant subset of a given world of assertions, which is of interest in this specific situation.

An information systems context dependent requirement is a pair (c, $\{R_1, \ldots, R_n\}$) where

- c is a context in which requirements $R_1, \ldots, R_n$ are relevant,
- c is one or more context graphs, and
- R_i is either a requirement graph r or a context dependent requirement (c', $\{R'_1, \ldots, R'_n\}$) for which c is a dominant context.

Information system requirements are represented by (T, $\{R_1, \ldots, R_m\}$) where T represents the most general context for the system. The context dependent requirements can be as deeply nested as required to represent the variety of viewpoints adopted by the system's stakeholders.

For example, let us consider the following system:

There are orders placed for parts. Orders are internal or external. An internal order is issued to a department. An external order has to have a supplier. External orders above $500 have to be approved by a manager. Internal orders above $1000 need to be approved by a manger.

In this example, there is one general requirement:

r_1: [ORDER]->(FOR)->[PART]

and requirements relevant in two different context

c_1: [ORDER]->(CHRC)->[ORDER_KIND: internal] and

c_2: [ORDER]->(CHRC)->[ORDER_KIND: external].

In context c_1 it is required that

r_2: [ORDER]->(TO)->[DEPARTMENT]

and in context c_2 it is required that

r_3: [ORDER]->(TO)->[SUPPLIER].

Additionally in context c_1 when

c_3: [ORDER]->(CHRC)->[AMOUNT > \$1000] then

r_4: [ORDER]<-(OBJ)<-[APPROVE]->(AGNT)->[MANAGER]

and in context c_2 when

c_4: [ORDER]->(CHRC)->[AMOUNT > \$500] then

r_5: [ORDER]<-(OBJ)<-[APPROVE]->(AGNT)->[MANAGER].

The above information system requirements can be represented as a hierarchy of nested contexts.

(T, { r_1,
    (c_1, {r_2,
        (c_3, {r_4})
        }
    ),
    (c_2, {r_3,
        (c_4, {r_5})
        }
)
    }
)

A hierarchy of context dependent requirements can be decomposed into individual requirements by using the following decomposition rules:

- $(c, \{R_1, \ldots, R_n\})$ can be decomposed into $(c, \{R_1\}), \ldots, (c, \{R_n\})$
- $(c, \{(c', \{R_1, \ldots, R_n\})\})$ can be decomposed into $(c.c', \{R_1, \ldots, R_n\})$ where $c.c'$ is a graph or graphs obtained by maximally joining graphs c and c' on identical concepts.

By using the rules, the hierarchy of nested context dependent requirements shown above can be decomposed into (T, $\{r_1\}$), (c_1, $\{r_2\}$), ($c_1.c_3$, $\{r_4\}$), (c_2, $\{r_3\}$) and ($c_2.c_4$, $\{r_5\}$). In this case $c_1.c_3$ is

$c_1.c_3$: [ORDER]-
    (CHRC)->[ORDER_KIND: internal]
    (CHRC)->[AMOUNT > \$1000],

and $c_2.c_4$ is

$c_2.c_4$: [ORDER]-
    (CHRC)->[ORDER_KIND: external]
    (CHRC)->[AMOUNT > \$500].

Note, that each of these decomposed context dependent requirements has its dominant context that can be derived from the requirements hierarchy. If requirement $(c, \{r\})$ has contexts $c_1, \ldots, c_n$ on the path between itself and the root of the context hierarchy tree, its dominant context is $c_1.c_2. \ldots .c_n$. For example, the dominant context of $(c_1, \{r_2\})$ is T while the dominant context of $(c_1.c_3, \{r_4\})$ is c_1.

3 Information System Model

An information system model is created from the requirements by successively unifying all the context dependent requirements into a common model. A resultant model is represented by conditional conceptual graphs. In conditional conceptual graphs every concept and relation is qualified by one or more condition graphs. A concept or relation is considered relevant only if its condition graphs are true. For example, requirement r_2 which is relevant in context c_1: [ORDER]->(CHRC)->[ORDER_KIND: internal] can be represented by a conditional conceptual graph as

cr_2: [ORDER|$N(c_1)$]->(TO|$N(c_1)$)->[DEPARTMENT|$N(c_1)$]

and requirement r_5 with context $c_2.c_4$ by a conditional graph like

cr_5: [ORDER|$N(c_2.c_4)$]<-(OBJ|$N(c_2.c_4)$)<-[APPROVE|$N(c_2.c_4)$]-
->(AGNT|$N(c_2.c_4)$)->[MANAGER|$N(c_2.c_4)$].

In this notation, function N means *normalization* of the graphs that are listed as its arguments. The normalization function takes the argument graphs and removes all graphs that are subsumed by one or more other argument graphs. In this way, all redundant graphs are removed. For example, $N(c_1, c_2)$, where c_1 is generalization of c_2, removes c_2 because it is redundant as c_1 already subsumes it.

To perform maximal join of two conditional graphs g_1 and g_2 the following procedure is used:

1° strip g_1 and g_2 of all conditions creating g'_1 and g'_2 respectively,

2° maximally join g'_1 and g'_2 on identical concepts creating an unconditional graph g,

3° restore original conditions in all concepts and relations of g that have not been joined with any other concept or relation,

4° if a concept or relation in g_1 with condition c_1 was joined with a concept or relation in g_2 with condition c_2, the resulting concept or relation in g is given condition $N(c_1, c_2)$.

For example, after maximally joining graphs cr'_2 and cr'_5 the following graph is produced:

cg: [ORDER|$N(c_1, c_2.c_4)$]-
(TO|$N(c_1)$)->[DEPARTMENT|$N(c_1)$]
(OBJ|$N(c_2.c_4)$)<-[APPROVE|$N(c_2.c_4)$]-
->(AGNT|$N(c_2.c_4)$)->[MANAGER|$N(c_2.c_4)$].

Using the above definitions the creation of information system model M out of the information system requirements can be described. Let us suppose that a hierarchy of context dependent requirements has been decomposed into individual requirements (c_i, $\{r_i\}$), and for each of them its dominant context c_{di} has been identified. Information system model M is created by using the following unification procedure:

1° **for** each of the individual requirements (c_i, $\{r_i\}$) with dominant context c_{di} **do**

1.1° remove all individual markers from c_i creating c'_i,

1.2° create conditional conceptual graph c''_i by adding condition c_{di} to all concepts and relations in c'_i,

1.3° maximally join graphs in M with c''_i on identical concepts creating M',

1.4° all graphs in c''_i that could not be joined should be added to M' creating M'',

1.5° create conditional conceptual graph cr_i by adding condition c_i to all concepts and relations in r_i,

1.6° maximally join graphs in M'' with cr_i on identical concepts creating M''',

1.7° if conditional graph cr_i could not be joined with any graph in M''', add it to M'''.

By following the above procedure the information system requirements of the example system produce the following information system model M:

[ORDER]-
(FOR)->[PART]
(CHRC)->[ORDER_KIND]
(TO|$N(c_1)$)->[DEPARMENT|{$N(c_1)$}]
(TO|$N(c_2)$)->[SUPPLIER|{$N(c_2)$}]
(CHRC|{$N(c_1, c_2)$})->[AMOUNT|{$N(c_1, c_2)$]
(OBJ|{$N(c_1, c_3, c_2.c_4$})<-[APPROVE|{$N(c_1.c_3, c_2.c_4)$}]-
->(AGNT|$N(c_1.c_3, c_2.c_4)$)->[MANAGER|$N(c_1.c_3, c_2.c_4)$]

Note that concepts ORDER, PART and ORDER_KIND do not have any condition attached. By default its condition is the general system context T. And since N(T, c) = T, step 4° of the conditional graphs maximal join procedure used in steps 1.3° and 1.6° of the unification procedure does not alter it.

4 Context Vocabulary

So far, it was assumed that all contexts and requirements were stated using the same set of type labels and the same type hierarchy. As it was suggested by Norman in [4], user local models can also differ from the global information system model in terminology used to describe elements of the system. For example, in the same company, while referring to final products marketing talks about merchandise, manufacturing about parts and warehousing about items.

In order to cater for the terminological differences, an information systems context dependent requirements definition would have to be extended into a tuple (c, t, {R_1, ..., R_n}) where t is a set of type definitions for all type labels used to state requirements R_1, ..., R_n in context c.

It is beyond the scope of this paper to discuss how the type definitions can be incorporated into the requirements unification process. However, some initial thoughts on this topic can be found in [5].

5 Extraction Function

Another useful operation on conditional conceptual graphs is *extraction*. Given a set of conditional conceptual graphs G and a graph c, $G|c$ means all the graphs that remain after removing from G all concepts and relations which conditions are not generalizations of c.

The extraction function allows retrieving all parts of the information system that are relevant in a given context c. Its primary role is requirements verification. In this case, the user can be presented with all the requirements that have been specified in a particular context. For example, for context c_2 the extraction function produces the relevant part of model M, which is

```
[ORDER]-
        (FOR)->[PART]
        (CHRC)->[ORDER_KIND]
        (TO)->[SUPPLIER]
        (CHRC)->[AMOUNT].
```

This graph can be presented to interested users for verification. And since it contains only relevant elements, the users can easily understand it and check its correctness and completeness.

Another useful function of the extraction function is system description completeness checking. This check determines whether all system definition elements have some useful purpose. To perform the check the extraction function has to be executed for all contexts used in the requirement specifications, and the results joined together into $\Sigma(c, M|c)$. If $\Sigma(c, M|c) = M$, it means that all elements in M have been used in some user context and therefore are necessary. Otherwise, there are some elements in M that are either of no use at all, or have not been analyzed as yet. However, $\Sigma(c, M|c)$ never contains elements not present in M since the requirements unification procedure always adds to M those elements that are needed to represent the requirements.

6 Conclusion

The paper described a method for formalizing context dependent information system requirements and showed how the requirements can be unified into an information system model. The unification preserves all the context information in conditional conceptual graphs. It is therefore always possible to extract from the model all the relevant elements for a given context. This feature can be used for requirements verification and completeness checking.

As suggested earlier an important and necessary extension to this approach is the introduction of type definition as a part of context definition. This will enable the users to have requirements stated in their own vocabulary making the requirements verification and completeness checking even more efficient.

Ultimately, it is foreseen to have the requirements and the system models integrated with a working information system and actively supporting user-computer interactions.

References

1. J. Sowa and J. A. Zachman. Extending and Formalizing the Framework for Information Systems Architecture. *IBM Systems Journal*, vol. 31, pp. 590-616, 1992.

2. J. F. Sowa. Peircean Foundations for a Theory of Context. In *Conceptual Structures: Fulfilling Peirce's Dream*, D. Lukose, H. Delugach, M. Keeler, L. Searle, and J. Sowa, Eds. Beriln: Springer Verlag, 1997, pp. 41-64.

3. G. W. Mineau and O. Gerbe. Contrexts: a Formal Definition of Worlds of Assertions. In *Conceptual Structures: Fulfilling Peirce's Dream*, D. Lukose, H. Delugach, M. Keeler, L. Searle, and J. Sowa, Eds. Beriln: Springer Verlag, 1997, pp. 81-94.

4. D. A. Norman. Some Observations on Mental Models. In *Mental Models*, D. Gentner and A. L. Stevens, Eds. Hillsdale, New Jersey: Lawrence Erlbaum Associates, Publishers, 1983.

5. R. Raban. Modelling Information Systems using Conceptual Graphs. In Proc. of the 1st Australian Conceptual Structures Workshop, Armidale, Australia, 1994.

Conceptual Structures Represented by Conceptual Graphs and Formal Concept Analysis

Guy Mineau[1], Gerd Stumme[2], Rudolf Wille[2]

[1] Département d'Informatique, Faculté des Sciences et de Génie, Université Laval, Cité universitaire, Québec, Canada, G1K 7P4; mineau@ift.ulaval.ca
[2] Technische Universität Darmstadt, Fachbereich Mathematik, D–64289 Darmstadt, Germany; {stumme, wille}@mathematik.tu-darmstadt.de

Abstract. Conceptual Graphs and Formal Concept Analysis have in common basic concerns: the focus on conceptual structures, the use of diagrams for supporting communication, the orientation by Peirce's Pragmatism, and the aim of representing and processing knowledge. These concerns open rich possibilities of interplay and integration. We discuss the philosophical foundations of both disciplines, and analyze their specific qualities. Based on this analysis, we discuss some possible approaches of interplay and integration.

1 Conceptual Structures in Knowledge Representation

Conceptual structures in knowledge representation are models (or artifacts) representing a perceived reality. With computer applications, these models support and delimit the kind of processing that a knowledge system will be able to carry out. Hence, the semantics they carry is of the outmost importance to the soundness and completeness of the applications that use them. With Artificial Intelligence related applications, these models represent human knowledge in a format available to an inference engine. To describe knowledge, a multitude of knowledge representation formalisms were devised. Among them, semantic networks aim at bringing together knowledge representation and graphical formalisms, hoping to alleviate the knowledge modeling and transfer problems pertaining to knowledge acquisition.

Through the years, different semantic network based formalisms were introduced, each having a different scope. Some dealt with linguistic knowledge, using the assumption that knowledge is necessarily expressible using natural language (cf. [33]). One of the underlying reasons for using this assumption was to facilitate the interfacing between the knowledge system and both the knowledge engineer and the end-user, since both could use a natural language based interface to communicate with the system. Furthermore, this would improve the interpretability of the achievements of the system at any given time.

Some other formalisms had preoccupations pertaining to complexity, as their applicability was often challenged by complexity results which discarded them from the spectrum of formalisms usable for the development of large-size applications (cf. [9], [23]). Applicability considerations must often restrict the kind of knowledge that can be represented by the formalism; it is referred to as the completeness/tractability trade-off. For instance, modal quantification, though very useful for many applications such as planning, is often omitted in order to simplify inference mechanisms.

Further formalisms put their emphasis on some logical aspects such as quantification and scope in order to address problems that require them (see [59]). The motivation behind this approach is to allow simple modeling of complex interrelations between sets of individuals. In brief, the nature of the knowledge captured by these formalisms varies greatly depending on the scope of the formalism: the same knowledge could be represented differently using different formalisms.

Twenty years ago, Ron Brachman surveyed the most popular semantic network based formalisms and identified five levels pertaining to knowledge representation [7]:

- *Implementational Level*: The primitives are nodes and links where links are merely pointers and nodes are simply destinations for links. On this level there are only data structures out of which to build logical forms.
- *Logical Level*: The primitives are logical predicates, operators, and propositions together with a structured index over those primitives. On this level logical adequacy is responsible for meaningfully factoring knowledge.
- *Epistemological Level*: The primitives are conceptual units, conceptual subpieces, inheritance and structuring relations. On this level conceptual units are determined in their inherent structure and their interrelationships.
- *Conceptual Level*: The primitives are word-senses and case relations, object- and action-types. On this level small sets of language-independent conceptual elements and relationships are fixed from which all expressible concepts can be constructed.
- *Linguistic Level*: The primitives are arbitrary concepts, words, and expressions. On this level the primitives are language-dependent, and are expected to change in meaning as the network grows.

Brachman discusses general criteria for judging the utility and formality of a given network language with respect to the described levels, namely *neutrality*, *adequacy*, and *semantics*. In a network, each level should be neutral (i.e.: not forcing any choice), and adequate in its support, to the level above it, and should have as definite as possible a well-defined semantics.

More recently, Randall Davis, Howard Schrobe, and Peter Szolovits stated five basic principles about knowledge representation in their critical review and analysis of the state of the art in knowledge representation [10]: A knowledge representation is (1) a surrogate, (2) a set of ontological commitments, (3) a fragmentary theory of intelligent reasoning, (4) a medium for efficient computation, and (5) a medium of human expression. These principles together with Brachman's level description and criteria for semantic networks shall be used as a framework for discussing the role of conceptual structures in knowledge representation performed by Conceptual Graphs and Formal Concept Analysis. The main aim of this paper is to investigate possibilities of interplay and integration of Conceptual Graphs and Formal Concept Analysis. For preparing this, we first give a brief introduction to both disciplines via illustrating examples in Section 2. Then their philosophical foundations are outlined in Section 3. In Section 4, methods of knowledge representation are described and evaluated, so that, finally in Section 5, possibilities of interplay and integration can be discussed.

2 Conceptual Graphs and Formal Concept Analysis: Illustrating Examples

To make this paper as much as possible self-contained, we give in this section brief introductions to Conceptual Graphs and Formal Concept Analysis via some illustrating examples. The reader who is familiar with those disciplines is recommended to skip the corresponding subsections, respectively.

2.1 Conceptual Graphs

Conceptual Graphs are a knowledge representation mechanism together with a reasoning mechanism. A *conceptual graph* is a labeled graph that represents the literal meaning of a sentence or even a longer text. It shows the concepts (represented by boxes) and the relations between them (represented by circles) (cf. [38]). As illustrating example, we show how the following text of instructions for decalcifying a coffee machine [1, p. 32] may be represented by a conceptual graph:

> In order to decalcify a coffee machine in an environment friendly way, one must fill it up with water and put in two teaspoons of citric acid (from the drugstore). Then one must turn it on and let the mixture go through the machine. Then, one must fill it up with clear water and let it go through the machine, twice.

Figure 1 shows a conceptual graph that represents the text and makes the text with its instructions tractable by a computer. For instance, it could be part of a document retrieval system for technical instructions. All concepts represented in the graph are *generic concepts*, because the text does not describe a specific situation, but instructions which are applicable to all kinds of coffee machines. The *concept types* (e. g., WATER) of conceptual graphs are usually organized in a type hierarchy. In this example, the concept types are all incomparable (with the exception ACTION < PROCESS). For a document retrieval system with many texts, however, the resulting type hierarchy will become larger and, as the texts treat similar topics, there will also arise comparable concept types. For keeping our introduction short, we consider all concept types as primitive types, i. e., we do not use type definitions.

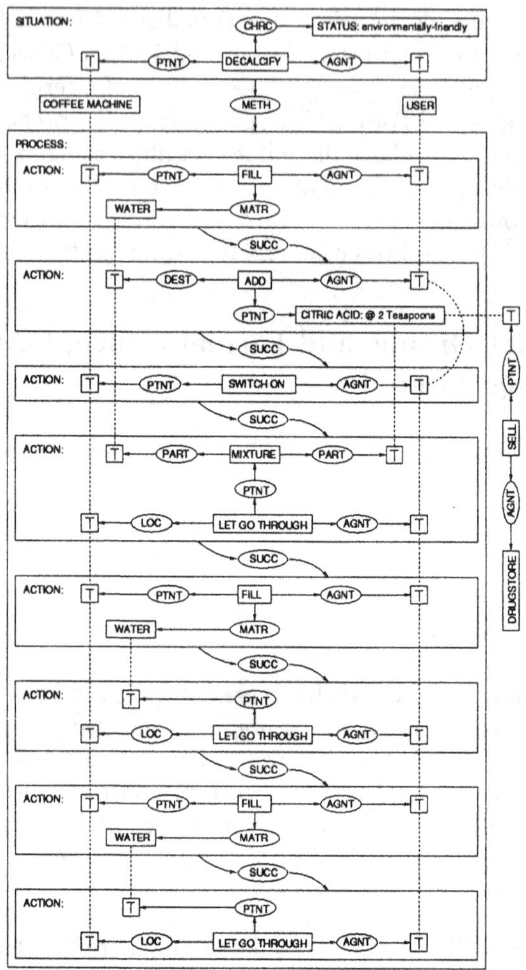

Fig. 1. A conceptual graph describing how to decalcify a coffee machine

In the conceptual graph, the process of decalcifying a coffee machine is splitted into succeeding actions. These actions are represented by larger boxes, called *contexts*, linked by the (SUCC)-relation. The sequel of all these actions is grouped together to a process which is stated as a method for decalcifying a coffee-machine in an environmentally-friendly way. Observe how the identity of the user and the coffee-machine is maintained throughout the whole process by dotted lines, called *coreference links*.

The conceptual graph in Figure 1 represents only the literal meaning of the instructions. It does not include background knowledge of a domain expert which is not provided by the text. For a document retrieval system, this approach is sufficient, because we assume that the end-user is familiar with the basic

STIFTUNG WARENTEST test KOMPASS — KAFFEEMASCHINEN MIT WARMHALTEKANNE (8 bis 10 Tassen)

test Ausgabe 12/98

	Mittlerer Preis in DM ca.	Preis für Ersatzkanne/ Glaseinsatz in DM ca.	Kaffeequalität	Technische Prüfung	Sicherheit	Handhabung	test-Qualitätsurteil
Gewichtung			35 %	30 %	10 %	25 %	
Neckermann Best.-Nr. 8628/409	40,–	35,–¹) / ❑	baugl. mit Otto Hanseatic Best.-Nr. 4327357				zufriedenst.
Otto Hanseatic Best.-Nr. 4327357	40,–	30,–²) / ❑	O	+	++	O	zufriedenst.
Quelle Privileg Best.-Nr. 7030720	40,–	24,50 / 17,50	baugl. mit Otto Hanseatic Best.-Nr. 4327357				zufriedenst.
Severin KA 9660	50,–	35,– / 23,–	baugl. mit Otto Hanseatic Best.-Nr. 4327357				zufriedenst.
Severin KA 4050	80,–	50,– / ❑	+	+	+	O	gut
Tchibo Aroma Garant Art.-Nr. 48469	80,–	27,50 / 19,50	+	+	+	O	gut
Ismet KM 582 starlight	84,–	47,– / 14,–	+	+	++	O	gut

Fig. 2. Part of a test report about coffee machines with integrated thermos flasks.

concepts of the domain. If there is the need of additional explanations, the set of documents must be extended.

This example shows that the decision to which extent domain knowledge should be formalized depends on the purpose. For a system with a human end-user one can rely on his background and common sense knowledge, while, for a knowledge base for a robot system which automatically decalcifies coffee machines, much more of the background knowledge has to be made explicit. For instance, we must let the system know that no coffee machine can be clean and dirty at the same time.

2.2 Formal Concept Analysis

Formal Concept Analysis is mainly used for analyzing data tables. We demonstrate this by a typical example: an investigation of coffee machines with integrated thermos flasks. Figure 2 shows part of a test report ([48]). This data table represents a so-called *many-valued context* in the sense of Formal Concept Analysis. [1] In order to obtain a concept lattice for investigating the data, we have to derive a *one-valued context*. Therefore we must decide which of the attributes are important, and how they shall be translated into one-valued attributes. The result of this so-called (*plain*) *conceptual scaling* is the *formal context* in Figure 3. The first four machines in Figure 2 are constructively indentical, so they are all represented by 'Otto Hanseatic' in the formal context. We emphasize that the choice of the conceptual scales depend on the purpose of the analysis. For instance, other attributes may be suitable for market analysis than for decision support. The attributes in Fig. 3 have been chosen in order to support a buyer in choosing a coffee machine.

[1] For the basic definitions of Formal Concept Analysis, see for instance [14] or [53].

	< 100 DM	< 125 DM	< 150 DM	< 200 DM	very high coffee quality	high coffee quality	sufficient coffee quality	deficient coffee quality	very high technical quality	high technical quality	sufficient technical quality	deficient technical quality	very high security	high security	sufficient security	deficient security	very good handling	good handling	sufficient handling	deficient handling
Otto Hanseatic	X	X	X	X			X			X	X		X	X	X				X	
Severin KA 4050	X	X	X	X		X	X			X	X			X	X				X	
Tschibo Aroma Garant	X	X	X	X		X	X			X	X			X	X				X	
Ismet KM 582	X	X	X	X		X	X			X	X		X	X	X				X	
Bosch TKA 2930			X	X		X	X		X	X	X			X	X				X	
Braun KF 170		X	X	X		X	X		X	X	X			X	X				X	
Krups 205 A		X	X	X		X	X			X	X			X	X				X	
Melitta 40001-89	X	X	X	X		X	X			X	X			X	X				X	
Moulinex AR 4		X	X	X		X	X		X	X	X			X	X				X	
Petra-electric KM 97.90	X	X	X	X		X	X			X	X			X	X				X	
Philips HD 7612				X		X	X		X	X	X			X	X				X	
Quelle Privileg		X	X	X		X	X			X	X		X	X	X				X	
Severin KA 5723				X		X	X			X	X			X	X				X	
Rowenta FT 774			X	X				X		X	X			X	X				X	

Fig. 3. Formal context derived from the table in Figure 2

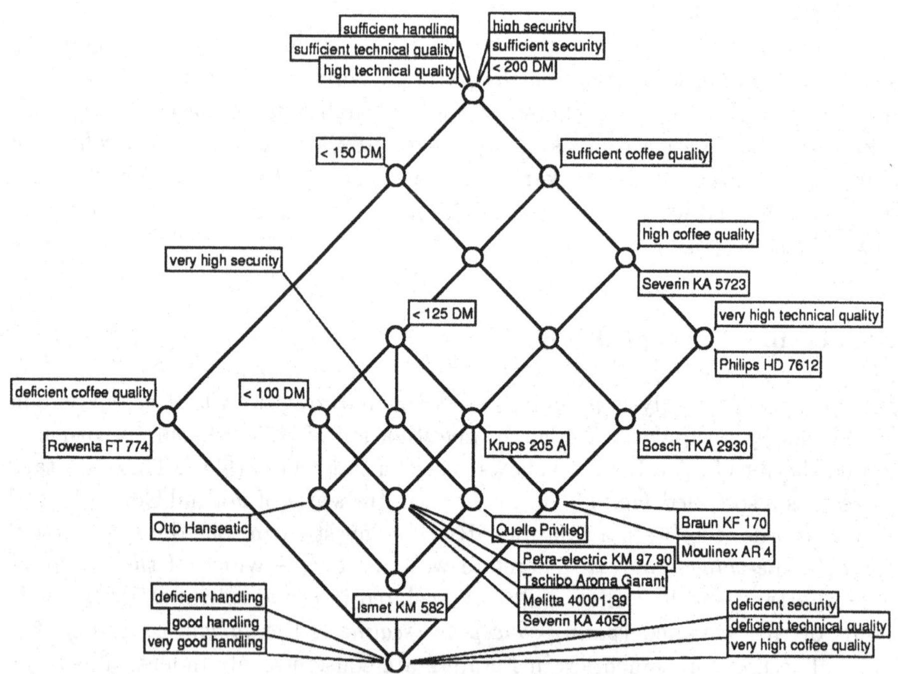

Fig. 4. The concept lattice derived from the formal context in Fig. 3

From the formal context, the concept lattice is computed, and visualized by a line diagram (cf. Fig. 4). We recall that each circle in the diagram represents a *formal concept*, i. e., a pair (A, B) where A is a subset of the set of objects, B

a subset of the set of attributes, and both are maximal such that each object has each attribute and vice versa. The *extent* A of a concept contains all objects whose labels are attached at concepts which are lower in the diagram, and the *intent* B contains all attributes whose labels are attached at concepts which are higher in the hierarchy. For instance, the rightmost concept has the objects 'Philips HD 7612', Bosch TKA 2930', 'Braun KF 170', and 'Moulinex AR 4' in its extent, and the attributes 'very high technical quality', 'high coffee quality', sufficient coffee quality', 'sufficient handling', ..., '< 200 DM' in its intent.

In the diagram, we can see that all machines have at least 'sufficient handling', 'high technical quality', 'high security', and cost '< 200 DM'. There is only one machine producing deficient coffee quality, 'Rowenta FT 774', and it does not even belong to the cheaper machines. On the other hand, 'Ismet KM 582', 'Braun KF 170', and 'Moulinex AR 4' have most of the positive attributes. Here, one has to decide between 'very high security' and '< 100 DM' on one side and 'very high technical quality' on the other. The fact that the label 'very high technical quality' is below the label 'high coffee quality' indicates that for all tested coffee machines the implication 'very high technical quality' $\rightarrow$ 'high coffee quality' holds.

In many applications, it is an advantage to limit the represented knowledge. Then the whole information can be visualized by one line diagram, which supports communication between knowledge engineers, experts, and novices. For instance, line diagrams of concept lattices have been used to a large extent for visualizing repertory grids tests in the treatment of patients suffering from bulimia ([41]). The diagrams were small enough to be understood by the patients, and at the same time they provided all information provided by the repertory grid. A similar application was the visualization of the water pollution of Lake Ontario for decision support ([42]).

For larger data, the attributes of the derived context are grouped to *conceptual scales* and, for each scale, its concept lattice is determined. By combining the concept lattices in *nested line diagrams*, information about the interplay of attributes is provided. Such applications are supported by the management system TOSCANA for Conceptual Information Systems. Examples of Conceptual Information Systems have been presented at ICCS '96 and ICCS '98 ([44], [15]). Other applications are, for instance, a conceptual information system about the medical nomenclature system SNOMED ([32]), a system for investigating international cooperations ([19]), and a retrieval system for a library ([20], [31]). The largest implemented system has 600,000 objects, and a system with 20,000,000 objects is planned for a basket analysis of a credit card company.

3 Philosophical Foundations

Conceptual Graphs and Formal Concept Analysis are each based on an elaborated philosophical understanding leading its development and application. For investigating the possibilities of their interplay and integration, it is necessary to bring the philosophical bases of both disciplines to mind. At first glance the philosophical foundations of Conceptual Graphs and Formal Concept Analysis seem to widely coincide: They both have conceptual structures as central theme of inquiry, both emphasize graphical representations for activating human thinking, both rely on Peirce's Pragmatism, and both aim to represent and process knowledge. But, since they were grown out of different roots and purposes in different scientific communities, each of the philosophical foundations has its special quality. This shall be outlined in this section to understand better how Conceptual Graphs and Formal Concept Analysis may be successfully combined.

According to [34], conceptual graphs arose in the Artificial Intelligence community when John Sowa started in 1968 to combine semantic networks and linguistic dependency graphs for forming a semantic representation of natural language. The used box and circle notation was influenced by the plastic templates used for computer flow charts. A systematic presentation of conceptual graphs which adapted also ideas of relational database theory was first published in [36]. Conceptual Graphs matured to a precise system of logic when John Sowa accommodated a new foundation based on Peirce's Logic of Existential Graphs in 1983. With this foundation, Sowa could present Conceptual Graphs as a rich and matured discipline in his book "Conceptual Structures: Information Processing in Mind and Machine" [37].

In this book Sowa views Conceptual Graphs as a knowledge representation language based on Linguistics, Psychology, and Philosophy, and in his "Conceptual Graphs Summary" [38] as a system of logic based on the existential graphs of Charles Sanders Peirce and the semantic networks of Artificial Intelligence. The key words "knowledge" and "logic" refer to a philosophical tradition of over two and a half millennia. Sowa extensively discusses this tradition in his new book "Knowledge Representation: Logical, Philosophical, and Computational Foundations" [39] to clarify the ontological basis of conceptual graphs. Starting with Heraclitus and Plato, he explains the systems of categories proposed by Aristotle, Kant, Peirce, Husserl, Whitehead, and Heidegger for deriving a new system of twelve categories which are obtained by associating the three major distinctions: Thing - Relation - Mediation; Physical - Abstract; Continuant - Occurrant. The categories provide a framework for classifying everything that exists, and can therefore be considered as a philosophical basis for hierarchies of concept types that are assumed for conceptual graphs.

In [39], the philosophical tradition for understanding the logical character of conceptual graphs is presented too. In sketching the history of logic, Sowa discusses Aristotle's syllogisms and their scholastic classification, first attempts toward automated reasoning by Lull and Leibniz, and the first major contributions to modern mathematical logic by Boole, Frege, Peirce, Schröder, Peano, Russell, and Whitehead. In the same way logic became more mathematical, the connections to common sense and natural language got weaker. Against this trend, conceptual graphs were designed as a logic with a human readable notation and a semantic basis of natural language for representing models of the real world or other possible worlds. In [37], Sowa describes this as follows:

– Rules of syntax map the graph to and from sentences in natural languages.

- Arcs of the graphs correspond to the *function words* and *case relations* of natural language.
- Nodes of the graphs are intensional concepts of individuals that may exist in the real world or some hypothetical world.
- Exact reasoning is based on Peirce's existential graphs [...]. The graphs are a complete notation for first-order logic with direct extensions to modal and higher-order logic.
- Plausible reasoning is based on schemata and prototypes, which codify the defaults and family resemblances that accommodate the variability of the real world.
- Model theory uses the same kinds of graph structures for both formal models and propositions about models.

Conceptual graphs have the ontological status of a model: they simulate significant structures and events in a possible world; a set of axioms, called "laws of the world", must at all times be true of the graphs; and certain graphs, called "schemata" and "prototypes", serve as patterns or frames that are joined to form the model. Understanding the knowledge of something as the ability to form a mental model that accurately represent the thing as well as the actions that can be performed by it and on it, it becomes clear why conceptual graphs as mental models are successful in knowledge representation, acquisition, and processing.

Formal Concept Analysis arose in the Fachbereich Mathematik of the TH Darmstadt out of efforts toward a better understanding of the relationships of mathematics to the world. These efforts led to the program of restructuring mathematics, an activity of reworking mathematical developments in order to integrate and to rationalize origins, connections, interpretations, and applications (cf. [54]). Besides several other attempts to restructure mathematical disciplines, the restructuring of mathematical order theory and lattice theory was successfully approached (see [49], [25]). The most fruitful result of this restructuring was the interpretation of complete lattices as concept lattices, which increasingly opened connections to new fields of applications. In this way, Formal Concept Analysis became established as a discipline of applied mathematics, based on a mathematization of concept and conceptual hierarchy, which activates mathematical thinking for conceptual data analysis and knowledge processing (see [14]).

Formal Concept Analysis is grounded on the philosophical understanding of a concept as a unit of thought constituted by its extension and its intension. This concept understanding has grown during centuries from Greek philosophy to late Scholastic and has finally found its modern formulation in the 17th century by the Logic of Port Royal (cf. [3], [16], [52]). Its actual use is exemplified by the German Standards DIN 2330 "Begriffe und Benennungen" [11]. Logically, Formal Concept Analysis fits in the tradition of the classical philosophical logic with its elementary doctrines of concepts, judgments, and conclusions (cf. [18], [54], [27]).

The mathematical notions of formal context, formal concept, and concept lattice - resulting from the mathematization of context, concept, and conceptual hierarchy - are not understood as realistic models, but as artifacts which may support human thinking, communication, and argumentation. Why such a support is possible, may be explained on the ground of Heidegger's characterization of Modern Science as guided by "das Mathematische" [17]. Martin Heidegger understands "das Mathematische" as the formal conception in which we take note of realities and acquire their use. Mathematics is then viewed as a specific culture of thinking in which "das Mathematische" is elaborated and conventionalized, caused by a complex process of communication within the respective community of mathematicians (cf. [58]). This explains that mathematics may support human thinking in general because mathematics is arising out of "das Mathematische" in human thought. The mathematizations used in Formal Concept Analysis are therefore understood as purpose-oriented constructions shaping "das Mathematische" in the thoughts which are mathematized.

As mathematical discipline, Formal Concept Analysis is semantically based on set theory as used in today's mathematics. On this basis, all its theoretical notions and results are of mathematical nature and have first of all to satisfy mathematical standards of adequacy and correctness. But as part of applied mathematics, Formal Concept Analysis also activates systematically connections of the mathematical theory to reality. Concerning such connections, human communication and argumentation must be considered since mathematical methods cannot grasp realities without an eventually serious loss of content. Therefore mathematizations have always to keep connections to their origins so that the consequences of mathematical treatments may be rationally analysed and interpreted in human communication. One way of establishing ties between mathematical theory and reality is to insert pieces of natural language into mathematical structures and their graphical representations. Formal Concept Analysis does this, for instance, by keeping the namings of objects and attributes and by using them in lattice diagrams to give conceptual meaning to structural relationships.

According to Peirce's Pragmatism, human knowledge is always incomplete and continuously requires intersubjective communication and argumentation for its formation. This understanding of knowledge is basic for Formal Concept Analysis (cf. [51]). Therefore its formalizations for the structural and computational treatment of data and knowledge are performed in such a way that they may support human communication and argumentation for establishing intersubjectively assured knowledge. For connecting all aspects of contents and formal settings, there is the vision of formally developing "conceptual landscapes of knowledge" which allow humans to commonly explore and acquire knowledge satisfying their specific requests (see [56]).

Although the above description shows quite different philosophical foundations, Conceptual Graphs and Formal Concept Analysis could join their specific qualities for obtaining a broader spectrum of methods of representing and processing conceptual structures. For instance, the world semantics of conceptual graphs and the set semantics of concept lattices could complement each other to increase the meaningfulness of representations and the mathematical tractability of structures; this would include an appropriate use of linguistic contexts and formal contexts for specific purposes in knowledge representation. More generally, Conceptual Graphs and Formal Concept Analysis could join to merge different developments of logics such as philosophical logics, mathematical logics, description logics, situational logics and others.

4 Conceptual Graphs and Formal Concept Analysis: Methods of Representation

The five levels of Brachman provide a framework for discussing the role of Conceptual Graphs and Formal Concept Analysis in knowledge representation. In the disussion, we also refer to the five principles of Davis, Schrobe, and Szolovits.

Implementational Level. The basic data structures of Conceptual Graphs are graphs in the usual mathematical understanding, as Sowa points out in [38]: "The dyadic relation LINK is the only primitive in the formal theory. All other conceptual relations may be defined in terms of it." However, virtually, all research in the Conceptual Graph community is done on the higher levels of representation. In Formal Concept Analysis, the basic data structures are formal (dyadic) contexts formed by a binary relation and many-valued (triadic) contexts formed by a ternary relation. The simplicity of these data structures makes clear that they are not understood as models imaging reality, but as surrogates or artifacts. On the implementational level, "there is nothing inherently 'semantic'" [7]; indeed, graphs and formal contexts are abstract mathematical structures without any content. Implementational issues of Formal Concept Analysis are discussed in [47].

Logical Level. In Conceptual Graphs and in Formal Concept Analysis, the logical level is activated in processing of the basic data structures. For Conceptual Graphs the logical primitives are the (abstract) concept and relation types and the connectives representing conjunctions, references, coreferences, contexts, actors, and type definitions. Reasoning is a major activity in processing Conceptual Graphs. Graph transformation rules (like projections) provide a fragmentary theory of intelligent reasoning which is inherent to Conceptual Graphs. Also the Peircing rules of deduction based on the context mechanism is also typical of the system of logic that the conceptual graph theory is. By mapping Conceptual Graphs to First Order Logic by the Φ-operator, other reasoning mechanisms can be activated for efficient computation as well. In Formal Concept Analysis, the (formal) object, attribute, and concept names are logical primitives which allow the composition of further predicates by logical connectives and quantifiers (cf. [5], [28]). Then the means for logical reasoning are taken from Predicate Logic. As logical operators, the numerous context constructions for one- and many-valued contexts may be considered; the mostly used context construction is the semiproduct which is basic for 'plain conceptual scaling' (described in Section 2.2).

Epistemological Level. The epistemological level addresses "the possibility of organizations of conceptual knowledge into units more structured than simple nodes and links or predicates and propositions, and the possibility of processing over larger units than single network links. The predominant use of concepts as intensional descriptions of objects hints that there is a class of relationship that is not accounted for by the [other] four levels [...]. This kind of relationship relates the parts of an intension to the intension as whole, and one intension to another" [7]. Brachman observed that for the semantic network representations of that time, this epistemological level is missing.

For Conceptual Graphs, the epistemological level is still not fully activated. The structures of type hierarchies belong to this level and also the expansion and contraction operators which are defined in the Conceptual Graph theory (see [37]). These operators allow multiple representations of the same knowledge because they provide a mechanism for the aggregation of lower-level constructs into higher abstractions (used at the conceptual level). Based on lambda expressions, these operators allow the representation and processing of the same knowledge at different levels of granularity. Type definitions implement these lambda expression definitions. They are solely based on extensional semantics and, in our opinion, they may not be sufficient to provide a sound justification for their existence. Beside representing the internal structure of concepts, another feature of the epistemological level is the representation of 'inheritance', which, according to [7], "is not a logical primitive; on the other hand, it is a mechanism assumed by all conceptual level nets [in particular Conceptual Graphs], but not accounted for as a 'semantic' (deep case) relation."

Concept lattices, the core structures of Formal Concept Analysis, are located on the epistemological level: Formal concepts are considered as "formal objects, with predetermined internal organization that is more sophisticated than sets of cases" [7]. Formal concepts bring together extensional and intensional views on 'concepts', and represent explicitly inheritance by refering to the set semantics of the intents (or extents) of the formal concepts. Concept lattice constructions also belong to the epistemological level. As Formal Concept Analysis is founded on lattice theory, lattice constructions and lattice decompositions can be activated for establishing more complex concept hierarchies out of simpler ones, and, vice versa, for reducing complex concept hierarchies to simpler ones. Constructions like direct products and tensor products of concept lattices and decompositions like subdirect and atlas decompositions have been successfully applied in data analysis. It is interesting to note that most concept lattice constructions (decompositions) have as counterpart a context construction on the logical level (see [14]). As formal contexts are only 'logarithmic in size' compared to the concept lattice, the knowledge representation on the logical level can be seen in the light of the fourthprinciple of Davis, Schrobe, and Szolovits as a medium of efficient computation.

The interplay between representations by formal contexts and by concept lattices supports also the fifth principle: The acceptance of Formal Concept Analysis as a medium of human expression by users from various domains (consider for instance the patients mentioned in Section 2.2 who have no higher mathematical education) results from the visualization of the data (as present on the logical level) on the epistemological level. In particular, the construction of complex lattices as (sub-)direct product of simpler lattices allows the visualization of complex data by nested line diagrams. The corresponding construction on the logical level is the semiproduct of conceptual scales, which is applied for efficiently querying the database.

Conceptual Level. The conceptual level is the major domain of Conceptual Graphs, just as it is for most semantical networks. Concept meanings are structured in type hierarchies according to levels of generality. They are linked by primitive semantic relations understood as those relations "that the verb in a sentence has with its subject, object, and prepositional phrase arguments in addition to those that underlie common lexical, classificational, and modificational relations" [35]. According to [7], "networks on this level can be characterized as having a small set of language-independent conceptual elements (namely, primitive object- and action-types) and conceptually primitive relationships (i. e., 'deep cases') out of which all expressible concepts can be constructed". In the theory of Conceptual Graphs, there is no predefined relation type other than LINK, which is the most general relation, being already formally located on the logical level. By applying concept expansion as described on the epistemological level, each concept type can be expanded such that all relationships are represented by the LINK relation. Hence case grammar based relations are in no way compulsory to the development of Conceptual Graph systems. However, the way Conceptual Graphs are used in practice identifies them clearly as being on the conceptual level: Although the LINK-relation is the only primitive relation *in theory*, in applications, the starter set of relation types is used as if consisting of primitive types *in practice*. The starter set plays the role of the small set of language-independent conceptually primitive relationships mentioned by Brachman. We conclude that the typical use of relation types identifies Conceptual Graphs as a knowledge representation formalism which is located mainly on the conceptual level.

In Formal Concept Analysis, word-senses are represented by the context attributes which lead to a contextual representation of concept intensions. As primitive case relations, there are defined four basic relations: an object has an attribute, an object belongs to a concept, a concept abstracts to an attribute, and a concept is a subconcept of another concept. These four relations are used as primitive relations in the mathematical model for Conceptual Knowledge Systems described in [22]. Other relations (especially meronomy) are discussed in [30].

Linguistical Level. Neither Conceptual Graphs nor Formal Concept Analysis provide (at least not in their core) language-specific primitives. The linguistical level is activated only indirectly. For Formal Concept Analysis, this activation is discussed in more detail in [2] from a pragmatic-semiotic point of view.

There the existence of qualified knowledge [anspruchsvolles Wissen] is claimed to depend from the existence of a community of discourse in which the knowledge is intersubjectively constituted.

Considering that the vocabulary (made of aggregations and primitive types) used to describe all Conceptual Graphs in a system is highly dependent on the type definition mechanism, we feel that Conceptual Graphs need to consider some complementary mechanism to ensure some epistemological soundness of the type definition mechanism. As Formal Concept Analysis is strong on the epistemological level, we are confident that Formal Concept Analysis can contribute to this. On the other hand, the experience of data analysis with Formal Concept Analysis has shown that there are applications with a need to enhance the expressiveness of Formal Concept Analysis on the conceptual level. We expect from the interplay of both theories a step in this direction.

5 Interplay and Integration

As already stated in Section 3, Conceptual Graphs and Formal Concept Analysis have basic concerns in common: the focus on conceptual structures, the use of diagrams for supporting communication, the orientation by Peirce's Pragmatism, and the aim of representing and processing knowledge. These concerns open rich possibilities of interplay and integration. Since both disciplines have their specific quality besides their common concerns, they may also complement each other. In this section, we discuss possible approaches of interplay and integration; but, because of the lack of space, we have to restrict to two themes: conceptual hierarchies and systems of logic.

5.1 Interplay: Deriving and Improving Conceptual Hierarchies

Conceptual hierarchies are basic in Conceptual Graphs and Formal Concept Analysis where they are mainly considered at the epistemological and the conceptual level. For conceptual graphs hierarchies of concept types are presumed as an ontological commitment. Often they are taken from conceptual taxonomies which, in general, are only ordered sets. But, for many purposes, it is desirable that the concept types form even a lattice. This can be effectively derived by methods of Formal Concept Analysis. In [13], Bernhard Ganter and Sergei Kuznetsov describe simple algorithms with satisfactory complexity bounds for minimally completing a finite ordered set to a lattice. For obtaining a meaningful type lattice, the minimal extension of a given ordered set of concept types might not be satisfactory. For finding useful completions, the construction of concept lattices of meaningfully deduced formal contexts can be offered as general method. The formal context need not contain all objects and attributes that belong to the concepts of the type hierarchy; it is sufficient to have enough typical objects and attributes. Then its concept lattice can be used as completed type hierarchy. The formal context can either be derived from an existing data set (if it provides enough typical objects and attributes), or it must be interactively acquired from some human expert.

For instance, there is a concept type COFFEE MACHINE in the type hierarchy for the document retrieval system in Section 2.1. This concept has in its extension all existing coffee machines and in its intension all attributes common to all coffee machines. In the knowledge acquisition process, however, it is sufficient to ask the user only for some few objects and attributes which distinguish this concept from the other involved concepts such as USER or CITRIC ACID.

Formal Concept Analysis provides different knowledge acquisition algorithms which can be used for that purpose: Attribute Exploration [14], Distributive Concept Exploration ([43]), and Concept Exploration ([46]). In [4], Franz Baader has used Attribute Exploration together with the Subsumption Algorithm of Description Logics for automatically deriving a $\bigwedge$-completion for an arbitrarily given type hierarchy. For a general overview concerning knowledge acquisition by methods of Formal Concept Analysis, see [50] and [45].

Methods of Formal Concept Analysis may also be applied to the generalization order on conceptual graphs. For instance, Gerard Ellis and Stephen Callaghan [12] have used conceptual scaling for improving the search in the generalization hierarchy of conceptual graphs. If the generalization order is represented as concept lattice, such a search might even be performed within a TOSCANA system (see [29]). Guy Mineau and Olivier Gerbé discuss in [24] context lattices in their definitional framework for contexts, based on Formal Concept Analysis.

As Formal Concept Analysis can be used for deriving and improving conceptual structures occuring in Conceptual Graphs, there may also be applications of Conceptual Graphs methods to Formal Concept Analysis. In particular, conceptual graphs could be very helpful in translating structural meaning of concept lattice into natural language to lead users of Formal Concept Analysis to a better understanding of possible interpretations. Another application would be to support knowledge acquisition for Conceptual Information Systems. In Conceptual Information Systems, the data are organized in a relational database. But if the data are only available in a less structured way (for instance from structured interviews), then Conceptual Graphs may support the formalization of the information, as they are close to natural language for being understood by the domain expert, and formal enough for being computationally tractable. Once the knowledge is represented in Conceptual Graphs, it can be transformed to formal contexts by using power context families (discussed in the next subsection).

5.2 Integration: Unifying Systems of Logic

An integration of developments in Conceptual Graphs and Formal Concept Analysis would mean to establish a common theoretical basis for the field of interest. An attractive field would be, as already indicated at the end of Section 3, knowledge-oriented systems of logic. In both disciplines, criticisms have been made about the inadequacy of mathematical logic concerning knowledge representation and processing. In [40], John Sowa lists a number of reasons why people who use a knowledge representation language may want to diverge from First Order Logic; those reasons are concerned with readability, computability, convenience, surprises, context dependence, and extended logics. All these issues have been actively explored in the Conceptual Graph community, and different approaches to them have been suggested, implemented, and published.

These activities might be combined with the recent development of "Contextual Logic" in Formal Concept Analysis (see [54], [55], [57], [26], [27]). Contextual Logic is understood as a mathematization of classical philosophical logic based on the elementary doctrines of concepts, judgments, and conclusions; Formal Concept Analysis yields the mathematization of the doctrine of concepts, and Conceptual Graphs is used for mathematizing the doctrines of judgments and conclusions. Contextual Logic is set-theoretically grounded on families of related contexts whose formal concepts allow a representation of the concepts and relations of conceptual graphs. Such representation of a conceptual graph is called a "Concept Graph" of the context family from which is is derived, and the family of related contexts is said to be a "Power Context Family". In [29] it is shown that the concept graphs of a power context family always form a lattice with respect to generalization, which can be represented as a concept lattice. Since power context families can always be derived from many-valued data contexts, the approach via Contextual Logic opens a large field of common applications for Conceptual Graphs and Formal Concept Analysis. This might stimulate to continue the process of integration of both disciplines and even further developments of logic systems.

References

1. ADAC: *Gewußt wie – 10000 praktische Tips für alle Tage.* ADAC Verlag, München 1993.
2. U. Andelfinger: Begriffliche Wissenssysteme aus pragmatisch-semiotischer Sicht. In: R. Wille, M. Zickwolff (eds.): *Begriffliche Wissensverarbeitung – Grundfragen und Aufgaben.* B.-I.-Wissenschaftsverlag, Mannheim 1994, 152–172.
3. A. Arnauld, P. Nicole: *La logique ou l'art de penser.* Amsterdam 1662.
4. F. Baader: Computing a minimal representation of the subsumption lattice of all conjunctions of concept defined in a terminology. In: *Proc. Intl. KRUSE Symposium*, August 11–13, 1995, UCSC, Santa Cruz 1995, 168–178.
5. H. Berg: *Terminologische Begriffslogik.* Diplomarbeit, TU Darmstadt 1997.
6. K. Biedermann: How triadic diagrams represent conceptual structures. LNAI **1257**. Springer, Heidelberg 1997, 304–317.
7. R. J. Brachman: On the epistemological status of semantic networks. In: N. V. Findler (ed.): *Associative networks: representation and use of knowledge by computers.* Academic Press, New York 1979, 3–50 (reprinted in [8]).
8. R. J. Brachman, H. L. Levesque (eds.): *Readings in Knowledge Represenation.* Morgan Kaufmann, Los Altos 1985.
9. R. J. Brachman, D. L. McGuinness, P. F. Patel-Schneider, L. Alperin, L. A. Resnick, A. Borgida: Living With Classic: When and how to use a KL-ONE-like language. In: J. F. Sowa (ed.): *Principles of Semantic Networks*, Morgan Kaufmann, Los Altos 1991, 401-456.

10. R. Davis, H. Schrobe, P. Szolovits: What is a knowledge representation? *AI Magazine* **14:1** (1993), 17–33.
11. Deutsches Institut für Normung: *DIN 2330; Begriffe und Benennungen: Allgemeine Grundsätze.* Beuth, Berlin-Köln 1979.
12. G. Ellis, S. Callaghan: Organization of knowledge using order factors. LNAI **1257**, Springer, Heidelberg 1997, 342–356.
13. B. Ganter, S. O. Kuznetsov: Stepwise construction of the Dedekind-MacNeille Completion. LNAI **1453**, Springer, Heidelberg 1998, 295–302.
14. B. Ganter, R. Wille: *Formal Concept Analysis: Mathematical Foundations.* Springer, Heidelberg 1999.
15. B. Groh, S. Strahringer, R. Wille: TOSCANA-Systems based on thesauri. LNAI **1453**. Springer, Heidelberg 1998, 127–138.
16. N. Hartmann: Aristoteles und das Problem des Begriffs. In: *Abh. Preuß. Akad. Wiss. Jg. 1939.* Phil.-hist. Kl. Nr.5. Verlag Akad. Wiss., Berlin 1939.
17. M. Heidegger: *Die Frage nach dem Ding.* Niemeyer, Tübingen 1962.
18. I. Kant: *Logic.* Dover, New York 1988.
19. B. Kohler-Koch, F. Vogt: *Normen und regelgeleitete internationale Kooperationen.* In: G. Stumme and R. Wille (eds.): *Begriffliche Wissensverarbeitung: Methoden und Anwendungen.* Springer, Heidelberg (to appear)
20. W. Kollewe, C. Sander, R. Schmiede,R. Wille: TOSCANA als Instrument der bibliothekarischen Sacherschließung. In: H. Havekost and H.J. Wätjen (eds.): *Aufbau und Erschließung begrifflicher Datenbanken.* (BIS)-Verlag, Oldenburg 1995, 95–114.
21. F. Lehmann, R. Wille. A triadic approach to formal concept analysis. LNAI **954**. Springer, Heidelberg 1995, 32–43.
22. P. Luksch, R. Wille: A mathematical model for conceptual knowledge systems. In: H.-H. Bock, P. Ihm (eds.): *Classification, data analysis, and knowledge organization.* Springer, Heidelberg 1991, 156–162
23. R. MacGregor: The evolving technology of classification-based knowledge representation systems. In: J. F. Sowa (ed.): *Principles of semantic networks*, Morgan Kaufmann, Los Altos 1991, 385–400.
24. G. W. Mineau, O. Gerbé: Contexts: A formal definition of worlds of assertions. LNAI **1257**. Springer, Heidelberg 1997, 80–94.
25. W. Poguntke, R. Wille: Zur Restrukturierung der mathematischen Ordnungstheorie. In: A. M. Kempf, F. Wille: *Mathematische Modellierung.* McGraw-Hill, Hamburg 1986, 283–293.
26. S. Prediger: Simple concept graphs: a logic approach. LNAI **1453**, Springer, Heidelberg 1998, 225–239.
27. S. Prediger: *Kontextuelle Urteilslogik mit Begriffsgraphen. Ein Beitrag zur Restrukturierung der mathematischen Logik.* Dissertation, TU Darmstadt. Shaker Verlag, Aachen 1998.
28. S. Prediger: Terminologische Merkmalslogik in der Formalen Begriffsanalyse. In: G. Stumme and R. Wille (eds.): *Begriffliche Wissensverarbeitung: Methoden und Anwendungen.* Springer, Heidelberg (to appear)
29. S. Prediger, R. Wille: The lattice of concept graphs of a relationally scaled context. Preprint, TU Darmstadt 1999. In this volume.
30. U. Priß: *Relational concept analysis: semantic structures in dictionaries and lexical databases.* Dissertation, TH Darmstadt 1996. Shaker Verlag, Aachen 1998.

31. T. Rock, R. Wille: Ein TOSCANA-System zur Literatursuche. In: G. Stumme and R. Wille (eds.): *Begriffliche Wissensverarbeitung: Methoden und Anwendungen.* Springer, Heidelberg (to appear)
32. M. Roth-Hintz, M. Mieth, T. Wetter, S. Strahringer, B. Groh, R. Wille: *Investigating SNOMED by Formal Concept Analysis.* FB4-Preprint, TU Darmstadt 1998.
33. R. C. Schank, C. J. Rieger: Inference and the computer understanding of natural language, Artificial Intelligence Journal **5** (4) (1974), 373–412.
34. L. Searle, M. Keeler, J. Sowa, H. Delugagh, D. Lukose: Fulfilling Peirce's dream: conceptual structures and communities of inquiry. LNAI **1257**. Springer, Heidelberg 1997, 1–11.
35. R. F. Simmons: Semantic networks: Their computation and use for understanding English sentences. In: R. C. Schank, K. M. Colby (eds.): *Computer models of thought and language.* Freeman, Can Francisco 1973, 63–113.
36. J. F. Sowa: Conceptual graphs for a data base interface. IBM Journal of Research and Development **20**(1976), 336–357.
37. J. Sowa: *Conceptual structures: information processing in mind and machine.* Adison-Wesley, Reading 1984.
38. J. F. Sowa: Conceptual graphs summary. In: T. E. Nagle, J. A. Nagle, L. L. Gerholz, P. W. Eklund (eds.): *Conceptual Structures: Current Research and Practice.* Ellis Horwood 1992, 3–51.
39. J. F. Sowa: *Knowledge representation: logical, philosophical, and computational foundations.* PWS Publishing Co., Boston (to appear)
40. J. F. Sowa: Conceptual Graph standard and extension. LNAI **1453**. Springer, Heidelberg 1998, 3–14.
41. N. Spangenberg, K. E. Wolff: Comparison between principal component analysis and formal concept analysis of repertory grids. In: W. Lex (ed.): *Arbeitstagung Begriffsanalyse und Künstliche Intelligenz*, Informatik–Bericht 89/3. TU Clausthal, 1991, 127–134.
42. S. Strahringer, R. Wille: Towards a structure theory for ordinal data. In: M. Schader (ed.): *Analyzing and Modeling Data and Knowledge.* Springer, Heidelberg 1992, 129–139.
43. G. Stumme: Knowledge acquisition by distributive concept exploration. *Suppl. Proc. 3rd ICCS*, Santa Cruz, CA, USA, 1995, 98–111.
44. G. Stumme: Local scaling in conceptual data systems. LNAI **1115**. Springer, Heidelberg 1996, 308–320.
45. G. Stumme: Exploration tools in Formal Concept Analysis. In: *Ordinal and Symbolic Data Analysis.* Studies in Classification, Data Analysis, and Knowledge Organization **8**, Springer, Heidelberg 1996, 31–44.
46. G. Stumme: Concept exploration – a tool for creating and exploring conceptual hierarchies. LNAI **1257**. Springer, Heidelberg 1997, 318–331.
47. F. Vogt: *Formale Begriffsanalyse mit C++: Datenstrukturen und Algorithmen.* Springer, Heidelberg 1996.
48. Stiftung Warentest: Kaffeemaschinen mit Warmhaltekannen: Wärme okay, Geschmack ade. *Test 12/98*, 71–75.
49. R. Wille: Restructuring lattice theory: an approach based on hierarchies of concepts. In: I. Rival (ed.): *Ordered sets.* Reidel, Dordrecht, Boston 1982, 445–470.
50. R. Wille: Knowledge acquisition by methods of formal concept analysis. In: E. Diday (ed.): *Data analysis, learning symbolic and numeric knowledge.* Nova Science Publishers, New York–Budapest 1989, 365–380.

51. R. Wille: Plädoyer für eine philosophische Grundlegung der Begrifflichen Wissensverarbeitung. In: R. Wille, M. Zickwolff (eds.): *Begriffliche Wissensverarbeitung: Grundfragen und Aufgaben.* B.I.-Wissenschaftsverlag, Mannheim 1994, 11-25.
52. R. Wille: Begriffsdenken: Von der griechischen Philosophie bis zur Künstlichen Intelligenz heute. In: Dilthey-Kastanie. Ludwig-Georgs-Gymnasium, Darmstadt 1995, 77–109.
53. R. Wille: Conceptual structures of multicontexts. LNAI **1115**. Springer, Heidelberg 1996, 23–39.
54. R. Wille: Restructuring mathematical logic: an approach based on Peirce's pragmatism. In: A. Ursini, P. Agliano (eds.): *Logic and Algebra.* Marcel Dekker, New York 1996, 267–281.
55. R. Wille: Conceptual Graphs and Formal Concept Analysis. LNAI **1257**. Springer, Heidelberg 1997, 290–303.
56. R. Wille: Conceptual landscapes of knowledge: a pragmatic paradigm for knowledge processing. In: G. Mineau and A. Fall (eds.): *Proc. 2nd Intl. KRUSE.* Simon Fraser University, Vancouver 1997, 2–13.
57. R. Wille: Triadic Concept Graphs. LNAI **1453**, Springer, Heidelberg 1998, 194–208.
58. R. Wille: *Bildung und Mathematik.* FB4-Preprint Nr. 2005, TU Darmstadt 1998.
59. W. Zadrozny: Logical dimensions of some graph formalisms. In: J. F. Sowa (ed.): *Principles of semantic networks.* Morgan Kaufmann, Los Altos 1991, 363–380.

A Simulation of Co-identity with Rules in Simple and Nested Graphs

Jean-François Baget

LIRMM
161, rue Ada, 34392 Montpellier - FRANCE
baget@lirmm.fr

Abstract. Equality of markers and co-reference links have always been a convenient way to denote that two concept nodes represent the same entity in conceptual graphs. This is the underlying cause of counterexamples to projection completeness with respect to these graphs FOL semantics. Several algorithms and semantics have been proposed to achieve completeness, but they do not always suit an application specific needs. In this paper, I propose to represent identity by relation nodes, which are first-class objects of the model, and I show that conceptual graphs rules can be used to represent and simulate reasonings defined by various semantics assigned to identity, be it in the case of simple or nested graphs. The interest of this method is that we can refine these rules to manage the identity needed by the application.

Introduction

Identity of concept nodes in conceptual graphs [10] has always been a crucial problem for completeness of projection with respect to these graphs logical semantics. According to the widely accepted FOL semantics Φ, two concept nodes sharing the same individual marker represent the same entity. Moreover, the necessity to express that some generic concept nodes represent the same entity has early led to the adoption of co-reference links which, however useful in simple graphs (SGs), proved to be of uttermost importance for positive nested graphs (NGs). These two representations of identity can be seen as an equivalence relation on concept nodes, which has been called *co-identity* [8].

But Chein and Mugnier [4], as well as Ghosh and Wuvongse [3] gave counterexamples to projection completeness with respect to Φ in the case of SGs, and proposed a normalization procedure to achieve completeness. In the case of NGs, Simonet has extended the normality condition into a k-normality condition [8], which achieves completeness with respect to a natural extension of the semantics Φ. Instead of forcing projection to conform to the semantics Φ, Simonet has proposed the semantics Ψ [7], defined as well for SGs as for NGs, which describes more precisely the graph as well as the projection mechanism. Projection is sound and complete with respect to Ψ, without any condition.

The problem is that the adoption of one semantics or another is a definitive and exclusive choice with respect to the reasonings allowed in the model. In the semantics Ψ, co-identical concept nodes are considered as different points of view on the same entity, and information on one of these nodes is considered to be meaningful only for the particular point of view represented by this node. On the other hand, the semantics Φ considers co-identical concept nodes as partial and complementary representations of an entity: information linked to one of the co-identical nodes is shared by all others. Now suppose we want to represent the sentence: *"A dark-haired man is looking at a photograph representing himself"*. We can represent it in a graph where two co-referent concept nodes represent the man, the man x looking at the photograph, and the man y represented by the photograph. We could want to deduce from this knowledge that y is dark-haired, but we do not want to deduce that he is looking at the photograph. With the semantics Φ, the two deductions are possible, and none of them are using Ψ. Since co-reference links are not first-class objects of the CG model, we cannot implement the desired behavior without rewriting the projection algorithm, with *ad hoc* procedures regarding to *sharable* or *non-sharable* relation types.

Instead, I propose in this paper to represent identity by relation nodes, which are first-class objects of the CG model. The semantics that can be assigned to identity will be simulated with conceptual graph rules [6]. I present the specific sets of rules which describe co-identity in SGs as defined by the semantics Φ and Ψ. In order to extend this work to NGs, I use a generalization of the NGs, that I call *boxed graphs*, where concept nodes in different contexts can be linked by relation nodes. These rules could easily be refined, using types restrictions, to conform to a specific semantics.

1 Co-Identity in Simple Graphs

The formal model for SGs used in this paper is similar to the one presented in [1]. Completeness can be achieved by normalization to conform to the semantics Φ [4], or without restriction with respect to Ψ [7], [9].

1.1 The Model

We consider a *knowledge base* where the basic ontological knowledge is coded in a *support* and the asserted facts are coded in a *simple graph* (SG). The basic operation for reasonings is the graph morphism classically called *projection*.

A support $\mathcal{S} = (T_C, T_R, \nu, \mathcal{I}, \tau)$ is defined by two partially ordered sets, the *concept types* T_C, the *relation types* T_R, a set of *individual markers* $\mathcal{I}$, and two mappings, the *valence*[1] ν and the *conformity relation* τ. T_C has a greatest element, $\top_C$. T_R is partitioned into partially ordered subsets $T_R^i, i \in \{1, \ldots, m\}$ of relation types of *valence* i, whose greatest element is $\top_R^i$. $\mathcal{I} \cup \{*\}$ is partially ordered, has a greatest element, the *generic marker* $*$, all others being pairwise incomparable. τ assigns a concept type to each individual marker.

A SG G, on a support $\mathcal{S}$, is a bipartite labelled multigraph (V_C, V_R, E, l), where the nodes in V_C are the *concept nodes*, those in V_R are the *relation nodes*.

[1] It is a simplified version of signature, which does not assign a maximum type to the arguments of a relation

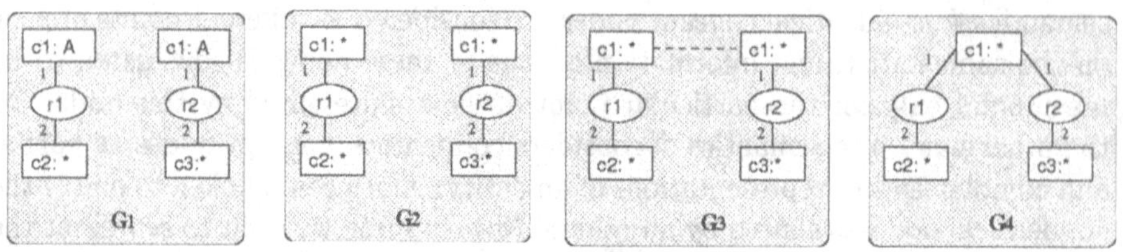

Fig. 1. *Very* **simple graphs ...**

l labels each concept node by a concept type and an individual or generic marker, called its referent. It is required that the type of an individual node c is exactly $\tau(\text{ref}(c))$. l labels each relation node r of degree i by a relation type of valence i. Edges of E incident on such a node are numbered from 1 to i.

SGs can be extended to simple conceptual graphs with co-reference links. A SG^{ref} is defined as a SG which is added an equivalence relation *co-ref* defined on the set of generic concept nodes. In order to avoid fusionning problems during normalization, only generic concept nodes having the same type can be declared co-referent. The equivalence relation ***co-ident*** is defined on all concept nodes by: $\forall c, c' \in V_C, \text{co-ident}(c, c') \Leftrightarrow \text{co-ref}(c, c')$ or $\text{ref}(c) = \text{ref}(c') \in \mathcal{I}$ The relation *co-ref* is traditionally represented in the drawing of the graph by a dotted line called a co-reference link[2] between concept nodes (see graph G_3 in fig. 1).

Let H and G be two SGs defined on a common support $\mathcal{S}$. A ***projection*** Π from H to G is given by a mapping from the nodes of H to the nodes of G, preserving edges and their numbering, and respecting the order defined on labels. In the case of SG^{ref}s, two co-referent generic concept nodes can only be projected into the same co-identity class. In fig. 1, the graph G_3 can be projected into the graph G_1, but G_3 cannot be projected into G_2. Instead of the traditional notation $H \geq G$, I note $H \sqsubseteq G$ if there exists a projection from H to G. This points out that all information in H is contained in G.

1.2 The Semantics Φ and SGs Normal Form

SGs and SG^{ref}s are assigned FOL semantics, Φ being the one presented in [10]. The support determines the vocabulary used in formulas, and is associated a formula $\Phi(\mathcal{S})$, translating the partial orderings on T_C and T_R. Each *co-ref* class in a graph G is assigned a unique variable. $\Phi(G)$ is the existential closure of the conjunction of the atoms associated with all nodes of the graph: $\forall c \in V_C$, $C(t)$ is defined by the unary predicate C assigned to its type and the term t, which is the variable associated to its co-reference class if c is generic, the constant assigned to its marker otherwise; $\forall r \in V_R$ such that $\nu(\text{degree}(r)) = i$, $R(t_1, ..., t_i)$ is defined by the i-ary predicate R assigned to its type, and the terms t_k associated with the kth neighbors of r. The graph G_3 in fig. 1 is interpreted by:

$$\Phi(G_3) = \exists xyz\ (\text{c1}(x) \wedge \text{c1}(x) \wedge \text{c2}(y) \wedge \text{c3}(z) \wedge \text{r1}(x, y) \wedge \text{r2}(x, z))$$

[2] A co-reference class of n concept nodes can be represented by $n - 1$ such links.

Projection is sound with respect to Φ [10]. Concerning completeness, Chein and Mugnier[4], and Ghosh and Wuvongse [3], pointed out that two co-identical concept nodes are assigned exactly the same atom by Φ, whereas projection treats them as two distinct nodes. By example, in fig. 1, assuming that all types involved are incomparable, there is no projection from G_4 into any of the other graphs, though $\Phi(G_4)$ can be deduced from $\Phi(G_1)$ or $\Phi(G_3)$.

A SG or SG^{ref} is said in *normal form* if the co-identity classes are restricted to the trivial ones: they only contain a single concept node. A graph G is put into its normal form $\mathcal{N}_{\mathcal{F}}(G)$ by fusionning all its co-identical nodes. In fig. 1, $G_4 = \mathcal{N}_{\mathcal{F}}(G_3)$. This normalization does not issue any problem since all concept nodes in the same co-identical class have the same type[3] and the same marker.

Theorem 1 (Soundness and Completeness [4]). *Let $\mathcal{S}$ be a support, and H and G be two SGs (or SG^{ref}s) defined on $\mathcal{S}$. Then $H \sqsubseteq \mathcal{N}_{\mathcal{F}}(G)$ iff $\Phi(\mathcal{S}), \Phi(G) \models \Phi(H)$.*

1.3 Adapting the Semantics to Projection: the Ψ Solution

The semantics Ψ was introduced in order to translate the whole information encoded in a graph. In this semantics, two terms are associated to each concept node. As in Φ, the first term represents the co-identity class of the node. The second term is a variable representing the node itself. Let G be a SG or SG^{ref} defined on a support $\mathcal{S}$. $\Psi(\mathcal{S})$ is obtained in the same way as $\Phi(\mathcal{S})$, excepted that the predicates associated with concept types become binary. $\Psi(G)$ is the existential closure of the conjunction of atoms obtained in the following way: If Φ associates an atom $C(t)$ to a concept node, then Ψ associates it an atom $C(t, v)$ where the term v is a new variable representing the node itself. Each relation node of degree i is assigned an atom $R(v_1, ..., v_i)$, where R is the predicate of arity i assigned to its type, and the v_k are the second terms associated with the kth neighbors of the node. The graph G_3 in fig. 1 is interpreted by:

$$\Psi(G_3) = \exists xyzabcd\ (\text{c1}(x, a) \wedge \text{c1}(x, b) \wedge \text{c2}(y, c) \wedge \text{c3}(z, d) \wedge \text{r1}(a, c) \wedge \text{r2}(b, d))$$

Theorem 2 (Soundness and Completeness [7], [9]). *Let $\mathcal{S}$ be a support, and H and G be two SGs or SG^{ref}s defined on $\mathcal{S}$. Then $H \sqsubseteq G$ iff $\Psi(\mathcal{S}), \Psi(G) \models \Psi(H)$.*

The co-reference as described by the semantics Ψ implements what I call a *"weak co-identity"*. Thanks to the additional constraint defined on projection, concept nodes belonging to the same co-reference class are assured to be projected into the same co-identity class. But the information on a concept node stays the property of this particular node. These nodes represent different and independent points of view on the same entity. On the other hand, the "normalization + projection" process implements a *"strong co-identity"*. Thanks to normalization, the whole information on a specific concept node is shared by every other node of its co-identity class.

[3] An explicit constraint for co-referent generic nodes, a consequence of the conformity relation for individual ones.

2 Implementing Co-Identity with Simple Graph Rules

As shown by the example in introduction, *"strong co-identity"* can be too strong, in the sense that too much information can be deduced from the graph. But *"weak co-identity"*, though conforming exactly to the projection mechanism, is definitively too weak. I propose to represent co-identity by relation nodes in the graph, and to simulate its desired behavior by conceptual graphs rules [6], [5]. I first give a modified (though exactly equivalent) version of rules, then show how they can be used to simulate weak and strong co-identities.

2.1 Conceptual Graphs Rules

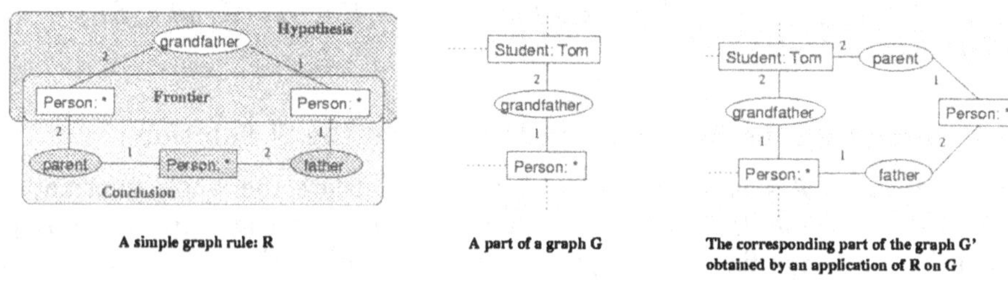

Fig. 2. A rule and its application

A simple graph rule (SGR) is defined as a simple graph which is associated a mapping *color* from $V_C \cup V_R$ into $\{0, 1\}$. A node v such that $\text{color}(v) = 0$ is called an *hypothesis* node, otherwise it is called a *conclusion* node. An hypothesis node having a conclusion node as a neighbor is called a *frontier* node. The subgraph of a SGR generated by the set of all hypothesis nodes is called the hypothesis of the rule. It must be a syntactically valid SG (see fig. 2). The subgraph generated by frontier nodes and conclusion nodes is called the conclusion.

Let $\mathcal{S}$ be a support, G a SG defined on $\mathcal{S}$, and R a SGR defined on $\mathcal{S}$. The rule R is *applicable* to G if there exists a projection from the hypothesis of R to G. In this case, the application of R on G following a projection Π is a SG G' constructed by making the disjoint union of G with the conclusion of R, then by fusionning[4] frontier nodes of the conclusion with the corresponding nodes in the projection of the hypothesis. The rule R in fig. 2 can be read *"If a person has a grandfather, then he has a parent whose father is this grandfather"*.

Let us consider a knowledge base $\mathcal{KB} = \{\mathcal{S}, \mathcal{R}\}$, where $\mathcal{S}$ is a support, and $\mathcal{R}$ a set of SGRs defined on $\mathcal{S}$. Let G and G' be two SGs defined on $\mathcal{S}$. We note $G \vdash G'$, if G' is obtained by an application of a rule of $\mathcal{R}$ on G. We say that G *derives* G' and we note $G \Vdash G'$, where $\Vdash$ is the reflexo-transitive closure of $\vdash$. We say that H is *deduced* from G and note $G \models H$ if there exists G' such that $G \Vdash G'$ and $H \sqsubseteq G'$.

[4] This fusion is not exactly similar to the one used in the normalization process, since projection can transform the label of a frontier node into a more specific one. The most specific label is kept for the resulting node.

Rules are given FOL semantics by extending Φ in the following way: each node is assigned the same atom as the one in the SG interpretation. $\Phi(R)$ is the universal closure of the formula $\Phi_H(R) \rightarrow \Phi_C(R)$ where $\Phi_H(R)$ is the conjunction of the atoms associated to the hypothesis nodes, and $\Phi_C(R)$ is obtained by existentially quantifying the variables of the conjunction of the other atoms, when these variables do not appear in $\Phi_H(R)$. The semantics Ψ can be extended to SGRs in the same way. The rule R in fig. 2 can be interpreted by the formulas:

$$\Phi(R) = \forall xy((\mathrm{Pe}(x) \wedge \mathrm{Pe}(y) \wedge \mathrm{gr}(x,y)) \rightarrow (\exists z(\mathrm{Pe}(z) \wedge \mathrm{pa}(z,y) \wedge \mathrm{fa}(x,z))))$$
$$\Psi(R) = \forall xyab((\mathrm{Pe}(x,a) \wedge \mathrm{Pe}(y,b) \wedge \mathrm{gr}(a,b)) \rightarrow (\exists zc(\mathrm{Pe}(z,c) \wedge \mathrm{pa}(c,b) \wedge \mathrm{fa}(a,c))))$$

Deduction is sound with respect to Φ [10]. To achieve completeness, we define a *normalizing derivation*. We note $G \vdash_{\mathcal{F}} G'$, when G' is the normal form of an application of a rule on G. It defines a derivation $\Vdash_{\mathcal{F}}$ and a deduction $\models_{\mathcal{F}}$.

Theorem 3 (Soundness and Completeness [6]). *Let $KB = \{\mathcal{S}, \mathcal{R}\}$ be a knowledge base, and G and H two SGs or SG^{ref}s defined on $\mathcal{S}$. Then $\mathcal{N}_{\mathcal{F}}(G) \models_{\mathcal{F}} H$ iff $\Phi(\mathcal{S}), \Phi(\mathcal{R}), \Phi(G) \models \Phi(H)$.*

Theorem 4 (Soundness and Completeness). *Let $KB = \{\mathcal{S}, \mathcal{R}\}$ be a knowledge base, and G and H two SGs or SG^{ref}s defined on $\mathcal{S}$. Then $G \models H$ iff $\Psi(\mathcal{S}), \Psi(\mathcal{R}), \Psi(G) \models \Psi(H)$.*

Proof. Thanks to completeness of projection with respect to the semantics Ψ, the proof is the same as the one given in [6] for the semantics Φ. □

2.2 The *Weak Co-Identity* Relation

Let us now represent co-identity by a new relation typed `co-ident` and associated rules. Given a support $\mathcal{S}$, $\mathcal{S}^{\mathcal{I}}$ is obtained by adding a new binary relation type `co-ident` and a new greatest element to $T^2_{\mathcal{R}}$, which covers both `co-ident` and $\top^2_{\mathcal{R}}$. Let $\mathcal{R}_{\mathcal{W}}$ be the set of rules defined in fig. 3. The first three rules indicate that `co-ident` is an equivalence relation, the last one is a set of rules, one for each individual marker in $\mathcal{S}$. These rules are obtained by replacing `Marker` and `Type` by m and $\tau(m)$, $\forall m \in \mathcal{I}$. Note that, since markers are not first-class objects of the graph, we cannot express that two nodes sharing the same individual marker represent the same entity in a single rule. The modified support and this set of rules define a knowledge base $\mathcal{KB}^{\mathcal{W}}(\mathcal{S}) = \{\mathcal{S}^{\mathcal{I}}, \mathcal{R}_{\mathcal{W}}\}$.

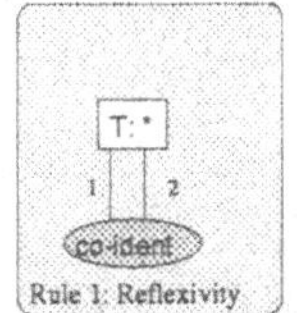

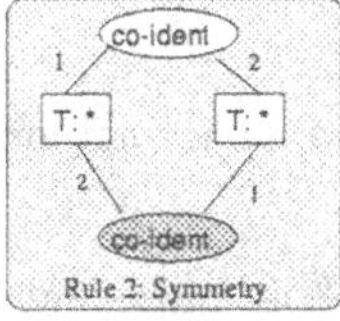

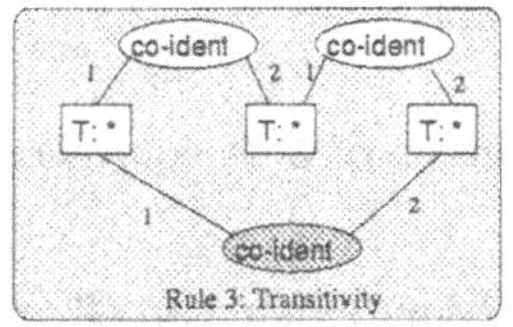

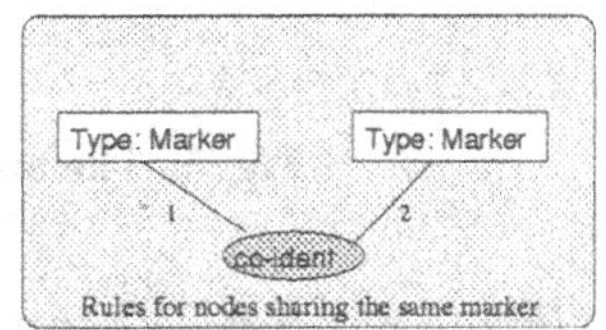

Fig. 3. Rules for weak co-identity

Proposition 1 (Equivalence). *Let $\mathcal{S}$ be a support, and G and H be two* SG^{ref}*s defined on $\mathcal{S}$. Let G' and H' be the* SGs *defined on $\mathcal{KB}^{\mathcal{W}}(\mathcal{S})$, obtained by replacing all co-reference links in G and H by relation nodes typed* `co-ident`. *Then $G' \models H'$ iff $\Psi(\mathcal{S}), \Psi(G) \models \Psi(H)$.*

Proof. We obtain a SG G'' by doing a complete irredundant expansion of G' using the rules in $\mathcal{R}^{\mathcal{W}}$. This operation is finite, since we create at most $|V_C(G)|^2$ relation nodes. Two concept nodes of G are co-identical iff there exists a relation node typed `co-ident` between these nodes in G''. □

2.3 The *Strong Co-Identity* Relation

Let $\mathcal{S}$ be a support. To simulate normalization, we now consider $\mathcal{KB}^{\mathcal{S}}(\mathcal{S}) = \{\mathcal{S}^{\mathcal{I}}, \mathcal{R}^{\mathcal{W}} \cup \mathcal{R}^{\mathcal{S}}\}$, where $\mathcal{R}^{\mathcal{S}}$ is the set of rules defined in fig. 4. There are in $\mathcal{R}^{\mathcal{S}}$ i rules of the form given in fig. 4 for each relation type of valence i. They are obtained by replacing the node typed `relation` by this relation type. These rules express that if a concept node C is linked by a relation node R to a concept node C', then every node indicated *co-ident* to C must be linked by R to C'.

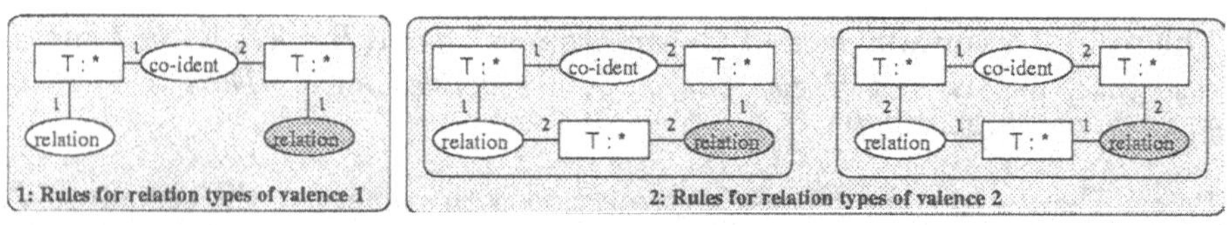

Fig. 4. Rules for strong co-identity

Proposition 2 (Equivalence). *Let $\mathcal{S}$ be a support, and G and H be two* SG^{ref}*s defined on $\mathcal{S}$. Let $G' = \vartheta(G)$ and $H' = \vartheta(H)$ be the* SGs *defined on $\mathcal{KB}^{\mathcal{S}}(\mathcal{S})$ obtained by replacing all co-reference links in G and H by relation nodes typed* `co-ident`. *Then $G' \models H'$ iff $\Phi(\mathcal{S}), \Phi(G) \models \Phi(H)$.*

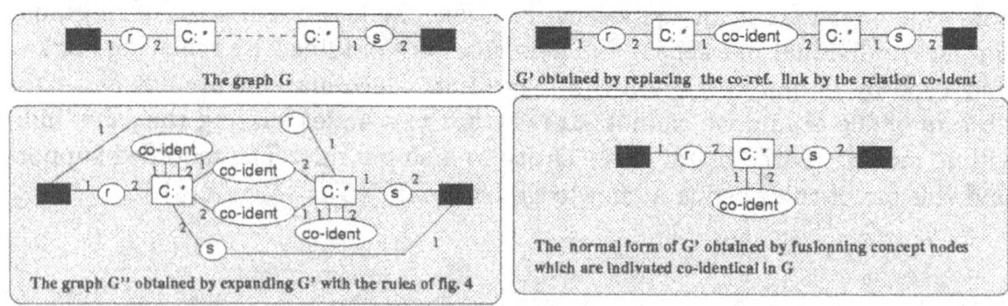

Fig. 5. The expansion of G' is equivalent to its normal form

Proof. The complete irredundant expansion G'' of G' is equivalent to $\vartheta(\mathcal{N}_{\mathcal{F}}(G'))$ (i.e. $G'' \sqsubseteq \vartheta(\mathcal{N}_{\mathcal{F}}(G')) \sqsubseteq G''$). This is proved by showing that co-identical nodes in G'' can be projected into the node C resulting from their fusion in $\mathcal{N}_{\mathcal{F}}(G')$, and that C can be projected into any of the co-identical nodes in G'' (fig. 5). □

3 From Nested Graphs to Boxed Graphs

The nested graphs (NGs) model allows to associate any concept node an internal description in the form of a NG. The formal model I use for NGs or NG^{ref}s is the one presented in [1] or [9]. Projection in this model is sound and complete without any restriction with respect to a natural extension of Ψ, but we have to consider a k-normality condition when using the semantics Φ [1], [9]. In order to extend the previous treatment of identity to NGs, I introduce the *boxed graphs* (BGs) model. BGs are a a generalization of NGs which allow relation nodes to link concept nodes in different descriptions. Moreover, I show that these boxed graphs are a "high-level representation" of a particular class of SGs, which are used to define BG rules.

3.1 Nested Graphs, k-Normality and the semantics Ψ

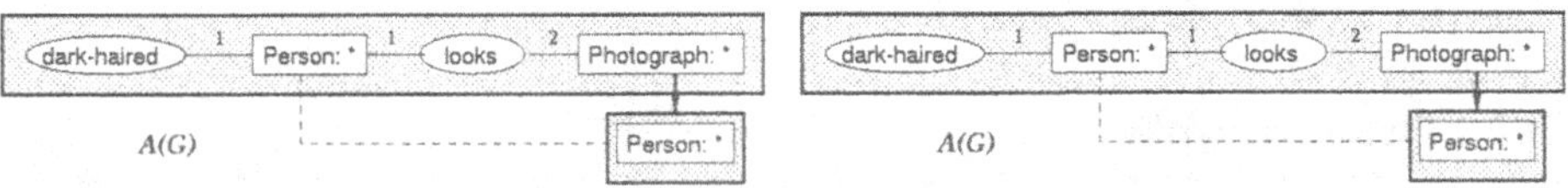

Fig. 6. A nested graph and its associated rooted tree

A NG is defined on a support identical to the one defined for SGs. A basic NG is obtained from a SG by adding to the label of each concept node a third field, called the *description* of the node, which is equal to ** (the empty description). A NG is obtained from a basic NG by replacing some of its descriptions by a NG. A NG^{ref} is a NG which is added an equivalence relation *co-ref* on the set of all its generic concept nodes, this relation is extended to *co-ident* as in SGs. Any NG is associated a rooted tree (fig. 6) whose nodes are the SGs used in its construction and edges (c, G) indicate that the concept node c is described by the NG whose root is G. Projection in the NG model can be defined on this tree: let H and G be two NGs, and $\mathcal{A}(H), \mathcal{A}(G)$ be their associated rooted trees. A projection from H to G is given by the projections (in the sense of simple graphs) of all nodes of $\mathcal{A}(H)$ into nodes of $\mathcal{A}(G)$ such that the root of $\mathcal{A}(H)$ is projected into the root of $\mathcal{A}(G)$, and the root of the description of a concept node can only be projected into the root of the description of its projection. Constraints on co-identical nodes must also be respected in the case of NG^{ref}s.

The semantics Φ and Ψ are extended to NG^{ref}s by associating another term to the atoms interpreting each node of the graph: n-ary predicates become $(n+1)$-ary. Every node of G in a node $\mathcal{X}$ of the rooted tree is associated the same additional term: the constant ρ if $\mathcal{X}$ is the root of $\mathcal{A}(G)$, otherwise the term identifying the concept node C such that $\mathcal{X}$ is the root of the description of C (the only term for Φ, the unique variable associated with the node in Ψ). By example, the graph G in fig. 6 is interpreted by:

$$\Phi(G) = \exists xy\ (\text{Pe}(x,\rho) \wedge \text{Ph}(y,\rho) \wedge \text{Pe}(x,y) \wedge \text{dh}(x,\rho) \wedge \text{lo}(x,y,\rho))$$
$$\Psi(G) = \exists xyabc\ (\text{Pe}(x,a,\rho) \wedge \text{Ph}(y,b,\rho) \wedge \text{Pe}(x,c,b) \wedge \text{dh}(a,\rho) \wedge \text{lo}(a,b,\rho))$$

Theorem 5 (Soundness and Completeness [7], [9]). *Let $\mathcal{S}$ be a support, G and H be two* NGs *or* $\mathtt{NG}^{ref}$*s. Then $H \sqsubseteq G$ iff $\Psi(\mathcal{S}), \Psi(G) \models \Psi(H)$*

In order to achieve completeness with respect to the semantics Φ, Simonet has defined a normal form for $\mathtt{NG}^{ref}$s, which is not always possible to compute, and a k-normal form, that I briefly recall here. A $\mathtt{NG}^{ref}$ G is said in k-normal form if every node of $\mathcal{A}(G)$ is a $\mathtt{SG}^{ref}$ in normal form, and for every concept node C such that its depth in $\mathcal{A}(G)$ is less than k, if C is co-identical to a concept node C' whose description is not empty, then the root of the description of C' must be exactly equivalent[5] to the root of the description of C. Putting a graph G into its k-normal form $\mathcal{N_F}^k(G)$ (by normalizing every node of $\mathcal{A}(G)$, and copying the roots of the descriptions of co-identical concept node as long as required by the k-normal form) does not change the semantics $\Phi(G)$.

Theorem 6 (Soundness and Completeness [8], [9]). *Let $\mathcal{S}$ be a support, G and H be two* NGs *or* $\mathtt{NG}^{ref}$*s, and $k \geq depth(\mathcal{A}(H))$ be a number. Then $H \sqsubseteq \mathcal{N_F}^k(G)$ iff $\Phi(\mathcal{S}), \Phi(G) \models \Phi(H)$*

3.2 Boxed Graphs

The main difference between NGs and *boxed graphs* (BGs) is that there can be relation nodes linking concept nodes which belong to different descriptions.

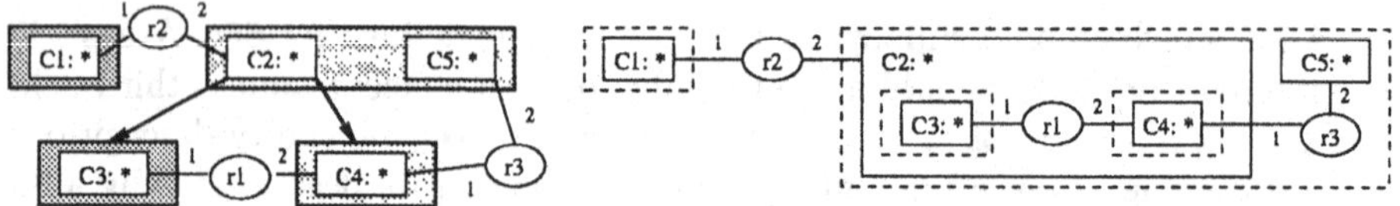

Fig. 7. A boxed graph and usual graphical representation

Definition 1 (Boxed Graphs). *Let $\mathcal{S}$ be a support, as defined for* SGs. *A* boxed graph *G is defined as a simple graph which is added a partition of V_C into* boxes *$\mathcal{B} = \{B_1, \ldots, B_k\}$, and a partial mapping* desc *from $\mathcal{B}$ into V_C, such that the oriented graph $\mathcal{A}(G) = \{\mathcal{B}, E\}$ defined by $(B_i, B_j) \in E$ iff $\exists x \in B_i, x = desc(B_j)$ is a collection of rooted trees.*

Boxes and *desc* are used to translate the nesting relation, and can be represented in the drawing of the graph by dotted rectangles drawn inside the concept node they describe (see fig. 7). Though this representation is similar to the one adopted for NGs, important differences must be noted. First, there can be several boxes describing the same concept node. Next, though there is a unique root in the rooted tree associated to NGs, the boxes for which *desc* is not defined are multiple *root boxes*. This will be of great utility for defining BG rules. Finally, there can be relation nodes linking concept nodes which belong to different boxes. This property will be used to represent the relation *co-ident* by relation nodes.

[5] Not only there is a projection from one to the other, and *vice-versa*, but these projections must map any generic concept node C into a generic concept node co-referent to C.

Projection on BGs is defined as on SGs, but two concept nodes in the same box must be projected into the same box ("a box is projected into a box"), and a box describing a concept node C must be projected into a box describing the image of C. Note the difference with projection as defined on NGs: a root box does not need to be projected into a root box.

A BG is said *nested* if it has only one root box, each concept node is described by at most one box, and there is no relation node linking concept nodes in different boxes. We can associate to each "nested BG" G the NG having the same graphic representation as G, and conversely, to each NG we associate a boxed graph which has the property of being nested. A *rooted projection* on a boxed graph is a projection such that root boxes can only be projected into root boxes. The proof of the next proposition is immediate, and it justifies the assertion that BGs are a generalization of NGs.

Proposition 3. *Let G and H be two NGs defined on $\mathcal{S}$, and G' and H' be their associated BGs. Then there is a projection from H to G iff there is a rooted projection from H' to G'.*

3.3 Associated Simple Graphs

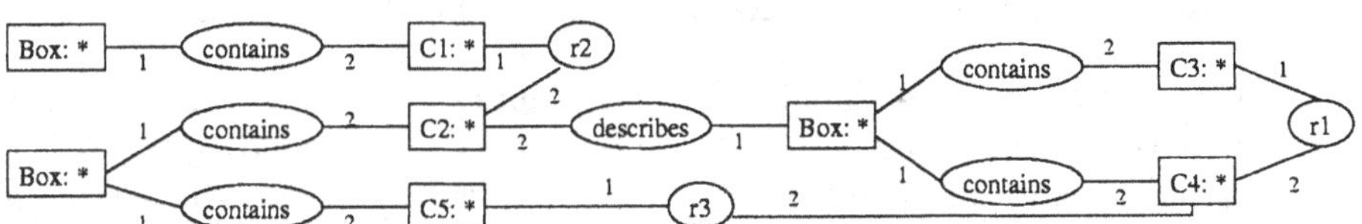

Fig. 8. The simple graph associated to the boxed graph of fig. 7

A problem with BGs is that boxes are not first-class objects of the model, but an assertion on some nodes of the graph which is represented in the drawing of the graph: boxes cannot be manipulated by rules. I define here the associated SG of a BG, where boxes and *desc* are reified into nodes of the SG.

Let $\mathcal{S}$ be a support, and G be a BG. The graph $\mathcal{S}_\mathcal{G}(G)$ associated to G is a SG defined on a support $\mathcal{S}_\mathcal{G}(\mathcal{S})$: T_C is added two types of concepts, $\top\top_C$, and *Box*, such that $\top\top_C$ covers $\top_C$ and *Box*. T_R is added three binary relation types, $\top\top^2_R$, *describes* and *contains*, such that $\top\top^2_R$ covers $\top^2_R$, *describes* and *contains*. $\mathcal{S}_\mathcal{G}(G)$ is the SG obtained by adding a generic concept node typed *Box* for every box in G, linking this node by a relation node typed *contains* to every concept node belonging to this box, and if this box describes a concept node C, linking the node typed *Box* to C by a relation node typed *describes*. The BG in fig. 7 is associated the SG in fig. 8.

Proposition 4. *Let $\mathcal{S}$ be a support, and G and H be two boxed graphs defined on $\mathcal{S}$. Let $\mathcal{S}_\mathcal{G}(\mathcal{S})$ be the support obtained from $\mathcal{S}$ as indicated above, and $\mathcal{S}_\mathcal{G}(G)$ and $\mathcal{S}_\mathcal{G}(H)$ the SGs defined on $\mathcal{S}_\mathcal{G}(\mathcal{S})$, associated with G and H. Then $H \sqsubseteq G$ iff $\mathcal{S}_\mathcal{G}(H) \sqsubseteq \mathcal{S}_\mathcal{G}(G)$.*

Again, the proof is immediate. I will now consider BGs (and NGs, thanks to prop. 3) as a high-level representation for a particular class of SGs whose support includes the types *Box*, *describes*, and *contains*. I will use indifferently the BG representation or the SG representation for these graphs. In particular, I will use this SG representation to define ***boxed graphs rules***. A BG rule is defined as a SG rule such that its hypothesis can be associated to a BG, the rule as a whole can be associated to a BG, and its subgraph generated by nodes typed *Box*, *describes*, and *contains* (the rooted forest representing nesting levels) is such that no conclusion node stands between an hypothesis node and its root. The consequence of this restriction is that BG rules only derive valid BGs. The result of the application of a BG rule R on a BG G is the boxed representation of the SG G' obtained by applying R (in a SGR sense) to the SG representation of G.

The semantics $\Psi_{\mathcal{B}}$ associated to a BG G or a BG rule R can then be defined as the semantics Ψ associated to $\mathcal{S}_{\mathcal{G}}(G)$, or R. Thanks to th. 6 and prop. 4, the proof of the following theorem is immediate.

Theorem 7 (Soundness and Completeness). *Let $KB = \{\mathcal{S}, \mathcal{R}\}$ be a knowledge base, where $\mathcal{R}$ is a set of* BG *rules, and G and H be two* BG*s. Then $G \models H$ iff $\Psi(\mathcal{S}), \Psi_{\mathcal{B}}(\mathcal{R}), \Psi_{\mathcal{B}}(G) \models \Psi_{\mathcal{B}}(H)$*

4 Rules Simulating Co-Identity in Boxed Graphs

I will now present the rules simulating the semantics Ψ and Φ in the particular class of boxed graphs (with *co-ident* relation nodes) corresponding to $\mathtt{NG}^{ref}$s. Assigning a semantics to co-identity by extending Φ or Ψ to any BG is beyond the scope of this paper.

Let G be a $\mathtt{NG}^{ref}$, defined on S. It is still required that only concept nodes sharing the same label can be declared co-identical. The BG $\mathcal{B}_{\mathcal{G}}(G)$ associated to G is defined on a support $\mathcal{S}^{\mathcal{I}}$ obtained by adding the *co-ident* relation type in $\mathcal{S}$, and $\mathcal{B}_{\mathcal{G}}(G)$ is represented by a graph obtained by replacing every co-reference link in G by a relation node typed *co-ident*. As these relation nodes can link concept nodes in different boxes, these *nested*^{ref} BGs are not nested BGs.

As I will now work only with nested *nested*^{ref} BGs, and in order to present more intuitive BG rules and equivalence results, I will use a slightly modified version of the associated SGs (note that these modifications only concern BGs, and not BG rules). First, in order to simulate a *rooted projection*, concept nodes typed *Box* representing root boxes will be labelled with the individual marker ρ. Next, for every concept node C such that there is no box describing C, we add a concept node typed *Box*, linked to C by a relation node typed *describes*. This feature will be used when defining rules in such a way that applying a rule does not create more than a box describing a single concept node.

4.1 The Semantics Ψ

Rules simulating co-identity in $\mathtt{SG}^{ref}$s are updated to conform to the boxed graphs syntax. The set of rules $\mathcal{R}_{\mathcal{B}}^{\mathcal{W}}$ is obtained from the rules presented in fig. 9 in the same way as in sec. 2.2. Since these rules only add *co-ident* relation nodes, they generate only nested^{ref} BGs.

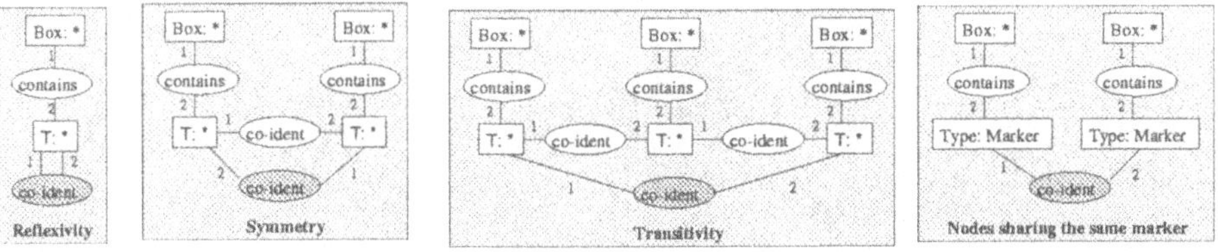

Fig. 9. Rules for weak-co-identity in boxed graphs

Proposition 5 (Equivalence). *Let $\mathcal{S}$ be a support, and G and H two* NG*s or* NG^{ref}*s defined on $\mathcal{S}$. Let G' and H' be their associated* BG*s, defined on the knowledge base $\mathcal{KB}_{\mathcal{B}}^{\mathcal{W}}(\mathcal{S}) = \{\mathcal{S}^{\mathcal{I}}, \mathcal{R}_{\mathcal{B}}^{\mathcal{W}}\}$. Then $G' \vDash H'$ iff $\Psi(\mathcal{S}), \Psi(G) \vDash \Psi(H)$.*

Proof. We must check that the *co-ident* classes can be computed regardless of the boxes containing the nodes. This can be done since the multiple roots in the hypothesis of the rules can be projected into any box of a given graph. □

4.2 The Semantics Φ

To simulate the semantics Φ in NG^{ref}s, I present a set of BG rules which mimic the operations used to put a graph into its k-normal form. The set of rules $\mathcal{R}_{\mathcal{B}}^{\mathcal{W}}$ presented in fig. 9 will be used to generate all *co-ident* relation nodes.

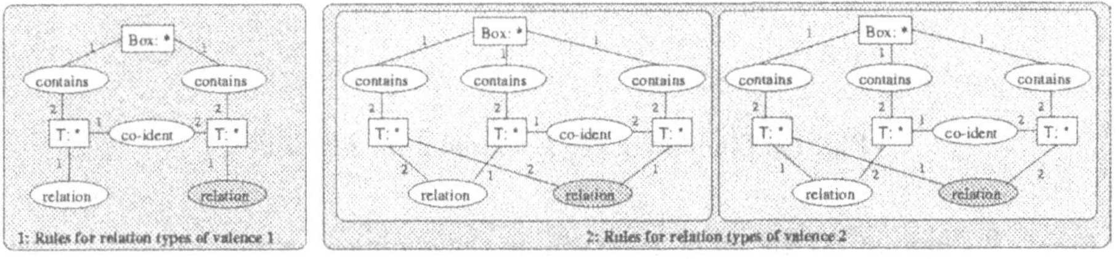

Fig. 10. Rules normalizing SGs in the same boxes

The set of rules $\mathcal{R}_{\mathcal{B}}^{\mathcal{S}1}$ presented in fig. 10 is a slightly updated version of the rules in fig. 4 used for simple graphs. Let G be a nested^{ref} BG, and G' its associated NG^{ref}. The application of these rules on G mimics normalization on every node of $\mathcal{A}(G')$. See that, as these rules only create relation nodes that link concept nodes in the same box, these rules only generate nested^{ref} BGs.

In order to simulate that two co-identical concept nodes must be described by the same box, the method used until now (duplicating the relation nodes) would create graphs which are not boxed. The set of rules $\mathcal{R}_{\mathcal{B}}^{\mathcal{S}2}$ in fig. 11 simulates a recursive copy of the contents of this description. Assuming that there are in $\mathcal{S}$ n_1 concept types and n_2 individual markers, the first rule drawn in fig. 11 represents $n_1 + n_2$ rules, the first n_1 being obtained by replacing Marker and Type by $*$ and t, $\forall t \leq \top_C$, the other n_2 by replacing them by m and $\tau(m), \forall m \in \mathcal{I}$. The second rule drawn also represents a set of rules, one for each relation type in $\mathcal{S}$. These rules simulate the fact that the descriptions of co-identical concept nodes must be exactly equivalent. I will now consider the set of rules $\mathcal{R}_{\mathcal{B}}^{\mathcal{S}} = \mathcal{R}_{\mathcal{B}}^{\mathcal{W}} \cup \mathcal{R}_{\mathcal{B}}^{\mathcal{S}1} \cup \mathcal{R}_{\mathcal{B}}^{\mathcal{S}2}$, necessary to simulate the semantics Φ.

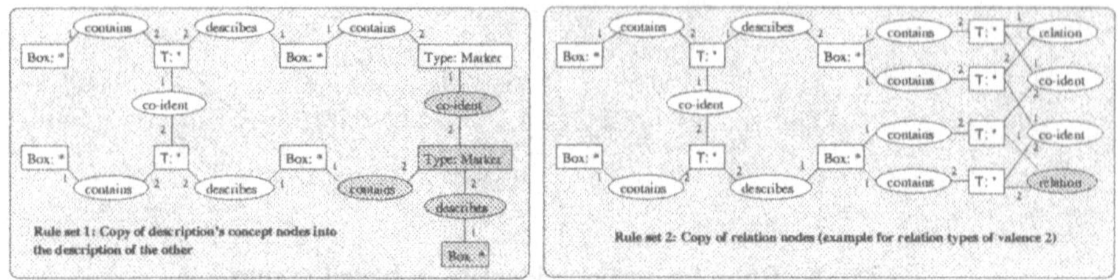

Fig. 11. Rules copying the description of co-identical nodes

Proposition 6 (Equivalence). *Let $\mathcal{S}$ be a support, and G and H two* NG*s or* NG^{ref}*s defined on $\mathcal{S}$. Let G' and H' be their associated* BG*s, defined on the knowledge base $\mathcal{KB}^{\mathcal{S}}_{\mathbb{B}}(\mathcal{S}) = \{\mathcal{S}^{\mathcal{I}}, \mathcal{R}^{\mathcal{S}}_{\mathbb{B}}\}$. Then $G' \models H'$ iff $\Phi(\mathcal{S}), \Phi(G) \models \Phi(H)$.*

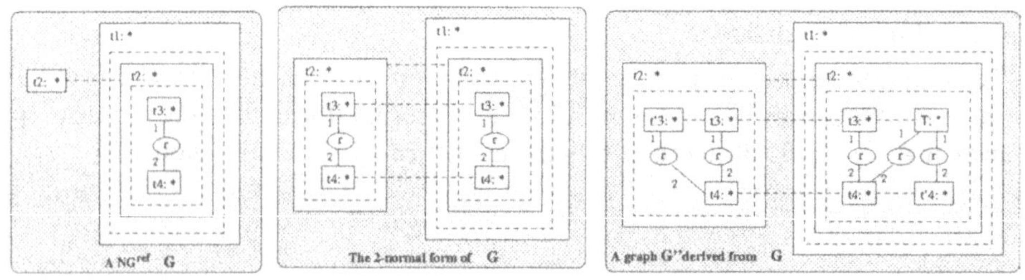

Fig. 12. Simulating the k-normal form with rules

Proof. We cannot, as for prop. 2, reason on the complete expansion of G'. Even with the simple example[6] in fig. 12, where any k-normal form of G is identical to its 2-normal form, we can derive from G a graph whose size is not bounded. But we can prove that, for any natural integer k, there exists a BG G'' such that $G' \Vdash G''$ and the BG associated with the k-normal form of G can be projected into G''. Conversely, we can prove that, for any BG G'' derived from G', there exists a natural integer $k = \text{depth}(\mathcal{A}(G''))$ such that G'' can be projected into the BG (with *co-ident* relation nodes) associated with the k-normal form of G.

A problem with the rules copying concept nodes in the descriptions of co-identical nodes (first rule of fig. 11) is that they do not generate valid nested^{ref} BGs. As shown in the graph G'' of fig. 12, for any concept node C labelled `(t, m)`, these rules can create an infinity of concept nodes co-identical to C, whose label can be a superlabel of `(t, m)`. As its co-identical concept nodes have different types, the graph obtained in such a way cannot be associated to a NG^{ref}. In order to solve this problem, we have to weaken the constraints on the relation *co-ident*: concept nodes sharing the same individual marker are in the same co-identity class, and the set of labels of concept nodes in the same co-identity class

[6] I adopted for these graphs a "nested graph representation", which is somewhat easier to read.

has a smallest element. Co-identical nodes are fusionned into a node having the most specific type and marker during normalizations. Adopting this weakened constraint solves this problem, without changing any of the preceding results. □

Conclusion

In this paper, I show that a model generalizing $\mathtt{NG}^{ref}$s can be exactly represented with SGs, in such a way that all projections are preserved when translating graphs from one model to the other. In order to simulate the various reasonings induced by co-identity, be it in SGs or NGs, interpreting identity with the semantics Φ or Ψ, I show that we only need the "SG + rules" model. At least, the rules presented can be seen as a "graphical illutration" of the different operations required by the co-identity relation. At most, this model can be seen as a "low-level layer" for the implementation of $\mathtt{NG}^{ref}$s, boxes being only a "man-machine interface" layer. I prefer to see this model as a "protyping tool", a way to rapidly define and test various semantics of identity, using a development platform such as CoGITaNT [2]. But, for an efficiency purpose, several reasonings that can be represented by rules must still be given a "hard-coded", specific algorithmic solution.

References

1. M. Chein, M-L. Mugnier, and G. Simonet. Nested Graphs : A Graph-based Knowledge Representation Model with FOL Semantics. In *Proceedings of the 6th International Conference "Principles of Knowledge Representation and Reasoning" (KR'98)*. Morgan Kaufmann Publishers, 1998.
2. D. Genest and E. Salvat. A Platform Allowing Typed Nested Graphs: How CoGITo Became CoGITaNT. In *Proceedings of the 6th International Conference on Conceptual Structures*, Lecture Notes in AI. Springer, 1998.
3. B.C. Ghosh and V. Wuvongse. Computational Situation Theory in the Conceptual Graph Language. In *Proceedings of ICCS'96*, Lecture Notes in AI. Springer, 1996.
4. M-L. Mugnier and M. Chein. Représenter des connaissances et raisonner avec des graphes. *Revue d'Intelligence Artificielle*, 10-1:7–56, 1996.
5. E. Salvat. *Raisonner avec des opérations de graphes : graphes conceptuels et règles d'inférence*. PhD thesis, Université de Montpellier II, 1997.
6. E. Salvat and M-L. Mugnier. Sound and Complete Forward and Backward Chainings of Graph Rules. In *Proceedings of the 4th International Conference on Conceptual Structures*, Lecture Notes in AI. Springer, 1996.
7. G. Simonet. Une autre sémantique logique pour les graphes conceptuels simples ou emboîtés. Research Report, 1996.
8. G. Simonet. Une sémantique logique pour les graphes conceptuels emboîtés. Research Report, 1996.
9. G. Simonet. Two FOL Semantics for Simple and Nested Conceptual Graphs. In *Proceedings of the 6th International Conference on Conceptual Structures*, Lecture Notes in AI. Springer, 1998.
10. J. F. Sowa. *Conceptual structures : Information processing in mind and machine.* Addison-Wesley, 1984.

Conceptual Graphs as Algebras - With an Application to Analogical Reasoning

Torben Braüner[1] and Jørgen Fischer Nilsson[2] and Anne Rasmussen[3]

[1] InterMedia
Aalborg University
Fredrik Bajers Vej 7 C, 9220 Aalborg East, Denmark
torbenb@intermedia.auc.dk

[2] Department of Information Technology
Technical University of Denmark
Building 344, 2800 Lyngby, Denmark
jfn@it.dtu.dk

[3] Department of Communication
Aalborg University
Langagervej 8, 9220 Aalborg East, Denmark
anne@hum.auc.dk

Abstract. The first part of this paper presents a logico-algebraic reconstruction of conceptual graph fundamentals using an appropriately extended binary relation algebra. The algebraisation comprises axioms which in a straightforward and instructive fashion admit deductive reasoning by term rewriting. As an application of the described logico-algebraic framework, the second part of the paper addresses formalisation of analogical reasoning. It is suggested that analogical reasoning be carried out as clausal resolution inference using axioms of the binary relation algebra together with the application specific axioms corresponding to the domains under consideration. This makes the analogical reasoning amenable to implementation in logic programming languages.

1 Introduction

Conceptual graphs (CGs) are data structures with accompanying operations advanced for representing concepts and relationships between concepts. As such, CGs offer themselves as a convenient alternative to traditional logical languages, they being closer to natural language by avoiding explicit use of variables. On the other hand, CGs fall short of the full syntactic apparatus and semantical notions such as logical consequence available, say, in predicate logic.

This paper seeks to combine the above mentioned advantages of CGs and logic in an algebraic logic put forward as a logico-algebraic reconstruction of CGs and similar representation forms such as frames. The first step is to establish the *isa* relationship in the logico-algebraic reconstruction as the partial ordering of a distributive lattice, for provision of the type structure. The lattice constitutes the skeleton in a conceptual structure for a given knowledge domain and each concept of the knowledge domain is situated in this taxonomy. The next step is to involve relations and attributed properties. This is done by enriching the distributive lattice such that it comprises a binary relation algebra. Such extended binary relation algebraic logics are two-sorted logics with a sort of concepts and a relation sort. Given a knowledge base in the form of ground algebraic equations, deductive reasoning may then be conducted as term rewriting using the axioms.

Analogical reasoning plays a central role in discovery and learning. The ability to reveal similarities between different concepts or conceptual domains can provide for extending knowledge as well as for a reorganisation of previously held knowledge. The most widespread use of analogy is reflected in the view that analogical reasoning deals with how existing similarities between two domains enable the uncovering and transfer of further similarities between them. As an application of the logico-algebraic formalism, we address formalisation of analogical reasoning. It is suggested that analogical reasoning be carried out as clausal resolution inference using equational axioms of an appropriately extended binary relation algebra together with the application specific axioms corresponding to the domains under consideration. This makes the analogical reasoning amenable to implementation in logic programming languages like PROLOG.

In the next section of this paper, section two, we give the logico-algebraic reconstruction of CG fundamentals. In section three, the formalisation of analogical reasoning is addressed as an example application of the logico-algebraic formalism. The final section, section four, makes some concluding remarks.

2 Conceptual Graphs as Algebras

2.1 Conceptual Graph Fundamentals

Conceptual graphs [17] in their kernel form are constituted by:

- Nodes representing either general or individual concepts.
- Labelled arcs representing binary relations between concepts.
- A distinguished binary relation, namely the *isa* concept or class inclusion relationship, which establishes the conceptual type structure.

Accordingly, in our conception a CG basically characterises a concept (or class) c by a bundle of binary relations $r_1, r_2, \ldots$ connecting to concepts $c_1, c_2, \ldots$. Diagrammatically, this can be depicted as

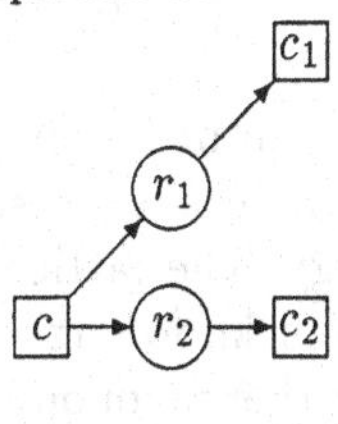

Fig. 1

This conforms with the notion of a frame

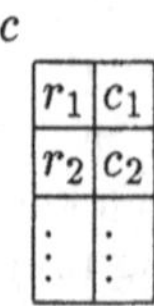

understood as an object named c comprising $0, 1$ or more entries in the form of relation-value pairs. The below algebraic reconstruction takes as starting point the distinguished *isa* inclusion relationship and then subsequently considers other relations.

2.2 Conceptual Lattices

The *isa* relationship in our logico-algebraic reconstruction of CGs is established as the partial ordering $\leq$ of a distributive lattice. For any concepts c_1 and c_2 in the lattice, the supremum $c_1 + c_2$ and infimum $c_1 \times c_2$ are assumed to exists, cf. [4], and the distributive lattice is bounded downwards by the null concept, lattice-bottom $\bot$, and upwards by the universal concept, lattice-top $\top$. The null concept serves to express disjointness, as in the lattice conceptual structure

$$\begin{aligned} pet &= cat + dog \\ cat \times dog &= \bot \end{aligned}$$

When conceptual structures are conceived as lattices they can be depicted as Hasse diagrams. Following the Hasse diagram convention, the *isa* relationships are unlabelled upwards directed arrows

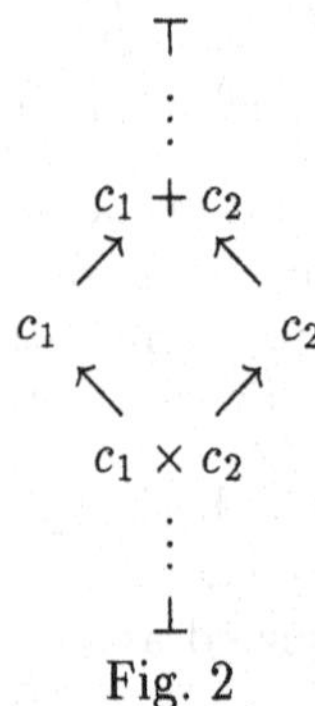

Fig. 2

Individual concepts are situated immediately above null as lattice-atomic elements. Formally, in the lattice there is no distinction between general and individual concepts as in mereology. The lattice constitutes the type skeleton in a conceptual structure for a given knowledge domain and each concept of the knowledge domain is situated in this taxonomy.

The two complementary lattice operations + (supremun, join) and $\times$ (infimum, meet), induced by the *isa* ordering relationship are axiomatised in lattice theory by the laws of idempotency, commutativity, associativity, absorption, and distributivity. Lattice-top $\top$ and lattice-bottom $\bot$ are axiomatised by boundary axioms. The fundamental result in lattice theory that

$$c_1 \leq c_2 \text{ iff } c_1 = c_1 \times c_2 \text{ iff } c_2 = c_1 + c_2$$

establishes the link between the order-theoretic and the equational definition of a lattice. The laws above are all well known from set-theory. As denotations $[\![c]\!]$ of ground terms c are taken sets, with definitions

$$\begin{aligned}
&[\![c_1 + c_2]\!] = [\![c_1]\!] \cup [\![c_2]\!] \\
&[\![c_1 \times c_2]\!] = [\![c_1]\!] \cap [\![c_2]\!] \\
&[\![\bot]\!] = \emptyset \\
&[\![\top]\!] = U
\end{aligned}$$

where U is an assumed universe, conforming with representation theorems for distributive lattices [4]. The algebraic conception of lattices enriches the taxonomy with means of calculation with concepts as further developed in the next section.

2.3 Concepts and Relations in Algebras

The conceptual lattice, cf. Fig 2., is to be augmented with relations, e.g. for attribution of properties as in Fig. 1. This is done by enriching the distributive lattice such that it comprises a binary relation algebra as originally proposed by Peirce and later revived by Tarski, [19]. Such binary relation algebraic extensions are two-sorted algebraic logics with a sort of concepts (conforming with unary predicates) and a relation sort (conforming with binary predicates). These extended relation algebras are axiomatised according to which operations they involve. Usually there are operations on both sorts as well as operations connecting the sorts, [1], see also [16].

In Boolean modules the two sorts are connected with the crucial concept-forming operation of Peirce product $(r : c)$ where r is a binary relation and c is a concept. Given a binary relation term r and a concept term c, the Peirce product $(r : c)$ is interpreted as

$$[\![r : c]\!] = \{x | \exists y \in [\![c]\!] \; \langle x, y\rangle \in [\![r]\!]\}$$

corresponding to the "some" predicate operator in description logic. Actually, these algebras may be conceived as algebraic counterparts of basic description logics [1] if extended with negation, that is complementation, leading to Heyting or Boolean algebras. See also [2], where further references to description logics can be found. On the other hand there seems to be no close affinity to Wille's concept lattices as described in [4]; see, however, [18] for a recent development.

The Peirce product now serves to attach properties in the form of attributed concepts as in the specification

$$blackcat = cat \times (color : black).$$

In general, a CG in our conception, cf. Fig.1, becomes represented by a ground algebraic term (frame term)

$$c \times (r_1 : c_1) \times ... \times (r_n : c_n).$$

The Peirce product enables us to handle specific relationships (between individuals) as well as general relationships (between non-individual concepts).

In the Concept Algebra formalism, [15, 16] a specialisation of the Peirce product is preferred where the term

$$(r_i : c)$$

is replaced by

$$a_i(c)$$

with a_i called an attribute. The attributes, which function as unary algebraic operators on concepts, are axiomatised in the distributive lattice simply as follows

$$a_i(X + Y) = a_i(X) + a_i(Y)$$
$$a_i(\bot) = \bot$$

Attributes are monotonic, that is, $c_1 \leq c_2$ implies $a(c_1) \leq a(c_2)$. This is easily proved from the axioms. The relations $[\![a_i]\!]$ denoted by the attributes a_i are usually assumed functional, that is, it is assumed that they are partial functions. This amounts to adding the axiom

$$a_i(X \times Y) = a_i(X) \times a_i(Y).$$

The price to pay for the simple axiomatisation above is abandoning of lattice operations and other operations on relations, notably composition, $r_1 \circ r_2$, and inversion, $r^{\smile}$.

2.4 Reasoning with Concepts and Relations in Algebras

Given a conceptual structure in the form of ground algebraic equations, deductive reasoning may be conducted as term rewriting using the axioms. For instance, given

$$greycat = cat \times (color : grey)$$

using the axioms and disjointness, that is,

$$grey \times black = \bot$$

and assuming *color* being functional, one gets

$$greycat \times blackcat = ... = cat \times (color : \bot) = \bot$$

telling that *greycat* and *blackcat* are disjoint. Similarly, given

$$darkcat = cat \times (color : (black + grey))$$

one gets

$$darkcat \times blackcat = ... = blackcat$$

which amounts to

$$blackcat \leq darkcat$$

telling that *blackcat* is included in *darkcat*. This latter example illustrates that properties inherit in the cruxing of terms. This means that $\times$ provides multiple - possibly conflicting - inheritance in the lattice taxonomy.

2.5 Relationship to Conceptual Graph Operations

Above it was explained how CGs in their kernel form can be reconstructed as algebraic terms by using the concept product (lattice-meet) $\times$ to attach a relation r to a concept c in combination with the Peirce product as in $c \times (r : c')$.

However, the lattice algebraic operations $+$ and $\times$ serve not only as constructors for CG terms; they further represent the algebraic counterparts of the CG operations [17]:

1. The lattice join $+$ corresponds to formation of the least common generalisation of two graphs.
2. The lattice meet $\times$ corresponds (with an ufortunate, confusing terminology) to formation of what is known in conceptual graph theory, [17], as the maximal join of two graphs, cf. also [14, 20].

Hence the algebraic operations with accompanying axioms may be considered as the logical counterpart of the graph data structure operations. Observe that lattice-meet, $\times$, is always defined: The case of incompatible graphs corresponds to disjoint concepts yielding the null concept, that is, lattice-bottom, $\perp$. In [15] is explained thow the meet, $\times$, generalises natural join in relational data bases.

3 An Application to Analogical Reasoning

3.1 Analogy as Similarity between Domains

Analogical reasoning plays a central role in discovery and learning. The ability to uncover similarities between different concepts or conceptual domains facilitate the extension of knowledge as well as the reorganisation of previously held knowledge. The most widespread use of analogy is reflected in the view that analogical reasoning deals with how existing similarities between two domains (a source and a target domain) provide for the uncovering and transfer of further similarities between them, [6, 9, 11]. This mechanism is for instance reflected in the problem solving strategy referred to by Gordon, [10], p. 98, as "making the strange familiar". Here, it is taken that one is able to establish an understanding of an unknown domain due to the existence of similarities between the unknown domain and another more well known one. Further, the existing similarities might suggest that other similarities among relations in the given domains hold as well. The probability of identifying an analogy increases in the case of simple analogical reasoning with surface similarity between the base and target. For instance one might rely on knowledge about how to drive a car when one is for the first time faced with the task of having to steer a boat.

However, more challenging examples is reflected in cases where the source domain is remote from the target domain. This sort of research into principles of analogical reasoning is for instance presented by Gentner et al. In the series of papers [5, 6, 7, 8] it is outlined how the meaning of an analogy can be derived from the meanings of its parts. Gentner's theoretical framework is corroborated by evidence from psychological studies, which seek to uncover the set of implicit rules by which people interpret analogy. It is suggested that the main characteristic of analogy is that it involves an alignment of relational structure.

> ... an analogy is an assertion that a relational structure that normally applies in one domain can be applied in another domain. ([6], p. 156)

As an example, the pioneering work in thermodynamics by the French scientist Sadi Carnot (1796–1832) provides an example of analogical reasoning in this sense. Carnot uses an analogy between heat (target domain) and water (source domain): The "fall" of heat from high to low temperature is compared to the fall of water from high to low elevation. By means of this analogy Carnot is able to organise ideas about the mechanical action of heat, for instance: The energy conferred when heat flows between two bodies varies with the difference in temperature between them. In this way the analogy conveys a system of connected knowledge and justify search for further similarities between two domains on behalf of already revealed similarities, cf. [5], p. 453. Thus, analogies can be used for learning purposes in order to support a student's conceptualisation of a given field. To make the analogy useful the learner must be able to focus on certain kinds of matches between two given domains. Furthermore, when the learner is able to reveal existing similarities between two domains this also indicates that there might be other similarities as well. The uncovering of relevant types of matches are described in Gentner's structure-mapping theory of analogy.

3.2 Analogy as Structure Mapping

The essence of Gentner's theoretical framework is reflected in two main principles.

1. The principle of relation mappings refers to similarity between relations in two domains (the base and the target domain), meaning that the formation and understanding of analogy has to be determined by comparing for similarity between relations instead of comparing for similarities among common features. An often considered example of this selectiveness of analogical mapping, is simple proportional analogies, which are characterised by relations such as "*a* is to *b* as *c* is to *d*", cf. [13], p. 30. As an example of this, Gentner mentions the following arithmetic analogy 3:6::2:4. Here, we do not care how many features 3 and 2 has in common, instead we focus on the relationship

"twice as great as" that holds between 3 and 6 as well as between 2 and 4, cf. [6], p. 156.
2. The so called systematicity principle stresses the fact that certain types of mappings are more preferable than others. Especially a base predicate that belongs to a mappable system of mutually interconnecting relations is more likely to be imported into the target than is an isolated predicate. Thus, the comparison process that takes part in analogical reasoning is seen as involving a rather sophisticated process of structural alignment and mapping over highly complex representations.

The ideas of analogical reasoning, viewed as comparison among relational structures, is by Gentner presented in a propositional network, where objects represent concepts and two types of predicates: The attributive predicate (predicate with one argument) states propositions about attributtes of concepts. The relational predicate (predicate with two or more arguments) is directed towards relations. Next, an analogy can be characterised by structure mapping between a base domain and the target domain one wishes to describe by employing an analogy. The structure mapping is carried out such that essentiel parts of relations in the base domain – in the form of relational predicates with more than one argument – are transferred to the target domain, while there is no or little transfer of attributtes (one place predicates) in analogical reasoning.

In this framework three types of similarity comparison can be considered. The distinction is not a sharply held one, rather the different types of similarity indicate a continuum of similarities stretching from simple kinds of similarity to more sophisticated kinds.

1. The literal similarity comparison refers to overlap in similarity between both relations and attributes. For example: The helium atom is like the neon atom. Here both attributtes and relations are similar, and all or most of the predicates are mapped.
2. In the case of overall similarity or mere appearance comparison one only deals with overlap between the attributtes of the given objects, meaning that relations do not enter into the comparison process at all: The planet is round as the ball.
3. On the other hand, a proper analogy uncovers similarities on the relational level between the base and target domains: The hydrogen atom is like the solar system. In this analogy, also referred to as the Rutherford-Bohr analogy, the comparison does not include any shared attributtes between the base (solar system) and the target (atom), while the comparison allows relational predicates to be mapped from base to target.

3.3 A Theoretical Framework for Analogy Based on Relations

Following Gentner, the interpretation of the Rutherford-Bohr analogy can be made explicit as follows. Let us assume that one's knowledge about the solar system includes propositions such as the following: The sun is yellow, the planets are colder than the sun, and since the planets are lighter than the sun, the planets revolve around the sun. These propositions can be represented as attributive and relational predicates:

yellow(sun)
colder–than(planet, sun)
lighter–than(planet, sun)
revolves–around(planet, sun)
cause(lighter–than(planet, sun), revolves–around(planet, sun))

Next, one may try to structure this knowledge by setting up the object correspondence between the two domains: *sun* is mapped to *nucleus* and *planet* is mapped to *electron*. Then object attributtes, such as *yellow(sun)* are skipped. Base relations such as *lighter–than(planet, sun)* are mapped onto the target domain: *lighter–than(electron, nucleus)*. Afterwards isolated relations, such as *colder–than(planet, sun)*, are discarded, while systems of relations, that are governed by higher-order constraining relations which can themselves be mapped, are preserved: *cause(lighter–than(planet, sun), revolves–around(planet, sun))*.

Changes in some of the above mentioned lower order relational predicates will cause changes in other predicates as well. If, for instance, we assume that the sun is lighter than the planet, this will cause inversion of the *revolves–around* predicate, implying that the sun revolves around the planet. The systematicity reflected by the mutually influences of the lower order relational predicates in question can be uncovered by means of higher order relations, which combine and constrain relations among lower order relational predicates.

Gentner takes systematicity to be the main selection principle. At the same time in the generel field of research into analogical reasoning there is a shared belief in the basic mechanism of mappings as based on comparisons among relational structures, whereas accounts differ on the nature of the selection principle. For instance some researchers use pragmatic context dependent information to control the analogical selection process (see for instance [21, 22]). Also Hofstadter and his colleagues, [12], only view systematicity as one of several mechanisms which influences on the mapping process. Despite differences in emphasis, the aspects of structure-mapping concerning mapping of objects and carry-over of predicates have received widespread support in investigations into analogical reasoning. In what follows we will therefore pay particular attention to the importance of structural alignment in connection to analogies.

3.4 Analogical Situations

On basis of the framework for analogy presented above is now considered the following scheme for analogies in the context of conceptual structures, here called an analogical situation logically characterised by

$$analogy(a, b, c, d, \ldots)$$

and depicted as

$$\begin{array}{ccc} a & \longleftrightarrow & c \\ \updownarrow & & \updownarrow \\ b & \longleftrightarrow & d \end{array}$$

Here a, b, c and d are (individual or general) concepts, cf. Subsection 3.2. Our goal is to formalise analogical reasoning of the following kind. We ask the question: What is the relationship between c and d, given say a and b and their relationships as explicated in a conceptual structure. And we solicit the answer telling that the relationship between c and d is such and such.

As an initial simple example, cf. [3] consider the analogical situation

$$\begin{array}{ccccc} & dark & \longleftrightarrow & night & \\ antonym & \updownarrow & & \updownarrow & antonym \\ & light & \longleftrightarrow & ? & \end{array}$$

with the given concept algebraic structure

$$\begin{array}{l} dark \leq (antonym : light) \\ night \leq (antonym : day) \end{array}$$

Here, obviously, the result $? = day$ is to be deduced, since the *antonym* relationship connects the concepts pairwise.

3.5 Analogical Reasoning with Extended Relation Algebras

We now outline a formal theoretical framework for analogical reasoning using our logico-algebraic reconstruction of CGs. More precisely, we thus make use of the notion of an equationally defined relation algebra, cf. above, with the two sorts: Concepts and binary relations. For the sake of simplicity we disregard operators which exclusively operate on relations. So there still is the Peirce product, $(r : c)$ but now the relation term, r is a constant. On concepts there are the operations join, $+$, meet, $\times$, the bottom concept, $\bot$, and the top concept, $\top$.

The relation algebra may be conceived to be defined in terms of many-sorted first-order predicate logic with equality. Thus, there are two linguistic levels:

1. The relation algebra.
2. Predicate logic with equality in which the relation algebra is defined.

Thus, in the second language we have predicate-logical atomic formulae

$$p(t_1, ..., t_n)$$

whose terms t_i are terms of appropriate sorts of the relation algebra. We suggest that analogical reasoning be formalised and conducted in the second language by carrying out clausal resolution inference using the equational axioms for the relation algebra together with the application specific equational axioms corresponding to the target and base domains under consideration. This makes the analogical reasoning amenable to implementation in logic programming languages like PROLOG or, preferably, an equational logic programming language.

In this set-up the analogical situation from previous subsection is characterised by

$$analogy(X, Y, U, V, A)$$

which asserts that X is to Y as U is V. The first four arguments of the predicate are concept terms in the relational algebra, and the term A is an analogy descriptor of an appropriate sort. The predicate *analogy* is defined by appropriate clauses, e.g.

$$analogy(X, Y, U, V, R) \leftarrow isa(X, (R : Y)) \wedge isa(U, (R : V))$$

where the algebraic order $\leq$ appears as the *isa* binary predicate at the clausal level. Here X and Y belong to the target domain, U and V to the base domain. These domains may coincide. In the simplest cases the analogy descriptor A is merely the name of the identified analogy relation (cf. the relation *antonym* in the above example). This relation need not be explicitly present, but may only be derivable in the conceptual structures. In more elaborate versions of this set-up the analogy descriptor A may be a maximal list of such relations as exemplified next.

3.6 The Rutherford-Bohr Analogy Revisited

Consider first the Rutherford-Bohr analogy in the simplified form where the general concept *planet* has been replaced with the individual concept *earth*. The Rutherford-Bohr analogy, in the described simplified form, involves the following propositions about the base domain: The earth is colder than, lighter than, and revolves around, the sun. This gives rise to the equational axioms

$$earth \leq (colder\text{--}than : sun)$$
$$earth \leq (lighter\text{--}than : sun)$$
$$earth \leq (revolves\text{--}around : sun)$$

Furthermore, the Rutherford-Bohr analogy involves the following propositions about the target domain under consideration: The electron is lighter than, and revolves around, the nucleus. This gives rise to the axioms

$$electron \leq (lighter\text{--}than : nucleus)$$
$$electron \leq (revolves\text{--}around : nucleus)$$

Given the axioms displayed, the query clause

$$\leftarrow analogy(electron, nucleus, U, V, A)$$

supposedly has as solution

$$U = earth$$
$$V = sun$$
$$A = [lighter\text{--}than, revolves\text{--}around]$$

So when we ask: What is the relationship between the electron and the nucleus like? We get the answer: The relationship between the electron and the nucleus is like the relationship between the earth and the sun in the sense that each of the relations *lighter–than* and *revolves–around* relates the electron to the nucleus as well as the earth to the sun.

Next, consider the Rutherford-Bohr analogy in the original form involving the general concept *planet* rather than the individual concept *earth*. We now take as base domain the sun together with all the planets of the solar system, not just the earth. This involves the following propositions: Venus is colder than, lighter than, and revolves around, the sun, the earth is colder than, lighter than, and revolves around, the sun, etc. Thus, we take as specification corresponding to the base domain a set of axioms for each planet analogous to the set of axioms for the earth which is displayed above, that is, we take the axioms

$$\begin{array}{lll} venus \leq (colder\text{–}than : sun) & earth \leq (colder\text{–}than : sun) & \ldots \\ venus \leq (lighter\text{–}than : sun) & earth \leq (lighter\text{–}than : sun) & \ldots \\ venus \leq (revolves\text{–}around : sun) & earth \leq (revolves\text{–}around : sun) & \ldots \end{array}$$

Furthermore, we add the axiom

$$planet = venus + earth + \ldots$$

So the general concept *planet* is defined to be the join of the individual concepts *venus*, *earth*, etc. The equational axioms corresponding to the target domain, that is, electron and nucleus, are left unchanged. Given the modified set of axioms, the query clause above also has as solution

$$\begin{array}{l} U = planet \\ V = sun \\ A = [lighter\text{–}than, revolves\text{–}around] \end{array}$$

because each of the equations

$$\begin{array}{l} planet \leq (lighter\text{–}than : sun) \\ planet \leq (revolves\text{–}around : sun) \end{array}$$

can be derived from the axioms using equational reasoning. So the question: What is the relationship between the electron and the nucleus like? Is now answered: The relationship is like the relationship between the planets and the sun in the sense that each of the relations *lighter–than* and *revolves–around* relates the electron to the nucleus as well as each planet to the sun.

4 Conclusion and Further Work

We have introduced an algebraic logic for CGs and applied it to simple analogical cases. The established framework for identification of analogies between conceptual structures may be extended and strengthened by relaxing the condition that relation terms in Peirce products be identifiers. This introduces the entire lattice of relations in the analogy identification process for example comprising conjunction of relations (meet in the lattice of relations) as in (*lighter–than* × *revolves–around*), cf. the example above with the Rutherford-Bohr analogy.

In less obvious analogical situations the formation of correspondences between conceptual structures calls for computation by inference. Consider the example

$$\begin{array}{ccccc} & & r' & & \\ & \textit{sight} & \longleftrightarrow & \textit{tv} & \\ r'' & \updownarrow & & \updownarrow & r'' \\ & \textit{hearing} & \longleftrightarrow & ? & \\ & & r' & & \end{array}$$

with the concept algebraic structure

$$\begin{array}{l} \textit{sight} \leq \textit{sense} \times (\textit{purp} : \textit{reception} \times (\textit{of} : \textit{pictures})) \\ \textit{hearing} \leq \textit{sense} \times (\textit{purp} : \textit{reception} \times (\textit{of} : \textit{sound})) \\ \textit{tv} \leq \textit{gadget} \times (\textit{purp} : \textit{transmission} \times (\textit{of} : \textit{pictures})) \\ \textit{radio} \leq \textit{gadget} \times (\textit{purp} : \textit{transmission} \times (\textit{of} : \textit{sound})) \end{array}$$

Here the result ? = *radio* is to be deduced. In this case the presumed relationship r'' is not explicitly present in in the conceptual structure, but may be established (computed), say, through a notion of congruence. Intuitively a congruence relationship exists between a and b if the specification for a is similar to the specification for b. A precise notion of congruence may be formalised for example by logical clauses which identifies the difference between two concepts. For instance, the difference between *sight* and *hearing* is accounted for by a relationship *different*(*pictures*, *sound*).

References

[1] C. Brink, K. Britz, and R. A. Schmidt. Peirce algebras. *Formal Aspects of Computing*, 6:339–358, 1994.

[2] P. Coupey and C. Faron. Towards correspondences between conceptual graphs and description logics. In M.-L. Mugnier and M. Chein, editors, *Proceedings of Sixth International Conference on Conceptual Structures*, volume 1453 of *LNAI*. Springer-Verlag, 1998.

[3] D. A. Cruse. *Lexical Semantics*. Cambridge University Press, 1986.

[4] B. A. Davey and H. A. Priestley. *Introduction to Lattices and Order*. Cambridge University Press, 1990.

[5] Gentner and Jeziorski. The shift from metaphor to analogy in western science. In A. Ortony, editor, *Metaphor and Thought*, pages 447–480. Cambridge University Press, 1993.

[6] D. Gentner. Structure-mapping: A theoretical framework for analogy. *Cognitive Science*, 7:155–170, 1983.

[7] D. Gentner and A. B. Markman. Structure mapping in analogy and similarity. *American Psychologist*, 1997.

[8] D. Gentner and C. Toupin. Systematicity and surface similarity in the developement of analogy. *Cognitive Science*, 10:277–300, 1986.

[9] M. L. Gick and K. J. Holyoak. Analogical problem solving. *Cognitive Psychology*, 12:306–355, 1980.

[10] W. J. J. Gordon. The metaphorical way of knowing. In G. Kepes, editor, *Education of Vision*. George Braziller Inc., University of Minnesota Press, 1965.

[11] M. B. Hesse. *Models and Analogies in Science*. University of Notre Dame Press, Notre Dame, Indiana, 1966.

[12] D. R. Hofstadter, T. M. Mitchell, and R. M. French. Fluid concepts and creative analogies: A theory and its computer implementation. Technical Report 87-1, Ann Arbor: University of Michigan, Fluid Analogies Research Group, 1987.

[13] B. Indurkhya. *Metaphor and Cognition - an Interactionist Approach*. Kluwer Academic Publishers, 1992.

[14] D. Leishman. Analogy as a constrained partial correspondence over conceptual graphs. In R. J. Brachman, H. J. Levesque, and R. Reiter, editors, *Proceedings of the First International Conference on Principles of Knowledge Representation and Reasoning*, pages 223–234. Morgan Kaufmann Publishers Inc., 1989.

[15] J. F. Nilsson. An algebraic logic for concept structures. In H. Jaakkola, H. Kangassalo, T. Kitahashi, and A. Markus, editors, *Proceedings of Information Modelling and Knowledge Bases V*, pages 75–84. IOS Press, 1994.

[16] J. F. Nilsson and J. Palomäki. Towards computing with extensions and intensions of concepts. In *Proceedings of Information Modelling and Knowledge Bases IX*. IOS Press, 1998.

[17] J. F. Sowa. *Conceptual Structures: Information Processing in Mind and Machine*. Addison-Wesley, Reading, 1984.

[18] G. Stumme. *Boolesche Begriffe*. Diplomarbeit, TH Darmstadt, 1994.

[19] A. Tarski. On the calculus of relations. *Journal of Symbolic Logic*, 6:73–89, 1941.

[20] E. C. Way. *Knowledge Representation and Metaphor*. Kluwer Academic Publishers, 1991.

[21] P. H. Winston. Learning and reasoning by analogy. *Communications of the ACM*, 23:689–703, 1980.

[22] P. H. Winston. Learning new principles from precedents and exercises. *Artificial Intelligence*, 19:321–350, 1982.

Unification over Constraints in Conceptual Graphs

Dan Corbett[1] and Robert Woodbury[2]

[1]University of Adelaide, Department of Computer Science
Adelaide, South Australia 5001
dcorbett@cs.adelaide.edu.au

[2]University of Adelaide, Department of Architecture
Adelaide, South Australia 5001
rw@arch.adelaide.edu.au

Abstract. The Conceptual Structures community has recently shown increased interest in methods and formalisms for the use of constraints in Conceptual Graphs (CGs), especially the definition of unification over constraints. None of the recent proposed constraint methods, however, are able to use simple unification methods, and still guarantee that a graph which is structurally valid under the canonical formation rules is also semantically valid in the knowledge domain. Our approach defines a method (and concept type) for constraining real values in the referent of a concept. The significance of our work is that a simple unification operation, using join and type subsumption, is defined which can be used to validate the constraints over an entire unified graph. A useful side-effect is that this constraint method can also be used to define real numbers in a referent.

1. Introduction

Until very recently, CGs have had no real-value constraint formalism. The standard method for representing and validating constraints has been to use type subsumption. One could constrain inputs and outputs to the system by forcing the concepts to be of the same type, or else to be subsumed by the same (more general) type. A similar method applies to relations.

However, there are some severe limitations on this type of constraint. First, it's not possible to represent any kind of real numbers with subsumption. By defining smaller integers as "more general" than larger integers, one can define "greater than" and "less than" over the integers. While this method (though quite trivial) does formally define these functions for CGs, this technique still falls flat for real numbers. A strict (or even partial) subsumption ordering can't be defined when numbers can be sandwiched in between already defined numbers.

Further, by adhering to a strict subsumption order on other types of concepts, it becomes difficult to represent certain domain knowledge. There are domains which rely on a continual construction and refining of the structures to find a structure which represents the solution to a problem in the domain. If a most general unifier is used to unify structures in these domains, it is sometimes possible (depending on the unification algorithm used) to lose some of the knowledge gained in the refinment process.

The method described here uses intervals to bound the value of an attribute, thus capturing the idea of a real number. Subsumption of an interval (defined on an interval lattice) is used to decide whether two concepts of the same type are still unifiable. The lattice operator "join" is used to decide subsumption. We are then able to define real-valued constraints in concepts, and use the standard join operation to decide whether a concept is valid according to the domain knowledge.

2. Interval constraints

We start by defining a lattice of real-value intervals, as shown in Figure 1.

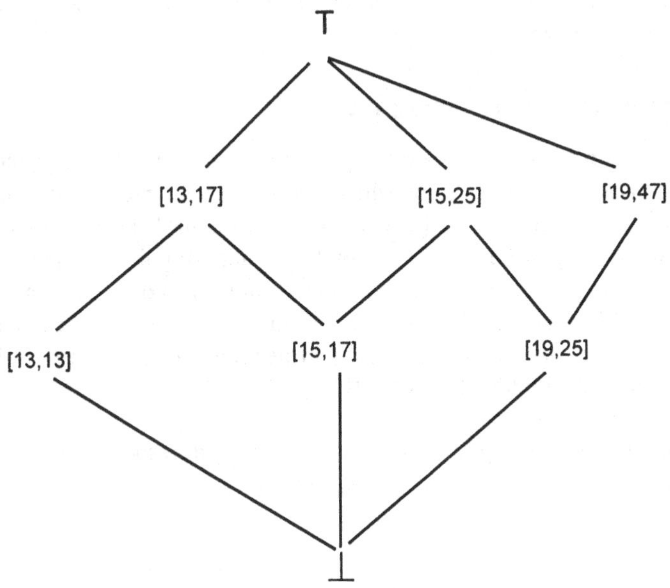

Fig. 1. An example of an interval lattice, ordered by interval inclusion.

Definition 1. Interval. An interval is defined as: Intr: [x, y] ● ¨ × ¨: x ≤ y.

This definition gives us T = [-,] and ⊥ is any absurd interval. Informally, the lattice is ordered on interval inclusion, such that two intervals have a join if there is some other interval which is in the "overlap" area of the two intervals. It is clear how such an operation on the lattice would be defined. A maximal join is the largest such overlap interval. These definitions give us the following desirable results:

$[13,17] \sqcap [15,25] \rightarrow [15,17]$

$[15,17] \sqcap [19,47] \rightarrow \perp$

$[15,11] \rightarrow \perp$

This lattice then defines the type hierarchy for an interval type for concepts. Two interval-type concepts can be joined if there is a join on the interval lattice for them. Constraints over real variables can then be expressed by specifying the interval into which the real value must be constrained. Note that a specific value can still be specified by using an interval: 13.7 can be expressed as the interval [13.7, 13.7]. Unifying two real-value concepts then becomes finding the join of the concepts on the interval lattice. The set of individual markers for concepts can now be expanded to include intervals as defined here.

3. Projection and constraints using intervals

The first step in defining constraints on concept markers is to define a projection operator, similar to Sowa's [1] operator, but which takes intervals into consideration. Sowa discusses his original definition in [8] but we follow the further formalized and refined versions presented by Willems [12] and by Chein and Mugnier [2, 6], and the following two definitions are adapted from their work. In these works, projection is defined as a mapping between two graphs, rather than a sequence of canonical derivations as Sowa defines it. The definitions of projection and specialization then become identical; with one implying the other [2].

Definition 2. Conceptual Graph. A conceptual graph with respect to a canon is a tuple $G = (C, q, R, type, referent, arg_1, \ldots, arg_m)$ where

C is the set of concepts, $type : C \rightarrow T$ indicates the type of a concept, and *referent* $: C \rightarrow I$ indicates the referent marker of a concept.

q is a distinguished member of C, the head or root node of the graph.

R is the set of conceptual relations, $type : R \rightarrow T$ indicates the type of a relation, and each $arg_i : R \rightarrow C$ is a partial function where $arg_i(r)$ indicates the i-th argument of the relation r. The argument functions are partial as they are undefined for arguments higher than the relation's arity.

We often write $c \in G$ instead of $c \in C$ when it is clear that c is a concept (similarly for relations $r \in G$). MŸller demonstrates in his work [7] that it is not always possible to find a common generalization for two arbitrary graphs in the same canon, and therefore it's not possible to guarantee that two graphs can be unified. If a subset of CGs is defined which is all graphs that have a designated head node, and the head nodes are of compatible types, then these graphs will necessarily have a most general unifier. We define q, the head node of a graph, in order to be able to guarantee unification of the graphs under MŸller's algorithms. This method allows us to combine the graphs while preserving the knowledge in both graphs, and still be able to use efficient unification methods as defined by Willems, MŸller, and others.

Definition 3. Projection. Consider two conceptual graphs $G = (C, q, R, type, referent, arg_1, \ldots, arg_m)$ and $G' = (C', q', R', type', referent', arg'_1, \ldots, arg'_m)$ with respect to the same canon. A projection from G to G' is a pair of functions $'_C : C \rightarrow C'$ and $'_R : R \rightarrow R'$, that are:

1. Type preserving: for all concepts $c \in C$ and $c' \in C'$, $'_C(c) = c'$ only if $type(c)$ ^{3} $type'(c')$, and $referent(c) = *$ or $referent(c)$ ^{3} $referent(c')$, and by extrapolation, q ^{3} q',
2. Type preserving: for all relations $r \in R$ and $r' \in R'$, $'_R(r) = r'$ only if $type(r)$ ^{3} $type'(r')$,
3. Structure preserving: for all relations $r \in R$ there holds that $arg'_i('_R(r)) = '_C(arg_i(r))$,
4. Non-empty: for all concepts $c \in C$ there is a concept $c' \in C'$ such that $'_C(c) = c'$.

This definition of projection is very similar to previous definitions, except that here, we specifically allow a referent marker to subsume another marker in the projection. The notion of an individual (not just the generic marker) subsuming another individual is not novel. The idea has wide acceptance in, for example, the use of conceptual graphs in Formal Concept Analysis (FCA) [11] . In FCA, an individual is lower on the concept lattice if it is more specified, or simply has more properties than another individual. In some domains in which FCA is used, it makes sense to derive an ordering on this hierarchy. This sense of an individual concept subsuming another which is more specified is the sense that we also employ in our work.

In our method, when two intervals are joined, their join may be another interval which is contained in both of the originals. This leads to a situation where, after joining two concepts, it may not be possible to find either individual marker exactly as it was before. It may be a joined interval, or similar concept type. The new marker would be a specialization of the previous markers. In our present research, an individual interval must be considered to have a more specific type, in order to be useful in the domains that we work in.

The other properties of the standard ^{1} operator remain the same as in previous work. We still guarantee that for each relation r in the graph in question, the old type must be equal to the new type (ie they must be the same type) and that the ith arc of the old relation must point to the same concept as the ith arc of the new relation.

The ^{1} operator applies to two graphs, one of which subsumes the other. In our work, we're combining the knowledge of two graphs into a more refined representation. We use the ^{1} operator to help define the unifier, and then to proceed with combining the two graphs into one.

4. Unification over constraints

The unification of two conceptual graphs with constraints now becomes the combination of two graphs which are compatible in corresponding concepts and relations, as defined by our definition of the projection operator and join. We again employ a definition from Willems [12] to complete the discussion of unification:

Definition 4. Unifier. Let G and G' be two conceptual graphs. A unifier for G and G' is a graph U such that:

1. projection: projections $U \rightarrow G$ and $U \rightarrow G'$ exist, and
2. compatibility: for any concept in U, the images $u \in U$ and $u' \in U$ must be compatible, i.e. *type* (u) ● *type*(u') - $\perp$.

Definition 5. Unification. The unification of G and G' on a unifier U is the graph G'' such that

1. commutativity: there exist morphisms $G \rightarrow G''$ and $G' \rightarrow G''$ such that $U \rightarrow G \rightarrow G'' = U \rightarrow G' \rightarrow G''$, and
2. universality: for any other graph Q satisfying the previous condition there exists a morphism $G'' \rightarrow Q$ such that $G \rightarrow Q = G \rightarrow G'' \rightarrow Q$ and $G' \rightarrow Q = G' \rightarrow G'' \rightarrow Q$.

We can now employ standard unification algorithms which recursively move down through the relations and concepts from the head node. Given compatible head nodes, our algorithm selects a relation from the first graph, and seeks a compatible relation in the second graph. If none is found, then the relation becomes part of the unified graph trivially. If a compatible relation is found in the second graph, then the unification algorithm is called recursively to check the concepts and relations which this relation points to. If these all prove to be compatible, then the two subgraphs are joined, and attached to the unified graph. If some part of the subgraphs are incompatible, then the unification fails. When all relations attached to the head concept in both graphs have been sucessfully processed in this way, the algorithm terminates successfully.

5. Examples of unification with constraints

The software to implement these ideas is currently being implemented, but preliminary results are encouraging. I present here some of the examples which have already been tested in the software in its preliminary state. The knowledge domain is that of the building architect. The architect will often try to reuse previous designs, not only to save time and resources, but also as solutions to problems previously encountered. Once a given problem has been solved, the design which represents the solution can be stored, and then retrieved when needed.

Consider in the domain of architectural design, a design for the kitchen of a custom-made house. In this design, the architect has specified some of the lighting design and that the floor area must be greater than 20 square meters. The architect has also retreived an old design, which specifies the remainder of the lighting design. The graphs specifying these two designs are shown in Figure 2. We assume that the portion of the graphs not shown in the diagram are compatible. The unification algorithm defined above combines these two graphs, with the result shown in Figure 3. In this graph, all the original knowledge of the first two graphs has been preserved, and the values in the concepts have been joined as specified.

Another example would have a design similar to the second in Figure 2, specifying most of the lighting design. Another graph would represent a kitchen design where only the plumbing design is specified. These two would unify since the two heads are compatible, and the remainder of the graphs would be included in the unified graph. All of the knowledge is represented in the unified graph, which would

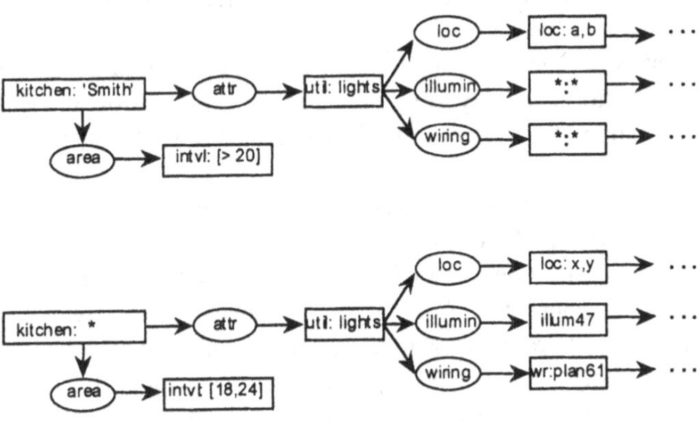

Figure 2. Two designs to be unified.

specify the design for the lighting and the plumbing.

These examples also illustrate how the interval type would allow real numbers to be represented in CGs. Any real number could be bounded inside an interval, similar to the concept of using floating point numbers to represent real numbers in software. Further, any concept containing a real value can be constrained in an interval. This allows the representation in CGs of constraint satisfaction problems. This use of interval constraints to represent real constraints has been used for some time in the Constraint Satisfaction Problem community. The work by van Hentenryck [9, 10] is a good example of intervals in CSP.

6. Previous work in constraints over CGs

There have been some interesting recent attempts to create a constraint system for conceptual graphs including the introduction of structural constraints on the graphs [4, 5] the use of fuzzy logic in the definition of concepts [1, 13], and enforcing domain semantics on the relations [3].

Work on structural constraint systems have included Mineau's system of representing topological constraints and domain constraints [5]. Mineau's topological constraints are produced by specifying graphs which are invalid in the specified domain, and then attempting to join them to graphs which are produced from the knowledge in the domain. The scheme relies on subsumption to find invalid graphs. Mineau specifies domain constraints by defining procedural attachments to graphs which are activated when their concepts are instantiated. These procedural attachments (actors) should always have a projection into the graphs where it is used. If a projection doesn't exist, then the graph violates the constraints, and is invalid.

Willems' unification algorithm [12] basically finds a segment of the CG which is common to both of the CGs being unified. In this approach, Willems finds a projection which is at least as (or possibly more) general than both of the graphs being unified. I.e. it's a graph in which every concept can subsume its corresponding concept in both graphs, and every relation can subsume its corresponding relation. This makes the unification algorithm efficient, as it takes advantage of the built-in CG attributes of subsumption and type hierarchies.

Willems' approach makes no attempt to implement a method for producing the unified graph, however. His effort is mainly to find the unifier. It is therefore unclear whether creating a generalization of a concept loses the essential knowledge that the user wants to retain. In our domain, the constructive nature of architectural design is a continual *refining* and *specifying* of the structure.

Willems' concept of a polyprojection allows for a certain amount of confusion in the implementation of joining concepts. A type of "crossover" of the relations attached to concepts could result from the generalization which is only useful in certain applications. While polyprojection does preserve the concept and relation pointers, the difference is that it's unclear which concept should be pointed to, if there is more than one possibility in the two graphs being unified. Essentially, he's defining a way to have no bindings on the variables, and still have a valid graph.

Kocura's approach [4] is another structural method, which essentially is to block certain parts of the graph away from the join; to disallow unification with certain parts of a graph. An illustrative example is one where he specifies that a man can marry a woman, but that a celibate man doesn't marry. He specifies which parts of a type hierarchy a graph can combine with, but excludes a certain concept, and then all of its specializations, including any join with other concepts or concept types (eg, the concept "celibate man" could be joined with the concept "professional man"). It basically is the idea of not allowing a match too far down a type hierarchy.

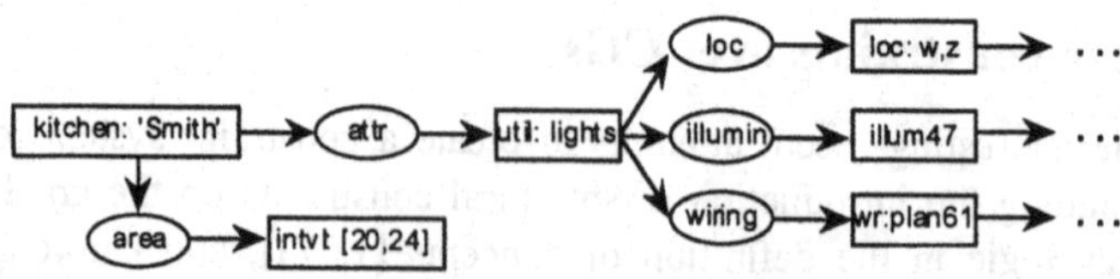

Figure 3. The unified design.

7. Discussion and Conclusion

We have defined a real-value constraint system over Conceptual Structures, and have partially created a software implementation of it. The system described in this paper allows real numbers to be represented in concepts, and also allows those values to be constrained by specifying valid intervals. In our software, we also use inequality relations and allow variables in the constraint specifications. The preliminary experiments with this software have shown these techniques to be useful and efficient.

In the methods discussed in this paper, we have attempted to improve on previous work in several ways. The major step is in formally specifying constraints for CGs, but this work also leads to defining the major constraint processing techniques from the Constraint Satisfaction Processing community. Further work in this area would allow the entire semantics of CSP to be included in the CG formalism. This would be a major benefit to both communities of researchers and users, as it would bring the entire Knowledge Representation formalism and structures of CGs (canonical formation rules, join, type subsumption) into the CSP work. The CG community would benefit by having formal definitions and algorithms for implementing standard constraint methods within the CG formalism.

Many of the projects discussed in the previous section are attempts to either constrain the structure of the CGs, or to specify what types of concepts are valid. One notable exception is the work of Cao et al [1] in the implementation of fuzzy values in the concepts. Cao's work is more of an implementation of fuzzy words and concepts, such as "fairly ripe" or "young." These implementations are useful to specifying boundaries for the values in concepts, but Cao has not pursued constraint processing technology. The possibility of implementing fuzzy membership functions in the value of a constraint is an intriguing idea which is waiting to be explored.

Mineau's work really is more of an attempt to define valid structures in a canon, while our work tries to constrain the values in concepts. A major difference in these approaches is that we rely only on projection and join, while Mineau has implemented actors to check the graphs after an attempt at a join. An interesting research direction might be to combine these two approaches to produce a complete constraint system. Using intervals to express the topological constraints that Mineau uses might lead to some short cuts in the processing time.

The formal definitions and the descriptions in this paper lead to a useful and efficient implementation of constraints over CGs. This work improves on previous work in allowing real numbers to be expressed in the concepts, and in allowing constraints to be placed on those values in the concepts. The constraints are defined as a concept type, and therefore can be used as a type in the normal way with conceptual graphs. The constraints are enforced in the unification and join operations, as defined in this paper. If a join operation violates the constraints on one of the concepts, the join fails. The unification algorithm used in this system is a standard algorithm, and is guaranteed to terminate since we restrict the CGs to a special subset, headed CGs.

8. Acknowledgment

The authors express their sincere gratitude to the anonymous reviewer from ICCS '98 who offered many detailed and helpful suggestions on the previous version of this paper. Those comments and suggestions helped to clarify many issues in our minds, and led to a much improved work.

References

1. Cao, T.H., P.N. Creasy, and V. Wuwongse. "Fuzzy Unification and Resolution Proof Procedure for Fuzzy Conceptual Graph Programs," in Proc. *Fifth International Conference on Conceptual Structures*, August, 1997. Seattle, Washington, USA: Springer-Verlag. Published as Lecture Notes in Artificial Intelligence #1257.
2. Chein, M. and M.-L. Mugnier, "Conceptual Graphs: Fundamental Notions," *Revue d'Intelligence Artificielle*, 1992. **6**(4): p. 365-406.
3. Dibie, J., O. Haemmerl , and S. Loiseau." A Semantic Validation of Conceptual Graphs," in Proc. *Sixth International Conference on Conceptual Structures*, August, 1998. Montpellier, France: Springer-Verlag. Published as Lecture Notes in Artificial Intelligence #1453.
4. Kocura, P. "Conceptual Graphs and Semantic Constraints," in Proc. *Fourth International Conference on Conceptual Structures*, August, 1996. Sydney, NSW, Australia: University of NSW Press.
5. Mineau, G.W. and R. Missaoui. " The Representation of Semantic Constraints in Conceptual Graph Systems," in Proc. *Fifth International Conference on Conceptual Structures*, August, 1997. Seattle, Washinton, USA: Springer-Verlag. Published as Lecture Notes in Artificial Intelligence #1257.
6. Mugnier, M.-L. and M. Chein, " Repr senter des Connaissances et Raisonner avec des Graphes," *Revue d'Intelligence Artificielle*, 1996. **10**(6): p. 7-56.
7. MŸller, T., *Conceptual Graphs as Terms: Prospects for Resolution Theorem Proving*, 1997, Masters Thesis, Department of Computer Science, Vrije Universiteit Amsterdam, Amsterdam, Netherlands.
8. Sowa, J.F., *Conceptual Structures: Information Processing in Mind and Machine*. 1984, Reading, Mass: Addison-Wesley.
9. Van Hentenryck, P., *Constraint Satisfaction in Logic Programming*. Logic Programming, ed. E. Shapiro. 1989, Cambridge, Massachusetts, USA: MIT Press.
10. Van Hentenryck, P., L. Michel, and Y. Deville, *Numerica*. 1997, Cambridge, Massachusetts, USA: MIT Press.
11. Wille, R. "Conceptual Graphs and Formal Concept Analysis," in Proc. *Fifth International Conference on Conceptual Structures*, August, 1997. Seattle, Washington, USA: Springer-Verlag. Published as Lecture Notes in Artificial Intelligence #1257.

12. Willems, M. "Projection and Unification for Conceptual Graphs," in Proc. *Third International Conference on Conceptual Structures*, August, 1995. Santa Cruz, California, USA: Springer-Verlag. Published as Lecture Notes in Artificial Intelligence #954.
13. Wuwongse, V. and T.H. Cao. "Towards Fuzzy Conceptual Graph Programs," in Proc. *Fourth International Conference on Conceptual Structures*, August, 1996. Sydney, Australia: Springer-Verlag. Published as Lecture Notes in Artificial Intelligence #1115.

Tractable and Decidable Fragments of Conceptual Graphs*

F. Baader, R. Molitor, and S. Tobies

LuFg Theoretical Computer Science,
RWTH Aachen, Germany.
E-mail: {baader,molitor,tobies}@informatik.rwth-aachen.de

Abstract. It is well-known that problems like validity and subsumption of general CGs are undecidable, whereas subsumption is NP-complete for simple conceptual graphs (SGs) and tractable for SGs that are trees. We will employ results on decidable fragments of first-order logic to identify a natural and expressive fragment of CGs for which validity and subsumption is decidable in ExpTime. In addition, we will extend existing work on the connection between SGs and description logics (DLs) by identifying a DL that corresponds to the class of SGs that are trees. This yields a tractability result previously unknown in the DL community.

1 Introduction

Conceptual graphs (CGs) are an expressive formalism for representing knowledge about an application domain in a graphical way. Since CGs can express all of first-order predicate logic (FO), they can also be seen as a graphical notation for FO formulae.

In knowledge representation, one is usually not only interested in *representing* knowledge, one also wants to *reason* about the represented knowledge. For CGs, one is, for example, interested in validity of a given graph, and in the question whether one graph subsumes another one. Because of the expressiveness of the CG formalism, these reasoning problems are undecidable for general CGs. In the literature [14, 16, 12] one can find complete calculi for validity of CGs, but implementations of these calculi have the same problems as theorem provers for FO: they may not terminate for formulae that are not valid, and they are very inefficient. To overcome this problem, one can either employ incomplete reasoners, or try to find decidable (or even tractable) fragments of the formalism. This paper investigates the second alternative.

The most prominent decidable fragment of CGs is the class of simple conceptual graphs (SGs), which corresponds to the conjunctive, positive, and existential fragment of FO (i.e., existentially quantified conjunctions of atoms). Even for this simple fragment, however, subsumption is still an NP-complete problem [5]. SGs that are trees provide for a tractable fragment of SGs, i.e., a class of simple

* This work was partially supported by the *Deutsche Forschungsgemeinschaft* Grant No. GRK 185/3-98 and Project No. GR 1324/3-1

conceptual graphs for which subsumption can be decided in polynomial time [13]. In this paper, we will, on the one hand, describe a decidable fragment of CGs that is considerably more expressive than SGs. On the other hand, we will identify a tractable fragment of SGs that is larger than the class of trees.

Instead of trying to prove new decidability or tractability results for CGs from scratch, our idea was to transfer decidability results from first-order logics [4] and from description logics [9, 10] to CGs. The goal was to obtain "natural" sub-classes of the class of all CGs in the sense that these sub-classes are defined directly by syntactic restrictions on the graphs, and not by conditions on the first-order formulae obtained by translating CGs into FO.

Although description logics (DLs) and CGs are employed in very similar applications (e.g., for representing the semantics of natural language sentences), it turned out that these two formalisms are quite different for several reasons: (1) conceptual graphs[1] are interpreted as closed FO formulae, whereas DL concept descriptions are interpreted by formulae with one free variable; (2) most DLs do not allow for relations of arity > 2; (3) SGs are interpreted by existential sentences, whereas almost all DLs considered in the literature allow for universal quantification; (4) because DLs use a variable-free syntax, certain identifications of variables expressed by cycles in SGs and by co-reference links in CGs cannot be expressed in DLs. As a consequence of these differences, we could not identify a natural fragment of CGs corresponding to an expressive DL whose decidability was already shown in the literature. We could, however, obtain a new tractability result for a DL corresponding to SGs that are rooted, arc- and node labeled trees. This correspondence result strictly extends the one in [7]. In addition, we have extended the tractability result from SGs that are trees to SGs that can be transformed into trees using a certain "cycle-cutting" operation.

An interesting decidable fragment of FO, which has recently been introduced by van Benthem [15], is the so-called *loosely guarded fragment* of FO. It contains (the first-order translations of) many modal logics and description logics, but is not restricted to unary and binary relations. We could identify a fragment of CGs corresponding to the loosely guarded fragment of FO in the sense that the first-order translation of CGs belonging to this fragment are equivalent (though not necessarily identical) to a loosely guarded formula, and every loosely guarded formula can be obtained in this manner. The characterization of this fragment is given by syntactic restrictions on the graphs, and for a given graph it is easily decidable whether it belongs to this fragment.

2 Preliminaries

To fix our notation, we recall basic definitions and results on conceptual graphs. Basic ontological knowledge from the application domain is coded in the **support**, which is a structure of the form $\mathcal{S} = \langle N_C, N_R, N_I \rangle$. Here N_C is the set of *concept types*, which is ordered by a partial order $\leq_C$ expressing the ***is-a-kind-of***

[1] Here, we restrict our attention to the first order fragment of CGs.

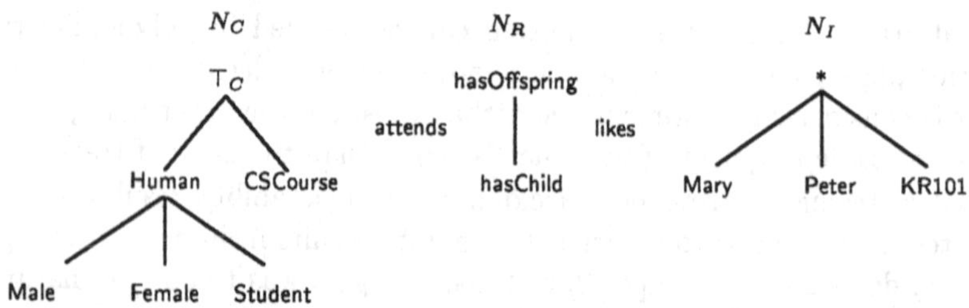

Fig. 1. An example of a support.

relation between concept types. We require N_C to contain a distinguished element $\top_C$ representing the entire domain, i.e., $\top_C$ is the greatest element w.r.t. $\leq_C$. Similarly, the set of *relation types*, N_R, is ordered by a partial order $\leq_R$. Each element of N_R has a fixed arity, and relation types with different arity are incomparable by $\leq_R$. The set of *individual markers* is denoted by N_I; an additional *generic marker* is denoted by $*$. We define a partial order $\leq_I$ on $N_I \cup \{*\}$ such that $*$ is the greatest element, and all other elements are pairwise incomparable.

Throughout the paper, we consider examples over the support $\langle N_C, N_R, N_I \rangle$ shown in Fig. 1, where all relation types are assumed to have arity 2.

A **simple graph** (SG) over the support $\mathcal{S}$ is a labeled bipartite graph of the form $g = \langle C, R, E, \ell \rangle$, where C and R are the node sets, called *concept nodes* and *relation nodes*, respectively, and $E \subseteq C \times R$ is the edge relation. The labeling ℓ labels g in the following way: Each concept node $c \in C$ is labeled by a pair $\ell(c) = (type(c), ref(c)) \in N_C \times (N_I \cup \{*\})$, called the *type* and the *referent* of c. If $ref(c) = *$, then c is called a *generic concept node*, otherwise c is called an *individual concept node*. Each relation node $r \in R$ is labeled with a relation type $\ell(r) \in N_R$. All edges that are incident to the same relation node are labeled by ℓ with sets of natural numbers in such a way that, if $\ell(r)$ has arity n, then all numbers from $\{1, \ldots, n\}$ appear exactly once in these labels. The concept node c linked to r by an edge with $j \in \ell(c, r)$ is called the jth neighbor of r and is denoted by $r(j)$. The set of all simple graphs over $\mathcal{S}$ is denoted by $SG(\mathcal{S})$.

SGs can be combined into more complex structures called **graph propositions**. A graph proposition p is a negated graph proposition, or a box that contains a SG (which may be empty) and finitely many graph propositions (see Fig. 2). These boxes are also called the *contexts* of the proposition. We require that all simple graphs appearing in a graph proposition have disjoint node sets. In a linear notation one can represent graph propositions as expressions generated by the EBNF grammar

$$p ::= [g\, p^*] \mid \neg p,$$

where g stands for a SG. For each SG g in p there is exactly one context p' which contains g at top level. This context is called *the context of* g, and we say that p' *contains* all nodes of g. We say that a context p *(strictly) dominates* a context q

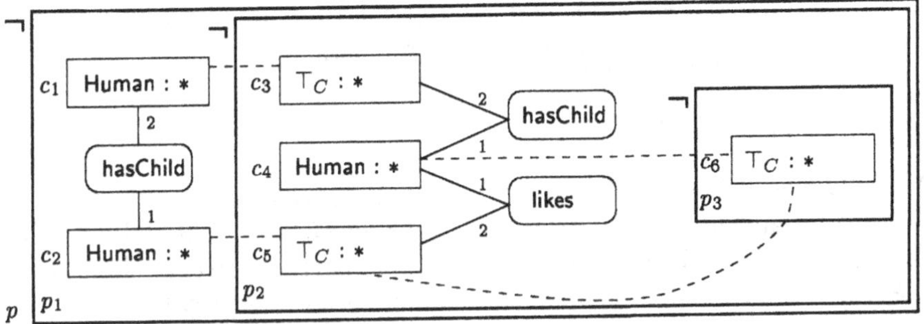

Fig. 2. An example of a conceptual graph.

iff q is (strictly) contained in p. The set of all concept nodes of all simple graphs occurring in p is denoted by $C(p)$.

A **conceptual graph** (CG) over the support $\mathcal{S}$ is a pair $G = \langle p, \mathit{coref} \rangle$, where p is a graph proposition over $\mathcal{S}$ and *coref*, the set of *coreference links*, is a symmetric binary relation over $C(p)$ satisfying the following property: for each pair of concept nodes $(c_1, c_2) \in \mathit{coref}$ contained in the contexts p_1 and p_2, respectively, either p_1 dominates p_2 or p_2 dominates p_1. We denote the set of all CGs over $\mathcal{S}$ by $CG(\mathcal{S})$.

An example of a CG built over the support from Fig. 1 is shown in Fig. 2 using the usual graphical notation. It asserts that for each parent and child there is another parent of that child who likes the first parent (see the translation of this graph into FO below). The extra markers p_i and c_i will be used in a later section of this paper to refer to the different parts of the CG.

Both SGs and CGs are given a semantics in FO by the **operator Φ**. Let $G = \langle p, \mathit{coref} \rangle \in CG(\mathcal{S})$ be the CG to be translated, and let $\mathcal{V}$ be a countably infinite set of variables. Firstly, we fix two mappings *id* and *links* as follows: *id* assigns a unique variable to each generic concept node $c \in C(p)$, and its individual marker to each individual concept node; we define $\mathit{links}(c)$ to consist of $\mathit{id}(c')$ of all concept nodes c' that are linked to c by a coreference link and are contained in a context dominating the context of c. The triple $\widehat{G} = \langle p, \mathit{id}, \mathit{links} \rangle$ obtained this way is translated by Φ into FO as follows:

- For a SG $g = \langle C, R, E, \ell \rangle$ we define $\phi(g) = \bigwedge_{c \in C} \phi(c) \wedge \bigwedge_{r \in R} \phi(r)$, where

$$\phi(c) := \bigwedge_{s \in \mathit{links}(c)} \mathit{id}(c) \doteq s \wedge \begin{cases} \mathit{id}(c) \doteq \mathit{id}(c) & \text{if } \mathit{type}(c) = \top_C \\ P(\mathit{id}(c)) & \text{if } \mathit{type}(c) = P \end{cases},$$

  and, for a relation node r with $\ell(r) = S$ of arity n,

$$\phi(r) := S(\mathit{id}(r(1)), \ldots, \mathit{id}(r(n))).$$

  The *quantifier prefix* of g is $\phi_p(g) := \exists x_1 \ldots \exists x_k$, where $\{x_1, \ldots, x_k\} := \{\mathit{id}(c) \mid c \in C\} \cap \mathcal{V}$.

– The *operator* Φ is defined by induction on the structure of graph propositions:
  - $\Phi[g\ p_1 \ldots p_m] := \phi_p(g).\big(\phi(g) \wedge \bigwedge_{j=1,\ldots,m} \Phi(p_j)\big)$,
  - $\Phi[\neg p] := \neg\Phi[p]$.

For a SG g we define its FO semantics $\Phi(g)$ by $\phi_p(g).\phi(g)$.

To translate the graph G from Fig. 2, we set $id(c_i) = x_i$ ($i = 1, \ldots, 6$). For *links*, this yields $links(c_3) = \{x_1\}$, $links(c_5) = \{x_2\}$, $links(c_6) = \{x_4, x_5\}$, and $links(c_i) = \emptyset$ for $i = 1, 2, 4$. After eliminating equalities of the form $x_i \doteq x_i$ we obtain the following FO formula:

$$\begin{aligned}\Phi(G) = \neg(\exists x_1 x_2.(&\mathsf{Human}(x_1) \wedge \mathsf{Human}(x_2) \wedge \mathsf{hasChild}(x_2, x_1) \wedge \\ &\neg(\exists x_3 x_4 x_5.(x_3 \doteq x_1 \wedge x_5 \doteq x_2 \wedge \mathsf{Human}(x_4) \wedge \\ &\quad \mathsf{hasChild}(x_4, x_3) \wedge \mathsf{likes}(x_4, x_5) \wedge \\ &\quad \neg(\exists x_6.(x_6 \doteq x_4 \wedge x_6 \doteq x_5))))))\end{aligned}$$

Note that the sub-formula $\neg(\exists x_6.(x_6 \doteq x_4 \wedge x_6 \doteq x_5)$ only expresses $x_4 \not\doteq x_5$.

We can also define the semantics of the order relations in the support by a FO formula. For a given support $\mathcal{S} = \langle N_C, N_R, N_I\rangle$, the partial orders $\leq_C$ and $\leq_R$ are interpreted as follows: $P_1 \leq_C P_2$ corresponds to the formula $\forall x.P_1(x) \rightarrow P_2(x)$, and for two relation types of arity n, $S_1 \leq_R S_2$ yields the formula $\forall x_1 \ldots x_n.S_1(x_1, \ldots, x_n) \rightarrow S_2(x_1, \ldots, x_n)$. We define $\Phi(\mathcal{S})$ to be the conjunction of all these formulae.

Validity with respect to a support $\mathcal{S}$ for a CG G can be defined with the help of the operator Φ: G is *valid* iff $\Phi(\mathcal{S}) \rightarrow \Phi(G)$ is a valid FO formula.

Subsumption with respect to a support $\mathcal{S}$ for two SGs or CGs G, H is defined as follows: G is *subsumed* by H ($G \sqsubseteq H$) iff $\Phi(\mathcal{S}) \wedge \Phi(G) \rightarrow \Phi(H)$ is a valid FO formula.

A SG g is said to be in **normal form** iff each individual marker $a \in N_I$ appears at most once as a referent of a concept node in g. Subsumption of two simple graphs g, h can be characterized by the existence of certain homomorphisms from h to g. To be more precise, if there exists a homomorphism from h to g, then $g \sqsubseteq h$ [14], and if $g \sqsubseteq h$ then there is such a homomorphism provided that g is in normal form [5].

Subsumption for SGs over a support $\mathcal{S}$ is an NP-complete problem [5]. Like Peirce's existential graphs, CGs are as expressive as FO formulae [14]. Thus, validity and subsumption for CGs are undecidable.

3 A Tractable Fragment of Simple Graphs

In this section, we introduce the description logic $\mathcal{ELIRO}^1$ as well as the class of rooted SGs. We will show that $\mathcal{ELIRO}^1$-concept descriptions can be translated into equivalent rooted SGs that are trees, and thus that subsumption in $\mathcal{ELIRO}^1$ can be decided in polynomial time. In addition, we extend the known tractability result for trees to a larger fragment of SGs.

Table 1. Syntax and semantics of $\mathcal{ELIRO}^1$-concept descriptions.

Construct name	Syntax	Semantics	
top-concept	$\top$	$x = x$	
primitive concept $P \in N_C$	P	$P(x)$	
conjunction	$C \sqcap D$	$\Psi_C(x) \wedge \Psi_D(x)$	$\mathcal{EL}$
existential restriction	$\exists r.C$	$\exists y.\Psi_r(x,y) \wedge \Psi_C(y)$	
constant $a \in N_I$	$\{a\}$	$x = a$	$\mathcal{O}^1$
primitive role $r \in N_R$	r	$r(x,y)$	
inverse role for $r \in N_R$	r^-	$r(y,x)$	$\mathcal{I}$
role conjunction	$r_1 \sqcap r_2$	$\Psi_{r_1}(x,y) \wedge \Psi_{r_2}(x,y)$	$\mathcal{R}$

Description Logics

In DLs, knowledge from an application domain is represented by so-called *concept descriptions*. Concept and role descriptions are inductively defined with the help of a set of *constructors*, starting with a set N_I of *constants*, a set N_C of *primitive concepts*, and a set N_R of *primitive roles*. The constructors determine the expressive power of the DL. In this paper, we consider concept descriptions built from the constructors shown in Table 1. The resulting DL is denoted by $\mathcal{ELIRO}^1$. Due to the fact that referents of individual concept nodes in SGs are single constants $a \in N_I$, we restrict ourselves to $\mathcal{ELIRO}^1$-concept descriptions in which *each conjunction contains at most one constant.*

The semantics of a concept description C (resp. a role description r) is defined by a FO formula $\Psi_C(x)$ with one free variable (resp. $\Psi_r(x,y)$ with two free variables): see Table 1 for the inductive definition of these formulae. Given an interpretation $\mathcal{I} = (\Delta, \cdot^{\mathcal{I}})$ of the signature $\langle N_C, N_R, N_I\rangle$, the concept description C is interpreted as $C^{\mathcal{I}} := \{\delta \in \Delta \mid \mathcal{I} \models \Psi_C(\delta)\}$.

For example, the concept description

$$D = \mathsf{Female} \sqcap \exists \mathsf{likes}.\mathsf{Male} \sqcap \exists \mathsf{has\text{-}child}.(\mathsf{Student} \sqcap \exists \mathsf{attends}.\mathsf{CScourse})$$

describes all women who like a man and have a child that is a student attending a CScourse. The semantics of D is given by the following FO formula:

$$\begin{aligned}\Psi_D(x_0) = {} & \mathsf{Female}(x_0) \wedge \exists x.(\mathsf{likes}(x_0,x) \wedge \mathsf{Male}(x)) \wedge \\ & \exists y.(\mathsf{has\text{-}child}(x_0,y) \wedge \mathsf{Student}(y) \wedge \exists z.(\mathsf{attends}(y,z) \wedge \mathsf{CScourse}(z))).\end{aligned}$$

In order to obtain a structured representation of the knowledge about the application domain one is interested in the subsumption hierarchy formed by the concept descriptions. Using their FO semantics, ***subsumption*** between concept descriptions is defined as $C \sqsubseteq D$ iff $\forall x_0.\Psi_C(x_0) \rightarrow \Psi_D(x_0)$ is valid.

Rooted Simple Graphs

We are interested in a class of SGs corresponding to $\mathcal{ELIRO}^1$-concept descriptions. On the one hand, we must restrict our attention to *connected* SGs over a support $\mathcal{S} = \langle N_C, N_R, N_I\rangle$ containing only *binary* relation types, because $\mathcal{ELIRO}^1$-roles correspond to binary relations and, as we will see, $\mathcal{ELIRO}^1$-concept descriptions always describe connected structures. Because of the restriction to binary relations, we can dispense with explicit relation nodes: instead we consider directed edges between concept nodes labeled by a relation type.

On the other hand, we must (1) deal with the different semantics of SGs and concept descriptions (closed formulae vs. formulae with one free variable), and (2) introduce conjunctions of types in SGs since conjunctions of primitive concepts may occur in $\mathcal{ELIRO}^1$-concept descriptions. In order to handle (2), we allow for concept nodes labeled by a set of concept types $\{P_1, \ldots, P_n\} \subseteq N_C$, where the empty set corresponds to $\top_C$. Due to (1), we extend the notion of SGs by introducing one distinguished concept node called the *root* of the SG.

Formally, we restrict the attention to *unordered* supports $\langle N_C, N_R, N_I\rangle$ where the orders on N_C and N_R are the identity relations.[2] Given such an unordered support $\langle N_C, N_R, N_I\rangle$ we define a *rooted SG* $\mathcal{G} = (V, E, c_0, \ell)$ over this support as a SG where V is a set of concept nodes, $E \subseteq V \times N_R \times V$ is a set of directed edges labeled by relation types from N_R, c_0 is the root of $\mathcal{G}$, and ℓ labels each $c \in V$ by a set of concept types $\{P_1, \ldots, P_n\} \subseteq N_C$ and a referent from $N_I \cup \{*\}$.

Given an interpretation $\mathcal{I} = (\Delta, \cdot^{\mathcal{I}})$ of $\langle N_C, N_R, N_I\rangle$, the semantics of a rooted SG $\mathcal{G}$ is given by $\{\delta \in \Delta \mid \mathcal{I} \models \Phi(\mathcal{G})(\delta)\}$, where Φ is an extension of the Φ operator from SGs to rooted SGs. To be more precise, the FO formula $\Phi(\mathcal{G})(x_0)$ with one free variable x_0 is obtained from $\mathcal{G}$ as follows. Let $id : V \rightarrow (\mathcal{V} \setminus \{x_0\}) \cup N_I$ be a mapping as defined in Section 2. Each concept node $c \in V$ with $type(c) = \{P_1, \ldots, P_n\}$ yields a conjunction $P_1(id(c)) \wedge \ldots \wedge P_n(id(c))$, and each edge $c_1 r c_2 \in E$ yields $r(id(c_1), id(c_2))$. Now, $\Phi(\mathcal{G})(x_0)$ is defined as the conjunction of $x_0 \doteq id(c_0)$ and the formulae corresponding to concept nodes and edges, where all variables except x_0 are existentially quantified.

For example, the rooted SG $\mathcal{G}_1$ with root c_0 depicted in Fig. 3 describes all women that are a daughter of Peter, and have a dear son that likes Peter and is a student attending the CScourse number KR101.

Just as for SGs, subsumption between rooted SGs can be characterized by the existence of a homomorphism. Here, the notion of a homomorphism between SGs w.r.t. a support $\mathcal{S}$ [6] must be adapted to rooted SGs. A *homomorphism* from $\mathcal{H} = (V_H, E_H, d_0, \ell_H)$ to $\mathcal{G} = (V_G, E_G, c_0, \ell_G)$ is a mapping $\varphi : V_H \rightarrow V_G$ such that (1) $\varphi(d_0) = c_0$, (2) $type_H(d) \subseteq type_G(\varphi(d))$ and $ref_H(d) \geq_I ref_G(\varphi(d))$ for all $d \in V_H$, and (3) $\varphi(d) r \varphi(d') \in E_G$ for all $d r d' \in E_H$.

The proof of the following theorem in [3] is similar to the proof of soundness and completeness of the characterization of subsumption in [6].

Theorem 1. *Let $\mathcal{G}$ be a rooted SG in normal form and $\mathcal{H}$ a rooted SG. Then $\mathcal{G} \sqsubseteq \mathcal{H}$ iff there exists a homomorphism from $\mathcal{H}$ to $\mathcal{G}$.*

[2] It should be noted that the restriction to unordered supports is without loss of generality since the order relation on N_C can be encoded into the type set labels, and the one on N_R into multiple edges between nodes. Vice versa, the introduction of sets of types is not a real extension since their effect can be simulated by an appropriately extended ordered support (see [3] for details).

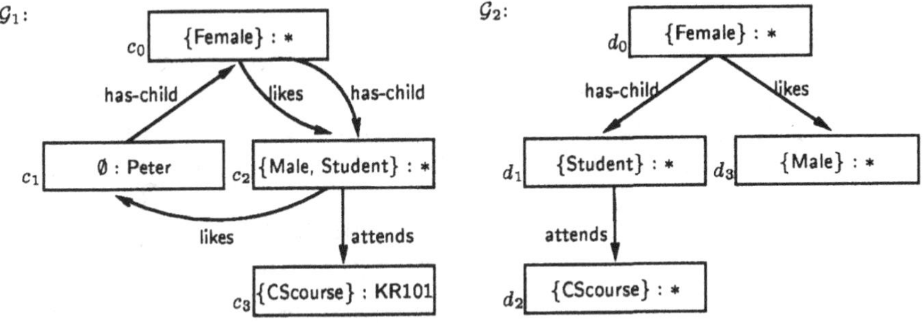

Fig. 3. Two rooted simple graphs.

For example, the rooted SG $\mathcal{G}_2$ with root d_0 depicted in Fig. 3 subsumes $\mathcal{G}_1$ because mapping d_0 onto c_0, d_1 and d_3 onto c_2, and d_2 onto c_3 yields a homomorphism from $\mathcal{G}_2$ to $\mathcal{G}_1$.

Unlike SGs over an arbitrary support, rooted SGs can be transformed (in polynomial time) into equivalent rooted SGs in normal form by identifying all concept nodes $c_1, \dots, c_n$ having the same referent $a \in N_I$ and defining the type set of the resulting concept node as $\bigcup_{1 \leq i \leq n} type(c_i)$.

For SGs over a support $\mathcal{S}$, subsumption is known to be an NP-complete problem. The known algorithms deciding $g \sqsubseteq h$ w.r.t. $\mathcal{S}$ are based on the characterization of subsumption by homomorphisms, and thus require the subsumee g to be in normal form. In order to obtain a subsumption algorithm for rooted SGs, we must simply adjust the conditions tested for nodes and edges according to the modified conditions on homomorphisms between rooted SGs. Conversely, subsumption of SGs w.r.t. $\mathcal{S}$ can be reduced to subsumption of rooted SGs [3]. This shows that subsumption for rooted SGs is also an NP-complete problem.

In [13], a polynomial-time algorithm is introduced that can decide $g \sqsubseteq \mathbf{t}$ w.r.t. a support $\mathcal{S}$ provided that $\mathbf{t}$ is a tree and g is a SG in normal form. In this context, a SG $\mathbf{t}$ is called a *tree* iff $\mathbf{t}$ contains no cycles of length greater than 2. The notion of a tree can be adapted to rooted SGs $\mathcal{T}$ by viewing $\mathcal{T}$ as an *undirected graph*. A simple modification of the algorithm in [13] yields a polynomial time algorithm deciding $\mathcal{G} \sqsubseteq \mathcal{T}$ for a rooted SG $\mathcal{T}$ that is a tree and a rooted SG $\mathcal{G}$ in normal form [3].

Now, we will show that this algorithm also yields a polynomial-time algorithm for subsumption of $\mathcal{ELIRO}^1$-concept descriptions.

Translating concept descriptions into rooted simple graphs
The main idea underlying the translation is to represent a concept description C as a tree T_C. Intuitively, C is represented by a tree with root c_0 where all atomic concepts and constants occurring in the top-level conjunction of C yield the label of c_0, and each existential restriction $\exists r.C'$ in this conjunction yields an r-successor that is the root of the tree corresponding to C'. For example, the concept description C below yields the tree T_C in Fig. 4:

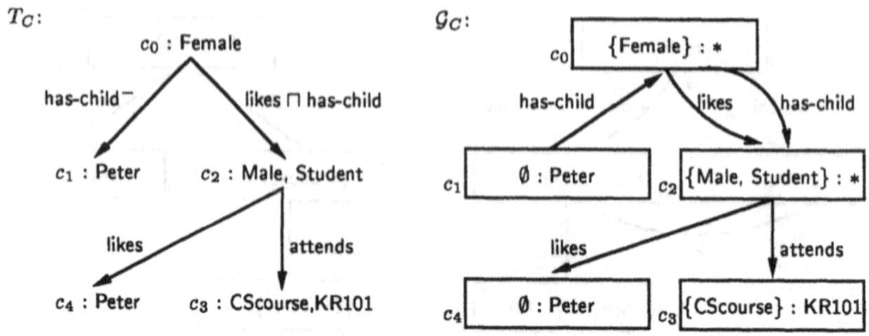

Fig. 4. Translating $\mathcal{ELIRO}^1$-concept descriptions into rooted simple graphs.

$$C := \mathsf{Female} \sqcap \exists \mathsf{has\text{-}child}^-.\{\mathsf{Peter}\} \sqcap \exists(\mathsf{likes} \sqcap \mathsf{has\text{-}child}).$$
$$(\mathsf{Male} \sqcap \mathsf{Student} \sqcap \exists \mathsf{attends}.(\mathsf{CScourse} \sqcap \{\mathsf{KR101}\}) \sqcap \exists \mathsf{likes}.\{\mathsf{Peter}\}).$$

Now, we can define the rooted SG $\mathcal{G}_C$ corresponding to C as follows. The nodes in T_C yield the set of concept nodes V of $\mathcal{G}_C$. The label $\ell(c)$ of a concept node $c \in V$ is determined by the label $\ell_T(c)$ of c in T_C, i.e., *type*(c) is the set of all atomic concepts occurring in $\ell_T(c)$ and, if there is a constant $a \in \ell_T(c)$, then *ref*$(c) := a$; otherwise *ref*$(c) := *$. Note that *ref*(c) is well-defined because we have restricted $\mathcal{ELIRO}^1$-concept descriptions to those containing at most one constant in each conjunction. Finally, the set of edges of $\mathcal{G}_C$ is obtained from the edges in T_C: conjunctions of roles are decomposed ($c(r_1 \sqcap \ldots \sqcap r_n)d$ yields n edges $cr_1d, \ldots, cr_nd$) and inverse roles are redirected (cr^-d yields the edge drc). In our example, we obtain the rooted SG $\mathcal{G}_C$ depicted in Fig. 4, which is a tree.

Using the recursive definition of the tree T_C, it can be shown [3] that C is equivalent to $\mathcal{G}_C$, i.e., $\forall x_0.\Psi_C(x_0) \leftrightarrow \Phi(\mathcal{G}_C)(x_0)$ is a valid FO formula. Conversely, any rooted SG $\mathcal{G}$ that is a tree can be translated into an equivalent concept description $C_\mathcal{G}$ [3]. Thus, there is a 1–1 correspondence between $\mathcal{ELIRO}^1$-concept descriptions and rooted SGs that are trees. Because of this correspondence, we can reduce subsumption in $\mathcal{ELIRO}^1$ to subsumption between rooted SGs, i.e., $C \sqsubseteq D$ iff $\mathcal{G}_C \sqsubseteq \mathcal{G}_D$. Since $\mathcal{G}_C$ is a tree and $\mathcal{G}_D$ can be transformed into normal form (in polynomial time), subsumption for $\mathcal{ELIRO}^1$-concept description is polynomial-time decidable by applying the polynomial-time algorithm mentioned above to the tree $\mathcal{G}_C$ and the normal form of $\mathcal{G}_D$. This yields the following tractability result for $\mathcal{ELIRO}^1$ [3]:

Theorem 2. *Subsumption $C \sqsubseteq D$ of $\mathcal{ELIRO}^1$-concept descriptions can be decided in time polynomial in the size of C and D.*

Strictly speaking, the above argument shows tractability only for concept descriptions where each conjunction contains only one constant. The result can, however, easily be extended to general $\mathcal{ELIRO}^1$-concept descriptions [3].

Extending the tractability result

We will now extend the tractability result from (rooted) SGs that are trees to (rooted) SGs that can be transformed into trees by "cutting cycles" of length greater than 2. For a given rooted SG $\mathcal{G}$, we can eliminate an (undirected) cycle $c_0, \dots, c_n$ where $c_0 = c_n$ in $\mathcal{G}$ by applying the ***split-operation*** on concept nodes as introduced for SGs in [5]. To be more precise, we (1) arbitrarily choose a node $c_i \in \{c_1, \dots, c_n\}$, (2) introduce a new node c labeled like c_i, and (3) replace all edges between c_{i-1} and c_i by edges between c_{i-1} and c. We then say that the cycle is *cut in* c_i. Obviously, any cyclic SG $\mathcal{G}$ can be transformed into an acyclic SG $\mathcal{G}^*$ by applying this operation a polynomial number of times. In general, however, the resulting SG $\mathcal{G}^*$ need not be equivalent to $\mathcal{G}$.

As an example, consider the rooted SG $\mathcal{G}_1$ in Fig. 3. On the one hand, we can eliminate the cycle c_1, c_0, c_2, c_1 by introducing a new node c_4 with label $(\emptyset, Peter)$ and replacing the edge $c_2 \mathsf{likes} c_1$ by $c_2 \mathsf{likes} c_4$. The resulting tree coincides with the tree $\mathcal{G}_C$ in Fig. 4, and it is equivalent to $\mathcal{G}_1$ because $\mathcal{G}_1$ is a normal form of $\mathcal{G}_C$. On the other hand, if we introduce a new node c labeled $(\{Female\}, *)$ and replace $c_1 \mathsf{has\text{-}child} c_0$ by $c_1 \mathsf{has\text{-}child} c$, then the resulting tree is not equivalent to $\mathcal{G}_C$ because the student's mother and Peter's child need no longer to be the same person.

The following proposition introduces a condition on rooted SGs that ensures that rooted SGs satisfying this condition can be transformed into equivalent trees by applying the split operation to individual concept nodes [3].

Proposition 1. *If each cycle of length greater than 2 in the rooted SG $\mathcal{G}$ contains at least one individual node c, then $\mathcal{G}$ can be transformed into an equivalent tree $\mathcal{G}^*$ in time polynomial in the size of $\mathcal{G}$.*

Consequently, $\mathcal{G} \sqsubseteq \mathcal{H}$ can be decided in polynomial time if $\mathcal{G}$ is a rooted SG in normal form and $\mathcal{H}$ satisfies the premise of the proposition. It is easy to see that this tractability result also applies to (non-rooted) SGs over a support.

4 The Loosely Guarded Fragment of Conceptual Graphs

Due to the expressiveness of the CG formalism, all the interesting reasoning problems (such as subsumption and validity) are undecidable for general CGs. We will identify a large class of CGs for which both validity and subsumption are decidable. This fragment, which we will call *loosely guarded fragment of CGs*, will be defined directly by syntactic restrictions on graphs. This allows for an efficient test for guardedness of graphs. The fragment corresponds to the so-called loosely guarded fragment of FO. In [1], the guarded fragment of FO was defined in an attempt to find a generalization of modal logics that still enjoys the nice properties of modal logics (like decidability, finite axiomatizability, etc.). In the same work, decidability of this fragment was shown. In [15], an even larger decidable fragment of FO was introduced, the loosely guarded fragment.

Definition 1. *Let Σ be a set of constant and relation symbols including equality (called the* signature*). The* loosely guarded fragment $LGF(\Sigma)$ *of first-order logic is defined inductively as follows:*

1. *Every atomic formula over Σ belongs to LGF(Σ).*
2. *LGF(Σ) is closed under the Boolean connectives $\neg, \wedge, \vee, \rightarrow$, and $\leftrightarrow$.*
3. *If $\mathbf{x}, \mathbf{y}$ are tuples of variables, if $\beta(\mathbf{x}, \mathbf{y})$ is a formula from LGF(Σ), and if $\alpha_1 \wedge \cdots \wedge \alpha_n$ is a conjunction of atoms, then*

$$\exists \mathbf{x}.((\alpha_1 \wedge \cdots \wedge \alpha_n) \wedge \beta(\mathbf{x}, \mathbf{y})) \quad \textit{and} \quad \forall \mathbf{x}.((\alpha_1 \wedge \cdots \wedge \alpha_n) \rightarrow \beta(\mathbf{x}, \mathbf{y}))$$

belong to LGF(Σ), provided that, for every variable x in $\mathbf{x}$ and every variable z in $\mathbf{x}$ or $\mathbf{y}$, there is an atom α_j (the guard*) such that x and z occur in α_j.*

An exact complexity result for the satisfiability problem of the loosely guarded fragment was shown by Grädel [11]. It turned out that the complexity of the satisfiability problem in *LGF*(Σ) depends on the arity of the relation symbols in the signature Σ. In general, the problem is 2-EXPTIME-complete. However, if the arity of all relation symbols in Σ is bounded by a constant, then the satisfiability problem for *LGF*(Σ) is "only" EXPTIME-complete; in particular, this is the case if Σ is finite.

The definition of the loosely guarded fragment of FO gives rise to the definition of a corresponding fragment of CGs, which we will call the *loosely guarded fragment of CGs.* The restrictions defining this fragment guarantee that all quantifiers in the FO translation of a loosely guarded graph can either be eliminated, or are loosely guarded in the sense of Def. 1. The same must apply to any variable appearing free in a sub-formula of the FO translation of a loosely guarded graph. To state the appropriate restrictions on the CGs, we identify the nodes representing free and bound variables in the contexts of a graph. These will be the *new* and *external* nodes introduced in the following definition.

Definition 2. *Let $G = \langle p, coref \rangle$ be a CG over $\mathcal{S}$. A concept node $c \in C(p)$ contained in a context q of p is called* external *iff it has a coreference link to a strictly dominating concept node. It is called* old *iff it satisfies one of the following conditions:*

- *c is an external or an individual concept node.*
- *c is linked by a coreference link to another old node in the same context q.*

Nodes that are not old are called new.

In the CG of Fig. 2, c_3, c_5, c_6 are external nodes while c_1, c_2, c_4 are new nodes. Note that c_4 is a new node even though it is linked by a chain of coreference links to the old node c_5. This is a desired effect of the definition since coreference links inside one context express equality of concept nodes, while the coreference links from c_4 and c_5 to c_6 are used to express inequality of c_4 and c_5.

Definition 3 (The loosely guarded fragment of conceptual graphs). *A CG $G = \langle p, coref \rangle \in CG(\mathcal{S})$ is called* loosely guarded *iff it satisfies the following:*

1. *If $(c_1, c_2) \in coref$ and the context p_1 of c_1 strictly dominates the context p_2 of c_2, then for each context q such that q lies between p_1 and p_2 (i.e. p_1 strictly dominates q and q strictly dominates p_2) it holds that q is labeled by a simple graph g containing no new nodes.*

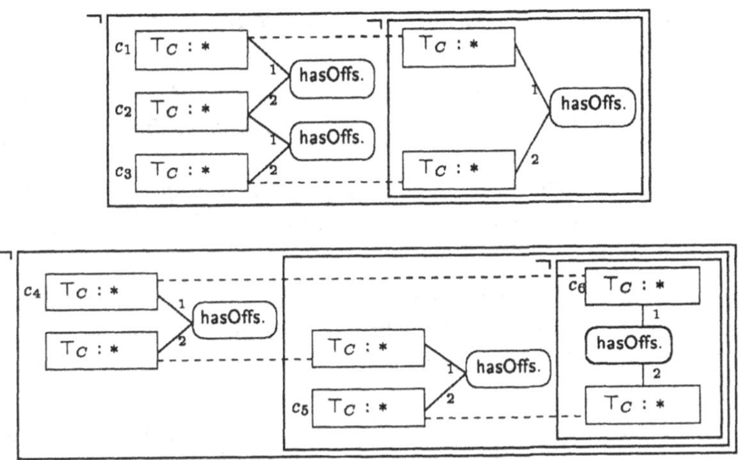

Fig. 5. Two graphs that are not loosely guarded.

2. For each simple graph $g = \langle C, R, E, \ell \rangle$ labeling a context of G, either g contains no new nodes, or g satisfies the following: if $C = \{c\}$, then $type(c) \neq \top_C$ or there is an $r \in R$ such that $(c, r) \in E$; if $|C| > 1$, then for each pair of distinct nodes $c, d \in C$ such that c is new and d is not an individual concept node, there is an $r \in R$ satisfying $\{(c, r), (d, r)\} \subseteq E$.

With $lgCG(\mathcal{S})$ we denote the set of all loosely guarded CGs over the support $\mathcal{S}$.

An example of a lgCG is the graph in Fig. 2. In fact, Property 1 is obviously satisfied since no coreference link crosses more than one context. Property 2 is satisfied as well; for example, in the context p_2, the new node c_4 shares a relation node with both c_3 and c_5. The nodes c_2 and c_4 need not be adjacent to the same relation node since both are old nodes.

CGs that violate one of the properties required by Def. 3 need not be equivalent to a loosely guarded FO formula. For example, transitivity of a binary relation symbol is an assertion that cannot be expressed by a loosely guarded formula [11]. Figure 5 shows two CGs that assert transitivity of the binary relation hasOffspring. The upper graph is not loosely guarded because it violates Property 2: c_1 and c_3 are both new nodes, but they are not adjacent to the same relation node. The lower graph is not loosely guarded because it violates Property 1: c_4 and c_6 are linked by a coreference link that spans a context containing the new node c_5.

Note that, even though the definition of lgCGs may look quite complex at first sight, it is a purely syntactic definition using easily testable properties of graphs. Indeed, it is easy to show that membership of a given CG over $\mathcal{S}$ in $lgCG(\mathcal{S})$ can be tested in polynomial time [2]. The name "loosely guarded fragment of CGs" is justified by the main theorem of this section:

Theorem 3. *Let $S = \langle N_C, N_R, N_I \rangle$ be a support and let Σ_S be the corresponding FO signature $\Sigma_S = N_C \cup N_R \cup N_I$.*

1. *For each $G \in lgCG(S)$ there exists a formula $\varphi_G \in LGF(\Sigma_S)$ such that φ_G is equivalent to $\Phi(G)$. In addition, φ_G is computable from G in polynomial time.*
2. *For each closed formula $\varphi \in LGF(\Sigma_S)$ there is a graph $G_\varphi \in lgCG(S)$ such that $\Phi(G_\varphi)$ is equivalent to φ. In addition, G_φ is computable from φ in polynomial time.*

A complete proof of this theorem can be found in [2]. Here, we will illustrate the main idea underlying the proof of the first part, using the example in Fig. 2. The transformation of $\Phi(G)$ into a loosely guarded formula works inductively over the structure of G. Hence, we start with the innermost context p_3 of G. The formula $\varphi_3(x_4, x_5) := \Phi(p_3) = \exists x_6.(x_6 \doteq x_6 \wedge x_6 \doteq x_4 \wedge x_6 \doteq x_5)$ is loosely guarded, and it is equivalent to the simpler loosely guarded formula $\varphi_3' = x_4 \doteq x_5$. The formula for the context p_2,

$$\varphi_2(x_1, x_2) := \Phi(p_2) = \exists x_3 x_4 x_5.(x_3 \doteq x_1 \wedge x_5 \doteq x_2 \wedge \mathsf{Human}(x_4) \wedge \\ \mathsf{hasChild}(x_4, x_3) \wedge \mathsf{likes}(x_4, x_5) \wedge \neg\varphi_3'(x_4, x_5)),$$

is not loosely guarded. In order to obtain a loosely guarded formula, we eliminate the identifiers of the old nodes (in this case x_3, x_5) together with their quantifiers, using the fact that φ_2 contains the conjuncts $x_3 \doteq x_1$ and $x_5 \doteq x_2$. Re-ordering the conjuncts yields the formula

$$\varphi_2'(x_1, x_2) = \exists x_4.(\mathsf{hasChild}(x_4, x_1) \wedge \mathsf{likes}(x_4, x_2) \wedge \mathsf{Human}(x_4) \wedge x_4 \not\doteq x_2),$$

which is loosely guarded and equivalent to φ_2. The necessary guards are given by the first two conjuncts, which correspond to the relation nodes adjacent to c_3, c_4, c_5. The existence of such nodes in a loosely guarded graph is guaranteed by Property 2 of Def. 3.

Since the context p_1 does not contain old nodes, the next steps (in which we also treat the two negation signs) directly yields the loosely guarded equivalent φ_G of $\Phi(G) = \Phi(p)$:

$$\varphi_G := \neg\exists x_1 x_2.(\mathsf{hasChild}(x_2, x_1) \wedge \mathsf{Human}(x_1) \wedge \mathsf{Human}(x_2) \wedge \neg\varphi_2'(x_1, x_2)).$$

Summing up, the techniques used to transform $\Phi(p)$ into its loosely guarded equivalent φ_G are: (1) Elimination of identifiers and the corresponding quantifiers for old nodes; and (2) using Property 2 of Def. 3 to find the appropriate guards for the remaining quantified variables. As has already been pointed out, Property 1 of Def. 3 is necessary to ensure that no free variable of a sub-formula escapes the guards (see Fig. 5).

The following theorem is an immediate consequence of part 1 of Theorem 3 and the known complexity results for the loosely guarded fragment of FO.

Theorem 4. *Let S be a finite support. Then subsumption and validity of loosely guarded CGs over S is decidable in deterministic exponential time.*

Because of part 2 of Theorem 3, the ExpTime-hardness result for $LGF(\Sigma_S)$ also transfers to $lgCG(S)$.

5 Conclusion

Although the characterization of the loosely guarded fragment of conceptual graphs may appear to be a bit complex, it can easily be checked whether a CG belongs to this fragment.It should also be easy to support the knowledge engineer in designing CGs belonging to this fragment by showing external and new nodes in different colors, and by pointing out new nodes that are not yet guarded. Another interesting point is that there are theorem-provers that are complete for FO, and behave as a decision procedure (i.e., always terminate) for the loosely guarded fragment [8]. If such a prover is used to prove validity of general CGs, then one automatically has a decision procedure if the CGs are loosely guarded.

References

1. H. Andréka, J. van Benthem, and I. Németi. Modal languages and bounded fragments of predicate logic. *J. of Philosophical Logic*, 27(3):217–274, 1998.
2. F. Baader, R. Molitor, and S. Tobies. The Guarded Fragment of Conceptual Graphs. LTCS-Report 98-10, available at http://www-lti.informatik.rwth-aachen.de/ Forschung/Papers.html.
3. F. Baader, R. Molitor, and S. Tobies. On the Relationship between Descripion Logics and Conceptual Graphs. LTCS-Report 98-11, available at http://www-lti.informatik.rwth-aachen.de/Forschung/Papers.html.
4. E. Börger, E. Grädel, and Y. Gurevich. *The Classical Decision Problem*. Perspectives in Mathematical Logic. Springer, 1997.
5. M. Chein and M. L. Mugnier. Conceptual graphs: fundamental notions. *Revue d'Intelligence Artificielle*, 6(4):365–406, 1992.
6. M. Chein, M. L. Mugnier, and G. Simonet. Nested graphs: a graph-based knowledge representation model with FOL semantics. In *Proc. KR'98*, 1998.
7. P. Coupey and C. Faron. Towards correspondences between conceptual graphs and description logics. In *Proc. ICCS'98*, LNCS 1453, 1998.
8. H. de Nivelle. A resolution decision procedure for the guarded fragment. In *Proc. CADE-15*, LNCS 1421, 1998.
9. F. M. Donini, M. Lenzerini, D. Nardi, and W. Nutt. The complexity of concept languages. *Information and Computation*, 134(1):1–58, 1997.
10. F. M. Donini, M. Lenzerini, D. Nardi, and A. Schaerf. Reasoning in description logics. In *Foundation of Knowledge Representation*, CSLI-Publications, 1996.
11. E. Grädel. On the restraining power of guards. *Journal of Symbolic Logic*, 1999. To appear.
12. G. Kerdiles and E. Salvat. A sound and complete CG proof procedure combining projections with analytic tableaux. In *Proc. ICCS'97*, LNCS 1257, 1997.
13. M. L. Mugnier and M. Chein. Polynomial algorithms for projection and matching. In *Proc. 7th Workshop on Conceptual Structures, 1992*, LNCS 754, 1993.
14. John F. Sowa. *Conceptual Structures*. Addison-Wesley, 1984.
15. J. van Benthem. Dynamic Bits and Pieces. Technical Report LP-1997-01, ILLC, University of Amsterdam, The Netherlands, 1997.
16. M. Wermelinger. Conceptual graphs and first-order logic. In *Proc. ICCS'95*, LNCS 954, 1995.

Dynamic Semantics for Conceptual Graphs

Gwen Kerdiles

LIRMM (University of Montpellier) and ILLC (University of Amsterdam)*
kerdiles@philo.uva.nl

Abstract. Dynamisation in Conceptual Graph Theory has been stimulated by mostly two trends: one inspired by computer science notions of actors and agents and the other one by computational semantics; we focus on the later, following John Sowa's parallel between Existential Graphs and Discourse Representation Theory or more generally Dynamic Semantics. CGs are usually interpreted by mean of a translation into FOL or directly but with a similar static impact, by mean of classical set-theoretic extensional semantics. This classical view in which meaning equals truth conditions does not capture contextual information. We propose to adopt the Dynamic Semantics shift in which the meaning of a graph is characterized by the change of information brought in an information state when it is updated with the graph.

1 Introduction

In [1], F. Giunchiglia and P. Bouquet metaphorically present a context in Artificial Intelligence as a sort of box which is part of the structure of an individual's representation of the world and which draws a sort of boundary between what is *in* and what is *out*. In J. McCarthy's pioneer work on the formalisation of context (see e.g. [2] and [3] for a recent survey), a such box is *a rich object* (a collection of parameters) upon which a representation depends. Typically, a representation can be true in some contexts and false in others. For instance, the sentence "It is raining" calls for a context of utterance to be interpreted and that context includes among the parameters, the time and place of utterance (In the context of Amsterdam, that sentence is often true and in particular on Sunday April 4, 1999). In this paper, we consider a very specific form of contextual information which corresponds to the discourse referents in Discourse Representation Theory (DRT) (see [4] and [5]). A context, as part of the cognitive state of an agent (the hearer), is used in interpreting new information. When a graph is asserted, it introduces two kinds of information: a relation between concepts provides some factual information and a concept itself introduces in the context an item that is available for further references. A Dynamic Semantics for Discourse Representation Structures (DRS), e.g. [6], takes into account these two kinds of information. To illustrate the presentation, some examples of graphical representation of natural language sentences are considered, in particular the use

* University of Amsterdam, Fac. der Geesteswetenschappen, Nieuwe Doelenstraat 15, 1012CP Amsterdam, The Netherlands. tel: (+31) 205254552, fax: (+31) 205254503

of anaphoric bindings, nevertheless this paper is in no way an attempt at tackling the problem of anaphora resolution, but rather a proposal for some formal semantical tools which are guided by some intuitions about anaphoric bindings and which might be useful in the problem of anaphora resolution. Our hearer can better be seen as a Conceptual Graph knowledge base (i.e., an agent representing a such knowledge base who speaks the language of Conceptual Graphs) which is updated with new graphical information.

In [7], J. Sowa argues that DRSs and Conceptual Graphs are isomorphic notations: they share the same primitives and the same scoping rules for quantifiers. Nevertheless, by inheriting Peirce's lines of identity and contrary to DRSs, CGs provide a notation of equivalent expressive power which is free of variables. This feature is relevant in theorem proving, indeed, efficient methods for constructing proofs like free-variable tableaux or resolution require *pure* representations (representations in which a variable is not quantified twice) and a renaming preprocess must be performed. By avoiding variable names, CGs are always pure. But the absence of variables is mostly relevant for an incremental construction of the representations: if it is assumed that the constituents of a text have a representation of their own, then building a representation of the text consists in merging the representations of the constituents. For instance, a DRS for "A man entered. A woman entered." can be obtained by the merging the following two DRSs:

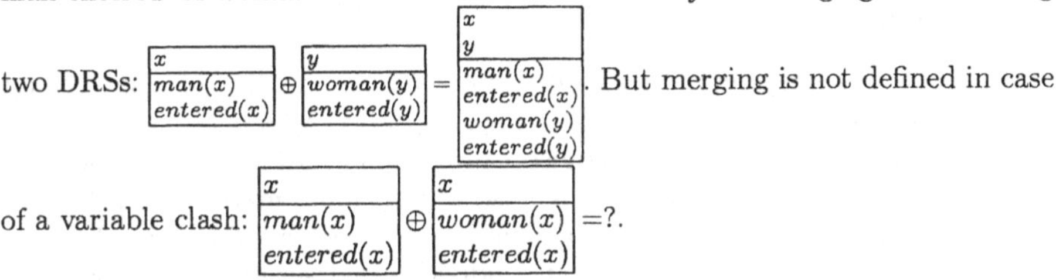

x
$man(x)$
$entered(x)$

$\oplus$

y
$woman(y)$
$entered(y)$

$=$

x y
$man(x)$
$entered(x)$
$woman(y)$
$entered(y)$

. But merging is not defined in case of a variable clash:

x
$man(x)$
$entered(x)$

$\oplus$

x
$woman(x)$
$entered(x)$

$=?$.

In Discourse Representation Theory, the problem of variable clashes is solved by always building the representation of a new sentence in the context of an existing DRS. Another solution consists in first renaming the variables occurring in different DRSs before merging them. CGs do not make use of variables, so variable clashes cannot occur and two CGs can always be merged by only juxtaposing them. The simplicity of this safe merging operation is an attractive feature of Conceptual Graphs: large knowledge bases can be gathered without any renaming.

To strengthen the parallel between DRT and CG Theory, we need to equip CGs with an adequate semantics which enables to distinguish factual information from contextual information: the factual information brought by the relation nodes can be verified in a (classical) model, whereas concepts introduce context items to which we may later make reference by means of coreference links (or lines of identity). We adapt Dynamic Semantics ([6]) to the language of Conceptual Graphs. CGs are viewed as programs (functions) that transform an information state into another information state. The meaning of a graph is not anymore its truth conditions in a model, but the change it brings to an information state defined with respect to a model. Entailment becomes dynamic: it allows coreference links between concepts in the premiss and concepts in the conclusion, e.g. a representation of "A farmer beats a donkey" entails a representation of "He beats it". This notion of entailment is not only interesting for the study of anaphoric bindings but it comes with an extended form of projection and provides as well a deduction theorem adapted to Conceptual Graph Rules ([8]): if G and H are two Simple Conceptual Graphs and f is a set of coreference links between G and H, then the rule $G \rightarrow_f H$ is valid iff $\boxed{\text{G} \oplus_f \boxed{\text{H}}}$[1] is valid iff G entails H under f iff there is a projection from H to G respecting f.

We first recall the construction rules for Simple Conceptual Graphs and presents the new semantics for this language. We then extend the definition of projection in order to cope with coreference links between the graph which is projected and the graph on which it is projected. Finally, in Section 3, we recall the language of CGs and extend the semantics to nested negation boxes.

2 Simple Conceptual Graphs

First concentrating on the negation-free fragment, will enable to focuss on the dynamic rôle of concepts. The following definitions are standard (e.g. [9]): a language is defined by a choice of ordered-sorted proper names (individual markers), ordered relation symbols and some construction rules.

2.1 Language

Definition 1 (Support). A support represents an ontology of a specific application domain. It is a 5-tuple $\Sigma = (T_C, T_R, \sigma, I, \tau)$ where T_C is a lattice of concept types. The infimum $\perp$ is also called the absurd type. T_R is a partition of posets of relation names. I is a set of individual markers such that for every concept type $t \neq \perp$, there are infinitely many individual markers (an enumerable set I_t). τ associates to any individual marker of I, the type of the individual represented which cannot be the absurd type. σ is a function which associates to any relation, its arity and the maximal type of each of its arguments: $\forall P \in T_R, \sigma(P) = (arity(P) > 0, t_1, \ldots, t_{arity(P)})$ where $\forall 1 \leq i \leq arity(P), t_i \in (T_C - \{\perp\})$; we note $\sigma_i(P) = t_i$, the maximal type of the ith-argument of any occurrence of the relation P.

Definition 2 (Ordering of concept labels). For a given support $\Sigma = (T_C, T_R, \sigma, I, \tau)$, we define a partial order on the pairs element of $\{(t, m)/t \in T_C$ and $m \in (I \cup \{*\})$ and if $m \in I$ then $t = \tau(m)\}$: $(t, m) \leq_\Sigma (t', m')$ iff (i) $t \leq_{T_C} t'$ and (ii) if $m \in I$ then either $m' = *$ or $m' = m$ and (iii) if $m = *$ then $m' = *$.

When there is no ambiguity on the considered support, we sometimes write $(t, m) \leq (t', m')$ instead of $(t, m) \leq_\Sigma (t', m')$.

As usual in Conceptual Graph theory and in classical First Order Logic(FOL), we adopt the assumption that two distinct concept nodes with the same individual marker do represent the same individual. However, we must stress that

[1] Where boxes represent negations.

this choice cannot be qualified as natural for Conceptual Graphs. Indeed, a linear notation, e.g. a language of FOL, is not equipped for distinguishing a single occurrence of a constant that occurs as argument of different relation symbols from distinct occurrences of the same constant: a linear notation, e.g. $HoldsAnInquiry(Holmes) \wedge HoldsAnInquiry(Holmes)$, does not differentiate

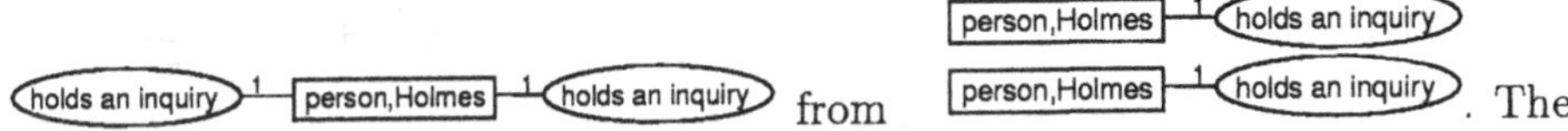

. The first graph represent redundant information, whereas the second could be the result of merging a graph about the detective Sherlock Holmes with one about the US Supreme Court Justice, Holmes. In a language of Conceptual Graphs, a coreference edge is used to state that two distinct concept nodes denote a single object. In this paper, we make the choice to simplify the presentation by reserving for existential nodes, the distinction between different occurrences of concept nodes with the same label: in a given model, an individual marker does always denote the same object.

Definition 3. A Simple Conceptual Graph (SCG) related to a support Σ, is a bipartite (relation nodes alternate with concept nodes), finite and not necessarily connected multigraph $G = (R, C, U, label, \equiv_{co})$ where R and C denote the two classes of relation and concept vertices, U is the set of edges such that the edges incident to each relation vertex are totally ordered, *label* is a mapping respecting σ and τ, which associates to a relation vertex its name, and to a concept vertex $c \in C$ a pair $(type(c), marker(c)) \in (T_C - \{\bot\}) \times (I \cup \{*\})$ and $\equiv_{co}$ is an equivalence relation on concept nodes, called the coreference relation[2]. G is inductively defined by:

1. An isolated concept node is a SCG ;e.g. [t,m]
   $R = \emptyset$, $C = \{c\}$, $U = \emptyset$, $\equiv_{co}$ is the identity on C.
   $label(c) = (type(c), marker(c))$ where $type(c) \in (T_C - \{\bot\})$ and $marker(c) \in (I \cup \{*\})$ and if $marker(c) \in I$ then $type(c) = \tau(marker(c))$. A concept node with marker $*$ is called an existential concept.
2. A star-graph, e.g.

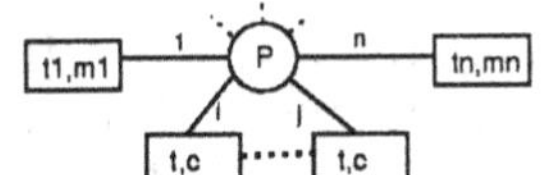

   $R = \{r\}$ where $P \in T_R$, $arity(P) = n$ and $label(r) = P$,
   $C = \{c_i / 1 \leq i \leq n\}$ and $\forall c \in C, label(c) = (type(c), marker(c))$ where $type(c) \in (T_C - \{\bot\})$ and $marker(c) \in (I \cup \{*\})$ and if $marker(c) \in I$ then $type(c) = \tau(marker(c))$,

[2] The adopted set notation of graphs presupposes that if $G^1 = (R^1, C^1, U^1, label^1, \equiv^1_{co})$ and $G^2 = (R^2, C^2, U^2, label^2, \equiv^2_{co})$ are two disjoint graphs, then $R^1 \cap R^2 = \emptyset$ and $C^1 \cap C^2 = \emptyset$. In order to lighten the drawings, we represent an equivalence relation on a set of concept nodes by an undirected graph covering the relation: $c \equiv_{co} c'$ iff there is a dashed path between c and c' or $marker(c) = marker(c') \in I$; Indeed, we do not need to represent the equivalence of individual concept nodes with the same marker as they are necessarily coreferent. We sometimes draw such links only to highlight them, e.g. in the example of a star-graph.

$U = \{(r, c_i, i)/1 \leq i \leq n\}$ and $type(c_i) \leq_{T_C} \sigma_i(P)$. $\forall c, c' \in C$, $c \equiv_{co} c'$ iff $marker(c) = marker(c') \in I$.
Note that the only $\equiv_{co}$-equivalent concept nodes are those which have the same individual marker.

3. The juxtaposition of two SCGs is a SCG

e.g.
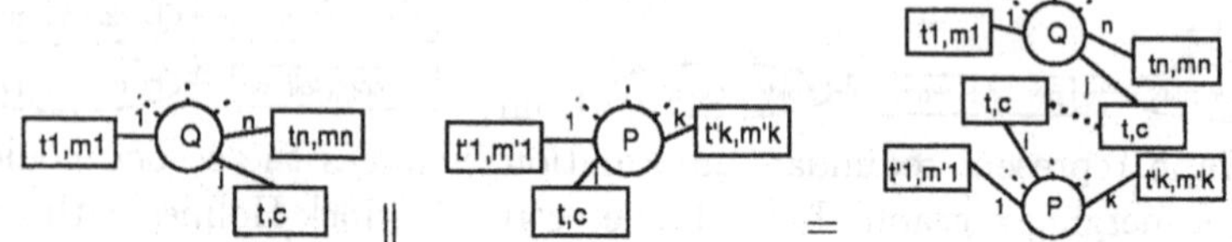

if $G^1 = (R^1, C^1, U^1, label^1, \equiv^1_{co})$ and $G^2 = (R^2, C^2, U^2, label^2, \equiv^2_{co})$ then $G^1 \parallel G^2 = (R^1 \cup R^2, C^1 \cup C^2, U^1 \cup U^2, label^1 \cup label^2, \equiv_{co})$ where $\forall c, c' \in (C^1 \cup C^2), c \equiv_{co} c'$ iff $c \equiv^1_{co} c'$ or $c \equiv^2_{co} c'$ or $marker(c) = marker(c') \in I$.
Equivalence classes of the two imported equivalence relations are preserved, the eventual new equivalences are those between concept nodes which have the same individual marker.

4. Internal linking: if $G' = (R, C, U, label, \equiv'_{co})$ is a SCG, then we may link two concept nodes c_1 and c_2 such that $label(c_1) = label(c_2)$. The result is a SCG in which the equivalence classes of the two nodes are merged[3]: $G = (R, C, U, label, \equiv_{co})$ such that
(i) $\forall x, y \in C$, if $x \equiv'_{co} y$ then $x \equiv_{co} y$.
(ii) $\forall x, y \in C$ if $x \equiv'_{co} c_1$ and $y \equiv'_{co} c_2$ then $x \equiv_{co} y$.

e.g.
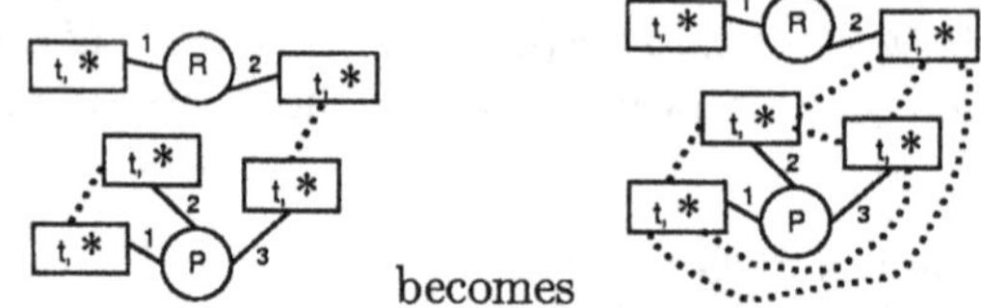

5. Concept merging: if $G' = (R, C', U', label', \equiv'_{co})$ is a SCG, then we may merge two concept nodes c_1 and c_2 such that $c_1 \equiv'_{co} c_2$ and the result is another SCG, $G = (R, C, U, label, \equiv_{co})$ such that
$C = C' - \{c_1\}$,
$\forall (r, c, i) \in U'$, if $c = c_1$ then $(r, c_2, i) \in U$ else $(r, c, i) \in U$,
label is the function of domain $(R \cup C)$ such that $\forall x \in (R \cup C), label(x) = label'(x)$ and $\equiv_{co}$ is the restriction of $\equiv'_{co}$ to C.

e.g.
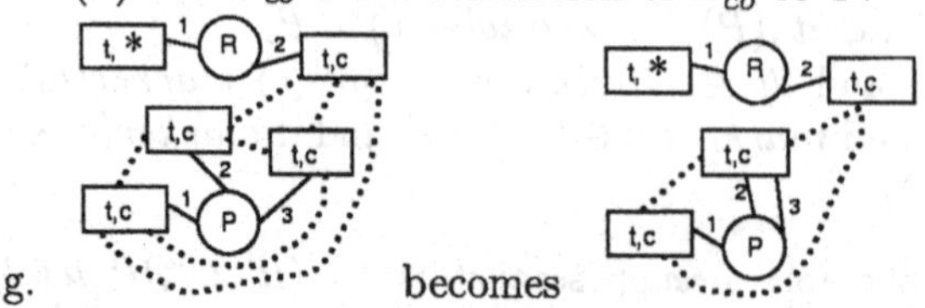

We note $G_\emptyset = (\emptyset, \emptyset, \emptyset, \emptyset, \emptyset)$, the empty SCG.

Extending a graph with new information, i.e. with another graph, can be seen either as a combination of juxtaposition of the two graphs and then internal

[3] This definition is deliberately not very constrained, indeed it allows the linking of two nodes that are already coreferent of each other which results in the same graph.

linking of the concepts representing the same objects or, and this is the course we will pursue in the rest of this paper, as an update of the first graph with the new representation. To update a piece of information, we first need to know which information is already available and in particular, to which concepts previously introduced we may refer in the information to add. We call *contextualisation*, an external (and directed from the representation to be added to the representation to be updated) notion of coreference links between two graphs.

Definition 4 (Contextualisation). For $G^1 = (R^1, C^1, U^1, label^1, \equiv^1_{co})$ and $G^2 = (R^2, C^2, U^2, label^2, \equiv^2_{co})$ two SCGs, a *contextualisation* of G^2 in G^1 is a relation $f \subseteq C^2 \times C^1$ such that the following conditions hold:

1. only existential nodes with the same type can be connected: for all pairs $(x, y) \in C^2 \times C^1$ if fxy then $\exists t \in T_C$ such that $label^2(x) = label^1(y) = (t, *)$.
2. the contextualisation is a function from equivalences classes to equivalence classes: $\forall x, x' \in C^2$ and $\forall y, y' \in C^1$:
$(x \equiv^2_{co} x' \& fxy)$ implies $fx'y$ and $(x \not\equiv^2_{co} x' \& fxy)$ implies $\neg(fx'y)$ and $(y \equiv^1_{co} y' \& fxy)$ implies fxy' and $(y \not\equiv^1_{co} y' \& fxy)$ implies $\neg(fxy')$.

The external counterpart for the combination of juxtaposition and then internal linking is a directed operation called *insertion*:

Definition 5 (Insertion). For two distinct SCGs, $G^1 = (R^1, C^1, U^1, label^1, \equiv^1_{co})$ and $G^2 = (R^2, C^2, U^2, label^2, \equiv^2_{co})$ and f a contextualisation of G^2 in G^1, the insertion of G^2 in G^1 under f, noted $G^1 \oplus_f G^2$ is the result of applying internal linking for every pair in f in the juxtaposition $G^1 \parallel G^2$.

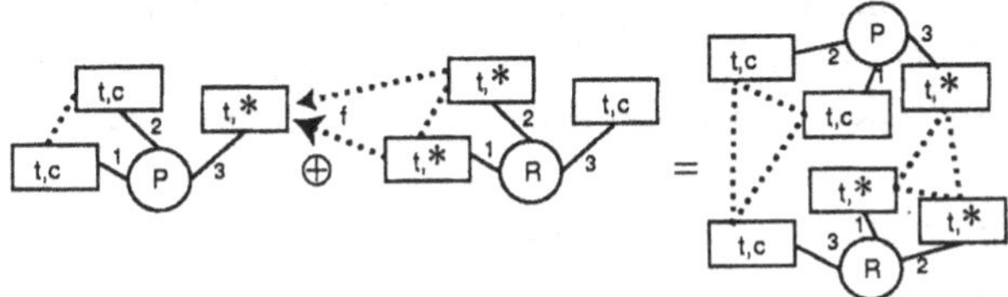

Example 1. Suppose that we have an expert building a simple-minded knowledge base about how the water of the canals in Amsterdam is cleaned. The knowledge base already contains: [lock,*]→(on)→[canal,*]←(connect)→[city,Amsterdam], [sea,NorthSea] representing "There is a canal connecting the city of Amsterdam to the North Sea and on this canal there is a lock". The expert can specialise the knowledge base by adding the information"It is daily opened". Anaphora resolution is not the problem at stake and the expert knows that the pronoun *it* refers to the concept corresponding to the lock. Furthermore, that concept is available for a connection and fulfills the requirements on the type of the argument of the relation *dailyOpened*. The expert can insert the new information under the pictured contextualisation f:

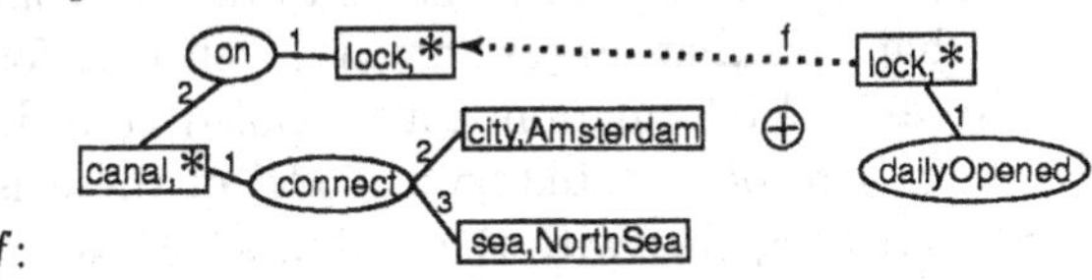

2.2 Dynamic Semantics

The semantics of simple Conceptual Graphs is usually given either by a translation into FOL, e.g. [9], or a classical set-theoretic extensional semantics in which the notion of entailment is static (there cannot be anaphoric bindings between the premiss and the conclusion), e.g. [10], [11] or [12]. A such semantics does not emphasize the notion of accessible concept whereas the essential rule for connecting two graphs, the insertion, is based on this notion of accessibility: added information can make reference (by means of coreference links) to the previously introduced concepts. For instance, in a classical semantics, the two different graphs [t,*]–1–(P) [t,*] and [t,*]–1–(P) have the same meaning, they are true in exactly the same models. In the proposed semantics, they have different meanings because the first graph introduces in the context two concepts to which we may later refer, whereas the second graph introduces only one. A graph is interpreted with respect to two different structures. The first one is a classical ensemblist model in which the factual information carried by the relations of the graph is verified. The second one, called an information state, is a record of the introduced concepts and their possible assignments to objects in the model.

A model is a structure which respects the ordering conveyed by the ontology, together with an interpretation of the relation labels and the individual markers:

Definition 6 (Models). A model, $M = (D, F)$, with respect to a support $\Sigma = (T_C, T_R, \sigma, I, \tau)$ consists of

- a set D such that for all concept types t and t' in $(T_C - \{\bot\})$, it holds that $D_t \subseteq D$ and $D_{t'} \subseteq D$ and if $t \leq_{T_C} t'$ then $D_t \subseteq D_{t'}$. These subdomains of interpretation respect the property of non-empty universe: $\forall t \in (T_C - \{\bot\}), D_t \neq \emptyset$.
- an interpretation function F on $(I \cup T_R)$ verifies:
  - if $m \in I$ then $F(m) \in D_{\tau(m)}$,
  - if $R \in T_R$ and $arity(R) = n$ then $F(R) \subseteq D_{\sigma_1(R)} \times \ldots \times D_{\sigma_n(R)}$,
  - $\forall R, R' \in T_R, R \leq_{T_R} R'$ implies that $F(R) \subseteq F(R')$.

The meaning of a graph is characterized by the change of information in an information state when it is acquainted with the knowledge represented by the graph. An information state is based on some model which fixes the interpretation of both relation labels and individual markers. Uncertainty is brought by existential nodes which do not have a predefined interpretation. When an existential node is encountered, it is added to the context and all possible assignments in the relevant domain (the set of objects in the model which have the same type as the existential node) are considered. Further (factual) information brought by the relation nodes enable to discard some of these possible assignments. If it happens that no assignment is left over for an element in the context, then the reached information state is called absurd. To summarize, an information state is a record related to a model of the existential nodes that have been encountered together with their admissible assignments:

Definition 7 (States). A state based on a model $M = (D, F)$ is a pair $\langle G, s \rangle$ such that G, called context, is a set of non-coreferent existential nodes; i.e. a SCG $G = (\emptyset, X, \emptyset, label, Id_X)$ such that $\forall x \in X, marker(x) = *$. s is a set of functions from X into D such that $\forall i \in s, \forall x \in X, i(x) \in D_{type(x)}$. Each element of s is called an assignment for X.

The state $\langle G_\emptyset, \{\emptyset\} \rangle$ is called the empty state. Any state $\langle G, \emptyset \rangle$ is called absurd.

Once more, we must stress that in this paper, the only piece of discourse information captured by an information state concerns existential concepts. A different perspective could consist in recording as well the full (normalised) structure of a graph which is the fuel in an update of an information state; updating would then be closer to a model construction.

Definition 8 (Interpretation of a SCG). Let $\langle X, s \rangle$ be a state based on $M = (D, F)$, $G = (R, C, U, label = (type, marker), \equiv_{co})$ be a SCG and f a contextualisation of G in X[4].

The update $\langle X, s \rangle_f[G]$ of $\langle X, s \rangle$ with G under f always exists and is defined as follows:

- $\langle X, s \rangle_\emptyset[G_\emptyset] = \langle X, s \rangle$: Updating with the empty graph does not change the state.
- $G = (R, C, U, label = (type, marker), \equiv_{co})$. Let $\langle c_1, \ldots, c_k \rangle$ be any ordering of C and $\langle r_{k+1}, \ldots, r_l \rangle$ be any ordering of R if $R \neq \emptyset$.
  We incrementally extend the context and the contextualisation of G:
  1. $\langle X_0, s_0 \rangle = \langle X, s \rangle$ and $f_0 = f$
  2. $\forall i$ such that $1 \leq i \leq k$
     (a) if either (i) $marker(c_i) \in I$
     or (ii) $marker(c_i) = *$ and $\exists x \in X_{i-1} / f_{i-1} c_i x$,
     then $\langle X_i, s_i \rangle = \langle X_{i-1}, s_{i-1} \rangle$ and $f_i = f_{i-1}$.
     (b) else, let N be a SCG containing a single node $c = \boxed{type(c_i), *}$.
     $X_i = X_{i-1} \parallel N$
     $f_i = f_{i-1} \cup \{(x, c) / x \in C \& x \equiv_{co} c_i\}$
     $s_i = \{a' / a \in s_{i-1} \& d \in D_{type(c_i)} \& a' : X_i \to D \& a'(c) = d \& \forall y \in X_{i-1}, a'(y) = a(y)\}$

  Updating with a concept node c does not change the state if the marker of c is individual or c is already linked to a node of the context. Otherwise, the existential concept node is added to the context and the set of assignments is expanded with all possible assignments for the new node:

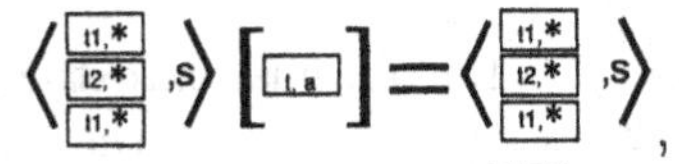

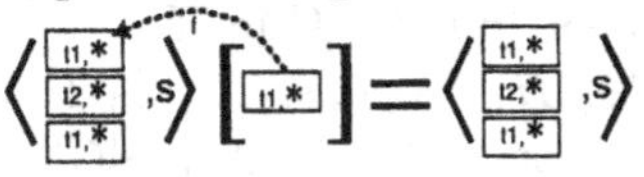

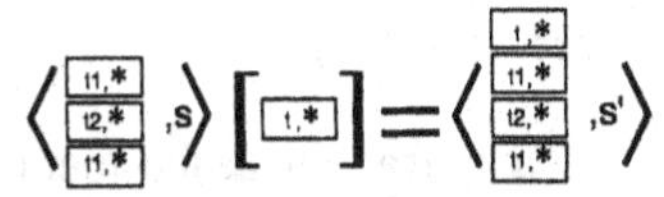

[4] f is the notation for those concepts that have already been linked to a context item. As the nodes in the context X are not coreferent to each other, f is a partial function from the set of existential nodes in G to X. Part of the updating process consists in building a total function.

3. if $R = \emptyset$ then $\langle X, s\rangle_f[G] = \langle X_k, s_k\rangle$,
   else $\forall i/(k+1) \leq i \leq l$, $m = arity(label(r_i))$, for $1 \leq p \leq m$, n_p refers to the p^{ith} neighbour of r_i in G and
   $s_i = \{a/a \in s_{i-1} \& ([\![n_1]\!]_a, \ldots, [\![n_m]\!]_a) \in F(label(r_i))\}$ where
   $[\![n]\!]_a = F(marker(n))$ if $marker(n) \in I$, else $[\![n]\!]_a = a(f_k(n))$.
   $\langle X, s\rangle_f[G] = \langle X_k, s_{k+l}\rangle$
   The assignments that cannot satisfy the factual information in the model are eliminated from the information state.

We call complete contextualisation of G in X_k, the total function f_k from the set of existential nodes in G to the context.

We chose to interpret a SCG all at once, but we could have done it connected compound by connected compound or relation node (and its neighborhood) by relation node as it is done in Dynamic Predicate Logic. The reason for this choice is that in Conceptual Graphs, the result of inserting a graph into another one has lost the information about which graph came first and which graph was inserted, whereas in a classical language of Predicate Logic, this information is encoded in a conjunction: $\exists x(Px \wedge Qx)$ and $\exists x(Qx \wedge Px)$ are represented by the same SCG, (P)–1–[*]–1–(Q).

Example 2. suppose that there is only one concept type, t, two individual markers a and b, and a binary relation P. Let $M = (D = \{A, B\}, F)$ be a model such that $F(a) = A, F(b) = B, F(P) = \{(B, B), (A, B)\}$.

⟨G∅,{∅}⟩[[t, a]–1–(P)–2–[t, *]] = ⟨ [t, *] –s→ A, B ⟩[[t, a]–1–(P)–2–[t, *]] = ⟨ [t, *] –s→ B ⟩

The context is expanded with a new existential node c of type t which is linked to the existential node of the graph. The set of assignments s is expanded with all possible assignments for c (i.e. $s = \{s_1 : c \to A, s_2 : c \to B\}$). Then, as $(A, A) \notin F(P)$, the assignment $s_1 : c \to A$ is eliminated from s.

The update of a state by a graph is equal to the result of successively updating with the elements of some partition of the graph:

Definition 9 (Composition of updates). Let $\langle X, s\rangle$ be a state based on $M = (D, F)$, G and H be some SCGs, g a contextualisation of H in G and f a contextualisation of $G \oplus_g H$ in X.

$\langle X, s\rangle_f[G \oplus_g H] = (\langle X, s\rangle_f[G])_{g'}[H]$ where g' is the obvious contextualisation of H in the context X' of $(\langle X, s\rangle_f[G]) = \langle X', s'\rangle$ (i.e., let f' be the complete contextualisation of G in X', $g'xy$ iff $\exists z \in G/gxz$ and $f'zy$).

Definition 10 (Entailment). Let $\langle X, s\rangle$ be a state, an assignment $i \in s$ *admits* an update with the SCG G under a contextualisation f iff $\langle X, \{i\}\rangle_f[G]$ is not absurd. The state $\langle X, s\rangle$ accepts G under f iff every $i \in s$ admits an update with G under f, e.g. in the previous example, the graph was accepted by the empty state.

1. For a support Σ and a SCG G, G is *valid*, $\models G$, iff for any model $M = (D, F)$ of Σ, the empty state accepts G (under the empty contextualisation).
   In other words, a graph is valid if for any model, the update of the empty state with the graph is not absurd. For instance, as the chosen models respect the property of non empty universe, a graph which does not contain any relation node is valid.
2. Given a support Σ, two SCGs G and H, and a contextualisation g of H in G, G *entails* H under g, $G \models_g H$, iff for any model $M = (D, F)$ of Σ, $\langle X, s\rangle$ accepts H under g'
   where $\langle X, s\rangle = \langle G_\emptyset, \{\emptyset\}\rangle_\emptyset[G]$, f is the complete contextualisation of G in X and g' is the contextualisation of H in X by: $g'xy$ iff $\exists z \in G/gxz$ and fzy.
   A graph entails another graph if for any model, every assignment resulting from the update of the empty state by the first graph is preserved in a further update by the second graph.

2.3 A projection for the dynamic interpretation of SCGs

A calculus for the language of simple Conceptual Graphs is the projection ([9]), a morphism from a graph to another one. Projection does not cope with anaphoric bindings as coreference links cannot occur between the two graphs. projection is extended in order to obtain a calculus sound and complete with respect to the dynamic entailment relation.

Definition 11. Let $G = (R, C, U, label, \equiv_{co})$ and $H = (R', C', U', label', \equiv'_{co})$ be two SCGs and f be a contextualisation H in G and let $C'' \subseteq C'$ be the set concept nodes that are connected to a relation node by an edge of U'. There is a projection of H to G under f, $G \vdash_f H$ iff there is a total function π from $(R' \cup C'')$ to $(R \cup C)$ such that:
$\forall r \in R'$, $\pi(r) \in R$ and $label(\pi(r)) \leq_{T_R} label'(r)$ and $\forall 1 \leq i \leq arity(label'(r))$, $\pi(H_i(r)) = G_i(\pi(r))$.
$\forall c \in C''$, $\pi(c) \in C$ and $label(\pi(c)) \leq label'(c)$ and $\forall c' \in C'', c \equiv'_{co} c'$ implies $\pi(c) \equiv_{co} \pi(c')$ and if $\exists c' \in C/fcc'$ then $\pi(c) \equiv_{co} c'$.

For completeness, isolated concept nodes are not projected; indeed if projection was defined as a total function from a graph to another one, there would not be any projection from [t,a]–1–(P) [t,b] to [t,a]–1–(P), whereas obviously [t,a]–1–(P) entails [t,a]–1–(P) [t,b] because the isolated concept [t,b] taken alone (as a graph) is valid. From an implementation point of view, it is very simple to adapt an algorithm for the classical projection; indeed, existential nodes in H that are linked to nodes in G behave exactly like individual nodes.

This extended notion of projection copes with anaphoric references between the premiss and the conclusion. It is not only essential when projection is part of a calculus for Conceptual Graphs as in [11] but also for redundancy checking: for instance, suppose that we know the following information "John beats a donkey. A farmer owns a donkey and he beats it": 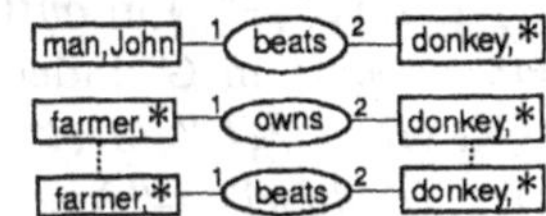
.

Is the information "he beats it" redundant with the rest of the information? No, because there is no projection of the bottom sub-graph to the rest of the graph under the contextualisation materialized by the dashed edges. In other words, "John beats a donkey and a farmer owns a donkey" does not entail that "he (the farmer) beats it (his donkey)", whereas classical projection does not deal with the coreferences between premiss and conclusion, e.g. the bottom sub-graph can be projected onto the top sub-graph.

Proposition 1 (Projection is sound and complete). *Let G and H be two SCGs and f be a contextualisation H in G, $G \vdash_f H$ iff $G \models_f H$.*

Proposition 2 (A deduction theorem). *Let G and H be two SCGs and f be a contextualisation H in G, the rule $G \rightarrow_f H$ is valid iff $G \models_f H$.*

3 Conceptual Graphs

The full expressive power of Conceptual Graphs is obtained by the introduction of negation boxes.

Definition 12. A CG is a graph $G = (R, C, U, label, \equiv_{co}, T, loc)$ where T is a tree of boxes and loc is a function which associates to every node (concept or relation) a node of T.

The root of T corresponds to Peirce's sheet of assertion, the outermost assertion zone, and the leaves the deepest assertion zones. $out(G)$ will refer to the SCG which is the subgraph of G whose nodes occur in the root of T. Note that the term 'context' is sometimes used for what we call an 'assertion zone': an assertion zone refers to the syntactical notion of graphical area delimited by a negation box whereas the term context refers to the semantic representation of a situation in which a graph is interpreted.

Definition 13. A Conceptual Graph G' is inductively defined by:

- A SCG is a CG: if $G = (R, C, U, label, \equiv_{co})$ is a SCG, then $G' = (R, C, U, label, \equiv_{co}, T, loc)$ is a CG where T is a single node r and $\forall x \in R \cup C, loc(x) = r$. $out(G') = G$.
- The result of enclosing a CG in a box is a CG: if $G = (R, C, U, label, \equiv_{co}, T, loc)$ is a CG then $G' = (R, C, U, label, \equiv_{co}, r(T), loc)$ is a CG where r is a new root for the tree of zones. $out(G') = G_\emptyset$. E.g. a representation of "it is false that a farmer beats a donkey" [farmer,*]—1—(beats)—2—[donkey,*]
- Insertion: let $G = (R, C, U, label, \equiv_{co}, r(T), loc)$ and $H = (R', C', U', label', \equiv'_{co}, r'(T'), loc')$ be two CGs, f be a contextualisation of $(R', C', U', label', \equiv'_{co})$ in $out(G)$.
  The insertion of H in G under the contextualisation f, noted $G \oplus_f H$, is another CG $(R'', C'', U'', label'', \equiv''_{co}, r''(T, T'), loc'')$ where
  - $(R'', C'', U'', label'', \equiv''_{co}) = (R, C, U, label, \equiv_{co}) \oplus_f (R', C', U', label', \equiv'_{co})$

- r'' is new and loc'' inherits loc and loc' as follows: $\forall x \in R'' \cup C''$ if $loc(x)$ is defined and $loc(x) \neq r$ then $loc''(x) = loc(x)$ else, (if $loc'(x)$ is defined and $loc'(x) \neq r'$ then $loc''(x) = loc'(x)$ else, $loc''(x) = r''$).
- $out(G'') = out(G) \oplus_f out(G')$.

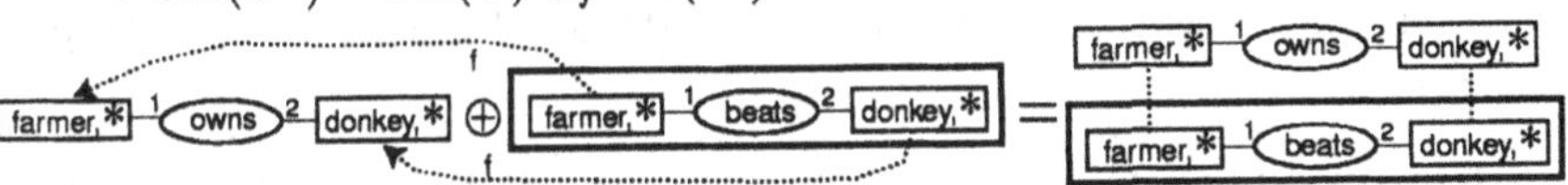

A CG G is the juxtaposition of a simple graph $out(G)$ and a (possibly empty) set of CGs $G_1, \ldots, G_n$, each of them being enclosed in a negation box. A concept occurring in one of the G_i may be coreferent to a concept in $out(G)$. To update a state with G, we first update it with $out(G)$ (concepts in this graph are inserted in the context and become available for further updates) and then successively with each graph $\boxed{G_i}$:

Definition 14 (Dynamic interpretation of CGs). Let $\langle X, s\rangle$ be a state based on $M = (D, F)$, $G = (R, C, U, label, \equiv_{co}, T, loc)$ be a CG and f be a contextualisation of G in X.

- If G is a SCG (i.e., T is a single node) then the update of the information state under f is given in definition 8
- If G is a CG G' enclosed in a negation box (i.e., $G = \boxed{G'}$) then $\langle X, s\rangle_f[G] = \langle X, s'\rangle$ where $i \in s'$ iff $i \in s$ and $\langle X, \{i\}\rangle_f[G]$ is absurd
  Updating a state with the negation of a graph G' does not introduce new context items; indeed, the construction rules forbid further information to refer to the concepts of G': if H is inserted in $\boxed{G'}$, then concepts in H cannot be coreferent to concepts in G'. The change in the information state which is updated with the negation of G' concerns the set of possible assignments: every assignment that would support an update with G' (that would not result in the absurd state) is eliminated.
- Else, $G = out(G) \oplus_{g_1} \boxed{G_1} \ldots \oplus_{g_n} \boxed{G_n}$, for $1 \leq n$.
  The state is first updated with the graph asserted in the outermost zone: $\langle X_0, s_0\rangle = \langle X, s\rangle_f[out(G)]$ and we call g'_0, the complete contextualisation of $out(G)$ in X_0. The resulting state is then successively updated with each of the negated graphs:
  For $1 \leq k \leq n$, we define a contextualisation g'_k of $\boxed{G_k}$ in X_0 by: $g'_k xy$ iff $\exists z \in out(G)/g_k xz$ and $g'_0 zy$.
  For $1 \leq k \leq n$, $\langle X_k, s_k\rangle = \langle X_{k-1}, s_{k-1}\rangle_{g'_k}[\boxed{G_k}]$.
  $\langle X, s\rangle_f[G] = \langle X_n, s_n\rangle$.

The notion of entailment remains the same as the one defined for the language of Simple Conceptual Graphs:

Definition 15 (Entailment). Let $\langle X, s\rangle$ be a state, an assignment $i \in s$ admits an update with the CG G under a contextualisation f iff $\langle X, \{i\}\rangle_f[G]$ is not absurd. The state $\langle X, s\rangle$ accepts G under f iff every $i \in s$ admits an update with G under f.

1. A CG, G, is valid, $\models G$, iff for any model, the empty state accepts G.
2. Given two CGs G and H, and a contextualisation g of H in $out(G)$, G entails H under g, $G \models_g H$, iff for any model, $\langle X, s\rangle$ accepts H under g' where $\langle X, s\rangle = \langle G_\emptyset, \{\emptyset\}\rangle_\emptyset[G]$, f is the complete contextualisation of G in X and g' is the contextualisation of H in X defined by: $g'xy$ iff $\exists z \in G/gxz$ and fzy.

Proposition 3 (deduction theorem). *Let G and H be two CGs and f be a contextualisation H in out(G),* $G \models_f H$ *iff* $\boxed{G \oplus_f \boxed{H}}$ *is valid.*

4 Conclusions

Construction rules and calculi for CGs make an extensive use of the notion of accessible concept, a concept introduced in some previously asserted graph. This notion should therefore be central to a semantics for CGs. We proposed an adaptation to CGs of the Dynamic Semantics for Discourse Representation Structures ([6]). A specific kind of contextual information becomes the cornerstone of the interpretation process: contexts are filled with (accessible) concepts associated to their possible interpretations and the meaning of a graph is given by the change its assertion brings into a context. This new semantics strengthen the links between CG Theory and theories for the study of discourse in which entailment is a dynamic notion (the conclusion can make reference to discourse items introduced in the premiss). A such perspective on the meaning of a graph is relevant to the field of knowledge bases: e.g. integration of two bases by inserting one into the other under a contextualisation, search of redundant representations or update of a knowledge base. Concerning this last point, we have scratched the possibility of choosing richer information states by for instance not only recording existential concepts and their possible assignments, but as well the structure of a graph serving in an update. A radical change would be to shift from the notion of test in a model to a notion of model construction; we can imagine that a such construction could start in an empty state (an empty knowledge base) and that new information would incrementally extend the model built (and extend its finite domain to another finite domain) resulting in a non-monotonic system: for instance, we could update an empty state with the information that *the farmer John owns the donkey Bram and every farmer beats every donkey he owns* and then update the resulting state with some new evidence like *the farmer Pete does not beat his donkey Boris.* In this final state, the information *every farmer beats every donkey he owns* is not anymore accepted. In this simplistic view on knowledge base update, as queries would be tested on a finite (but growing with the updates) domain, entailment would be decidable.

We considered contexts with a flat structure (contexts are represented by Simple Conceptual Graphs), but the semantics could be extended to more expressive and structured knowledge representations like Nested Conceptual Graphs, graphs in which a concept is itself described by a graph; Contexts would then inherit these nested structures. A modal (but static) perspective on contextual reasoning is proposed in [13] and [3], and [14] studies this modal view in the framework of Nested Conceptual Graphs. Further work should integrate the notion of dynamic entailment in the modal framework of Nested CGs.

Acknowledgements: I would like to thank the anonymous referees for their helpful comments that I could not entirely exploit in this paper, but suggest some future work in the study of contextual reasoning in CG theory.

References

1. Fausto Giunchiglia and Paolo Bouquet. Introduction to Contextual Reasoning. An Artificial Intelligence Perpective. Technical Report 970519, IRST, 1997.
2. John McCarthy. Generality in Artificial Intelligence. In *ACM Turing Award Lectures, The First Twenty Years.* ACM Press, 1987.
3. John McCarthy and Saša Buvač. Formalizing Context (Expanded Notes). In A. Aliseda, R. van Glabbeek, and D. Westerståhl, editors, *Computing Natural Language*, volume 81 of *CSLI Lecture Notes*, pages 13–50. Center for the Study of Language and Information, Stanford University, 1997.
4. Hans Kamp. A theory of truth and semantic representation. In Jeroen Groenendijk, Theo Janssen, and Martin Stokhof, editors, *Formal Methods in the Study of Language.* Mathematical Centre, Amsterdam, 1981.
5. Hans Kamp and Uwe Reyle. *From Discourse to Logic.* Kluwer, Dordrecht, 1993.
6. Jeroen Groenendijk and Martin Stokhof. Dynamic Predicate Logic. *Linguistics and Philosophy*, 14:39–100, 1991.
7. John F. Sowa. Peircean Foundations for a Theory of Context. In Lukose et al. [15], pages 41–64.
8. Eric Salvat and Marie-Laure Mugnier. Sound and Complete Forward and Backward Chaining of Graph Rules. In P.W. Eklund, G. Ellis, and G. Mann, editors, *Conceptual Structures: Knowledge Representation as Interlingua (Proceedings of ICCS'96, Sydney, Australia)*, volume 1115 of *LNAI*, pages 248–262. Springer-Verlag, 1996.
9. John F. Sowa. *Conceptual Structures, Information Processing in Mind and Machine.* Addison Wesley, 1984.
10. Marie-Laure Mugnier and Michel Chein. Représenter des connaissances et raisonner avec des graphes. *RIA*, 10.1:7–56, 1996.
11. Gwen Kerdiles and Eric Salvat. A Sound and Complete CG Proof Procedure Combining Projections with Analytic Tableaux. In Lukose et al. [15], pages 371–385.
12. Susanne Prediger. Simple Concept Graphs: A Logic Approach. In Marie-Laure Mugnier and Michel Chein, editors, *Conceptual Structures: Theory, Tools and Applications (Proceedings of ICCS'98, Montpellier, France)*, volume 1453 of *LNAI*, pages 225–239. Springer-Verlag, 1998.
13. Saša Buvač. Quantificational logic of context. In *proceedings of the Thirteenth National Conference on Artificial Intelligence*, 1996.
14. Gwen Kerdiles. Graph matching in contextual reasoning. Technical Report 99034, LIRMM, 1999.
15. Dikson Lukose, Harry Delugach, Mary Keeler, Leroy Searle, and John F. Sowa, editors. *Conceptual Structures: Fulfilling Peirce's Dream (Proceedings of ICCS'97, Seattle, USA)*, volume 1257 of *LNAI.* Springer-Verlag, 1997.

A Case for Variable-Arity Relations: Definitions and Domains

Dan Corbett

Department of Computer Science
University of Adelaide
Adelaide, South Australia
dcorbett@cs.adelaide.edu.au

1. Why Variable Arity?

There are some domains which currently do not fit easily into the standard definitions and formalisms of Conceptual Structures, due to the dynamic nature of the knowledge being represented. Some simple examples include a variable number of students in a class, or several exits from a room. In these cases, the number of concepts to be "attached" to the relation is not known beforehand. The use of variable arity relations will allow representation of knowledge where a relation may not point to the same number of concepts in every situation.

2. Definitions and Explanations on Variable Arity Relations

To make variable arity relations consistent with the rest of CG formalism, it is first necessary to modify the standard definition of a CG slightly.

Definition 1. A conceptual graph is a tuple $G = (C, R, type, referent, arg_1, \ldots, arg_m)$ where:

C is the set of concepts, $type : C \rightarrow T$ indicates the type of a concept.

$R \# T \_ C \{\_ C \_ \ldots \_ C\}$ is the set of relations between the concepts. T is the set of types and $(t, c_i, \{c_j, \ldots, c_m\}) \in R$ yields an arc of type t from concept i to concept j, and possibly also to other concepts.

In this definition, there are optional arguments to each relation, which may be used in a given circumstance, thus creating relations of variable arity.

3. How Does This Change Joins, and the [1] Operator?

The objective of a join operation is to find a graph which represents knowledge which is *more specific* than the two graphs being joined. It isn't difficult to define a join operation which includes variable arity, if we let *higher* arity mean *more* specific. For example, a kitchen design specifies some lighting but only relates to a few plumbing concepts. Another kitchen design specifies different aspects of the plumbing. The join of these two graphs is a graph with both lighting and plumbing completely specified. The joined graph is more specific because it's more completely specified, and the arity of the relations is higher in the joined graph.

The ^{1} operator expects that each relation in the previous graphs will still point to the same concepts in the new graph. We can guarantee that if a relation points to a given concept in a previous graph, it will still point to it in the new, joined graph, but we don't guarantee that the order of the arcs is preserved. We can guarantee that the semantics of the relation is still the same, though, ie that even though a concept has changed places, its meaning is still intact. This obviously would not work for all domains, and the canonical formation rules would specify which relations were to have a variable arity in a given domain.

Another issue which arises from the definition of variable arity is the definition of the mgu in a unification process. This definition follows on in a straightforward manner from the modification to the ^{1} operator discussed above. I leave the formalisation of the ^{1} operator and the mgu to future papers.

4. Examples of Variable Arity Relations in Use

The design and architecture domains can benefit from the use of variable-arity relations. Designers often attempt to reuse previous designs, and apply them to new problems. In the case of designing a building, we want joined or unified designs to reflect all of the knowledge contained in the originals, even if the number of exits (or the rooms which are adjacent, or the exact floor area) isn't an exact match. For example, a retrieved design may show the kitchen adjacent to the living room and dining room, while the customer requirements specify that the kitchen must be adjacent to the dining room and laundry. The adjacent relation is of arity 2 in these instances. Once unified, however, the new graph of the kitchen must show it as adjacent to the laundry, living room and dining room. This change in arity preserves the knowledge from both of the original graphs, and is in keeping with the designer's intent.

In the natural language domain, a grammar is used to specify that adjectives can precede nouns (in English). However, it is not known in advance how many adjectives will be used in a noun phrase. A variable-arity relation " adjective" would allow for any number of adjectives. Since word order is essential in language understanding, and given that our join operator doesn't guarantee the order of concepts after a join, a modification to the join operator would be necessary for the natural language domain.

Graph Structures in Parametric Spaces for Representation of Verbs

Valery Solovyev

Kazan State University, Kremlevskaia str., 18, 420018, Russia
solovyev@tatincom.ru

Spaces of word attributes are investigated for different semantic classes of verbs. These spaces are designated to collate the semantics of verbs with close meaning. It is an alternative to dynamic maps proposed in [1].

For example, let us consider a group of verbs: 'wander', 'prowl', 'stray', 'meander', 'ramble', 'roam', 'rove'. We shall collate them by the next attributes: *distance, purpose, internal state of subject, direction, speed, appearance, figure of way.*

Attributes can be binary (for the verb 'meander' it means the *direction* of movement be specified or not) or possess several values (e.g. for the purpose attribute: aimless, purposefully, for pleasure, with reprehensible purpose). Attribute values can also be fuzzy (*distance* - large).

Possibly, each semantic class contains a word with the most general meaning. Such word has no fixed values of attributes. In the given example the verb 'wander' satisfies this notion. Attribute space dimensions may be reduced by using the next factorization procedure. If some attributes take the same value for all the words in the current class, they are allowed to be changed by the single integral attribute without losing the discerning force of attribute system.

Word possession of attributes can be expressed in the form of table or tree. To draw the tree form of representation, we have to apply a certain taxonomic classification on words from the chosen group. Then the structure of such tree must be investigated. Two main cognitive principles exerting influence on the tree structure maintenance are stated below.

1. Any word is obliged not to have many attributes (otherwise, it looks too special).

2. The space is obliged not to contain many words (economy principle).

This approach can be applied to the study of semantic peculiarities of the words in many fields. For example, it can help to define more precisely the semantics of the words, which mean emotions, similar to [2].

References

1. Cohen P. R. Dynamic Maps as Representations of Verbs. Pros. of the 13th Biennial European Conference on Artificial Intelligence, John Wiley Sons, 1998.

2. Wierzbicka A. Emotion, Language and "Cultural Scripts". In "Emotion and Culture: Empirical Studies of Mutual Influence", eds.Kitayama, H.Markus, Washington, 1994.

PORT: Peirce Online Resource Testbeds

Mary Keeler and William Tepfenhart

Center for Advanced Research and Technology in the Arts and Humanities,
University of Washington, Seattle, WA 98117
mkeeler@u.washington.edu
Monmouth University, Software Engineering Dept.,
West Long Branch, NJ, USA
btepfenh@monmouth.edu

Abstract. This overview of PORT covers the goals to achieve, the institutions involved, and the status of the project. Proposals for funding PORT development are pending, but this preliminary report provides URLs to sites where the work has begun.

PORT has proposed to build a communications network environment for investigating and improving collaborative research to support the production of digital library resources based on scanned images of manuscript archives. In selecting C.S. Peirce's manuscripts as the base for this model project, we have two purposes: 1) to preserve a legacy of archived work that is on the brink of decay and 2) to provide researchers in many disciplines significantly better access to material that is valuable in a wide range of intellectual pursuits. Currently, PORT has 100 test images of the Peirce manuscript collection at Harvard, scanned and available online through the World Wide Web.

PORT's initial development has two primary goals: 1) application of scanning techniques and technologies that preserve the character and condition of the original archived documents and 2) development of an environment that enables scholars to collaborate in pursuing investigation of the document content and subject matter. An example of the challenges to which PORT's goals must respond is that over 10,000 pages of the manuscript collection have been misplaced or lost. Replacement of these scanned pages in the scanned collection of available manuscripts will require high-quality images that provide researchers with the evidence required to determine where a particular page belongs. In one recent test, a high-resolution scan clearly showed the lines of a page and marks on its underside which did not match the notebook context to which the page had been assigned. Collaborative work in an effective "show-and-tell" environment, would make recovery of the archive's order as efficient as possible.

The need to collaborate across international boundaries adds another challenging dimension to PORT. Most of the widely available, Web-based applications provide only rudimentary mechanisms for viewing imaged material remotely, and commenting on its condition and content. Often, only a simple form to fill out is available for such communications. And effective collaboration among a large group of researchers requires the support of a NewsNet-type facility, but with more discrete links to the original material than simple text mechanism affords. A number of sites in the U.S. and Europe currently host the Peirce manuscript test images and related documentation. These testbed sites represent the first stage of PORT's effort to build what the U.S. National Science Foundation calls a "collaboratory," as a model for digital library resource development. The individuals involved, their affiliations, and the URLs related to this effort are listed below.

Dr. Harry Delugach
Computer Science Department and the Information Technology and Systems Center, Univ. Alabama in Huntsville
http://concept.cs.uah.edu/PORT/

Dr. Ulrich Thiel (Said Kutschekmanesch)
Integrated Publication and Information Systems Institute
GMD - German National Research Center for Information Technology
http://delite.darmstadt.gmd.de/Peirce/

Dr. Peter Oehrstroem, Ph.D., D.Sc.
Department of Communication, Aalborg University
http://www.hum.auc.dk/~poe
http://www.hum.auc.dk/~magnus/tiff/demo.html

Dr. Uta Priss
School of Library and Information Science, Indiana University
http://php.indiana.edu/~upriss/peirce/Peirce145.htm

Prof. Tony Jappy,
Department of English and American Studies,
Research Institute in Semiotic, Communication and Education Studies (IRSCE),
University of Perpignan, France
http://capcir.univ-perp.fr/peirce/default.htm

Prof. J. Jay Zeman
Department of Philosophy, University of Florida
http://web.clas.ufl.edu/users/jzeman/

Prof. Jaime Nubiola
Dept. Filosofia, Universidad de Navarra
http://www.unav.es/gep/

Aldo de Moor
Infolab, Tilburg University
http://infolab.kub.nl/people/ademoor

Prof. William M. Tepfenhart
Software Engineering Department, Monmouth University
URL: http://peirce.monmouth.edu/

Assuring Computer Agent Communications

William Tepfenhart and Mark Holder

Software Engineering, Monmouth University, West Long Branch, NJ
btepfenh@monmouth.edu
Computer Sciences Corporation
mholder@csc.com

Abstract. In this position paper, the author identifies a research direction currently being taken concerning communications among computer agents (not necessarily intelligent agents). Initial efforts are focused on miscommunication -- that is how two independent programs can fail to communicate accurately based on the individual interpretations of the data exchanged between them.

1. Introduction

When two programs communicate they actually exchange a string of bits. Programs that obey the same set of protocols will overlay an identical structure on the bits within a message so that type information and values can be extracted from the message. In many cases, several protocols will be stacked one atop the other. The lowest protocol is used to first impose one kind of structure (e.g. treat it as a character stream) and then next is used to impose a second kind of structure upon that (e.g. treat it as a set of attribute type value triplets). Usually the highest protocol is imposed to allow the receiver to act upon the stream (e.g., use it as a postscript message to drive a printer). The highest level usually has a semantics that is well defined -- that is, some tokens are defined as to their functional meaning and, hence, place upon the arguments some functional role such that the message can be interpreted without ambiguity.

For two programs to communicate effectively both programs must be based upon the same semantic model -- that such and such a token has a very specific functional semantic. Unfortunately, the use of such well defined and widely distributed protocols may become the exception rather than rule for future interacting software agents. Recent introductions into main stream computing (such as the Dynamic Invocation Interface (DII) of CORBA) suggest that interacting programs will lack a common semantic model and this practice is going to be more common. As an example, DII allows CORBA programs to access functionality of a server object without requiring the client to have been developed using the IDL definition of the object. This means that the client program will not be built incorporating the explicit static semantics and semantic labeling typical of past practices. The semantics of individual tokens will have to be established dynamically -- that is, at run-time.

Dynamically establishing a common semantics between two independently developed programs is, in many ways, a more difficult task than understanding natural language text. At least with natural language text there is usually some well defined set of tokens and a dictionary by which all people who use the language are more or less bound. Unfortunately, with computer systems and service definitions, programmers tend to shove several words together (hence creating new tokens) on an as needed basis and with little forethought. Different programmers, in different environments will create different tokens for the same functionality (e.g., print_document(), sendDocumentToPrinter() and printDocument()). A human reading the name can often figure out what service is being provided, but can a machine? There is no dictionary of allowed tokens to fall back upon and it isn't always clear as to where the string should be broken into parts for application of conventional text processing.

In our laboratory, research is currently focused on ways by which two computer systems may fail to inter-operate because of a lack of a universal high level semantic protocol. We accept from the start that the basic protocols (e.g., TCP, IP, HTTP, IIOP, etc.) function as they are intended. That is, they cast the bit stream into the appropriate structures for use within the program. The real problem is assuring that the tokens within the structures are interpreted correctly. With respect to this problem several key assumptions are being made:

- A single formal semantics for service interfaces will never exist;
- Interacting programs will have to dynamically synchronize their semantics; and
- Synchronizing semantics will require a special purpose protocol.

The investigation regarding the consequences of these assumptions is the initial step in this investigation. At present, an explicit identification of the various ways in which a token (passed syntactically correct) can be misinterpreted by the receiving program has been undertaken and is soon to be completed. This work will continue in the future to establish ways in which the mechanisms which lead to misinterpretation can be corrected. The work is incorporating past efforts in ontology as providing an initial mechanism for agents to synchronize their individual interpretations of tokens to assure common meanings. This work will utilize the current standards established by Sowa for conceptual graphs.

Author Index

Author Index

Springer and the environment

At Springer we firmly believe that an international science publisher has a special obligation to the environment, and our corporate policies consistently

We also expect our business partners – paper mills, printers, packaging manufacturers, etc. – to commit themselves to using materials and production processes that do not harm the environment. The paper in this book is made from low- or no-chlorine pulp and is acid free, in conformance with international standards for paper permanency.

Lecture Notes in Artificial Intelligence (LNAI)

Vol. 1489: J. Dix, L. Fariñas del Cerro, U. Furbach (Eds.), Logics in Artificial Intelligence. Proceedings, 1998. X, 391 pages. 1998.

Vol. 1495: T. Andreasen, H. Christiansen, H.L. Larsen (Eds.), Flexible Query Answering Systems. Proceedings, 1998. IX, 393 pages. 1998.

Vol. 1501: M.M. Richter, C.H. Smith, R. Wiehagen, T. Zeugmann (Eds.), Algorithmic Learning Theory. Proceedings, 1998. XI, 439 pages. 1998.

Vol. 1502: G. Antoniou, J. Slaney (Eds.), Advanced Topics in Artificial Intelligence. Proceedings, 1998. XI, 333 pages. 1998.

Vol. 1504: O. Herzog, A. Günter (Eds.), KI-98: Advances in Artificial Intelligence. Proceedings, 1998. XI, 355 pages. 1998.

Vol. 1510: J.M. Zytkow, M. Quafafou (Eds.), Principles of Data Mining and Knowledge Discovery. Proceedings, 1998. XI, 482 pages. 1998.

Vol. 1515: F. Moreira de Oliveira (Ed.), Advances in Artificial Intelligence. Proceedings, 1998. X, 259 pages. 1998.

Vol. 1527: P. Baumgartner, Theory Reasoning in Connection Calculi. IX, 283 pages. 1999.

Vol. 1529: D. Farwell, L. Gerber, E. Hovy (Eds.), Machine Translation and the Information Soup. Proceedings, 1998. XIX, 532 pages. 1998.

Vol. 1531: H.-Y. Lee, H. Motoda (Eds.), PRICAI'98: Topics in Artificial Intelligence. XIX, 646 pages. 1998.

Vol. 1532: S. Arikawa, H. Motoda (Eds.), Discovery Science. Proceedings, 1998. XI, 456 pages. 1998.

Vol. 1534: J.S. Sichman, R. Conte, N. Gilbert (Eds.), Multi-Agent Systems and Agent-Based Simulation. Proceedings, 1998. VIII, 237 pages. 1998.

Vol. 1535: S. Ossowski, Co-ordination in Artificial Agent Societies. XVI, 221 pages. 1999.

Vol. 1537: N. Magnenat-Thalmann, D. Thalmann (Eds.), Modelling and Motion Capture Techniques for Virtual Environments. Proceedings, 1998. IX, 273 pages. 1998.

Vol. 1544: C. Zhang, D. Lukose (Eds.), Multi-Agent Systems. Proceedings, 1998. VII, 195 pages. 1998.

Vol. 1545: A. Birk, J. Demiris (Eds.), Learning Robots. Proceedings, 1996. IX, 188 pages. 1998.

Vol. 1555: J.P. Müller, M.P. Singh, A.S. Rao (Eds.), Intelligent Agents V. Proceedings, 1998. XXIV, 455 pages. 1999.

Vol. 1562: C.L. Nehaniv (Ed.), Computation for Metaphors, Analogy, and Agents. X, 389 pages. 1999.

Vol. 1570: F. Puppe (Ed.), XPS-99: Knowledge-Based Systems. VIII, 227 pages. 1999.

Vol. 1571: P. Noriega, C. Sierra (Eds.), Agent Mediated Electronic Commerce. Proceedings, 1998. IX, 207 pages. 1999.

Vol. 1572: P. Fischer, H.U. Simon (Eds.), Computational Learning Theory. Proceedings, 1999. X, 301 pages. 1999.

Vol. 1574: N. Zhong, L. Zhou (Eds.), Methodologies for Knowledge Discovery and Data Mining. Proceedings, 1999. XV, 533 pages. 1999.

Vol. 1582: A. Lecomte, F. Lamarche, G. Perrier (Eds.), Logical Aspects of Computational Linguistics. Proceedings, 1997. XI, 251 pages. 1999.

Vol. 1585: B. McKay, X. Yao, C.S. Newton, J.-H. Kim, T. Furuhashi (Eds.), Simulated Evolution and Learning. Proceedings, 1998. XIII, 472 pages. 1999.

Vol. 1599: T. Ishida (Ed.), Multiagent Platforms. Proceedings, 1998. VIII, 187 pages. 1999.

Vol. 1609: Z. W. Raś, A. Skowron (Eds.), Foundations of Intelligent Systems. Proceedings, 1999. XII, 676 pages. 1999.

Vol. 1611: I. Imam, Y. Kodratoff, A. El-Dessouki, M. Ali (Eds.), Multiple Approaches to Intelligent Systems. Proceedings, 1999. XIX, 899 pages. 1999.

Vol. 1612: R. Bergmann, S. Breen, M. Göker, M. Manago, S. Wess, Developing Industrial Case-Based Reasoning Applications. XX, 188 pages. 1999.

Vol. 1617: N.V. Murray (Ed.), Automated Reasoning with Analytic Tableaux and Related Methods. Proceedings, 1999. X, 325 pages. 1999.

Vol. 1620: W. Horn, Y. Shahar, G. Lindberg, S. Andreassen, J. Wyatt (Eds.), Artificial Intelligence in Medicine. Proceedings, 1999. XIII, 454 pages. 1999.

Vol. 1621: D. Fensel, R. Studer (Eds.), Knowledge Acquisition Modeling and Management. Proceedings, 1999. XI, 404 pages. 1999.

Vol. 1632: H. Ganzinger (Ed.), Automated Deduction – CADE-16. Proceedings, 1999. XIV, 429 pages. 1999.

Vol. 1634: S. Džeroski, P. Flach (Eds.), Inductive Logic Programming. Proceedings, 1999. VIII, 303 pages. 1999.

Vol. 1638: A. Hunter, S. Parsons (Eds.), Symbolic and Quantitative Approaches to Reasoning and Uncertainty. Proceedings, 1999. IX, 397 pages. 1999.

Vol. 1640: W. Tepfenhart, W. Cyre (Eds.), Conceptual Structures: Standards and Practices. Proceedings, 1999. XII, 515 pages. 1999.

Vol. 1650: K.-D. Althoff, R. Bergmann, L.K. Branting (Eds.), Case-Based Reasoning Research and Development. Proceedings, 1999. XII, 598 pages. 1999.

Lecture Notes in Computer Science

Vol. 1603: V. Vitek, C.D. Jensen (Eds.), Secure Internet Programming. X, 501 pages. 1999.

Vol. 1605: J. Billington, M. Diaz, G. Rozenberg (Eds.), Application of Petri Nets to Communication Networks. IX, 303 pages. 1999.

Vol. 1606: J. Mira, J.V. Sánchez-Andrés (Eds.), Foundations and Tools for Neural Modeling. Proceedings, Vol. I, 1999. XXIII, 865 pages. 1999.

Vol. 1607: J. Mira, J.V. Sánchez-Andrés (Eds.), Engineering Applications of Bio-Inspired Artificial Neural Networks. Proceedings, Vol. II, 1999. XXIII, 907 pages. 1999.

Vol. 1608: S. Doaitse Swierstra, P.R. Henriques, J.N. Oliveira (Eds.), Advanced Functional Programming. Proceedings, 1998. XII, 289 pages. 1999.

Vol. 1609: Z. W. Raś, A. Skowron (Eds.), Foundations of Intelligent Systems. Proceedings, 1999. XII, 676 pages. 1999. (Subseries LNAI).

Vol. 1610: G. Cornuéjols, R.E. Burkard, G.J. Woeginger (Eds.), Integer Programming and Combinatorial Optimization. Proceedings, 1999. IX, 453 pages. 1999.

Vol. 1611: I. Imam, Y. Kodratoff, A. El-Dessouki, M. Ali (Eds.), Multiple Approaches to Intelligent Systems. Proceedings, 1999. XIX, 899 pages. 1999. (Subseries LNAI).

Vol. 1612: R. Bergmann, S. Breen, M. Göker, M. Manago, S. Wess, Developing Industrial Case-Based Reasoning Applications. XX, 188 pages. 1999. (Subseries LNAI).

Vol. 1613: A. Kuba, M. Šámal, A. Todd-Pokropek (Eds.), Information Processing in Medical Imaging. Proceedings, 1999. XVII, 508 pages. 1999.

Vol. 1614: D.P. Huijsmans, A.W.M. Smeulders (Eds.), Visual Information and Information Systems. Proceedings, 1999. XVII, 827 pages. 1999.

Vol. 1615: C. Polychronopoulos, K. Joe, A. Fukuda, S. Tomita (Eds.), High Performance Computing. Proceedings, 1999. XIV, 408 pages. 1999.

Vol. 1617: N.V. Murray (Ed.), Automated Reasoning with Analytic Tableaux and Related Methods. Proceedings, 1999. X, 325 pages. 1999. (Subseries LNAI).

Vol. 1619: M.T. Goodrich, C.C. McGeoch (Eds.), Algorithm Engineering and Experimentation. Proceedings, 1999. VIII, 349 pages. 1999.

Vol. 1620: W. Horn, Y. Shahar, G. Lindberg, S. Andreassen, J. Wyatt (Eds.), Artificial Intelligence in Medicine. Proceedings, 1999. XIII, 454 pages. 1999. (Subseries LNAI).

Vol. 1621: D. Fensel, R. Studer (Eds.), Knowledge Acquisition Modeling and Management. Proceedings, 1999. XI, 404 pages. 1999. (Subseries LNAI).

Vol. 1622: M. González Harbour, J.A. de la Puente (Eds.), Reliable Software Technologies – Ada-Europe'99. Proceedings, 1999. XIII, 451 pages. 1999.

Vol. 1625: B. Reusch (Ed.), Computational Intelligence. Proceedings, 1999. XIV, 710 pages. 1999.

Vol. 1626: M. Jarke, A. Oberweis (Eds.), Advanced Information Systems Engineering. Proceedings, 1999. XIV, 478 pages. 1999.

Vol. 1627: T. Asano, H. Imai, D.T. Lee, S.-i. Nakano, T. Tokuyama (Eds.), Computing and Combinatorics. Proceedings, 1999. XIV, 494 pages. 1999.

Col. 1628: R. Guerraoui (Ed.), ECOOP'99 - Object-Oriented Programming. Proceedings, 1999. XIII, 529 pages. 1999.

Vol. 1629: H. Leopold, N. García (Eds.), Multimedia Applications, Services and Techniques - ECMAST'99. Proceedings, 1999. XV, 574 pages. 1999.

Vol. 1631: P. Narendran, M. Rusinowitch (Eds.), Rewriting Techniques and Applications. Proceedings, 1999. XI, 397 pages. 1999.

Vol. 1632: H. Ganzinger (Ed.), Automated Deduction – Cade-16. Proceedings, 1999. XIV, 429 pages. 1999. (Subseries LNAI).

Vol. 1633: N. Halbwachs, D. Peled (Eds.), Computer Aided Verification. Proceedings, 1999. XII, 506 pages. 1999.

Vol. 1634: S. Džeroski, P. Flach (Eds.), Inductive Logic Programming. Proceedings, 1999. VIII, 303 pages. 1999. (Subseries LNAI).

Vol. 1636: L. Knudsen (Ed.), Fast Software Encryption. Proceedings, 1999. VIII, 317 pages. 1999.

Vol. 1638: A. Hunter, S. Parsons (Eds.), Symbolic and Quantitative Approaches to Reasoning and Uncertainty. Proceedings, 1999. IX, 397 pages. 1999. (Subseries LNAI).

Vol. 1639: S. Donatelli, J. Kleijn (Eds.), Application and Theory of Petri Nets 1999. Proceedings, 1999. VIII, 425 pages. 1999.

Vol. 1640: W. Tepfenhart, W. Cyre (Eds.), Conceptual Structures: Standards and Practices. Proceedings, 1999. XII, 515 pages. 1999. (Subseries LNAI).

Vol. 1644: J. Wiedermann, P. van Emde Boas, M. Nielsen (Eds.), Automata, Languages, and Programming. Proceedings, 1999. XIII, 720 pages. 1999.

Vol. 1649: R.Y. Pinter, S. Tsur (Eds.), Next Generation Information Technologies and Systems. Proceedings, 1999. IX, 327 pages. 1999.

Vol. 1650: K.-D. Althoff, R. Bergmann, L.K. Branting (Eds.), Case-Based Reasoning Research and Development. Proceedings, 1999. XII, 598 pages. 1999. (Subseries LNAI).

Vol. 1653: S. Covaci (Ed.), Active Networks. Proceedings, 1999. XIII, 346 pages. 1999.